THE
SCIENTIFIC PAPERS OF
SIR GEOFFREY INGRAM TAYLOR

VOLUME IV

MECHANICS OF FLUIDS:
MISCELLANEOUS PAPERS

SIR GEOFFREY TAYLOR, F.R.S.

Reproduced from a portrait by Ruskin Spear
commissioned by friends in commemoration of
Sir Geoffrey Taylor's eightieth birthday
on 7 March 1966

THE
SCIENTIFIC PAPERS OF
SIR GEOFFREY INGRAM TAYLOR

ORDER OF MERIT

FELLOW OF TRINITY COLLEGE, CAMBRIDGE

FELLOW AND FORMERLY YARROW RESEARCH PROFESSOR OF
THE ROYAL SOCIETY

VOLUME IV
MECHANICS OF FLUIDS:
MISCELLANEOUS PAPERS

EDITED BY

G. K. BATCHELOR, F.R.S.
TRINITY COLLEGE, CAMBRIDGE

CAMBRIDGE
AT THE UNIVERSITY PRESS
1971

CAMBRIDGE UNIVERSITY PRESS
Cambridge, New York, Melbourne, Madrid, Cape Town, Singapore,
São Paulo, Delhi, Dubai, Tokyo, Mexico City

Cambridge University Press
The Edinburgh Building, Cambridge CB2 8RU, UK

Published in the United States of America by Cambridge University Press, New York

www.cambridge.org
Information on this title: www.cambridge.org/9780521158794

First published 1971
First paperback edition 2010

A catalogue record for this publication is available from the British Library

Library of Congress Catalogue Card Number: 58-14675

ISBN 978-0-521-07995-2 Hardback
ISBN 978-0-521-15879-4 Paperback

EDITOR'S PREFACE

This is the last of four volumes containing the scientific papers written by Sir Geoffrey Taylor, and is being published, by happy chance, on his 85th birthday.

The papers have been prepared for the present volumes with as little change of their original form as possible. Printing errors and small mathematical slips in the first publication have been rectified without comment, but any substantial change in the text as originally published has been disclosed in a footnote.

The papers have been grouped into volumes according to subject, although the divisions between the different aspects of fluid mechanics in Volumes II, III and IV are not sharp. Within each volume the papers have been arranged in chronological order, and each volume has its own numbering system based on that order. A consolidated list of the papers contained in all four volumes in chronological order is provided at the end of this volume.

There will also be found at the end of this volume a list of papers and articles by Sir Geoffrey Taylor which have not been reproduced in these volumes. Some are purely expository articles prepared for delivery as lectures or for publication in collective works, some are papers which duplicate work included in these volumes, and some are articles for a general audience. A few other articles of this kind may well have escaped my notice.

The publication of these volumes was made possible by the receipt of a financial guarantee from the Ministry of Technology (through the good offices of the Fluid Motion Sub-Committee and its parent body, the Aeronautical Research Council) and a gift from the Master and Fellows of Trinity College.

Paper 15 is Crown copyright material and is published with the permission of the Controller of H. M. Stationery Office. The other papers herein are reprinted with the kind permission of the original publishers.

G. K. BATCHELOR

September 1970

CONTENTS

PAPERS BY G. I. TAYLOR

1

INTERFERENCE FRINGES WITH FEEBLE LIGHT

REPRINTED FROM

Proceedings of the Cambridge Philosophical Society, vol. xv (1909), pp. 114–15

The phenomena of ionisation by light and by Röntgen rays have led to a theory according to which energy is distributed unevenly over the wave-front.* There are regions of maximum energy widely separated by large undisturbed areas. When the intensity of light is reduced these regions become more widely separated, but the amount of energy in any one of them does not change; that is, they are indivisible units.

So far all the evidence brought forward in support of the theory has been of an indirect nature; for all ordinary optical phenomena are average effects, and are therefore incapable of differentiating between the usual electromagnetic theory and the modification of it that we are considering. Sir J. J. Thomson however suggested that if the intensity of light in a diffraction pattern were so greatly reduced that only a few of these indivisible units of energy should occur on a Huygens zone at once the ordinary phenomena of diffraction would be modified. Photographs were taken of the shadow of a needle, the source of light being a narrow slit placed in front of a gas flame. The intensity of the light was reduced by means of smoked glass screens.

Before making any exposures it was necessary to find out what proportion of the light was cut off by these screens. A plate was exposed to direct gas light for a certain time. The gas flame was then shaded by the various screens that were to be used, and other plates of the same kind were exposed till they came out as black as the first plate on being completely developed. The times of exposure necessary to produce this result were taken as inversely proportional to the intensities. Experiments made to test the truth of this assumption showed it to be true if the light was not very feeble.

Five diffraction photographs were then taken, the first with direct light and the others with the various screens inserted between the gas flame and the slit. The time of exposure for the first photograph was obtained by trial, a certain standard of blackness being attained by the plate when fully developed. The remaining times of exposure were taken from the first in the inverse ratio of the corresponding intensities. The longest time was 2000 hours or about 3 months. In no case was there any diminution in the sharpness of the pattern although the plates did not all reach the standard blackness of the first photograph.

In order to get some idea of the energy of the light falling on the plates in these experiments a plate of the same kind was exposed at a distance of two metres from

---

* J. J. Thomson, *Proc. Camb. phil. Soc.* xiv (1907), 417.

a standard candle till complete development brought it up to the standard of blackness. Ten seconds sufficed for this. A simple calculation will show that the amount of energy falling on the plate during the longest exposure was the same as that due to a standard candle burning at a distance slightly exceeding a mile. Taking the value given by Drude for the energy in the visible part of the spectrum of a standard candle, the amount of energy falling on 1 cm.^{2} of the plate is 5×10^{-6} ergs/sec. and the amount of energy per cm.^{3} of this radiation is $1 \cdot 6 \times 10^{-16}$ ergs.

According to Sir J. J. Thomson this value sets an upper limit to the amount of energy contained in one of the indivisible units above.*

* *Editor's note:* The three photographs in pl. 1 were not a part of the original paper, but are included here in view of their historical interest.

PLATE 1

A general view of the apparatus.

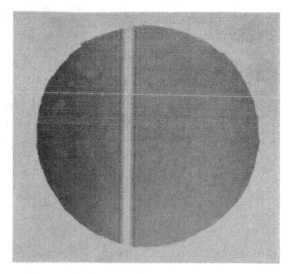

Exposure time 4 minutes.

Exposure time 24 hours.

2

MOTION OF SOLIDS IN FLUIDS WHEN THE FLOW IS NOT IRROTATIONAL

REPRINTED FROM

Proceedings of the Royal Society, A, vol. XCIII (1917), pp. 99–113

The chief interest in the results obtained in the following pages lies in the fact that a mathematical result has been obtained concerning the motion of solids in fluids which is verified accurately when recourse is had to experiment, with real solids moving in real fluids. This is so exceptional a circumstance that it is hoped that the interest which it gives to the mathematical work will serve to extenuate, to a certain extent, the clumsiness of the methods employed.

The problem solved is two-dimensional. An infinite cylindrical body of any cross-section moves in a uniformly rotating fluid with its generators parallel to the axis of rotation. The stream lines and the reaction between the solid and the fluid are found.

Suppose that a stream function ψ' has been found which represents the irrotational motion of an incompressible fluid when a cylindrical solid (or several cylindrical solids) of the required cross-section is moved in an assigned manner starting from rest in a fluid which has a given boundary or has a given irrotational motion at infinity. ψ' is a function of x and y, the co-ordinates of a point in a plane perpendicular to the axis of rotation, and of t, the time.

Since the motion is irrotational ψ' satisfies the relation $\nabla^2\psi' = 0$ everywhere, and $-\partial\psi'/\partial s = V_n'$ at the solid boundaries, where V_n' represents the velocity normal to the boundary of a point on the surface of a cylindrical solid moving in the fluid, and $\partial\psi'/\partial s$ represents the rate of change in ψ' measured in a direction along the solid boundary. These, together with the conditions at infinity, if the fluid is unenclosed, are the necessary and sufficient conditions for determining ψ'. The components of velocity of the fluid are then

$$u' = -\partial\psi'/\partial y \quad \text{and} \quad v' = \partial\psi'/\partial x.$$

Now consider the function

$$\psi = \psi' + \tfrac{1}{2}\omega(x^2 + y^2), \tag{1}$$

where ω is a constant both in regard to space and to time. It satisfies the dynamical equation of motion

$$D(\nabla^2\psi)/Dt = 0$$

for $\nabla^2\psi = 2\omega$, which is constant; and it is the stream function of the fluid motion obtained when the whole system represented by ψ' is rotated with uniform angular velocity ω about the origin. The boundary conditions of the rotating system are evidently satisfied if the cylindrical solids move relative to the rotating system in the same way that they moved relative to fixed axes in the case of the motion represented by ψ'. Hence it appears that the system consisting of the cylindrical solids and the fluid in which they move may be rotated uniformly without affecting the motion of the fluid relative to the rotating system, provided the cylinders are constrained to move, relative to the rotating system, in the same way that they moved, relative to fixed axes, when the system was not rotating.* If, however, the solids are free to move under the action of their own inertia and of the pressure of the fluid, the rotation will make a considerable difference to the relative motion of the solids and the fluid. If therefore becomes important to find the pressure at any point.

Let p' be the pressure at the point (x, y) in the irrotational case, and let p be the pressure when the whole system is rotated.

The equations for p and p' are

$$\left.\begin{aligned} -\frac{1}{\rho}\frac{\partial p'}{\partial x} &= \frac{\partial u'}{\partial t} + u'\frac{\partial u'}{\partial x} + v'\frac{\partial u'}{\partial y} \\ -\frac{1}{\rho}\frac{\partial p'}{\partial y} &= \frac{\partial v'}{\partial t} + u'\frac{\partial v'}{\partial x} + v'\frac{\partial v'}{\partial y} \end{aligned}\right\}, \tag{2}$$

$$\left.\begin{aligned} -\frac{1}{\rho}\frac{\partial p}{\partial x} &= \frac{\delta u}{\delta t} + u\frac{\partial u}{\partial x} + v\frac{\partial u}{\partial y} \\ -\frac{1}{\rho}\frac{\partial p}{\partial y} &= \frac{\delta v}{\delta t} + u\frac{\partial v}{\partial x} + v\frac{\partial v}{\partial y} \end{aligned}\right\}, \tag{3}$$

where u and v are the components of velocity in the rotational motion.

The symbol $\delta u/\delta t$ has been used to represent the rate of change, at a point fixed in space, in the component of velocity parallel to a fixed direction which momentarily coincides with the axis of x.

This is not the same thing as $\partial u/\partial t$. Since u may be regarded as being known in terms of the co-ordinates (x and y), referred to rotating axes, and t the time, $\partial u/\partial t$ represents the rate of change in the component of the velocity of the fluid which is parallel to the rotating axis of x at a point which moves with the axes. It is evident that $\partial u/\partial t = \partial u'/\partial t$ and $\partial v/\partial t = \partial v'/\partial t$.

To find the value of $\delta u/\delta t$ and $\delta v/\delta t$, consider the rate of change, at a point fixed in space, in the component of velocity parallel to a fixed direction which momentarily makes an angle β with the axis of x.

The component of the velocity of the fluid parallel to this direction is

$$u\cos\beta + v\sin\beta.$$

After a short interval of time, δt, the co-ordinates of the fixed point relative to the moving axes are
$$x + \omega y\,\delta t \quad \text{and} \quad y - \omega x\,\delta t.$$

---

* It will be shown later that this proposition cannot be extended to the case of the three-dimensional motion.

The components of velocity parallel to the rotating axes (which now make an angle $\omega\, \delta t$ with their previous positions) are

$$u+\left(\frac{\partial u}{\partial t}+\omega y\,\frac{\partial u}{\partial x}-\omega x\,\frac{\partial u}{\partial y}\right)\delta t \quad \text{and} \quad v+\left(\frac{\partial v}{\partial t}+\omega y\,\frac{\partial v}{\partial x}-\omega x\,\frac{\partial v}{\partial y}\right)\delta t.$$

The component of velocity parallel to the fixed direction is therefore

$$\left[u+\left(\frac{\partial u}{\partial t}+\omega y\,\frac{\partial u}{\partial x}-\omega x\,\frac{\partial u}{\partial y}\right)\delta t\right]\cos\left(\beta-\omega\,\delta t\right)+\left[v+\left(\frac{\partial v}{\partial t}+\omega y\,\frac{\partial v}{\partial x}-\omega x\,\frac{\partial v}{\partial y}\right)\delta t\right]\sin\left(\beta-\omega\,\delta t\right).$$

The rate of change in velocity parallel to the fixed direction is therefore

$$\left(\frac{\partial u}{\partial t}+\omega y\,\frac{\partial u}{\partial x}-\omega x\,\frac{\partial u}{\partial y}-\omega v\right)\cos\beta+\left(\frac{\partial v}{\partial t}+\omega y\,\frac{\partial v}{\partial x}-\omega x\,\frac{\partial v}{\partial y}+\omega u\right)\sin\beta.$$

Putting $\beta = 0$ we find

$$\frac{\delta u}{\delta t}=\frac{\partial u}{\partial t}+\omega y\,\frac{\partial u}{\partial x}-\omega x\,\frac{\partial u}{\partial y}-\omega v$$

and putting $\beta = \frac{1}{2}\pi$,

$$\frac{\delta v}{\delta t}=\frac{\partial v}{\partial t}+\omega y\,\frac{\partial v}{\partial x}-\omega x\,\frac{\partial v}{\partial y}+\omega u.$$

Substituting these values in (3), subtracting equations (2) and substituting for u and v, it will be found that

$$-\frac{1}{\rho}\frac{\partial}{\partial x}(p-p')=-\omega^{2}x-2\omega v' \quad \text{and} \quad -\frac{1}{\rho}\frac{\partial}{\partial y}(p-p')=-\omega^{2}y+2\omega u'.$$

These equations may be integrated in the form

$$(p-p')/\rho=\tfrac{1}{2}\omega^{2}(x^{2}+y^{2})+2\omega\psi'. \tag{4}$$

At this stage it is easy to prove that the proposition proved on the previous page cannot be extended to the case of three-dimensional motion.

Let u', v', w' be the components of the velocity of a fluid in irrotational motion. Suppose that the motion defined by

$$u=u'-\omega y, \quad v=v'+\omega y, \quad w=w',$$

is possible.

Proceeding as before, it will be found that the pressure equations can be reduced to

$$-\frac{1}{\rho}\frac{\partial}{\partial x}(p-p')=-\omega^{2}x-2\omega v',$$

$$-\frac{1}{\rho}\frac{\partial}{\partial y}(p-p')=-\omega^{2}y+2\omega u',$$

$$-\frac{1}{\rho}\frac{\partial}{\partial z}(p-p')=0.$$

These are not consistent unless u' and v' are independent of z; that is, unless the motion is two-dimensional.

Let us now apply (4) to find the resultant force and couple which the fluid pressure exerts on a solid moving in a rotating fluid.

Let F_x', F_y', and G' be the resultant forces and couple due to fluid pressure on the solid in the case when the system is not rotating. F_x' and F_y' are supposed to act at the centre of gravity C of the area of the cross-section of the solid. Let F_x, F_y, and G be the corresponding quantities in the case when the system is rotating.

If χ represents the angle between the normal to the surface of the solid and the axis of x, then

$$F_x - F_x' = -\int (p - p') \cos \chi \, ds,$$

$$F_y - F_y' = -\int_s (p - p') \sin \chi \, ds,$$

$$G - G' = \int_s (p - p')(\eta \cos \chi - \xi \sin \chi) \, ds,$$

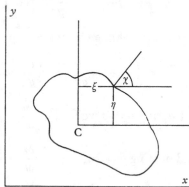

where ξ and η are the co-ordinates of a point on the surface referred to axes parallel to the axes of x and y, and passing through C; and the integrals are taken round the surface of the solid. Substituting the value of $p - p'$ given by (4) these may be integrated.

Thus

$$-\frac{F_x - F_x'}{\rho} = \frac{1}{\rho}\int_s (p - p') \cos \chi \, ds = \tfrac{1}{2}\omega^2 \int (x^2 + y^2) \cos \chi \, ds + 2\omega \int \psi' \cos \chi \, ds. \quad (5)$$

Now $\int_s y^2 \cos \chi \, ds$ vanishes since $\cos \chi \, ds = dy$. Also

$$\tfrac{1}{2}\omega^2 \int_s x^2 \cos \chi \, ds = \omega^2 A x_0,$$

where A is the area of cross-section of the solid, and (x_0, y_0) are the co-ordinates of its centroid.

$2\omega \int_s \psi' \cos \chi \, ds$ may be integrated by parts. It then becomes

$$-2\omega \int_s y \frac{d\psi'}{ds} \, ds = -2\omega \int_s (y_0 + \eta)\frac{d\psi'}{ds} \, ds = -2\omega \int_s \eta \frac{d\psi'}{ds} \, ds, \quad (6)$$

since $\int_s y_0 \frac{d\psi'}{ds} \, ds$ evidently vanishes.

Now $-\partial \psi'/\partial s$ represents the velocity of the fluid normal to the surface of the solid. The boundary condition which must be satisfied by ψ' is

$$-\partial \psi'/\partial s = (\dot{x}_0 - \Omega \eta) \cos \chi + (\dot{y}_0 + \Omega \xi) \sin \chi,$$

where Ω is the angular velocity of the body.

Substituting in (6) and remembering that $\cos\chi\,ds = d\eta$ and $\sin\chi\,ds = -d\xi$, it will be found that

$$2\omega\int_s \psi'\cos\chi\,ds = 2\omega\int_s \eta(\dot{x}_0 - \Omega\eta)\,d\eta - 2\omega\int_s (\dot{y}_0 + \Omega\xi)\,\eta\,d\xi.$$

The first of these integrals vanishes and the second may be written

$$-2\omega\dot{y}_0\int_s \eta\,d\xi - 2\omega\Omega\int_s \eta\xi\,d\xi,$$

Now $\int_s \eta\,d\xi = -A$ and $\int_s \eta\xi\,d\xi = 0$, since C is the centroid of the area of cross-section.

Hence from (5) $\qquad -(F_x - F'_x)/\rho = \omega^2 A x_0 + 2\omega A\dot{y}_0.$

Similarly it will be found that

$$-(F_y - F'_y)/\rho = \omega^2 A y_0 - 2\omega A\dot{x}_0.$$

It will be noticed that $\omega^2 A x_0$ and $\omega^2 A y_0$ are the components of a force $\omega^2 AR$ acting radially, R being the distance of C from the centre of rotation. Also $2\omega A\dot{y}_0$ and $-2\omega A\dot{x}_0$ are the components of a force $2\omega AQ$ acting at right angles to the direction of motion of C relative to the rotating axes, Q being the relative velocity of C.

Now consider the couple $G - G'$ due to the rotation

$$\frac{G - G'}{\rho} = \int_s \frac{p - p'}{\rho}(\eta\,d\eta + \xi\,d\xi).$$

Substituting from (4)

$$G - G' = \frac{\omega^2}{2}\int_s \{(x_0 + \xi)^2 + (y_0 + \eta)^2\}(\eta\,d\eta + \xi\,d\xi) + 2\omega\int_s \psi'(\eta\,d\eta + \xi\,d\xi).$$

Neglecting all terms which contain only powers of ξ or of η* and integrating the second integral by parts, this becomes

$$\frac{G - G'}{\rho} = \frac{\omega^2}{2}\left[2x_0\int_s \xi\eta\,d\eta + 2y_0\int_s \eta\xi\,d\xi + \int_s \xi^2\eta\,d\eta + \int_s \eta^2\xi\,d\xi\right] - 2\omega\int_s \frac{\xi^2 + \eta^2}{2}\frac{d\psi'}{ds}\,ds.$$

Now $\qquad\qquad \int_s \xi\eta\,d\eta = \int_s \eta\xi\,d\xi = 0$

and $\qquad\qquad \int_s \xi^2\eta\,d\eta + \int_s \eta^2\xi\,d\xi = \tfrac{1}{2}\int_s d(\xi^2\eta^2) = 0;$

also since $\qquad -\frac{\partial\psi'}{\partial s}\,ds = (\dot{x}_0 - \Omega\eta)\,d\eta + (\dot{y}_0 + \Omega\xi)\,d\xi,$

$(G - G')/\rho$ reduces to

$$2\omega\left[\dot{x}_0\int_s \tfrac{1}{2}\xi^2\,d\eta - \dot{y}_0\int_s \tfrac{1}{2}\eta^2\,d\xi - \Omega\int_s \xi^2\eta\,d\eta - \Omega\int_s \eta^2\xi\,d\xi\right] = 0.$$

* For they vanish when integrated round a closed contour.

The forces due to fluid pressure, which act on a body moving in an assigned manner in a rotating fluid, may therefore be regarded as being made up as follows:

(1) The forces F'_x, F'_y, and the couple G' which would act on the body if it moved in the same way relatively to the fluid at rest.

(2) A force equivalent to $\rho\omega^2 AR$ acting towards the centre of rotation through C.

(3) A force $2\rho\omega AQ$ acting at C in a direction perpendicular to the relative motion of C and the rotating axes, and directed to the left if the rotation of the fluid is anti-clockwise.

We can therefore solve any problem on the motion of cylindrical solids in a rotating fluid if we can obtain a solution of a similar problem respecting the motion of the solids in a fluid at rest.

Now, consider the forces and the couple which it is necessary to apply to a solid body of mass M, in order that it may move in an assigned manner relatively to rotating axes. Suppose that a force F' and a couple G' must be applied at its centre of gravity, in order that it may move in the assigned manner relatively to fixed axes. The additional force which it is necessary to apply when the system is rotating uniformly with angular velocity ω may be shown to consist of a force $2M\omega Q$ perpendicular to the direction of the velocity Q of the centre of gravity relative to the rotating system, together with a force $M\omega^2 R$ acting through the centre of gravity towards the centre of rotation.

It will be noticed that, if the position of the centre of gravity of the solid coincides with the centroid of its cross-section, and if the mass per unit length of the solid is equal to ρA, that is to say, if the mass and the centre of gravity of the solid are the same as those of the fluid displaced, then these forces are the same as those which act on the solid, owing to the additional pressures in the fluid due to its rotation.

These considerations lead to the conclusion that, if a solid of the same density as the fluid be moved along a certain path by certain assigned external forces, then a uniform rotation of the whole system, including the external force, makes no difference to the path which the solid pursues relative to the system.

This theorem applies only to the case of two-dimensional motion. In the case of a finite cylinder, for instance, its seems almost obvious that the pressures due to the rotation must fall off towards its ends. It is natural to suppose, therefore, that the reaction of the fluid would not be sufficient to hold a finite cylinder in its path when the whole system is rotated.

The case of a sphere moving in a rotating fluid presents considerable mathematical difficulties, but the initial motion has been investigated by Mr J. Proudman, who has kindly consented to allow the author to make use of his results, though they are not yet published.* He finds that, if a sphere of volume V starts from rest in the rotating fluid and moves with uniform velocity along a straight line relative to the rotating system, it is acted on initially by a force $V\rho\omega^2 R$ directed towards the centre of the rotation (which is at a distance R from the centre of the sphere) and by a

---

* Since the above was written Mr Proudman has published his results. They appeared in *Proc. Roy. Soc.*, A, xcii (1916), 408–24.

force $\frac{1}{2}V\rho Q\omega$ acting in a direction perpendicular to its path. But in order that a sphere of the same density as the fluid, that is, one whose mass is $V\rho$, may move along a straight path relative to the rotating system, it must be acted on by a force $V\rho\omega^2 R$ directed towards the centre of rotation and by a force $2V\rho\omega Q$ perpendicular to its path.

The forces due to fluid pressure are not sufficient to supply the second of these. If, therefore, the sphere were drawn through the rotating fluid by means of a string, it would not move in the direction the string was pulling it, but would be deflected to the left if the fluid were rotating clockwise, and to the right if it were rotating anticlockwise. On the other hand, if a cylinder of the same density were drawn through the rotating fluid, the force necessary to hold it in its straight path would be supplied by the fluid pressure. The cylinder would therefore move straight through the fluid in the direction the string was pulling it.

These conclusions have been tested and completely verified by means of experiments made by the author in the Cavendish Laboratory with water in a rotating tank.

Experiments made with a Rotating Tank of Water

A glass tank full of water was mounted so that it could be rotated about a vertical axis at various speeds by means of an electric motor. The speeds varied from 2 to 6 seconds per revolution. Two bodies were prepared, one cylindrical and the other spherical. The former consisted of a piece of thin-walled brass tube about 6 in. × $\frac{3}{4}$ in. stopped at the end with waxed cork, while the other was a spherical glass bulb. They were weighted until they would fall very slowly through water, and the positions of the weights were adjusted till they would stay almost at rest in any position in the water. The centres of gravity of the bodies were then coincident with the centres of gravity of the water displaced by them.

A simple mechanism was next devised to tow them through the tank from one end to the other. It consisted of a wood pulley about 4 inches in diameter, mounted on a vertical spindle which was driven into a wood bridge, fixed to the tank over the middle of it. This spindle coincided with the axis of rotation of the tank. Cotton was then wound round the pulley, passed through some small rings screwed into a board fixed to one end of the tank, and led horizontally along the tank to the cylinder or sphere, which was fixed at the other end.

The body was held in a holder while the tank and water were being brought to a state of uniform rotation. A device was arranged so that the holder could release the body and at the same moment the wood pulley on which the cotton was wound could be fixed in space. As the tank was then rotating round the pulley the cotton wound up round it, and pulled the bodies along the middle of the tank from one end to the other.

Result. It was found that the cylinder moved straight through the middle of the tank. Even when the tank was rotating very rapidly the cylinder always passed over the central line. The sphere, however, was violently deviated to the left (the tank was rotating clockwise). When the tank was rotated quite slowly, about once in 6 seconds, the sphere would not quite touch the side, though it never came up to the stop at the other end from a direction less than 45° away from the central line. When the tank rotated more rapidly the complete path could never be seen, because the sphere always hit the side of the tank before it had gone more than a few inches in the direction along which the cotton was trying to pull it. After striking the side of the tank the sphere would follow the side along, touching all the time, till it got to a position close to the other end where the string was pulling in a direction making an angle of about 50° with the side of the tank. It would leave the side and approach the point towards which the cotton was pulling it along a curved path.

The accuracy with which the experiments just described verify the hydro-dynamical theory of rotating fluids is at first sight most surprising. Besides the fact that there is apparently no other case in which experiments made with real solids moving in real fluids agree with the predictions of hydrodynamics, it is known that the stream lines of a real fluid round a circular cylinder in particular bear no resemblance to the stream lines used in the ordinary hydrodynamical theory. It will be noticed, however, that in order that there may be agreement between theory and experiment in the particular respect to which attention has been drawn, it is unnecessary that the actual flow pattern shall be the same as the flow pattern contemplated in the ordinary hydrodynamical theory. All that is necessary is that the flow pattern in the case of the cylinder shall be two-dimensional, while that in the case of the sphere shall be three-dimensional.

Experiments with Vortex Rings in a Rotating Fluid

The theory explained on pp. 8, 9 leads to the conclusion that if a homogeneous solid, which is not cylindrical, be projected in a rotating fluid of the same density as itself it will be deviated to the left if the rotation is clockwise, and to the right if the rotation is anti-clockwise, of the path it would pursue through the fluid if the whole system were not rotating. Now, a vortex ring affects the fluid round it in much the same way as a solid ring of the same dimensions as the cyclic portion of the flow system. If it is projected through a fluid at rest it travels along a straight line. We should expect, therefore, that if a vortex ring were projected through a rotating fluid it would follow a curved path relative to the fluid, being deviated to the left if the fluid were rotating clockwise.

This conclusion was tested experimentally and found to be correct. A small vortex box with a rubber top and a circular hole in the side was made. This was filled with a solution of fluorescene and placed in one end of the tank, which was filled with water and held fixed. On striking the rubber lightly a vortex ring was produced which travelled straight down the tank and struck the middle of the opposite end.

The same experiment was repeated when the tank and vortex box were rotating. On tapping the box, rings started out in the same direction as before, but were deflected in a curved path, so that they hit the side instead of the end of the tank. By tapping the box quite lightly and rotating the tank fairly rapidly the rings could be made to turn in such small circles that they came round and struck the vortex box again without touching the side of the tank on the way. They would, in fact, turn in a circle whose diameter was only about four times the diameter of the rings.

It was pointed out by Dr F. W. Aston, to whom the writer was showing this experiment, that the rings appeared to remain parallel to a plane fixed in space, while the rest of the fluid rotated. He suggested that the gyroscopic action prevented the ring from being deviated from this plane, and that in order that the ring might move relative to the fluid in a direction perpendicular to its plane it would have to move through the fluid along a curved path.

MOTION OF A CIRCULAR CYLINDER IN A FLUID WHICH HAS A STEADY ROTATIONAL MOTION AT INFINITY BUT DOES NOT NECESSARILY ROTATE AS A WHOLE

The results given in the rest of this paper have no immediate practical interest. The author entered on the investigation with a view to getting an idea on how the instability which is known to exist in a uniformly shearing laminar flow would be likely to manifest itself, and to find out whether the characteristics of the motion of solids in rotating fluids, which have been discussed in the first part of this paper, have any counterpart in the case of solids moving in a fluid whose undisturbed motion is a uniform laminar flow.

The problem of finding the motion of a circular cylinder in a rotationally moving fluid divides itself naturally into two parts, that of finding the stream function for a given motion of the cylinder, and that of finding the force which the pressure associated with that stream function exerts on the cylinder. The stream function for a certain type of rotational flow in which the vorticity is uniform will now be found.

Let (r, θ) be the polar co-ordinates of a point referred to axes through the centre of the cylinder, and let (x_0, y_0) be the co-ordinates of the centre of the cylinder referred to fixed axes, so that the equation $\theta = 0$ represents a line parallel to the axis of x at a distance y_0 from it.

Consider the stream function

$$\psi = \tfrac{1}{2}\zeta r^2 + \left(Ar + \frac{B}{r}\right)\cos\theta + \left(Cr + \frac{D}{r}\right)\sin\theta + \left(Er^2 + \frac{F}{r^2}\right)\cos 2\theta + \left(Gr^2 + \frac{H}{r^2}\right)\sin 2\theta.$$

$$(7)$$

It satisfies the equation $\nabla^2\psi = 2\zeta$ everywhere.

If, therefore, the constants A, B, C, etc., be so chosen that the boundary condition

$$\frac{1}{r}\frac{\partial\psi}{\partial\theta} + \dot{x}_0\cos\theta + \dot{y}_0\sin\theta = 0 \tag{8}$$

is satisfied where $r = a$, a being the radius of the cylinder, then ψ is the stream function which represents the motion of a fluid which, if the cylinder were removed, would be moving in accordance with the velocities given by the stream function

$$\psi_1 = \tfrac{1}{2}\zeta r^2 + Ar\cos\theta + Cr\sin\theta + Er^2\cos 2\theta + Gr^2\sin 2\theta. \tag{9}$$

Now (8) must be satisfied for all values of θ; hence we may equate coefficients of $\cos\theta$, $\sin\theta$, $\cos 2\theta$, and $\sin 2\theta$, separately to zero. In this way the following relations between the constants are determined:

$$A + \frac{B}{a^2} - \dot{y}_0 = 0, \quad C + \frac{D}{a^2} + \dot{x}_0 = 0, \quad Ea + \frac{F}{a^3} = 0, \quad Ga + \frac{H}{a^3} = 0. \tag{10}$$

It will be noticed that ψ_1, the stream function of the fluid before the introduction of the cylinder, is expressed in terms of co-ordinates referred to moving axes. In order to find the motion of a cylinder in a fluid whose undisturbed motion before the introduction of the cylinder is known with reference to fixed axes, we must transform (9) so as to give ψ_1 in terms of co-ordinates x and y referred to the fixed axes used to fix the position of the cylinder. The transformation is performed by putting

$$r\cos\theta = x - x_0, \quad r\sin\theta = y - y_0.$$

ψ_1 then becomes

$$\tfrac{1}{2}\zeta\{(x-x_0)^2 + (y-y_0)^2\} + A(x-x_0) + C(y-y_0)$$
$$+ E\{(x-x_0)^2 - (y-y_0)^2\} + 2G(x-x_0)(y-y_0). \tag{11}$$

If the motion of the fluid before the introduction of the cylinder be given by the function

$$\psi_1 = \tfrac{1}{2}\zeta(x^2+y^2) + A'x + C'y + E'(x^2-y^2) + 2G'xy, \tag{12}$$

where A', C', E', G' are given constants, we find, by equating coefficients of x, y, x^2, xy, y^2 in (11) and (12), the following relations determining A, C, E, G, in terms of A', C', E', G', x_0 and y_0,

$$-\zeta x_0 + A - 2Ex_0 - 2Gy_0 = A', \quad -\zeta y_0 + C + 2Ey_0 - 2Gx_0 = C', \left.\vphantom{\begin{matrix}a\\b\end{matrix}}\right\}$$
$$E = E', \quad G = G'. \tag{13}$$

Solving (10) and (13) we obtain the following values of A, B, C, D, E, F, G, H,

$$\left.\begin{aligned}
A &= A' + \zeta x_0 + 2E'x_0 + 2G'y_0,\\
B &= -a^2(-\dot{y}_0 + A' + \zeta x_0 + 2E'x_0 + 2G'y_0),\\
C &= C' + \zeta y_0 - 2E'y_0 + 2G'x_0,\\
D &= -a^2(\dot{x}_0 + c' + \zeta y_0 - 2E'y_0 + 2G'x_0),\\
E &= E', \quad F = -E'a^4,\\
G &= G', \quad H = -G'a^4.
\end{aligned}\right\} \tag{14}$$

Hence the stream function is obtained for the motion of a cylinder in a fluid whose undisturbed motion may be expressed by a stream function of the form ψ_1. The two particular cases which are of the greatest interest are those of uniform rotation, for which $\psi_1 = \frac{1}{2}\omega(x^2 + y^2)$, and uniformly shearing laminar flow, for which $\psi_1 = -\frac{1}{2}\alpha y^2$, α being the rate of shear. Before discussing these cases, however, it is necessary to find an expression in terms of ψ for the force on the cylinder.

In general there does not appear to be a simple pressure integral like Bernoulli's for the case of irrotational motion, or the expression given in equation (4) for the pressure in a rotating fluid. It is necessary to go back to the original equations of motion of the fluid.

If the rate of change in pressure along a direction which makes an angle χ with the axis of x be represented by the symbol dp/ds_χ, ds_χ representing an element of length in the direction χ, then the equation of motion is

$$-\frac{1}{\rho}\frac{dp}{ds_\chi} = \frac{Dv_\chi}{Dt},$$

where v_χ represents the component of velocity of the fluid in the direction χ. Its value may be found in terms of ψ by the equation

$$v_\chi = \frac{\partial\psi}{\partial r}\sin(\chi - \theta) - \frac{1}{r}\frac{\partial\psi}{\partial\theta}\cos(\chi - \theta). \tag{15}$$

Now $\dfrac{Dv}{Dt}$ may be written

$$\frac{\delta v_\chi}{\delta t} - \frac{\partial\psi}{r\,\partial\theta}\frac{\partial v_\chi}{\partial r} + \frac{\partial\psi}{\partial r}\frac{\partial v_\chi}{r\,\partial\theta} \tag{10}$$

where $\delta v_\chi/\delta t$ represents, as before, the rate of change in v_χ at a point fixed in space.

If δr, $\delta\theta$ are the changes in the co-ordinates of a fixed point in time δt,

$$\frac{\delta v_\chi}{\delta t} = \frac{\partial v_\chi}{\partial t} + \frac{\partial v_\chi}{\partial\theta}\frac{\delta\theta}{\delta t} + \frac{\partial v_\chi}{\partial r}\frac{\delta r}{\delta t}, \tag{17}$$

where $\partial v_\chi/\partial t$ represents the rate of change in v_χ at a point fixed relative to the moving axes. The value of $\partial v_\chi/\partial t$ may be obtained by differentiating the expression (15) with respect to time, which occurs in all terms which contain $x_0, y_0, \dot{x}_0$ or $\dot{y}_0$.

The values of δr and $\delta\theta$ may be found by resolving the velocity of c, the centre of the cylinder, along and perpendicular to r.

Thus $\quad \delta r = -(\dot{x}_0\cos\theta + \dot{y}_0\sin\theta)\,\delta t, \quad r\,\delta\theta = (\dot{x}_0\sin\theta - \dot{y}\cos\theta)\,\delta t;$

substituting in (17),

$$\frac{\delta v_\chi}{\delta t} = \frac{\partial v_\chi}{\partial t} + (\dot{x}_0\sin\theta - \dot{y}_0\cos\theta)\frac{\partial v_\chi}{r\,\partial\theta} - (\dot{x}_0\cos\theta + \dot{y}_0\sin\theta)\frac{\partial v_\chi}{\partial r};$$

substituting this in (16),

$$-\frac{1}{\rho}\frac{\partial p}{\partial s_\chi} = \frac{Dv_\chi}{Dt} = \frac{\partial v_\chi}{\partial t} - \left(\frac{\partial\psi}{r\,\partial\theta} + \dot{x}_0\cos\theta + \dot{y}_0\sin\theta\right)\frac{\partial v_\chi}{\partial r} + \left(\frac{\partial\psi}{\partial r} + \dot{x}_0\sin\theta - \dot{y}_0\cos\theta\right)\frac{\partial v_\chi}{r\,\partial\theta}.$$

Now $-\partial\psi/r\,\partial\theta - \dot{x}_0\cos\theta - \dot{y}_0\sin\theta$ represents the component, normal to the surface, of the relative velocity of the fluid and the cylinder. It must therefore vanish.

Hence

$$-\frac{1}{\rho}\frac{\partial p}{\partial s_\chi} = \frac{\partial v_\chi}{\partial t} + \left(\frac{\partial \psi}{\partial r} + \dot{x}_0 \sin\theta - \dot{y}_0 \cos\theta\right)\frac{\partial v_\chi}{r\,\partial\theta}$$

and substituting for v_χ from (15),

$$-\frac{1}{\rho}\frac{\partial p}{\partial s} = \sin(\chi-\theta)\left[\frac{\partial^2\psi}{\partial r\,\partial t} + \left(\frac{\partial\psi}{\partial r} + \dot{x}_0\sin\theta - \dot{y}_0\cos\theta\right)\left(\frac{\partial^2\psi}{r\,\partial r\,\partial\theta} - \frac{\partial\psi}{r^2\,\partial\theta}\right)\right]$$

$$+ \cos(\chi-\theta)\left[-\frac{1}{r}\frac{\partial^2\psi}{\partial\theta\,\partial t} + \left(\frac{\partial\psi}{\partial r} + \dot{x}_0\sin\theta - \dot{y}_0\cos\theta\right)\left\{-\frac{\partial\psi}{r\,\partial\theta} - \frac{1}{r}\frac{\partial}{r\,\partial\theta}\left(\frac{\partial\psi}{r\,\partial\theta}\right)\right\}\right].$$

If χ be put equal to $\frac{1}{2}\pi + \theta$, we obtain the variation in pressure round the cylinder in the form

$$-\frac{1}{\rho}\left[\frac{\partial p}{r\,\partial\theta}\right]_{r=a} = \left[\frac{\partial^2\psi}{\partial r\,\partial t} + \left(\frac{\partial\psi}{\partial r} + \dot{x}_0\sin\theta - \dot{y}_0\cos\theta\right)\frac{\partial}{\partial r}\left(\frac{1}{r}\frac{\partial\psi}{\partial\theta}\right)\right]_{r=a}. \qquad (18)$$

If F_x and F_y represent the components of the resultant force acting on the cylinder due to fluid pressure,

$$F_x = -\int_0^{2\pi} p\cos\theta\,a\,d\theta, \qquad F_y = -\int_0^{2\pi} p\sin\theta\,a\,d\theta.$$

These may be integrated by parts.

F_x then becomes $a^2 \int_0^{2\pi}\left[\dfrac{\partial p}{r\,\partial\theta}\right]_{r=a}\sin\theta\,d\theta$ and $F_y = -a^2 \int_0^{2\pi}\left[\dfrac{\partial p}{r\,\partial\theta}\right]_{r=a}\cos\theta\,d\theta.$ (19)

By substituting the value obtained for $(\partial p/r\,\partial\theta)_{r=a}$ in (18) we can find the force exerted by fluid pressure when the cylinder has any assigned motion for which a stream function can be found.

This method will be applied to two particular cases. In Case (1) the general motion of the fluid is one of uniform rotation. This problem has been solved already in the first part of this paper, but it seems worth while to verify the calculation. In Case (2) the general motion of the fluid is one of uniform shearing.

Case 1. The stream function of the general motion of the fluid is $\psi_1 = \frac{1}{2}\omega(x^2 + y^2)$. In this case, then, $\zeta = \omega$ and $A' = C' = E' = G' = 0$. The stream function of the motion round the cylinder is

$$\psi = \omega\frac{r^2}{2} + \left\{\omega x_0 r - \frac{a^2}{r}(-\dot{y}_0 + \omega x_0)\right\}\cos\theta + \left\{\omega y_0 r - \frac{a^2}{r}(\dot{x}_0 + \omega y_0)\right\}\sin\theta.$$

Substituting in (18) the value of $(\partial p/r\,\partial\theta)_{r=a}$ may be found, and substituting this value in (19) it will be found that

$$-F_x = \pi\rho a^2(\ddot{x}_0 + 4\omega\dot{y}_0 - 2\omega^2 x_0), \qquad F_y = \pi\rho a^2(\ddot{y}_0 + 4\omega\dot{x}_0 + 2\omega^2 y_0).$$

If these expressions be transformed by the transformation

$$x_0 = R\cos(\phi + \omega t), \qquad y_0 = R\sin(\phi + \omega t),$$

so that R, ϕ, are the polar co-ordinates of a point referred to axes which rotate with the fluid, it will be found that the forces F_x, F_y, may be resolved into components F_R along R, and F_ϕ perpendicular to it where

$$F_R = \pi \rho a^2 \{ -\ddot{R} + R\dot{\phi}^2 - 2R\omega\dot{\phi}^2 - R\omega^2 \},$$

$$F_\phi = \pi \rho a^2 \{ -R\ddot{\phi} - 2\dot{R}\dot{\phi} + 2\omega\dot{R} \}.$$

This agrees with the results obtained on p. 8, for the force whose components are F_R and F_ϕ may be regarded as being made up in the following way: (1) a force $\pi \rho a^2 \times$ (acceleration of the cylinder relative to the rotating axes); (2) a force $\pi \rho a^2 \omega^2 R$ acting towards the centre of rotation; and (3) a force $2\pi \rho a^2 \omega \times$ (velocity of the cylinder relative to the rotating axes) acting at right angles to the direction of relative motion. These are evidently the same as the three forces discussed on p. 8.

Case 2. The general motion of the fluid is one of uniform shearing. The fluid moves parallel to the axis of x with velocity αy, which increases at a uniform rate as y increases. In this case $\psi_1 = -\frac{1}{2}\alpha y^2$, which may be written

$$\psi_1 = -\tfrac{1}{4}\alpha(x^2+y^2) + \tfrac{1}{4}\alpha(x^2-y^2).$$

Comparing this with (12) it appears that $\zeta = -\frac{1}{2}\alpha$ and $E' = \frac{1}{4}\alpha$, while

$$A' = C' = G' = 0.$$

Hence from (7) and (14)

$$\psi = -\tfrac{1}{4}\alpha r^2 + \frac{a^2 \dot{y}_0}{r}\cos\theta - \left\{\alpha y_0 r + \frac{a^2}{r}(\dot{x}_0 - \alpha y_0)\right\}\sin\theta + \tfrac{1}{4}\alpha\left(r^2 - \frac{a^4}{r^2}\right)\cos 2\theta.$$

Hence differentiating and putting $r = a$,

$$\left[\frac{\partial^2 \psi}{\partial r\, \partial t}\right]_{r=a} = -\ddot{y}_0\cos\theta + \ddot{x}_0\sin\theta - 2\alpha\dot{y}_0\sin\theta,$$

$$\left[\frac{\partial\psi}{\partial r} + \dot{x}_0\sin\theta - \dot{y}_0\cos\theta\right]_{r=a} = \alpha(\cos 2\theta - \tfrac{1}{2}) - 2\dot{y}_0\cos\theta + 2(\dot{x}_0 - \alpha y_0)\sin\theta,$$

$$\left[\frac{\partial}{\partial r}\left(\frac{\partial\psi}{r\,\partial\theta}\right)\right]_{r=a} = \frac{1}{a}\{2\dot{y}_0\sin\theta + (\dot{x}_0 - \alpha y_0)\cos\theta - 2a\alpha\sin 2\theta\}.$$

Hence from (18)

$$-\frac{1}{\rho}\left(\frac{\partial p}{r\,\partial\theta}\right)_{r=a} = \ddot{x}_0\sin\theta - \ddot{y}_0\cos\theta - 2\alpha\dot{y}_0\sin\theta$$

$$+ \frac{2}{a}\{\dot{y}_0\sin\theta + (\dot{x}_0 - \alpha y_0)\cos\theta - \alpha a\sin 2\theta\}\{2(\dot{x}_0 - \alpha y_0)\sin\theta - 2\dot{y}_0\cos\theta + a\alpha(\cos 2\theta - \tfrac{1}{2})\}.$$

Hence from (19)

$$F_x = -\pi\rho a^2\{\ddot{x}_0 - \alpha\dot{y}_0\}, \quad F_y = -\pi\rho a^2\{\ddot{y}_0 + 2\alpha(\dot{x}_0 - \alpha y_0)\}. \tag{20}$$

This result will now be applied to find the motion of a cylinder of the same density as the fluid when it is projected from the origin with velocity whose components are U and V.

The equations of motion of the cylinder are

$$\pi\rho a^2 \ddot{x}_0 = -\pi\rho a^2\{\ddot{x}_0 - 2\alpha\dot{y}_0\}, \quad \pi\rho a^2 \ddot{y}_0 = -\pi\rho a^2\{\ddot{y}_0 + 2\alpha(\dot{x}_0 - \alpha y_0)\},$$

or

$$\ddot{x}_0 - \alpha\dot{y}_0 = 0, \quad \ddot{y}_0 = -\alpha(\dot{x}_0 - \alpha y_0).$$

The first of these may be integrated in the form

$$\dot{x}_0 - \alpha y_0 = \text{constant.}$$

That is to say, the component parallel to the axis of x of the relative velocity of the cylinder and the fluid is constant and equal to U. The acceleration of the cylinder in the direction of the axis of x is constant and equal to $-\alpha V$. If $U = 0$, that is to say, if the cylinder is shot off in a direction perpendicular to the direction of shear, then the component of velocity parallel to the axis of y is constant, and the fluid pressure is just sufficient to give the cylinder the acceleration αV, which is necessary in order that the velocity of the cylinder relative to the fluid round it may remain constant. This property of uniformly shearing fluids appears to be analogous to a certain extent to the property of rotating fluids discussed on p. 8.

3

EXPERIMENTS WITH ROTATING FLUIDS

REPRINTED FROM

Proceedings of the Royal Society, A, vol. c (1921), pp. 114–21

It is well known that predictions about fluid motion based on the classic hydro-dynamical theory are seldom verified in experiments performed with actual fluids. The explanation of this want of agreement between theory and experiment is to be found chiefly in the conditions at the surfaces of the solid boundaries of the fluid.

The classical hydrodynamical theory assumes that perfect slipping takes place, whereas in actual fluids the surface layers of the fluid are churned up into eddies. In the case of motions which depend on the conditions at the surface, therefore, no agreement is to be expected between theory and experiment. This class of a fluid motion, unfortunately, includes all cases where a solid moves through a fluid which is otherwise at rest.

On the other hand, there are types of fluid which only depend to a secondary extent on the slip at the boundaries. For this reason theoretical predictions about waves and tides, or about the motion of vortex rings, are in much better agreement with observation than predictions about the motion of solids in fluids. Some time ago the present writer* made certain predictions about the motion of solids in rotating fluids, or rather about the differences which might be expected between the motion of solids in a rotating fluid and those in a fluid at rest. These predicted features of the motion did not depend on conditions at the boundaries. It was therefore to be anticipated that they might be verified by experiment. The experiments were carried out and the predictions were completely verified.

In view of the interest which attaches to any experimental verification of theoretical results in hydrodynamics, and more particularly to verifications of those concerning the motion of solids in fluids, it seems worth while to publish photographs showing the experiments in progress. In the second and third part of the paper further experiments are described in which theoretical predictions are verified in experiments with water.

MOTION OF CYLINDER AND SPHERE IN ROTATING FLUID

In these experiments a solid cylinder and a solid sphere, of the same density as water, were drawn through a rotating vessel containing water. The threads by means of which these solids were dragged, passed through small rings attached to the vertical wall of the circular rotating glass vessel in which the water was contained. The solids were initially attached to the opposite point of this vessel to that at which

* Paper 2 above.

the rings were attached. Under these circumstances, if the vessel were not rotating, a symmetrical body like a sphere or cylinder would evidently pass along a diameter through the centre of the apparatus when towed by the threads.

When the vessel was rotating, however, the theoretical prediction* was that the cylinder would pass through the centre of the apparatus just as if the whole system were not rotating, while a sphere, or any symmetrical three-dimensional body, would be deflected and would pass to one side of the centre. The verification of this prediction was first made in an apparatus with which it was difficult to obtain photographs owing to the difficulty of throwing a light through the water towards a camera placed on the axis of rotation above the apparatus. A new apparatus was therefore devised, in which the vertical central spindle used in the previous apparatus was done away with.

Two dishes in the form of circular cylinders were made. Each had a thin plate-glass bottom, about 1 mm. thick. The diameter of one of them was made about $\frac{1}{8}$ in. larger than the other, so that the smaller one would fit inside the larger one, leaving a space of about $\frac{1}{16}$ in. all round. The inner dish was filled about two-thirds full of water, and the outer one was filled quite full. The inner cylinder would then float in the outer one. It was driven round a vertical axis by means of a jet of water projected at its outside surface. When the apparatus was set up so as to run truly, it was found that a very uniform rate of rotation would be obtained by this method. The whole apparatus stood in a trough with a plate-glass bottom, in which the water which overflowed could be collected. This apparatus is shown in fig. 1.

In order to take instantaneous photographs, the apparatus was illuminated from underneath, and a camera with a lens of 15 in. focus was fixed about 6 ft. above it. Two methods of illumination were used. In the first method the direct light from a spark between an aluminium wire and a hole in an aluminium plate was focused, by means of a condenser, on the lens of the camera. In the second method use was made of diffused light, a mercury vapour spark being used to illuminate a ground-glass plate placed close under the apparatus. The photographs (a), (b), (e) and (f) on pl. 1, were taken by the first method, while (c) and (d) were taken by the second. It is worth pointing out that the apparatus is very simple, and that it is easy to project all the experiments to be described in this paper on to a screen by means of a lantern.

It was thought at first that it might be difficult to arrange to tow the solids through the rotating basin in such a way that no appreciable variation in the speed of rotation would occur. To avoid this difficulty, the threads used to tow the solids were led from the edge of the apparatus back to the centre, and were pulled upwards from that point. The method adopted for doing this was to fix a transparent celluloid bridge across the basin in such a position that its centre line ran from the initial position of the solid to the point towards which the threads were towing it. This transparent bridge shows up clearly as a broad band in pl. 1 (a), (b). The threads passed through a small hole in the centre of the bridge, and were attached to a small wire ring, which rested on it in the middle.

* Paper 2 above.

To perform the experiment, the solid, which was usually made of wax or boxwood, was placed on small pins projecting from the side of the trough. The threads were then stretched across the basin to the opposite side, and led from there up to a ring close under the celluloid bridge. From this point they passed along the underside of the bridge to the centre, where they passed up through the central hole.

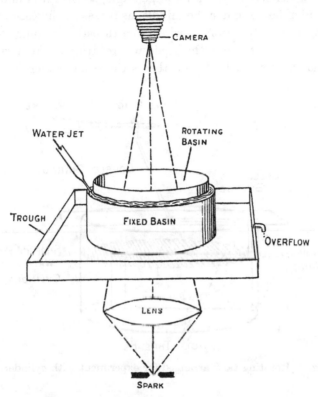

Fig. 1. General arrangement of apparatus.

The tank was then set rotating, and, when the whole system had attained a constant speed, a steel pricker was inserted into the wire ring, and raised approximately vertically. The sketch (fig. 2) shows the apparatus in action. In the case of the cylinder two threads were used to ensure that the axis should remain vertical. This is the case shown in the sketch. With the sphere, only one thread was used.

The results in the two cases are shown in the photographs (*a*) and (*b*) on pl. 1. In pl. 1 (*a*), the cylinder is shown at the middle point of its path. Since its axis is nearly vertical it is seen end-on, and appears therefore as a black circle in the photograph. In the particular experiment shown in pl. 1 (*a*) the axis of the cylinder had accidentally got tilted very slightly away from the vertical so that both the upper and lower threads show as two distinct threads. The pricker and wire ring are seen at the bottom of the photograph; they are naturally rather out of focus. The lines in the photograph which are not straight are threads used for drawing the solids in other experiments. They have nothing to do with the present experiment. It will be seen that the predicted result is verified. The cylinder is moving practically straight across under the middle line of the celluloid bridge.

Pl. 1 (*b*) shows the result in the case of the sphere. In this case it will be seen that the sphere is being deflected through a large angle, so that it is under the edge of the bridge instead of being under the middle, as it was in the case of the cylinder. It is not moving in the direction in which the thread is pulling it, but is being deflected to the right. In this case the liquid was rotating in the direction opposite to that of the hands of a clock. Again, this is the result which was predicted by theory.

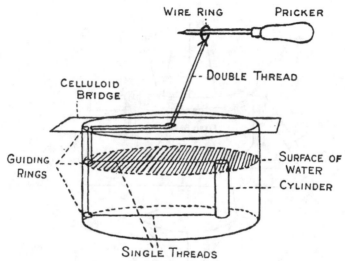

Fig. 2. Rotating tank arranged for experiment with cylinder.

MOTION OF VORTEX RINGS IN A ROTATING FLUID

The difference between two- and three-dimensional fluid motion which has been discussed must apply to all fluid motions. It must therefore apply to the propagation of a vortex ring, as well as to the motions of a spnere and cylinder.

It is impossible, apparently, to produce a two-dimensional analogue of a vortex ring. If such a thing could be produced, it should, according to this theory, propagate itself in a straight line through a fluid, whether the fluid is rotating or not. In other words, since a rotation of the whole system should make no difference to any two-dimensional flow, it should make no difference to the two-dimensional analogue of a vortex ring.

The flow in a vortex ring, however, is three-dimensional, and a rotation of the whole system should affect it. For reasons explained in the paper already referred to,* it was anticipated on theoretical grounds that if a vortex ring can be propagated

* Paper 2 above.

in a rotating fluid, it will not move in a straight line relative to the rotating system, but will move in a circle in the *opposite direction* to that in which the whole system is rotating.

In pl. 1 (c) is shown a photograph of a vortex ring which has been projected from a small vortex box immersed in the water.* The basin was not rotating in this experiment. It will be seen that the ring, which is being projected along a diameter of the basin, moves in a straight line, as was to be expected. A short wire was attached to the vortex box, so as to point in the direction in which the ring was aimed. This will be seen in the photograph pl. 1 (c).

Pl. 1 (d) is a photograph of a vortex ring projected when the whole system is rotating. In this photograph, the predicted curved path traversed by the ring is shown up clearly by the track of coloured fluid left behind during its flight.

The jet which drives the apparatus can be seen on the right-hand side of the photograph. It will be noticed that the basin is being driven in the counter-clockwise direction, and that the vortex ring is going round its circle in a clockwise direction, as predicted theoretically.

One point which the experiments bring out is that the direction of the axis of the ring appears to be fixed in space, so that the ring would go round in a circle once during each revolution of the system. It would therefore be possible to consider the motion as being a steady motion relative to axes whose directions are fixed in space, but whose origin moves and is situated at the centre of the path of the vortex ring.

Slow Motions in a Rotating Fluid

The investigations described above naturally led to enquiries as to whether rotating liquids possess any other properties which can be predicted from hydrodynamical theory. The following striking peculiarity of rotating fluids was discovered in the course of this work. If any small motion is communicated to a fluid which is initially rotating steadily like a solid body, the resulting flow must be two-dimensional, though small oscillations about this state of slow motion are possible.† This may be proved as follows:

Let u, v, w be the components of velocity of any particle of fluid relative to a system which is rotating uniformly with velocity ω about the axis of z.

The circulation round any circuit in the fluid is

$$I = \int \{(u - \omega y)\, dx + (v + \omega x)\, dy + w\, dz\},$$

the integral being taken round the circuit.

---

* The liquid eventually used was not pure water, but an acid solution of ferrous sulphate. The rings consisted of a solution of permanganate of potash made up to the same density as the ferrous sulphate solution by mixing with a heavy neutral salt. The rings then dissolved after they had broken up and a large number could be projected without discolouring the solution. For the suggestion to use this solution I am indebted to Dr A. A. Robb.

† This is practically the same thing as the fact previously noted by Proudman, that small steady motions of a rotating fluid are two-dimensional.

This may be divided into two parts

(a) $$I' = \int \{u\,dx + v\,dy + w\,dz\}$$

which may be called the circulation due to the relative motion, and

(b) $$\int (-\omega y\,dx + \omega x\,dy)$$

which can be expressed in the form $2\omega A$, where A is the area of the projection of the circuit, on a plane perpendicular to the axis of rotation.

In a non-viscous fluid the circulation round any circuit which always consists of the same ring of particles, is constant. Hence it will be seen that $I = I' + 2\omega A$ is constant.

Evidently if the motion relative to the rotating system is small this means that the variations in A during the whole motion are small. The liquid must therefore move in such a way that the area of the projection of any ring of particles on a plane perpendicular to the axis of rotation is nearly constant.

Let us now enquire how this geometrical condition may be expected to reveal itself during the motion. First, we shall see what types of motion are possible in a fluid for which the areas of the projections on a given plane of all possible circuits of particles remain constant during the motion. This condition may be expressed mathematically by writing down an expression for the rate at which the area of the projection increases and equating it to 0. In this way it is found that

$$\int (v\,dx - u\,dy) = 0,$$

the integral being taken round the circuit.

This expression may be transformed by Stokes' Theorem into the form

$$\int\int \left\{ l\frac{\partial u}{\partial z} + m\frac{\partial v}{\partial z} - n\left(\frac{\partial u}{\partial x} + \frac{\partial v}{\partial y}\right)\right\} dS = 0,$$

where dS is an element of surface of any surface which is bounded by the circuit in question and l, m, n are the direction cosines of the normal to that surface.

Since this relation holds for all possible circuits,

$$\partial u/\partial z = 0, \quad \partial v/\partial z = 0, \quad \text{and} \quad \partial u/\partial x + \partial v/\partial y = 0. \tag{1}$$

Hence, since the fluid is incompressible.

$$\partial w/\partial z = 0. \tag{2}$$

The conditions (1) show that any two particles which are originally in a line perpendicular to the given plane, will always remain in a line perpendicular to it. The condition (2) shows that they also remain at a constant distance apart throughout the motion.

If therefore any small motion be communicated to a rotating fluid the resulting motion of the fluid must be one in which any two particles originally in a line parallel to the axis of rotation remain so, except for possible small oscillations about that position. This property of rotating fluids is found to be true experimentally.

PLATE 1

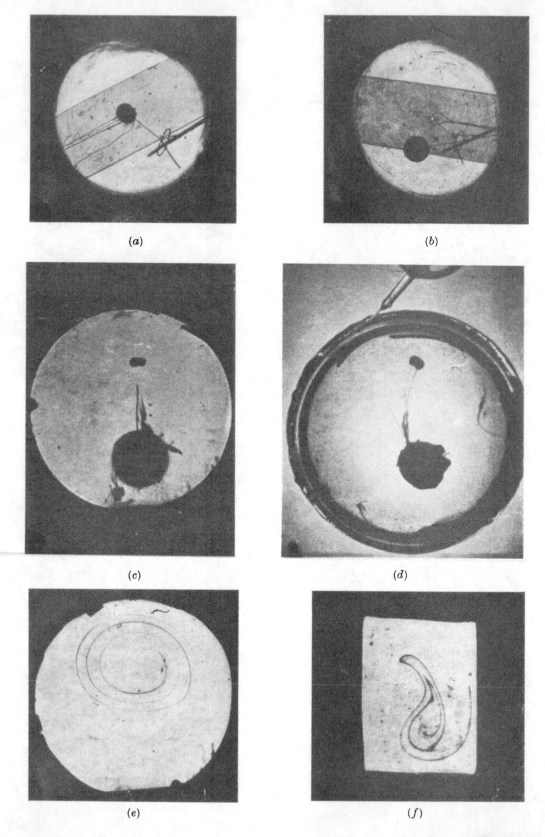

(a)

(b)

(c)

(d)

(e)

(f)

It can be demonstrated in a very striking way by means of the apparatus described in the first part of this paper. The liquid is first made to rotate steadily as a solid body. A small motion is then communicated to it, and a few drops of coloured liquid are inserted. However, carefully these drops are inserted the volume occupied by coloured water necessarily measures at least half a centimetre in every direction.

The slow movement of the fluid then draws this coloured portion of the fluid out into sheets. These sheets remain always parallel to the axis of rotation. They go on spreading almost indefinitely till they may perhaps be twenty or thirty times as long as the diameter of the basin, their thickness decreases correspondingly till they are only a small fraction of a millimetre thick.

The accuracy with which they remain parallel to the axis of rotation is quite extraordinary. After the motion has been going on for some time it is only possible to see that the colouring matter is not uniformly diffused through the liquid by placing one's eye on, or near, the axis of rotation. The portion of the fluid which is passing immediately beneath the eye then appears to be filled with fine lines which are, of course, in reality, thin sheets seen edgewise.

The photographs shown in pl. 1 (e), (f) were taken by a camera placed accurately on the axis of rotation of the basin. It will be seen that the lines are extremely fine. The sheets into which the originally diffuse patch of coloured liquid are drawn are therefore extremely thin, and are moreover accurately parallel to the axis of rotation. It appears therefore that in this case also the theoretical prediction is completely verified by experiments with real fluids.

In pl. 1 (e) the small motion was communicated to the fluid by changing the speed of the rotating basin temporarily. Pl. 1 (f) shows an experiment in which a rectangular boundary was placed in the rotating basin in order to alter the effect produced by a change of speed, and so produce a different pattern.

In a future paper the author hopes to discuss what happens in the case when the boundaries of the fluid move slowly in such a way that three-dimensional motion must take place.

DESCRIPTION OF PLATE 1

(a) Cylinder passing under centre of tank, seen from point above the tank and on the axis of rotation.

(b) Sphere being deflected, seen from the same point as the cylinder shown in fig. 1.

(c) Vortex ring being projected in a non-rotating tank. The large black disc is the vortex box, seen end-on. The wire seen projecting from the box indicates the direction of projection of the ring. The ring will be seen clear of the end of the wire.

(d) Vortex ring projected in rotating tank. The curved path is shown by the trail of coloured liquid left by the ring in its flight. The jet which drives the inner basin is seen at the top of the photograph.

(e) Sheet of coloured liquid seen from a point on the axis of rotation in a rotating liquid.

(f) Another sheet of coloured liquid. In this case the liquid is contained in a rectangular boundary.

4

THE MOTION OF A SPHERE IN A ROTATING LIQUID

REPRINTED FROM

Proceedings of the Royal Society, A, vol. CII (1922), pp. 180–9

In some recent papers* the author has drawn attention to certain general properties of rotating fluids, especially to the differences which may be expected between two- and three-dimensional motion. Unfortunately, mathematical difficulties have so far prevented the solution of any three-dimensional problem in a rotationally moving fluid from being obtained, except in one case, when the motion is very slow. In this case, Professor Proudman has shown how it is possible to approximate to the solution of the problem of the slow motion of a sphere in a rotating fluid.† Even in this case the analysis is very complicated.

There seems little prospect of obtaining a more general solution of the problem when the inertia terms which Proudman neglected are taken into account. On the other hand, it is shown in the following pages that a solution can be obtained in the case when the sphere moves *steadily* along the axis of rotation of the fluid. The limitation imposed by considering only a steady motion necessarily excludes the case considered by Proudman, for all slow steady motions of a rotating fluid are two-dimensional.

In the case when the sphere moves steadily with velocity U along the axis of a fluid rotating with angular velocity Ω, it is possible to reduce the flow to a steady motion by superposing a velocity $-U$ on the whole system. Since the motion is symmetrical about the axis, only two independent co-ordinates specifying positions are necessary, namely, r, the distance of any point from the centre of the sphere, and θ, the angle between the radius from the centre and the axis of symmetry.

Let u be the component of velocity of the fluid along a radius from the centre of the sphere, v the component in an axial plane and perpendicular to the radius, w the component perpendicular to the axial plane. The scheme is shown in fig. 1. Since the motion is symmetrical, the equation of continuity is satisfied if

$$u = -\frac{1}{r^2 \sin\theta}\frac{\partial\psi}{\partial\theta}, \quad v = \frac{1}{r \sin\theta}\frac{\partial\psi}{\partial r}, \tag{1}$$

where ψ is Stokes' stream function.

---

* Papers 2 and 3 above, and *Proc. Camb. phil. Soc.* XX (1921), 326 (not reproduced in these volumes).

† *Proc. Roy. Soc.* A, XCII (1916), 408. Proudman's work has recently been extended by Mr S. F. Grace, *Proc. Roy. Soc.* A, CII (1922), 89.

Since the motion is symmetrical, the equations of motion are

$$\frac{Du}{Dt} - \frac{v^2}{r} - \frac{u^2}{r} = -\frac{1}{\rho}\frac{\partial p}{\partial r}, \tag{2}$$

$$\frac{Dv}{Dt} - \frac{w^2 \cot\theta}{r} + \frac{uv}{r} = -\frac{1}{\rho r}\frac{\partial p}{\partial\theta}, \tag{3}$$

$$\frac{Dw}{Dt} + \frac{uw}{r} + \frac{vw \cot\theta}{r} = 0, \tag{4}$$

where p is the pressure and ρ the density of the fluid.

Since the motion is steady $\dfrac{D}{Dt}$ is $\quad u\dfrac{\partial}{\partial r} + \dfrac{v}{r}\dfrac{\partial}{\partial\theta}.$

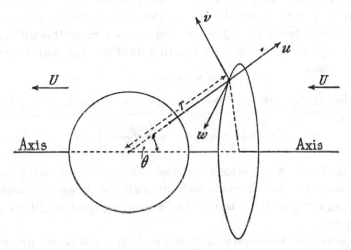

Fig. 1. Scheme of co-ordinates.

It is easy to verify that $w = A\psi/(r\sin\theta)$ satisfies the equation (4), and if the motion at infinity consists of a flow with uniform velocity $-U$ parallel to axis and a rotation about this axis with angular velocity Ω, the constant A is evidently equal to $2\Omega/U$, so that

$$w = 2\Omega\psi/(Ur\sin\theta). \tag{5}$$

This equation evidently expresses the fact that the circulation in a ring of fluid, which is symmetrical with respect to the axis, remains constant during the motion.

Let us now search for possible solutions of the form

$$\psi = f\sin^2\theta, \tag{6}$$

where f is a function of r only.

The components of velocity are then

$$u = -\frac{2f\cos\theta}{r^2}, \quad v = \frac{f'\sin\theta}{r}, \quad w = \frac{2\Omega}{U}\frac{f\sin\theta}{r}, \tag{7}$$

where f' is df/dr.

Eliminating p between (2) and (3), it will be found that the form assumed for ψ is legitimate, provided that f satisfies the equation

$$r^3 f''' - 2r^2 f'' - 2rf' + 8f + (4\Omega^2/U^2)(rf' - 2f) = 0. \tag{8}$$

The complete solution of this equation is

$$f = Cz^2 + A\left(\cos z - \frac{\sin z}{z}\right) + B\left(\sin z + \frac{\cos z}{z}\right),$$

or alternatively $\qquad f = Cz^2 + D\left\{\cos(z+\epsilon) - \frac{\sin(z+\epsilon)}{z}\right\}, \tag{9}$

where $\qquad\qquad\qquad\qquad k = 2\Omega/U, \quad z = kr,$

and A, B, C, D, ϵ are arbitrary constants. When z becomes infinite, the first term in (9) becomes large compared with the others. The motion at a great distance from the sphere is therefore represented by $f = Cz^2$, and, on comparing this with (7), it will be seen that this represents a uniformly rotating fluid moving with velocity $-2Ck^2$ along the axis. Hence

$$C = U/2k^2. \tag{10}$$

The condition at the surface of the sphere, $r = a$, is $u = 0$. Writing $\mu = ka$, this condition gives

$$\tfrac{1}{2}U\frac{\mu^2}{k^2} + D\left\{\cos(\mu+\epsilon) - \frac{\sin(\mu+\epsilon)}{\mu}\right\} = 0. \tag{11}$$

Any values of D and ϵ which satisfy (11) lead to a possible solution of the problem.

It appears therefore, that there are an infinite number of possible motions round a sphere moving steadily along the axis of a rotating fluid, and that all of them vanish at infinity.

I have not been able to discover how the motion could be set up. Perhaps the different solutions represent the stream-lines due to different ways of starting the motion. It is clear, however, that some of the possible motions represented by (9) could not be set up by starting the sphere from rest; it would be impossible for instance to set up a motion in this way for which f vanished or changed sign anywhere except at the surface of the sphere, because a negative value of f corresponds with a reversed rotation of the fluid about the axis. Such a motion is dynamically possible however, and it corresponds with a case in which all the liquid inside a certain sphere, concentric with the solid sphere, moves with it, forming a kind of sheath of liquid which possesses a rotation about the axis opposite to the rotation at infinity.

It is possible that these considerations may be extended so as to differentiate between the various possible motions corresponding to the various possible pairs of values of D and ϵ which are consistent with (11). The circulation round any symmetrical ring of fluid remains constant during any motion of the fluid. It is equal to $I = 2\pi\Omega y_0^2$, where y_0 is the initial distance of a particle of the ring considered from the axis. If the sphere starts from rest, the total deficiency of fluid in the field which possesses circulation lying between I and $I + (dI/dy_0)\,\delta y_0$ below that which the fluid would possess in the absence of the sphere, is the volume of fluid displaced by the part of the sphere which is contained between cylinders concentric with the axis, whose radii are y_0 and $y_0 + \delta y_0$. This volume is

$$2\pi y_0 (a^2 - y_0^2)^{\frac{1}{2}} \, \delta y_0 \quad \text{if} \quad y_0 < a, \quad \text{or} \quad 0 \quad \text{if} \quad y_0 > a.$$

When the sphere moves in steady motion with velocity U, ψ is connected with y_0 by the equation

$$y_0^2 = 2\psi/U.$$

The total deficiency of fluid possessing circulation lying between I and

$$I + (dI/dy_0) \, \delta y_0$$

is, therefore, found by calculating the deficiency of fluid possessing circulation between I and $I + (dI/d\psi) \, \delta\psi$ (where $\delta\psi = U y_0 \delta y_0$), by integrating the total volume of fluid between the stream-lines ψ and $\psi + \delta\psi$ through the whole field. If this deficiency is not equal to $2\pi(a^2 - y_0^2) \, \delta y$, when $y_0 < a$ and 0 when $y_0 > a$, then the motion could not be produced by starting a sphere from rest in a rotating fluid, unless there is a finite motion of the fluid at infinity along the stream-lines close to the axis during the time the motion is being established. This reasoning makes it appear that there is very little chance that any of the motions represented by (9) would be started by moving a sphere from rest in a rotating fluid.

It is interesting to notice that it is possible to find solutions in which $u = v = w = 0$ at the surface of the sphere, so that there is no slipping between the fluid and the surface of the sphere. The condition $U = 0$ at $r = a$ leads to the equation $0 = [f']_{z=\mu}$,

or

$$\frac{\mu U}{k^2} + D \left\{ \left(\frac{1}{\mu^2} - 1 \right) \sin(\mu + \epsilon) - \frac{1}{\mu} \cos(\mu + \epsilon) \right\} = 0. \tag{12}$$

This together with (11) yields the following values for D and ϵ

$$D = \tfrac{1}{2} U (\mu^4 + 3\mu^2 + 9)/k^2, \tag{13}$$

$$\tan(\mu + \epsilon) = 3\mu(3 - \mu^2)^{-1}, \tag{14}$$

so that the motion represented by

$$\frac{2k^2}{U} f = z^2 + (\mu^4 + 3\mu^2 + 9)^{\frac{1}{2}} \left\{ \cos(z + \epsilon) - \frac{\sin(z + \epsilon)}{z} \right\}, \tag{15}$$

is one for which the disturbance due to the sphere vanishes at infinity, and it is characterized by the fact that there is no slipping at the surface of the sphere.

This may be a point of some importance because it is the assumption that there is slipping at the surface of a solid body moving in a liquid which vitiates all the ordinary hydrodynamical theories of the motion of solids in fluids. It is possible, therefore, that the solution given above may represent the motion of a sphere in a rotating liquid more closely than the ordinary irrotational solution for a sphere moving in an infinite fluid at rest represents the actual flow in that case.

The surfaces ψ = constant are surfaces of revolution and the stream-lines are spirals wrapped on these surfaces. The sections of the surfaces ψ = constant by an axial plane may be called the stream-lines of the motion in the axial plane. These stream-lines are shown for a particular case in fig. 2. The case chosen is that of a sphere moving along the axis of a rotating fluid at such a speed that it travels a distance equal to its diameter (i.e. $\mu = 2\pi$) during each revolution of the liquid, and the particular solution for which there is no slipping between the sphere and the liquid is chosen. It will be seen that the stream-lines are very different from those which surround a sphere moving in a non-rotating liquid.

Fig. 2. Stream-lines due to the motion of a sphere in a rotating fluid.
Case when $\mu = 2\pi$.

PROPAGATION OF AN ISOLATED DISTURBANCE ALONG THE AXIS OF A ROTATING FLUID

An interesting point about the motion represented by (15) is that it is possible to reduce the radius of the sphere to zero while still retaining a finite velocity in the disturbed motion.

Taking a very small, μ becomes small, and (14) becomes tan $\epsilon = 0$, so that $\epsilon = 0$, (15) then becomes

$$f = \frac{U}{2k^2}\left[z^2 + 3\left(\cos z - \frac{\sin z}{z}\right)\right].$$

(16)

It can be verified, by substituting in the original equations of motion, that

$$\psi = \frac{U}{2k^2}\left[k^2r^2 + 3\left(\cos kr - \frac{\sin kr}{kr} \right) \right] \sin^2 \theta, \tag{17}$$

is a solution of the equations of motion, (17) also represents a motion for which the velocity and the pressure are finite and continuous at the origin. The motion consists of a kind of non-rotating core* of liquid propagated with velocity U along the axis of a rotating liquid. It is analogous to the motion produced by a vortex ring, but in an inverse sense.

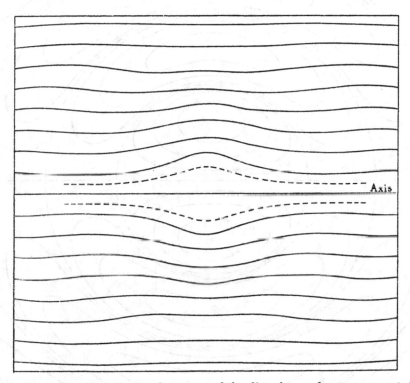

Fig. 3. Stream-lines relative to the centre of the disturbance due to a non-rotating core travelling along the axis of a rotating fluid.

The stream-lines, due to the motion of a non-rotating core in a rotating fluid, are shown in figs. 3 and 4. The stream-lines of the steady motion, relative to the moving core, are shown in fig. 3. It will be seen that there are no closed stream-lines, so that the disturbance does not carry any fluid with it.

The stream-lines, relative to the main body of the liquid, are shown in fig. 4. It will be seen that the central part of the disturbance resembles Hill's spherical vortex†, and that it is surrounded by spherical waves which travel with it. The

* Region on the axis where the rotation about the axis is small compared with the rotation of the fluid as a whole.

† M. J. M. Hill, 'On a Spherical Vortex', *Phil. Trans. Roy. Soc.* A (1894).

analogy between the present disturbance and a spherical vortex is only superficial, for the vortex ring is a mass of rotating fluid which can move through a non-rotating fluid. The present disturbance is a type which could only be propagated in a rotating fluid, and it consists of a core which rotates more slowly than the surrounding fluid and moves parallel to the axis of rotation.

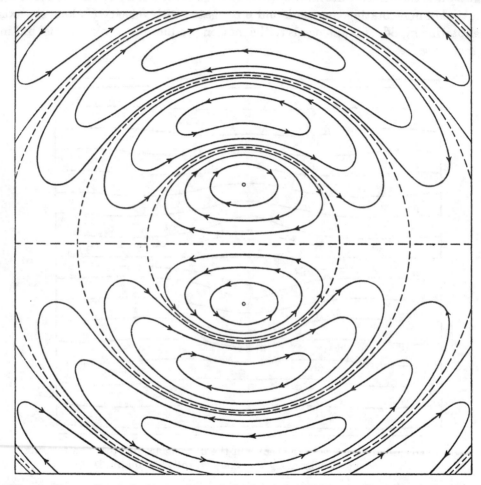

Fig. 4. Stream-lines relative to the main body of the fluid in the disturbance due to a non-rotating core propagated along the axis of a rotating fluid.

WAVE SYSTEMS IN A ROTATING FLUID

An essential feature of the motions described above is the system of spherical waves which accompany the moving sphere or moving disturbance. That a system of waves would accompany a body moving in a rotating fluid is to be expected. It has been pointed out by Lord Kelvin* that rotation confers on a fluid certain

* Kelvin's *Mathematical and Physical Papers*, IV, 152, 170.

properties resembling those of an elastic solid, and in particular, a rotating fluid can transmit waves.

In order that a system of waves may accompany a disturbance which moves with velocity U along any axis, it is necessary that the velocity of the wave along a normal to the wave-front should be $U \cos \alpha$, where α is the angle between this normal and the direction of the axis along which the disturbance is moving. It is easy to show that a plane wave, of given wave-length, moves in a rotating fluid with velocity proportional to $\cos \alpha$, so that the required condition is satisfied.

Writing the equations of motion of a fluid relative to rotating rectangular axes in the form

$$\frac{\partial u}{\partial t} + u\frac{\partial u}{\partial x} + v\frac{\partial u}{\partial y} + w\frac{\partial u}{\partial z} - 2\Omega u = -\frac{\partial}{\partial x}\left(\frac{p}{\rho} - \Omega^2 r^2\right)$$

$$\frac{\partial v}{\partial t} + u\frac{\partial v}{\partial x} + v\frac{\partial v}{\partial y} + w\frac{\partial v}{\partial z} + 2\Omega u = -\frac{\partial}{\partial y}\left(\frac{p}{\rho} - \Omega^2 r^2\right) \Bigg\}, \qquad (18)$$

$$\frac{\partial w}{\partial t} + u\frac{\partial w}{\partial x} + v\frac{\partial w}{\partial y} + w\frac{\partial w}{\partial z} = -\frac{\partial}{\partial z}\left(\frac{p}{\rho} - \Omega^2 r^2\right)$$

and the equation of continuity as

$$\frac{\partial u}{\partial x} + \frac{\partial v}{\partial y} + \frac{\partial w}{\partial z} = 0,$$

it will be seen at once that these equations are satisfied by

$$(u, v, w, p/\rho - \Omega^2 r^2) = (A, B, C, P)\, e^{i(ax+by+cz+nt)}$$

provided*

$$\begin{aligned} niA - 2\Omega B &= -i\alpha P \\ niB + 2\Omega A &= -ibP \\ niC &= icP \\ Aa + Bb + Cc &= 0 \end{aligned} \Bigg\} . \qquad (19)$$

Eliminating A, B, C, P it will be found that

$$a^2 + b^2 + c^2 = 4\Omega^2 c^2/n^2. \qquad (20)$$

Hence
$$n/2\Omega = c(a^2 + b^2 + c^2)^{-\frac{1}{2}};$$

but $c(a^2 + b^2 + c^2)^{-\frac{1}{2}} = \cos \alpha$, hence

$$n/2\Omega = \cos \alpha. \qquad (21)$$

The velocity of the waves represented by (18) is $n\lambda/2\pi$, where λ is the wave-length. Hence, from (21), the velocity is $\Omega\lambda \cos\alpha/\pi$, which is proportional to $\cos \alpha$ if λ is constant. It appears, therefore, that a spherical system of waves of length $U\pi/\Omega$ can be propagated with velocity U parallel to the axis of a rotating fluid.

It is an unusual feature of both the plane and the spherical types of wave that the amplitudes are not limited to small motions.

---

* The terms involving products of u and v all vanish in this solution on account of the relation $Aa + Bb + Cc = 0$, a condition which implies that the motion of the fluid is confined to the plane of the wave front.

Experimental Demonstration of the Existence of a Non-rotating Sheath of Fluid Round a Sphere Moving in a Rotating Fluid

It is difficult to realise a practical demonstration that any of the types of motion, considered above, actually exist in any real fluid; on the other hand, I have been able to show experimentally that, when a sphere moves along the axis of a rotating fluid, it is surrounded by a sheath of fluid which does not rotate with the rest of the fluid. That this result is to be expected is clear from equations (7), for at the surface of the sphere $f = 0$, so that $w = 0$, i.e. the fluid immediately in contact with the sphere has no tendency to rotate it.

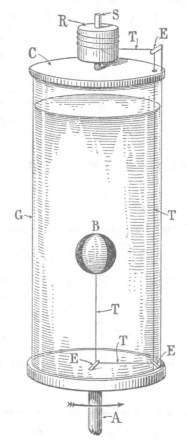

The apparatus with which this was demonstrated is shown in fig. 5. In this diagram G is a glass cylinder full of water, A is the axis about which the cylinder is made to rotate uniformly, B is the sphere which consists of a light celluloid ball, of the type used in the game of ping-pong. T is a thread which holds the ball down. The ball can be moved at a uniform speed along the axis by unrolling the thread, T, at a uniform speed from a reel R. This reel is mounted on a spindle fixed in the middle of a lid, C, which is fitted over the glass cylinder G. The thread passed from R through a series of small eyes, E, the last of which was in the centre of the bottom of the cylinder, and the ball B, was painted with black and white quadrants, so as to make it easy to see whether it was rotating.

The apparatus was first rotated uniformly for some time till the ball and liquid were both rotating as a solid body with the glass cylinder. To perform the experiment, the reel, R, was suddenly fixed so that the thread, T, unwound from the reel at a uniform rate. This gave the ball, B, a uniform speed along the axis of the cylinder.

Fig. 5. Apparatus intended to demonstrate that a sphere is surrounded by a non-rotating sheath of liquid when it travels along the axis of a rotating fluid.

It was found that the ball stopped rotating directly it started moving along the axis. As soon as the reel was released, so that the ball stopped moving along the axis, it quickly picked up the rotation of the rest of the system once more. To perform the experiment successfully, it was found necessary to use the lightest possible type of ball, to use a thin single or plaited silk thread, and to take great care that it was not twisted at the time of the experiment. To ensure success, it was found necessary also to make the ball move at a rate greater than about one diameter per revolution of the system (i.e. $\mu < 2\pi$). If the ball travelled more slowly than this it was found that it did not stop rotating, and investigation of the stream-lines with coloured water, showed that a column of liquid, of the same diameter as the sphere, was apparently pushed along in front of the sphere. This observation suggests that the explanation of the rotation of the sphere, when it moves slowly along the axis, is that the stream-line, $\psi = 0$, does not keep to the surface of the sphere.

In the course of these experiments, it was noticed that if the sphere was stopped suddenly when half-way up the cylinder, and if there was some colouring matter present to show up the motion, a mass of liquid appeared to detach itself from the sphere, and to continue moving along the axis of rotation with the same velocity as that with which the sphere had been moving. The impression produced on the author's mind was that the flow was similar to that shown in fig. 3. This conclusion, however, should be treated with reserve.

5

STABILITY OF A VISCOUS LIQUID CONTAINED BETWEEN TWO ROTATING CYLINDERS

REPRINTED FROM

Philosophical Transactions of the Royal Society, A, vol. CCXXIII (1923), pp. 289–343

PART I. THEORETICAL

Introduction

In recent years much information has been accumulated about the flow of fluids past solid boundaries. All experiments so far carried out seem to indicate that in all cases* steady motion is possible if the motion be sufficiently slow, but that if the velocity of the fluid exceeds a certain limit, depending on the viscosity of the fluid and the configuration of the boundaries, the steady motion breaks down and eddying flow sets in.

A great many attempts have been made to discover some mathematical representation of fluid instability, but so far they have been unsuccessful in every case. The case, for instance, in which the fluid is contained between two infinite parallel planes which move with a uniform relative velocity has been discussed by Kelvin, Rayleigh, Sommerfeld, Orr, Mises, Hopf, and others. Each of them came to the conclusion that the fundamental small disturbances of this system are stable. Though it is necessarily impossible to carry out experiments with infinite planes, it is generally believed that the motion in this case would be turbulent, provided the relative velocity of the two planes were sufficiently great.

Various suggestions have been made to account for the apparent divergence between theory and experiment. Among the most recent is that of Hopf, who points out that the flow would be unstable if the two infinite planes were flexible, so that the pressure could remain constant along them. There seems little to recommend this theory as an explanation of the observed turbulent motion of fluids, for there is no experimental evidence that instability is in any way connected with want of rigidity in the solid boundaries of the fluid. The more generally accepted view that infinitely small disturbances are stable, but that disturbances of finite size tend to increase, seems to be more in accordance with the experimental evidence, for it has been shown by Osborne Reynolds that the velocity at which water flowing through a pipe becomes turbulent depends to a very large extent on the amount of initial disturbance in the reservoir from which the water originally came.

On the other hand it has not yet been shown that disturbances of small but finite

---

* All cases where there is relative motion between the fluid and the boundaries, thus excluding the case of steady rotation of a liquid in a rotating vessel.

size do increase in such a manner as to give rise to the large disturbances observed in cases of turbulent motion.

So far all attempts to calculate the speed at which any type of flow would become unstable have failed. The most promising was that of Osborne Reynolds who assumed an arbitrary disturbance in the flow and determined whether it would tend to increase or decrease initially. As applied by Reynolds himself this method does not lead to any definite result. It does not determine an upper limit to the speed of flow which must be stable because some other type of disturbance might exist which would increase initially at a lower speed of the fluid. Neither does it determine a lower limit to the speeds at which the flow must be unstable, because the assumed disturbance which initially increases might decrease indefinitely at some later stage of the motion. It has been shown in fact that certain types of initial disturbance exist for which this actually is the case.*

The method of Osborne Reynolds has been modified by Orr, who has determined in two cases† the highest speed of flow at which all small disturbances initially decrease. At this speed evidently any initial small disturbance will decrease indefinitely.

Orr's method gives the only definite result which has yet been obtained in the subject. The result, however, is merely a negative one, in that it affords no indication as to whether flow at high speeds would be unstable. Orr's result, for instance, in the case of flow through a pipe of circular cross-section is that when the mean speed, W, of the fluid is less than the value given by $W = 180\nu/D$, D being the diameter of the pipe and ν the kinematic viscosity, the motion will be stable. The value of W so obtained is less than $\frac{1}{70}$ of the highest speed at which the flow has been observed to be stable under suitable experimental conditions. Orr's method therefore is of very little assistance in understanding the observed instability of fluid flow.

Indeed, Orr remarks in the introduction to his paper: 'It would seem improbable that any sharp criterion for stability of fluid motion will ever be arrived at mathematically.'

Scope of the Present Work

It seems doubtful whether we can expect to understand fully the instability of fluid flow without obtaining a mathematical representation of the motion of a fluid in some particular case in which instability can actually be observed, so that a detailed comparison can be made between the results of analysis and those of experiment. In the following pages a special type of fluid instability is discussed and experiments are described in which the results of analysis are subjected to numerical verification.

The attention of mathematicians has been concentrated chiefly on the problem of the stability of motion of a liquid contained between two parallel planes which move relatively to one another with a uniform velocity. This problem has been

* Orr, 'Stability or Instability of Motions of a Viscous Fluid', *Proc. Roy. Ir. Acad.* (1907), p. 90.

† *Loc. cit.*, p. 134.

chosen because it seemed probable that the mathematical analysis might prove comparatively simple; but even when the discussion is limited to two-dimensional motion it has actually proved very complicated and difficult. On the other hand it would be extremely difficult to verify experimentally any conclusions which might be arrived at in this case, because of the difficulty of designing apparatus in which the required boundary conditions are approximately satisfied.

It is very much easier to design apparatus for studying the flow of fluid under pressure through a tube, or the flow between two concentric rotating cylinders. The experiments of Reynolds and others suggest that in the case of flow through a circular tube, infinitely small disturbances are stable, while larger disturbances increase, provided the speed of flow is greater than a certain amount. The study of the fluid stability when the disturbances are not considered as infinitely small is extremely difficult. It seems more promising therefore to examine the stability of liquid contained between concentric rotating cylinders. If instability is found for infinitesimal disturbances in this case it will be possible to examine the matter experimentally.

Previous Work on the Stability of a Viscous Liquid contained between Two Concentric Rotating Cylinders

The stability of an *inviscid* fluid moving in concentric layers has been studied by the late Lord Rayleigh. Perfect slipping was assumed to take place at the two bounding cylinders. If the motion is confined to two dimensions his conclusion is that the motion is stable if the liquid is initially flowing steadily with the same distribution of velocity which a viscous liquid would have if confined between two concentric rotating cylinders. All two-dimensional motions of an incompressible fluid, which do not involve change in area of internal boundaries are unaffected by a rotation of the whole system, so that this result merely depends on the existence of a *relative* angular velocity of the two cylinders.

In the case when the disturbances are assumed to be symmetrical about the axis, Lord Rayleigh* developed an analogy with the stability of a fluid of variable density under the action of gravity. In this analogy the varying centrifugal force of the different layers of fluid plays the part of gravity and the resulting condition for stability is that the square of the circulation must increase continuously in passing from the inner to the outer cylinder, just as the density of a fluid under gravity must decrease continuously with height in order that it may be in stable equilibrium. This condition leads to the conclusion that if the initial flow of the inviscid fluid is the same as that of a viscous fluid in steady motion, this flow will be unstable if the two cylinders are rotating in opposite directions. If they rotate in the same direction then the motion is stable or unstable according as $\Omega_2 R_2^2$ is greater or less than $\Omega_1 R_1^2$. Ω_1 and Ω_2 are the angular velocities of the inner and outer cylinders respectively, while R_1 and R_2 are their radii.

* 'On the Dynamics of Revolving Fluids', *Proc. Roy. Soc.* A (1916), pp. 148–54.

The investigations of Kelvin, Orr, Sommerfeld, Mises and Hopf on the stability of a viscous fluid shearing between two planes have not been extended to the cylindrical case, but recently W. J. Harrison* has extended Orr's method to find the maximum relative speed which the two cylinders can possess in order that the energy of *all* possible types of initial disturbance may initially decrease. In this work Harrison assumes that the motion is two-dimensional. His value for Reynolds' criterion therefore contains only the *relative* speeds of the two cylinders. It is unaltered if the whole system is rotated uniformly at any speed. His criterion is therefore the same whether the inner cylinder is fixed and the outer one rotated or *vice versa*.

The question has been investigated experimentally by Couette† and Mallock.‡

In the experiments of Couette the inner cylinder was fixed while the outer one rotated. It was found that the moment of the drag which the fluid exerted on the inner cylinder was proportional to the velocity of the outer cylinder, provided that velocity was less than a certain value. As the speed of the outer cylinder increased above this value the drag increased at a greater rate than the velocity. The change was attributed to a change from steady to turbulent motion. Rayleigh's theory of stability of an inviscid fluid for symmetrical disturbances makes the case when the inner cylinder is fixed stable at all speeds.

Mallock's experiments yielded the same result as Couette's, but in this case the value of Reynolds' criterion was higher than that obtained by Couette.§

Mallock extended his experiments to cover the case in which the inner cylinder rotated and the outer one was at rest. In this case he found instability at all speeds of the inner cylinder. This result is in accordance with Lord Rayleigh's theoretical prediction for the case of an inviscid fluid, but on the other hand it seems certain, in fact Lord Rayleigh‖ has proved, that at very low speeds all steady motions of a viscous fluid must be stable.

In spite of these differences between theory and experiment there is one point in which Rayleigh's 'inviscid fluid' theory is in agreement with Mallock's experiments, namely the large difference in regard to stability between the cases when the inner and when the outer cylinder is fixed. This shows clearly that in the case when the outer cylinder is fixed at any rate, the disturbance is not two-dimensional in character. Whether it is actually of a symmetrical type as contemplated by Rayleigh, or whether it is of some other three-dimensional form, remains to be seen.

The most striking feature of Lord Rayleigh's theory for inviscid fluids is the criterion for stability when both cylinders are rotating in the same direction, namely, $\Omega_2 R_2^2 > \Omega_1 R_1^2$. Owing to the construction of their apparatus no information as to

---

* W. J. Harrison, *Proc. Camb. phil. Soc.* (1921), p. 455.

† *Annls. chim. Phys.* 6me sér., XXI (1890).

‡ *Phil. Trans. Roy. Soc.* A (1896), p. 41.

§ See Orr, *Proc. Roy. Ir. Acad.* XXVII (1907–9), 78.

‖ Rayleigh, *Phil. Mag.* XXVI (1913), 776–86.

the correctness of this criterion of stability is obtainable from the experiments of either Mallock or Couette.

Author's Preliminary Experiments

For this reason I decided to construct a rough apparatus in which the two cylinders could be rotated separately. The experiments performed with this apparatus are described in a preliminary paper.* The results appeared to show that the criterion $\Omega_2 R_2^2 > \Omega_1 R_1^2$ is approximately satisfied in a viscous fluid, but that Rayleigh's result is not true for the case when the two cylinders are rotating in opposite directions. The experiments also indicated that the type of disturbance which is formed when instability occurs is symmetrical. These results encouraged me to embark on the complicated problem of trying to calculate the possible symmetrical disturbances of a viscous liquid contained between two rotating cylinders, and at the same time I started to construct an apparatus for observing as accurately as possible the conditions under which instability arises.

The complexity of the mathematical problem arises from the fact that it is necessary to obtain a three-dimensional solution of the equations of motion in which all three components of velocity vanish at both the cylindrical boundaries.

1. STABILITY FOR SYMMETRICAL DISTURBANCES

Before proceeding to the details of the solution of the problem it may be helpful to readers to give a list of the symbols employed. In table 1 the number of the page on which each symbol is defined is given.

Table 1. *List of symbols used, with number of page on which they first appear or are first defined*

$(V, \Omega_1, \Omega_2, R_1, R_2, A, B, r, \mu)$, p. 39; (z, t, u, v, w), p. 39;
$(p, \rho, \nu, \nabla_1^2, u_1, v_1, w_1, \lambda, \sigma)$, p. 40; $(J_1 (\kappa_s r), W_1 (\kappa_s r), C_1, C_2)$, p. 40;
$(B_0 (\kappa_s r), B_1 (\kappa_s r), \kappa_1, \kappa_2, \kappa_3, \ldots)$, p. 41; (H_s), p. 41;
$(a_m, C_3, C_4, \lambda', b_m, i)$, p. 43; (C_5, C_6, C_7, c_m), p. 43;
(C_5', C_6', C_7', d_m), p. 44; $({}_s c_m)$, p. 45; (L_m', Δ_1), p. 46;
(d), p. 48; $(x, C_1', U_2', \epsilon')$, p. 49; (y, κ), p. 50;
$(\alpha, \beta, \gamma, \eta)$, p. 51; (θ, L_m), p. 52; (Δ_2), p. 52; (P), p. 53; (P'), p. 54;
(Δ_3, ϵ), p. 58; (f_1), p. 59; $(f_2, \delta_1 P, P_1)$, p. 60;
$(\Delta_{11}, \Delta_{22}, \ldots \Delta_{88})$, p. 62; (ψ), p. 64;
$M_1, M_2, \ldots M_8)$, p. 67.

Steady Motion

Let V be the velocity at any point of an incompressible viscous fluid in steady motion between two infinitely long concentric rotating cylinders of radii R_1 and R_2 ($R_2 > R_1$). If r is the distance of a point from the axis then it is known that

$$V = Ar + B/r, \tag{1·0}$$

* Taylor, *Proc. Camb. phil. Soc.* xx (1921), 326 (not reproduced in these volumes).

when A and B are constants which are connected with the angular velocities Ω_1 and Ω_2 of the two cylinders by the relations

$$\left.\begin{array}{l} \Omega_1 = A + B/\mathrm{R}_1^2 \\ \Omega_2 = A + B/\mathrm{R}_2^2 \end{array}\right\}. \tag{1·1}$$

Solving these equations A and B can be expressed in terms of R_1, R_2, Ω_1 and Ω_2 and

$$A = \frac{\mathrm{R}_1^2\Omega_1 - \mathrm{R}_2^2\Omega_2}{\mathrm{R}_1^2\mathrm{R}_2^2} = \frac{\Omega_1(1 - \mathrm{R}_2^2\mu/\mathrm{R}_1^2)}{1 - \mathrm{R}_2^2/\mathrm{R}_1^2}, \tag{1·2}$$

$$B = \frac{\mathrm{R}_1^2\Omega_1(1 - \mu)}{1 - \mathrm{R}_1^2/\mathrm{R}_2^2}, \tag{1·3}$$

where $\mu = \Omega_2/\Omega_1$.

Specification of Symmetrical Disturbance

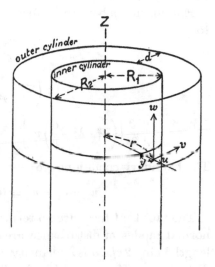

Fig. 1. Scheme of co-ordinates.

Let u, $V + v$, w, be the components of velocity in a disturbed motion, u is the component in an axial plane and perpendicular to the axis, $V + v$ is the component perpendicular to the meridian plane and to the axis—that is, in the direction of the undisturbed motion— w is the component parallel to the axis. The scheme is represented in fig. 1.

We shall assume that u, v and w are small compared with V, and that the disturbance is symmetrical, so that they are functions of r, z and t only; z is the co-ordinate parallel to the axis and t is the time.

2. Equations of Motion

Neglecting terms containing products or squares of u, v, w, the equations of disturbed motion may be written

$$\frac{1}{\rho}\frac{\partial p}{\partial r} - \frac{V^2}{r} = -\frac{\partial u}{\partial t} + 2\left(A + \frac{B}{r^2}\right)v + \nu\left(\nabla_1^2 u + \frac{\partial^2 u}{\partial z^2} - \frac{u}{r^2}\right), \tag{2·0}$$

$$0 = -\frac{\partial v}{\partial t} - 2Au + \nu\left(\nabla_1^2 v + \frac{\partial^2 v}{\partial z^2} - \frac{v}{r^2}\right), \tag{2·1}$$

$$\frac{1}{\rho}\frac{\partial p}{\partial z} = -\frac{\partial w}{\partial t} + \nu\left(\nabla_1^2 w + \frac{\partial^2 w}{\partial z^2}\right), \tag{2·2}$$

where p represents pressure, ρ density, ν is kinematic viscosity, and ∇_1^2 represents the operator

$$\frac{\partial^2}{\partial r^2} + \frac{1}{r}\frac{\partial}{\partial r}.$$

The equation of continuity is

$$\frac{\partial u}{\partial r} + \frac{u}{r} + \frac{\partial w}{\partial z} = 0. \tag{2·3}$$

The six boundary conditions which must be satisfied are

$$u = v = w = 0 \quad \text{at} \quad r = R_1 \quad \text{and} \quad r = R_2. \tag{2·4}$$

Assume as a solution

$$\left.\begin{aligned} u &= u_1 \cos \lambda z\, e^{\sigma t}, \\ v &= v_1 \cos \lambda z\, e^{\sigma t}, \\ w &= w_1 \sin \lambda z\, e^{\sigma t}, \end{aligned}\right\} \tag{2·5}$$

where u_1, v_1 and w_1 are functions of r only.

Eliminating p between (2·0) and (2·2) equations (2·0), (2·1), (2·2) and (2·3) reduce to

$$\frac{u_1}{r} + \frac{\partial u_1}{\partial r} + \lambda w_1 = 0, \tag{2·6}$$

$$\nu\left(\nabla_1^2 - \frac{1}{r^2} - \lambda^2 - \frac{\sigma}{\nu}\right) v_1 = 2A u_1, \tag{2·7}$$

$$\frac{\nu}{\lambda}\frac{\partial}{\partial r}\left\{\left(\nabla_1^2 - \lambda^2 - \frac{\sigma}{\nu}\right) w_1\right\} = -2\left(A + \frac{B}{r^2}\right) v_1 - \nu\left(\nabla_1^2 - \frac{1}{r^2} - \lambda^2 - \frac{\sigma}{\nu}\right) u_1. \tag{2·8}$$

The boundary conditions are

$$u_1 = v_1 = w_1 = 0 \quad \text{at} \quad r = R_1 \quad \text{and} \quad r = R_2. \tag{2·9}$$

The fact that there are no terms containing z in these equations shows that the normal modes of disturbance are simple harmonic with respect to z, the wavelength being $2\pi/\lambda$. σ is a quantity which determines the rate of increase in a normal disturbance. If σ is positive the disturbance increases and the motion is unstable. If σ is negative the disturbance decreases and the motion is stable. If σ is zero the motion is neutral. It will be seen from the way in which σ enters into the equations that it cannot be imaginary or complex unless u_1, v_1, w_1 are complex.

3. Bessel Functions used in the Solution

The solution of equations (2·6), (2·7) and (2·8) is developed by means of a type of Bessel functions of order 1 which vanish at $r = R_1$ and $r = R_2$. Let $J_1(\kappa_s r)$ and $W_1(\kappa_s r)$ be two independent solutions of the Bessel equation,

$$\left(\nabla_1^2 + \kappa_s^2 - \frac{1}{r^2}\right) f = 0. \tag{3·0}$$

The general solution of (3·0) is

$$f = C_1 J_1(\kappa_s r) + C_2 W_1(\kappa_s r), \tag{3·01}$$

where C_1 and C_1 are constants.

Let us now choose C_1 and C_2 so that f vanishes at $r = R_1$ and $r = R_2$; we then obtain two equations which suffice to determine C_1/C_2.

The equation for κ_s is

$$\frac{J_1(\kappa_s R_1)}{W_1(\kappa_s R_1)} = \frac{J_1(\kappa_s R_2)}{W_1(\kappa_s R_2)}. \tag{3.10}$$

Let the roots of this equation be $\kappa_1, \kappa_2, \kappa_3 \ldots$ in ascending order of magnitude. The equation for C_1/C_2 is

$$-\frac{C_2}{C_1} = \frac{J_1(\kappa_s R_1)}{W_1(\kappa_s R_2)}. \tag{3.11}$$

Writing $B_1(\kappa_s r)$ for

$$C_1 J_1(\kappa_s r) + C_2 W_1(\kappa_s r), \tag{3.12}$$

and $B_0(\kappa_s r)$ for the corresponding Bessel function of zero order, namely,

$$B_0(\kappa_s r) = C_1 J_0(\kappa_s r) + C_2 W_0(\kappa_s r), \tag{3.13}$$

we notice that

$$\frac{1}{\kappa_s} \frac{\partial}{\partial r} B_0(\kappa_s r) = -B_1(\kappa_s r), \tag{3.14}$$

and that $B_0(\kappa_s r)$ does not vanish at R_1 and R_2.

In order to develop any function of r in Bessel–Fourier expansions valid between the limits R_1 and R_2 the following formulae will be used*:

$$\int_{R_1}^{R_2} J_n(\kappa_s r) J_n(\kappa_t r) r \, dr = \frac{R_2}{\kappa_s^2 - \kappa_t^2} \{ \kappa_t J_n(\kappa_s R_2) J_n'(\kappa_t R_2) - \kappa_s J_n(\kappa_t R_2) J_n'(\kappa_s R_2) \}$$
$$- \frac{R_1}{\kappa_s^2 - \kappa_t^2} \{ \kappa_t J_n(\kappa_s R_1) J_n'(\kappa_t R_1) - \kappa_s J_n(\kappa_t R_1) J_n'(\kappa_s R_1) \} \tag{3.20}$$

and

$$\int_{R_1}^{R_2} J_n^2(\kappa_s r) r \, dr = \frac{R_2^2}{2} \left\{ J_n'^2(\kappa_s R_2) + \left(1 - \frac{n^2}{\kappa_s^2 R_2^2} \right) J_n^2(\kappa_s R_2) \right\}$$
$$- \frac{R_1^2}{2} \left\{ J_n'^2(\kappa_s R_1) + \left(1 - \frac{n^2}{\kappa_s^2 R_1^2} \right) J_n^2(\kappa_s R_1) \right\}, \tag{3.21}$$

where $J_n(\kappa_s r)$ is any Bessel function of order n, and κ_s and κ_t may be real or complex numbers. Particular cases of (3.20) and (3.21) when $n = 0$ and $n = 1$ and κ_s and κ_t are roots of (3.10) are

$$\int_{R_1}^{R_2} B_0(\kappa_s r) B_0(\kappa_t r) r \, dr = 0, \tag{3.22}$$

$$\int_{R_1}^{R_2} B_1(\kappa_s r) B_1(\kappa_t r) r \, dr = 0, \tag{3.23}$$

$$\int_{R_1}^{R_2} B_0^2(\kappa_s r) r \, dr = \tfrac{1}{2} \{ R_2^2 B_0^2(\kappa_s R_2) - R_1^2 B_0^2(\kappa_s R_1) \} = H_s, \tag{3.24}$$

$$\int_{R_1}^{R_2} B_1^2(\kappa_s r) r \, dr = \tfrac{1}{2} \{ R_2^2 B_0^2(\kappa_s R_2) - R_1^2 B_0^2(\kappa_s R_1) \} = H_s. \tag{3.25}$$

* Gray and Mathews, 'Bessel Functions', p. 53.

Any continuous function $f(r)$ of r which vanishes at R_1 and R_2 may be developed in a Bessel–Fourier series

$$f(r) = \sum_{s=1}^{\infty} a_s B_1(\kappa_s r). \tag{3.30}$$

This series is valid between the limits R_1 and R_2 and

$$a_s = \frac{1}{H_s} \int_{R_1}^{R_2} f(r)\, B_1(\kappa_s r)\, r\, dr. \tag{3.31}$$

On the other hand any continuous function $F(r)$ of r may be developed in a Bessel–Fourier series

$$F(r) = b_0 + \sum_{s=1}^{\infty} b_s B_0(\kappa_s r). \tag{3.32}$$

This series is also valid between the limits R_1 and R_2 and

$$b_s = \frac{1}{H_s} \int_{R_1}^{R_2} F(r)\, B_0(\kappa_s r)\, r\, dr. \tag{3.33}$$

It will be noticed that a constant term occurs in (3.32). At first sight this appears surprising. In most Bessel–Fourier expansions this extra term does not appear because it is possible to express a constant as a Bessel–Fourier expansion containing all the other terms. In the case of the expansion (3.32) it will be found that it is not possible to do this. The functions $B_0(\kappa_1 r)$, $B_0(\kappa_2 r)$... do not form a complete set of normal functions without the constant.

4. Development of Solution of (2.6), (2.7) and (2.8) in Bessel Functions

It is found convenient to express u_1, v_1 and w_1 as series of types (3.30) and (3.32) because when these series are introduced into the equations (2.6), (2.7) and (2.8) they yield linear relations between the coefficients of the various expansions. At the same time the form (3.30) is specially convenient because a series of that form automatically satisfies the boundary conditions at R_1 and R_2.

Integral of (2.7)

Assume the following series for u_1

$$u_1 = \sum_{m=1}^{\infty} a_m B_1(\kappa_m r). \tag{4.0}$$

This satisfies the conditions $u_1 = 0$ at R_1 and R_2. Substituting this in equation (2.7) it will be seen that the complete solution of (2.7) is

$$v_1 = C_3 J_1(i\lambda' r) + C_4(i\lambda' r) + \sum_{m=1}^{\infty} b_m B_1(\kappa_m r), \tag{4.10}$$

where i is $\sqrt{-1}$ and C_3 and C_4 are the two arbitrary constants occurring in the complementary function,

$$\lambda'^2 = \lambda^2 + \sigma/\nu, \tag{4.11}$$

and

$$b_m = -\frac{2A a_m}{\nu(\kappa_m^2 + \lambda^2 + \sigma/\nu)}. \tag{4.12}$$

The boundary condition $v_1 = 0$ at R_1 and R_2 gives

$$C_3 = C_4 = 0. \tag{4.13}$$

Integral of (2.8)

The complete solution of the equation (2.8) may be written in the form

$$w_1 = C_5 + C_6 J_0(i\lambda' r) + C_7 W_0(i\lambda' r) + \sum_{m=1}^{\infty} c_m B_0(\kappa_m r), \tag{4.20}$$

where C_5, C_6, C_7 are the three arbitrary constants which occur in the complementary function—that is, in the solution of

$$\frac{\nu}{\lambda} \frac{\partial}{\partial r} \left(\nabla_1^2 - \lambda^2 - \frac{\sigma}{\nu} \right) w_1 = 0, \tag{4.201}$$

$J_0(i\lambda' r)$ and $W_0(i\lambda' r)$ are two independent solutions of

$$(\nabla_1^2 - \lambda'^2) J_0(i\lambda' r) = 0.$$

Substituting (4.20) in (2.8) the following equation is obtained to determine the coefficients c_m

$$\sum_{m=1}^{\infty} \frac{\nu}{\lambda} (\kappa_m^2 + \lambda'^2) c_m \frac{\partial}{\partial r} B_0(\kappa_m r) = 2\left(A + \frac{B}{r^2} \right) \sum_{m=1}^{\infty} b_m B_1(\kappa_m r)$$

$$- \nu \sum_{m=1}^{\infty} (\kappa_m^2 + \lambda'^2) a_m B_1(\kappa_m r). \tag{4.21}$$

Substituting for b_m from (4.12) and using the relation

$$\frac{\partial}{\partial r} B_0(\kappa_m r) = -\kappa_m B_1(\kappa_m r)$$

(4.21) becomes

$$\sum_{m=1}^{\infty} c_m \left(\frac{\nu \kappa_m}{\lambda} \right) (\lambda'^2 + \kappa_m^2) B_1(\kappa_m r) = \sum_{m=1}^{\infty} \nu(\kappa_m^2 + \lambda'^2) a_m B_1(\kappa_m r)$$

$$+ 2\left(A + \frac{B}{r^2} \right) \sum_{m=1}^{\infty} \frac{2A a_m}{\nu(\kappa_m^2 + \lambda'^2)} B_1(\kappa_m r). \tag{4.22}$$

Treatment of the Equation of Continuity (2.6)

Substituting for u_1 from (4.0) and for w_1 from (4.20)

$$\frac{u_1}{r} + \frac{\partial u_1}{\partial r} \quad \text{becomes} \quad \Sigma \kappa_m a_m B_0(\kappa_m r), \tag{4.30}$$

so that (2.6) becomes

$$0 = \sum_{m=1}^{\infty} (\kappa_m a_m + \lambda c_m) B_0(\kappa_m r) + \lambda \{ C_5 + C_6 J_0(i\lambda' r) + C_7 W_0(i\lambda' r) \}. \tag{4.31}$$

In order that we may equate coefficients of $B_0(\kappa_m r)$ for all values of m it is necessary to expand the terms inside the second bracket in (4·31) in a Bessel–Fourier series of the form (3·32).

Expansion of $C_5 + C_6 J_0(i\lambda' r) + C_7 W_0(i\lambda' r)$.

Let
$$C_5 + C_6 J_0(i\lambda' r) + C_7 W_0(i\lambda' r) = C_5' + \sum_{m=1}^{\infty} d_m B_0(\kappa_m r), \qquad (4\cdot40)$$

then from (3·33)

$$d_m = \frac{1}{H_m} \int_{R_1}^{R_2} B_0(\kappa_m r)\{C_6 J_0(i\lambda' r) + C_7 W_0(i\lambda' r)\}\, r\, dr. \qquad (4\cdot41)$$

This integral is a particular case of (3·20). Remembering that

$$B_0'(\kappa_m R_1) = B_0'(\kappa_m R_2) = 0,$$

it will be seen from (3·20) that

$$d_m = \frac{i\lambda'}{H_m(\kappa_m^2 + \lambda'^2)}\left[\begin{array}{l} C_6\{R_2 B_0(\kappa_m R_2) J_0'(i\lambda' R_2) - R_1 B_0(\kappa_m R_1) J_0'(i\lambda' R_1)\} \\ + C_7\{R_2 B_0(\kappa_m R_2) W_0'(i\lambda' R_2) - R_1 B_0(\kappa_m R_1) W_0'(i\lambda' R_1)\} \end{array}\right]. \qquad (4\cdot42)$$

The constant term is

$$C_5' = C_5 + \frac{2i}{(R_2^2 - R_1^2)\lambda'}[C_6\{R_2 J_0'(i\lambda' R_2) - R_1 J_0'(i\lambda' R_1)\}$$
$$+ C_7\{R_2 W_0'(i\lambda' R_2) - R_1 W_0'(i\lambda' R_1)\}]. \qquad (4\cdot43)$$

Writing
$$C_6' = i\lambda' R_2\{C_6 J_0'(i\lambda' R_2) + C_7 W_0'(i\lambda' R_2)\}, \qquad (4\cdot44)$$

$$C_7' = -i\lambda' R_1\{C_6 J_0'(i\lambda' R_1) + C_7 W_0'(i\lambda' R_1)\}. \qquad (4\cdot45)$$

The expansion may now be written

$$C_5 + C_6 J_0(i\lambda' r) + C_7 W_0(i\lambda' r) = C_5' + C_6' \sum_{m=1}^{\infty} \frac{B_0(\kappa_m R_2)}{H_m(\kappa_m^2 + \lambda'^2)} B_0(\kappa_m r)$$

$$+ C_7' \sum_{m=1}^{\infty} \frac{B_0(\kappa_m R_1)}{H_m(\kappa_m^2 + \lambda'^2)} B_0(\kappa_m r). \qquad (4\cdot46)$$

Since C_5, C_6 and C_7 are entirely arbitrary constants, and the coefficients of C_6' and C_7' are independent functions of r, we can regard the right-hand side of (4·46) as being the complementary function of (4·201), the three arbitrary constants now being C_5', C_6' and C_7'.

We are now in a position to make effective use of the equation of continuity (4·31). Substituting (4·46) in (4·31) we can equate coefficients of $B_0(\kappa_m r)$. In this way

$$\left. \begin{array}{l} C_5' = 0 \\[2mm] 0 = \dfrac{\kappa_m a_m}{\lambda} + c_m + C_6' \dfrac{B_0(\kappa_m R_2)}{H_m(\kappa_m^2 + \lambda'^2)} + C_7' \dfrac{B_0(\kappa_m R_1)}{H_m(\kappa_m^2 + \lambda'^2)}. \end{array}\right\} \qquad (4\cdot47)$$

These equations give c_m in terms of a_m and C_6' and C_7'.

Equations for Determining Coefficients a_m

Next substitute in (4·22) the value of c_m given by (4·47).

There results an equation containing only $a_1, a_2, \ldots, a_m, \ldots$ and C_6' and C_7'. It is not possible, however, to equate coefficients of $B_1(\kappa_m r)$ directly on account of the factor $A + B/r^2$ which occurs on the right-hand side. In order to equate coefficients it is necessary to expand every term of type $(A + B/r^2) B_1(\kappa_m r)$ in a Bessel–Fourier series of type (3·30).

Let
$$(A + B/r^2) B_1(\kappa_m r) = {}_1c_m B_1(\kappa_1 r) + {}_2c_m B_1(\kappa_2 r) + \ldots + {}_sc_m B_1(\kappa_s r), \tag{4·50}$$

so that
$$\,_sc_m = \frac{1}{H_s} \int_{R_1}^{R_2} (A + B/r^2) B_1(\kappa_m r) B_1(\kappa_s r)\, dr. \tag{4·51}$$

Substituting these series in (4·22) and also substituting for c_m from (4·47), (4·22) becomes

$$- \sum_{m=1}^{\infty} \left\{ \left(\frac{\nu \kappa_m}{\lambda} \right) (\lambda'^2 + \kappa_m^2) \left(C_6' \frac{B_0(\kappa_m R_2)}{H_m(\kappa_m^2 + \lambda'^2)} + C_7' \frac{B_0(\kappa_m R_1)}{H_m(\kappa_m^2 + \lambda'^2)} + \frac{\kappa_m a_m}{\lambda} \right) + \nu(\kappa_m^2 + \lambda'^2) a_m \right\}$$

$$\times B_1(\kappa_m r) = 4A \sum_{m=1}^{\infty} \left(\frac{a_m}{\nu(\kappa_m^2 + \lambda'^2)} \sum_{s=1}^{\infty} \,_sc_m B_1(\kappa_s r) \right). \tag{4·52}$$

We can now equate coefficients of $B_1(\kappa_m r)$ in (4·52). The result is

$$0 = \frac{4A}{\nu} \left[\frac{a_1\,{}_mc_1}{\kappa_1^2 + \lambda'^2} + \frac{a_2\,{}_mc_2}{\kappa_2^2 + \lambda'^2} + \frac{a_3\,{}_mc_3}{\kappa_3^2 + \lambda'^2} + \ldots \right]$$

$$+ C_6' \frac{\nu \kappa_m}{\lambda H_m} B_0(\kappa_m R_2) + C_7' \frac{\nu \kappa_m}{\lambda H_m} B_0(\kappa_m R_1) + \frac{\nu}{\lambda^2} (\kappa_m^2 + \lambda^2)(\kappa_m^2 + \lambda'^2) a_m. \tag{4·53}$$

We have now a system of linear equations connecting $a_1, a_2, \ldots, a_m, C_6'$ and C_7'. It will be noticed that there are two more unknowns than there are equations. There are, however, two more conditions of which no account has yet been taken which must be satisfied by the solution: w_1 must vanish at R_1 and at R_2.

Using the equation of continuity (2·6), it will be seen from (4·30) that the conditions that w_1 vanishes at R_1 and R_2 are

$$\sum_{m=1}^{\infty} \kappa_m a_m B_0(\kappa_m R_1) = 0 \tag{4·54}$$

and
$$\sum_{m=1}^{\infty} \kappa_m a_m B_0(\kappa_m R_2) = 0. \tag{4·55}$$

Determination of σ

We have now used all the boundary conditions and differential equations available. We have also the same number of equations as unknowns, and since the equations are homogeneous, all the unknowns can be eliminated from them. The resulting equation takes the form of an infinite determinant equated to zero. It can be regarded as an equation to determine σ. It is

$$
0 = \begin{vmatrix}
0 & 0 & \kappa_1 B_0(\kappa_1 R_1) & \kappa_2 B_0(\kappa_2 R_1) & \kappa_3 B_0(\kappa_3 R_1) & \cdots \\
0 & 0 & \kappa_1 B_0(\kappa_1 R_2) & \kappa_2 B_0(\kappa_2 R_2) & \kappa_3 B_0(\kappa_3 R_2) & \cdots \\
\dfrac{\nu^2 \kappa_1}{4A\lambda H_1} B_0(\kappa_1 R_1) & \dfrac{\nu^2 \kappa_1}{4A\lambda H_1} B_0(\kappa_1 R_2) & \dfrac{\nu^2}{4A\lambda^2}(\kappa_1^2+\lambda^2)(\kappa_1^2+\lambda'^2)+\dfrac{{}_1 c_1}{\kappa_1^2+\lambda'^2} & \dfrac{{}_1 c_2}{\kappa_2^2+\lambda'^2} & \dfrac{{}_1 c_3}{\kappa_3^2+\lambda'^2} & \cdots \\
\dfrac{\nu^2 \kappa_2}{4A\lambda H_2} B_0(\kappa_2 R_1) & \dfrac{\nu^2 \kappa_2}{4A\lambda H_2} B_0(\kappa_2 R_2) & \dfrac{{}_2 c_1}{\kappa_1^2+\lambda'^2} & \dfrac{\nu^2}{4A\lambda^2}(\kappa_2^2+\lambda^2)(\kappa_2^2+\lambda'^2)+\dfrac{{}_2 c_2}{\kappa_2^2+\lambda'^2} & \dfrac{{}_2 c_3}{\kappa_3^2+\lambda'^2} & \cdots \\
\cdots & \cdots & \cdots & \cdots & \cdots & \cdots
\end{vmatrix}
\tag{4.60}
$$

This may be written

$$
0 = \Delta_1 = \begin{vmatrix}
0 & 0 & \dfrac{\kappa_1}{H_1} B_0(\kappa_1 R_1) & \dfrac{\kappa_1}{H_1} B_0(\kappa_1 R_2) & \cdots \\
0 & 0 & \dfrac{\kappa_2}{H_2} B_0(\kappa_2 R_1) & \dfrac{\kappa_2}{H_2} B_0(\kappa_2 R_2) & \cdots \\
\kappa_1(\kappa_1^2+\lambda'^2) B_0(\kappa_1 R_1) & \kappa_1(\kappa_1^2+\lambda'^2) B_0(\kappa_1 R_2) & L_1' & {}_1 c_1 & \cdots \\
\kappa_2(\kappa_2^2+\lambda'^2) B_0(\kappa_2 R_1) & \kappa_2(\kappa_2^2+\lambda'^2) B_0(\kappa_2 R_2) & {}_2 c_1 & L_2' & \cdots \\
\cdots & \cdots & \cdots & \cdots & \cdots
\end{vmatrix}
\tag{4.61}
$$

where

$$
L_m' = \frac{\nu^2}{4A\lambda^2}(\kappa_m^2+\lambda^2)(\kappa_m^2+\lambda'^2)^2 + {}_m c_m,
\tag{4.62}
$$

Δ_1 is written to represent the whole determinant, and $\sigma/\nu = \lambda^2 - \lambda'^2$.

Stability of Symmetrical Disturbances

Equation (4·61) may be regarded as a criterion for the stability of given initial disturbances of the type specified by equations (2·5). If the value of σ determined from (4·61) is real, then the motion is stable or unstable according as σ is negative or positive. If σ is complex the motion is unstable if the real part of σ is positive. The motion is then an oscillation of increasing amplitude. A complete discussion of stability necessitates a search for complex roots of (4·61) as well as real ones.

Reasoning on the lines of Rayleigh's analogy it will be noticed that the type of instability which ensues when a liquid whose density increases with height is disturbed from its position of unstable equilibrium cannot be an oscillation of increasing amplitude. Though Rayleigh's analogy cannot be applied without modification to viscous fluids, it seems unlikely that unstable oscillations of this type can exist when the disturbance is symmetrical. It will be seen moreover in Part II, that careful experiments over a wide range of speeds have failed to detect them. It does not seem worth while, therefore, to embark on the extremely laborious and difficult work which a search for complex roots of (4·61) would entail. I have, therefore, limited the work which follows to a discussion of the real roots of (4·61).

Direction in which it is Profitable to Continue the Discussion of (4·61)

The object with which this work was undertaken was to search for a mathematical solution of some case of fluid instability which can conveniently be subjected to experimental investigation.

It is known that all possible types of steady motion of a viscous fluid are stable at very low speeds.[*] If, therefore, one is examining experimentally the stability of any type of steady motion which is dynamically possible at all speeds, it is convenient, in carrying out the experiment, to start the flow at a slow speed and to increase the speed slowly. If the motion is ever unstable it will become so at some definite speed, and the experiment would naturally involve measuring that speed. The instability which then sets in is that particular type of instability which occurs at the lowest speed, and evidently for this type of instability $\sigma = 0$.

If σ be put equal to 0, so that $\lambda = \lambda'$, (4·61) may be regarded as an equation giving the point at which instability will first appear when the speed of the initial steady motion is slowly increased. Equation (4·61), however, gives us more information than that. Up to the present the wave-length of the disturbance which is equal to $2\pi/\lambda$ has been considered as entirely arbitrary. Equation (4·61) determines the speed at which instability of arbitrary wave-length λ first appears. One particular value of λ will correspond with the minimum speed at which instability can appear. In experiments made with viscous fluids this value would be the one which would be observed when the instability first appeared. Probably it is the only one which could ever be observed.

---

[*] Rayleigh, 'On the Motion of a Viscous Fluid', *Phil. Mag.* xxvi (1913), 776–86.

It will be seen that equation (4·61) can therefore be used to predict the dimensions and form of the disturbance as well as the speed at which it will appear. Accordingly in the numerical work which follows, when $\sigma = 0$ (4·61) is regarded as an equation in two variables. The ratio of the speed of the outer cylinder to that of the inner is regarded as a constant, μ; the speed of the flow is then proportional at all points to the speed of the inner cylinder, Ω_1, which is taken as one of the variables. This enters into the equation in the quantities A and $_rc_s$. The other variable is λ. To determine the instability which will first appear with any particular value of μ, i.e. Ω_2/Ω_1, various values of λ are inserted in (4·61) and the one which yields the minimum value of Ω_1 is taken.

To prove that the steady motion is unstable at slightly higher speeds, and stable, so far as real roots of (4·61) are concerned, at slightly lower speeds of the cylinders, it is necessary to show that a slight increase in Ω_1 gives rise to a small positive value of σ, while a slight decrease in Ω_1 gives rise to a small negative value. It is shown later that this is the case.

5. APPROXIMATE FORMULAE

If any particular values of R_1 and R_2 be taken, and also a particular value of μ, it would certainly be possible to find the corresponding numerical values of Ω_1 and λ from (4·61). The labour involved would, however, be so great that it might take months to perform the computation in a single case. To complete the investigation would necessitate finding solutions for various values of R_2/R_1 and for a complete range of μ from large negative to large positive values.

These considerations show that it would be practically impossible to undertake a complete numerical discussion of the problem. On the other hand it will be shown in the second part of this paper that the dimensions of the apparatus which was constructed to investigate the problem, impose a limitation on the range of ratios R_2/R_1 with which it is possible to perform satisfactory experiments. In that apparatus it was found that if the radius of the inner cylinder was much less than half that of the outer one, effects due to the ends of the apparatus began to be appreciable and difficult to eliminate, so that the initial motion ceased to be the same as that between two infinite cylinders. Most of the experiments were therefore conducted with cylinders for which $R_2 - R_1$ was considerably smaller than either R_1 or R_2, that is to say the thickness of the space between the cylinders was small compared with their radii.

Under these conditions it is possible to reduce (4·61) to an approximate form which can be used effectively for numerical calculation. Writing d for $R_2 - R_1$ the work can conveniently be carried to the second approximation, so as to include small terms involving the first power of d/R_1, but not those involving $(d/R_1)^2$.

Approximate Expressions for $B_0(\kappa_m r)$ and $B_1(\kappa_m r)$

Writing $\kappa_m r = x$ it will readily be seen that in order that $B_1(\kappa_m r)$ may vanish at $r = R_1$ and $r = R_2$ when d/R_1 is small x must be a large number. Hence the ordinary asymptotic expressions for Bessel functions can be used.

The asymptotic expansions used are correct to the second approximation, i.e. they include terms in $1/x$ but not terms in $1/x^2$. They are*

$$\sqrt{(\tfrac{1}{2}\pi x)}\, J_0(x) = \cos\left(x - \frac{\pi}{4}\right) + \frac{1}{8x}\sin\left(x - \frac{\pi}{4}\right),$$

$$\sqrt{(\tfrac{1}{2}\pi x)}\, W_0(x) = \sin\left(x - \frac{\pi}{4}\right) - \frac{1}{8x}\cos\left(x - \frac{\pi}{4}\right),$$

$$\sqrt{(\tfrac{1}{2}\pi x)}\, J_1(x) = \sin\left(x - \frac{\pi}{4}\right) + \frac{3}{8x}\cos\left(x - \frac{\pi}{4}\right),$$

$$\sqrt{(\tfrac{1}{2}\pi x)}\, W_1(x) = -\cos\left(x - \frac{\pi}{4}\right) + \frac{3}{8x}\sin\left(x - \frac{\pi}{4}\right).$$

$$(5{\cdot}0)$$

Let
$$\sqrt{\kappa_m}\, B_1(x) = \{C_1' J_1(x) + C_1' W_1(x)\}\sqrt{(\tfrac{1}{2}\pi)}, \qquad (5{\cdot}11)$$

and let the constants C_1' and C_2' be chosen so that

$$\sqrt{\kappa_m}\, B_1(x) = (x)^{-\frac{1}{2}} \sin(x - \tfrac{1}{4}\pi + \epsilon'). \qquad (5{\cdot}12)$$

Then from $(5{\cdot}0)$
$$\left. \begin{array}{l} C_1' + \dfrac{3}{8x} C_2' = \cos\epsilon' \\[2mm] -C_2' + \dfrac{3}{8x} C_3' = \sin\epsilon'. \end{array} \right\} \qquad (5{\cdot}13)$$
and

Solving $(5{\cdot}13)$ and neglecting terms in $1/x^2$, ϵ' can be regarded as constant over the range of values of x corresponding with the space between the cylinders, and the following expressions are obtained for C_1' and C_2'

$$C_1' = \cos\epsilon' + \frac{3}{8x}\sin\epsilon', \quad C_2' = -\sin\epsilon' + \frac{3}{8x}\cos\epsilon'. \qquad (5{\cdot}14)$$

To find the corresponding expression for $B_0(x)$, note that

$$\sqrt{\kappa_m}\, B_0(x) = \{C_1' J_0(x) + C_2' W_0(x)\}\sqrt{(\tfrac{1}{2}\pi)},$$

substituting from $(5{\cdot}14)$ and $(5{\cdot}0)$

$$\sqrt{\kappa_m}\, B_0(x) = x^{-\frac{1}{2}}\left\{\cos(x - \tfrac{1}{4}\pi + \epsilon') + \frac{1}{2x}\sin(x - \tfrac{1}{4}\pi + \epsilon')\right\}. \qquad (5{\cdot}15)$$

Next replace x by $\kappa_m r$ and put $r = R_1 + y$. y is then the distance of any point from the inner cylinder. Choose ϵ' so that

$$\kappa_m R_1 - \tfrac{1}{4}\pi + \epsilon' = 0. \qquad (5{\cdot}16)$$

* See Jahnke and Emde, *Functionen Tafeln*, p. 99.

$B_1(\kappa_m r)$ then vanishes at $r = R_1$, and (5·12) becomes

$$B_1(\kappa_m r) = (R_1 + y)^{-\frac{1}{2}} \sin \kappa_m y, \tag{5·17}$$

and (5·15) becomes

$$B_0(\kappa_m r) = (R_1 + y)^{-\frac{1}{2}} \left\{ \cos \kappa_m y + \frac{1}{2(R_1 + y)\,\kappa_m} \sin \kappa_m y \right\}. \tag{5·18}$$

The values of κ_m are found by putting $B_1(\kappa_m R_2) = 0$, i.e. $\sin(\kappa_m d) = 0$. Hence evidently

$$\kappa_m = m\pi/d, \tag{5·20}$$

where m is a positive integer. The successive values of κ_m are

$$\pi/d, \quad 2\pi/d, \quad 3\pi/d, \dots$$

Writing κ for π/d so that $\kappa_m = m\kappa$ the asymptotic expression for the Bessel functions up to and including first order small terms are

$$B_1(\kappa_m r) = (R_1 + y)^{-\frac{1}{2}} \sin m\kappa y, \tag{5·21}$$

$$B_0(\kappa_m r) = (R_1 + y)^{-\frac{1}{2}} \{ \cos m\kappa y + [2m\kappa(y + R_1)]^{-1} \sin m\kappa y \}. \tag{5·22}$$

It will be noticed that if we had attempted to proceed beyond the second approximation it would not have been found that $\kappa_m = m\kappa$, and the work would have been much more complicated.

Approximate Expressions for the Terms in (4·61)

In the first two rows of determinant we can replace

$$B_0(\kappa_m R_1) \quad \text{by} \quad (R_1)^{-\frac{1}{2}} \quad \text{and} \quad B_0(\kappa_m R_2) \quad \text{by} \quad (-1)^m R^{-\frac{1}{2}}. \tag{5·30}$$

$\kappa_m = m\kappa$ and λ' is the same as λ since $\sigma = 0$.

In the first two columns of (4·61) appears H_m. From (5·30) and (3·24)

$$H_m = \tfrac{1}{2}(R_2 - R_1) = \tfrac{1}{2}d. \tag{5·31}$$

It remains to find the approximate expressions for $_sc_m$ and L'_m. At this stage some care is necessary. On referring to equation (4·51) it will be seen that $_sc_m$ is an integral containing the expression $A + B/r^2$ which represents the angular velocity of any annulus of liquid in the undisturbed state. When d is small compared with R_1 and when neither Ω_1 nor Ω_2 are very large the quantities A and B/r^2 are both large and nearly equal in magnitude, but of opposite sign. For this reason therefore it is necessary to express $A + B/r^2$ in terms of Ω_1, μ, d and R_1. This has been done, the expansion being carried to the second approximation so as to include terms containing the first power of d/R_1. It is

$$A + B/r^2 = \Omega_1 \left\{ 1 - \frac{y}{d}(1 - \mu)\left(1 + \frac{3d}{2R_1}\right) + \frac{3}{2}\frac{y^2}{d^2}(1 - \mu)\frac{d}{R_1} \right\}, \tag{5·32}$$

where $\mu = R_2/R_1$.

This expression is now substituted in (4·51) which becomes

$$_sc_m = \frac{1}{H_s} \int_0^d \left(\alpha + \beta \frac{y}{d} + \gamma \frac{y^2}{d^2}\right) \sin m\kappa y \sin s\kappa y \; \kappa \, dy, \tag{5·33}$$

where

$$\alpha = \frac{\Omega_1}{\kappa}, \quad \beta = -\frac{\Omega_1}{\kappa}(1-\mu)\left(1 + \frac{3}{2}\frac{d}{R_1}\right), \quad \gamma = \frac{3}{2}\frac{\Omega_1 d}{\kappa R_1}(1-\mu). \tag{5·34}$$

Let $\kappa y = \eta$, then

$$_sc_m = \frac{1}{H_s} \int_0^\pi \left(\alpha + \beta \frac{\eta}{\pi} + \gamma \frac{\eta^2}{\pi^2}\right) \sin m\eta \sin s\eta \, d\eta. \tag{5·35}$$

If s is not equal to m

$$\left.\begin{aligned}
\int_0^\pi \sin m\eta \sin s\eta \, d\eta &= 0, \\[2mm]
\int_0^\pi \eta \sin m\eta \sin s\eta \, d\eta &= 0, \quad \text{when} \quad m+s \text{ is even.} \\[2mm]
&= \frac{-4ms}{(m^2-s^2)^2}, \quad \text{when} \quad m+s \text{ is odd,} \\[2mm]
\int_0^\pi \eta^2 \sin m\eta \, s\eta \, d\eta &= \frac{4\pi ms}{(m^2-s^2)^2}, \quad m+s \text{ even,} \\[2mm]
&= \frac{-4\pi ms}{(m^2-s^2)^2}, \quad m+s \text{ odd.}
\end{aligned}\right\} \tag{5·36}$$

If $s = m$

$$\left.\begin{aligned}
\int_0^\pi \sin^2 m\eta \, d\eta &= \tfrac{1}{2}\pi, \\[2mm]
\int_0^\pi \eta \sin^2 m\eta \, d\eta &= \tfrac{1}{4}\pi^2, \\[2mm]
\int_0^\pi \eta^2 \sin^2 m\eta \, d\eta &= \frac{\pi^3}{3} - \frac{\pi}{4m^2}.
\end{aligned}\right\} \tag{5·37}$$

Inserting the values of α, β, γ, H_m, it will be found after some reduction that

$$\left.\begin{aligned}
_sc_m &= \frac{8ms\Omega_1(1-\mu)}{(m^2-s^2)^2\pi^2}, \quad m+s \text{ odd,} \\[2mm]
_sc_m &= \frac{8ms\Omega_1(1-\mu)}{(m^2-s^2)^2\pi^2}\left(\frac{3d}{2R_1}\right), \quad m+s \text{ even,}
\end{aligned}\right\} \tag{5·38}$$

and

$$_mc_m = \tfrac{1}{2}\Omega_1(1+\mu) - \Omega_1(1-\mu)\frac{d}{R_1}\left(\tfrac{1}{4} + \frac{3}{4m^2\pi^2}\right).$$

Approximate Expression for Determinants

On replacing the terms in the first two columns and the first two rows of the determinant Δ_1 by their approximate values, some reductions can be made immediately. The first column can be divided by $2\kappa d^{-1}R_1^{-\frac{1}{2}}$, the second by $2\kappa d^{-1}R_2^{-\frac{1}{2}}$, the first row by $\kappa^3 R_1^{-\frac{1}{2}}$ and the second by $\kappa^3 R_2^{-\frac{1}{2}}$. If θ be written for λ/κ (4·6) becomes

$$0 = \begin{vmatrix} 0 & 0 & 1(1^2+\theta^2) & 2(2^2+\theta^2) & 3(3^2+\theta^2) & \dots \\ 0 & 0 & -1(1^2+\theta^2) & 2(2^2+\theta^2) & -3(3^2+\theta^2) & \dots \\ 1 & -1 & L_1' & {}_1c_2 & {}_1c_3 & \dots \\ 2 & 2 & {}_2c_1 & L_2' & {}_2c_3 & \dots \\ 3 & -3 & {}_3c_1 & {}_3c_2 & L_3' & \dots \\ \multicolumn{6}{c}{\dotfill} \end{vmatrix}, \qquad (5\cdot40)$$

Next perform the following operations on this determinant:

(1) Divide the $(n+2)$th column and the $(m+2)$th row by m.

(2) Add and subtract the first two rows and the first two columns. This reduces every alternate term to zero.

(3) Multiply all terms by $\pi^2\{8\Omega_1(1-\mu)\}^{-1}$ but divide these factors out again from the first two rows and columns. The equation $\Delta_1 = 0$ may now be written

$$0 = \Delta_2 = \begin{vmatrix} 0 & 0 & 1^2+\theta^2 & 0 & 3^2+\theta^2 & \dots \\ 0 & 0 & 0 & 2^2+\theta^2 & 0 & \dots \\ 1 & 0 & L_1 & \dfrac{1}{(1^2-2^2)^2} & \dfrac{3d}{2R_1(1^2-3^2)^2} & \dots \\ 0 & 1 & \dfrac{1}{(2^2-1^2)^2} & L_2 & \dfrac{1}{(2^2-3^2)^2} & \dots \\ 1 & 0 & \dfrac{3d}{2R_1(3^2-1^2)} & \dfrac{1}{(3^2-2^2)^2} & L_3 & \dots \\ 0 & 1 & \dfrac{1}{(4^2-1^2)^2} & \dfrac{3d}{2R_1(4^2-2^2)^2} & \dfrac{1}{(4^2-3^2)^2} & \dots \\ \multicolumn{6}{c}{\dotfill} \end{vmatrix}, \qquad (5\cdot41)$$

where Δ_2 is used as a symbol to represent the determinant and

$$L_m = \frac{\pi^2}{8\Omega_1(1-\mu)\,m^2}\left\{\frac{\pi^2\nu^2(m^2+\theta^2)^3}{4Ad^4\theta^2} + \Omega_1\left(\frac{1+\mu}{2}\right) - (1-\mu)\frac{\Omega_1 d}{R_1}\left(\frac{1}{4}+\frac{3}{4m^2\pi^2}\right)\right\}.$$

Remembering that $\qquad A = \Omega_1\left(1-\frac{R_2^2}{R_1^2}\mu\right)\left(1-\frac{R_2^2}{R_1^2}\right)^{-1},$

it will be found that

$$L_m = \frac{\pi^2}{16m^2}\left\{\frac{1+\mu}{1-\mu} - \frac{d}{R_1}\left(\frac{1}{2} + \frac{3}{2m^2\pi^2}\right) - \frac{P(m^2+\theta^2)^3}{\theta^2}\right\} \tag{5·42}$$

where

$$P = \frac{\pi^4\nu^2(R_1+R_2)}{2\Omega_1^2 d^3 R_1^2(1 - R_2^2\mu/R_1^2)(1-\mu)}. \tag{5·43}$$

Since Ω_1 only occurs in Δ_2 through the term P, P may be regarded as the variable. It is required therefore to find the maximum possible value of P consistent with (5·41). To do this it is necessary to insert a number of different trial values of θ in Δ_2 and then to solve (5·41) to find P in each case. The value of θ for which P is a maximum determines the dimensions of the eddies into which the flow will resolve itself when instability sets in. At first sight this seems to be a very complicated piece of work, but it is possible to perform certain operations on Δ_2 which greatly increase the rapidity with which its roots converge to definite values. These operations will now be explained.

Limiting Case when μ is nearly equal to 1

When μ is nearly equal to 1 the diagonal terms of Δ_2 which contain the factor $(1-\mu)^{-1}$ become large compared with all the other terms. Consider the determinant obtained by taking the first $m+2$ rows and columns of Δ_2. If this determinant be expanded each term will contain $m+2$ factors, and the greatest terms will be those containing the maximum number of factors L_m from the diagonal. In the limit when $\mu \rightarrow 1$ these terms will become infinitely great compared with all the others. Since two of the factors of each term must come from the first two rows and two from the first two columns, neither of which contained any of the L_m terms, it follows that no term can contain more than $m-2$ factors of type L_m. The limiting value of the determinant will therefore be found by taking all terms which can be obtained by choosing a term from each of the first two rows, a term from each of the first two columns and $m-2$ diagonal terms.

Each term is of the form

$$\left(\frac{L_1 L_2 \ldots L_m}{L_s L_t}\right)(s^2+\theta^2)(t^2+\theta^2),$$

where s and t are two integers, one of which is even and the other odd. It is evident therefore that the limiting value of Δ_2 can be expressed in the form

$$\underset{\mu\rightarrow 1}{Lt}\ \Delta_2 = (L_1 L_2 L_3 \ldots L_m \ldots)\left(\frac{1^2+\theta^2}{L_1} + \frac{3^2+\theta^2}{L_3} + \frac{5^2+\theta^2}{L_5} + \ldots\right)\left(\frac{2^2+\theta^2}{L_2} + \frac{4^2+\theta^2}{L_4} + \ldots\right). \tag{5·50}$$

Evaluation of the Greatest Root of (5·41)

It is clear that the greatest value of P consistent with the equation $\Delta_2 = 0$* is the greatest root of the equation

$$\frac{1^2+\theta^2}{L_1} + \frac{3^2+\theta^2}{L_3} + \ldots = 0. \tag{5·51}$$

* The greatest root of $(2^2+\theta^2)/L_2 + \ldots = 0$ can easily be shown to be less than the greatest root of (5·51).

Writing
$$P' = \frac{1-\mu}{1+\mu} P \tag{5.52}$$

and neglecting $\dfrac{d}{R_1}\left(\dfrac{1-\mu}{1+\mu}\right)$ which $\to 0$ as $\mu \to 1$ (5.51) becomes

$$\frac{1^2(1^2+\theta^2)}{1-(P'/\theta^2)\,(1^2+\theta^2)^3} + \frac{3^2(3^2+\theta^2)}{1-(P'/\theta^2)\,(3^3+\theta^2)^3} + \cdots = 0. \tag{5.53}$$

For any particular value of θ it is a simple matter to approximate to the greatest root of (5.53), which evidently lies between $P' = \theta^2(1^2+\theta^2)^{-3}$ and $P' = \theta^2(3^2+\theta^2)^{-3}$. After a few terms the 1 in the denominator becomes small compared with

$$\frac{P'}{\theta^2}\,(m^2+\theta^2)^3.$$

Neglecting it, the $\frac{1}{2}(m+1)$th term is then

$$-\frac{m^2\theta^2}{P'\,(m^2+\theta^2)^2}.$$

The rest of the series including this term is then

$$-\frac{\theta^2}{P'}\left\{\frac{m^2}{(m^2+\theta^2)^2} + \frac{(m+2)^2}{((m+2)^2+\theta^2)^2} + \cdots\right\}. \tag{5.54}$$

After a few more terms it will be possible to neglect the θ^2 which occurs in the denominator of each term. If the first term inside the bracket of (5.54) for which it is possible to do this is

$$\frac{s^2}{(s^2+\theta^2)^2},$$

the remainder of the series including this term is

$$\frac{1}{s^2} + \frac{1}{(s+2)^2} + \frac{1}{(s+4)^2} + \cdots.$$

This series can be summed exactly.

Proceeding in this way it was found in a rough calculation that the greatest roots of (5.53) are associated with values of θ^2 in the neighbourhood of 1. Accordingly the values of P' were calculated to three significant figures for the series of values $\theta^2 = 0.8,\ 0.9,\ 1.0,\ 1.1,\ 1.2$. The corresponding values of P' are 0.0562, 0.0569, 0.0571, 0.0569, 0.0563. The variation of P' with θ is shown in a curve in fig. 2. On looking at that curve it will be seen that the maximum value of P' is 0.0571. It occurs when $\theta^2 = 1.00$. There is no reason to suppose that the correct value of θ^2 is exactly 1, but it almost certainly lies between 0.98 and 1.02.

Stability when the Cylinders are rotating in the same direction with slightly different velocities

We are now in a position to make some definite predictions about the stability of the flow when μ is nearly 1—that is, when the cylinders are rotating in the same direction with slightly different velocities. In the first place the motion changes from being stable to being unstable when Ω_1 passes through the value given by[*]

$$\left(\frac{\pi^2\nu^2(R_1+R_2)}{2\Omega_1^2 d^3 R_1^2(1-R_2^2\mu/R_1^2)(1-\mu)}\right)\left(\frac{1-\mu}{1+\mu}\right) = 0.0571. \tag{5.6}$$

It seems evident that at speeds below this the motion must be stable while at higher speeds it must be unstable, but it is perhaps worth while to prove that this is the case by writing down the equation for σ and showing that it changes from a negative to a positive value as Ω_1 increases through the value given by (5.6).

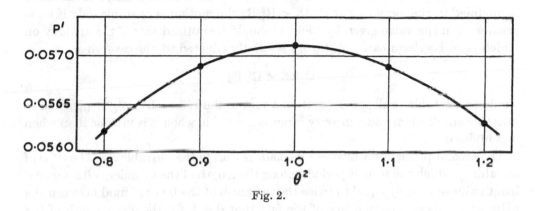

Fig. 2.

Retaining terms in σ from (4.61) the equation equivalent to (5.53) is

$$0 = \sum_{m \text{ odd}} \frac{m^2(m^2+\theta^2+x)}{1-(P'/\theta^2)(m^2+\theta^2)(m^2+\theta^2+x)^2}, \tag{5.61}$$

where $x = \sigma/\kappa\nu$. If P' differs from the value it would have if σ were 0 by a small quantity $\delta P'$, x will also be small and from (5.61) it will be found that

$$x\left[\sum_{m \text{ odd}} \frac{m^2\{1+(P'/\theta^2)(m^2+\theta^2)^3\}}{\{1-(P'/\theta^2)(m^2+\theta^2)\}^2}\right] = -\frac{\delta P'}{P'}\left[\sum_{m \text{ odd}} \frac{m^2(m^2+\theta^2)}{\{1-(P'/\theta^2)(m^2+\theta^2)^3\}^2}\right]. \tag{5.62}$$

Since the series inside the square brackets are positive when P' is positive, it appears that x is positive when $\delta P'$ is negative. Also from (5.43) and (5.52)

$$\frac{\delta P'}{P'} = -2\frac{\delta\Omega_1}{\Omega_1},$$

[*] From (5.43) and (5.52).

where $\delta\Omega_1$ is the change in Ω_1 corresponding with δP_1. Hence when $\delta\Omega_1/\Omega_1$ is positive, i.e. when the speed increases above the speed for which $\sigma = 0$, σ is positive and the motion is unstable. When the speed is slightly less than the speed for which $\sigma = 0$, the steady motion is stable for symmetrical disturbances.

Equation (5·6) gives a positive value for Ω_1^2 if μ is less than R_1^2/R_2^2. If therefore, $\Omega_2/\Omega_1 < R_1^2/R_2^2$ the motion is unstable for values of Ω_1 greater than that given by (5·6). If $\Omega_2/\Omega_1 > R_1^2/R_2^2$ the value of Ω_1 given by (5·6) is imaginary and the motion is stable unless a negative root of (5·53) can be found. It is obvious that there are no negative roots of (5·53) because a negative value of P' makes every term in the series on the left-hand side of the equation positive. The sum of the series, therefore, cannot vanish.

Equation (5·6) shows that Lord Rayleigh's criterion for stability of an inviscid fluid is a limiting case of the criterion for a viscous fluid. Lord Rayleigh's criterion was that a fluid would be stable at all speeds if $\Omega_2/\Omega_1 > R_1^2/R_2^2$, and unstable if $\Omega_2/\Omega_1 < R_1^2/R_2^2$. The former of these is equally true for viscous fluids, but the latter is modified in the sense that if $\Omega_2/\Omega_1 < R_1^2/R_2^2$ the motion is unstable only if Ω_1 is greater than the value given by (5·6). It should be noticed that if the analogy on which Lord Rayleigh based his theory is strictly adhered to, the case when

$$\Omega_2/\Omega_1 < R_1^2/R_2^2$$

would be unstable at all speeds, even in a viscous fluid; for a heterogeneous fluid in unstable equilibrium under gravity is not more stable when it is viscous than when it is inviscid.

The second prediction which can be made is that in the unstable case the type of instability which will form is periodic along the length of the cylinder, with a wave-length almost exactly equal to twice the thickness of the layer of fluid between the cylinders. This is a consequence of the fact that $\theta = 1$, for the wave-length of the unstable disturbance is $2\pi/\lambda = 2\pi/\kappa\theta = 2d/\theta$ and this is

$$2d \quad \text{when} \quad \theta = 1. \tag{5·7}$$

It will be shown in the second part of this paper that the symmetrical type of instability does actually occur under experimental conditions, and that both these predictions are verified with considerable accuracy.

6. Numerical Approximations when $1-\mu$ is not small

When $1-\mu$ is not small the first few diagonal terms may be of the same order of magnitude as the neighbouring terms which are of the form $(m^2-n^2)^{-2}$, but on passing down the diagonal of Δ_2 the $(m+2)$th term contains a factor of order of magnitude Pm^4, while the other terms in the neighbourhood of the diagonal decrease with the factor $(m^2-n^2)^{-2}$. This is largest when $m-n = 1$, when it is of order m^{-2}. The diagonal terms, therefore, rapidly become very great compared with all other terms.

It appears, therefore, that in this case the effect of the parts of the determinant which are situated far from the top left-hand corner may be expected to be of the same kind as that of the same terms in the case when $1 - \mu$ is small. The difference is that in the case when $1 - \mu$ is small *all* the diagonal terms are large, whereas in the case where $1 - \mu$ is not small, all except a few terms near the top left-hand corner are large.

We have seen that in the case when $\mu \to 1$ no terms are of importance except those in the first two columns, the first two rows and the diagonal terms.

Hence in this case

$$\underset{\mu \to 1}{Lt} \Delta_2 = \begin{vmatrix} 0 & 0 & 1^2+\theta^2 & 0 & 3^2+\theta^2 & \ldots \\ 0 & 0 & 0 & 2^2+\theta^2 & 0 & \ldots \\ 1 & 0 & L_1 & 0 & 0 & \ldots \\ 0 & 1 & 0 & L_2 & 0 & \ldots \\ 1 & 0 & 0 & 0 & L_3 & \ldots \\ \hdotsfor{6} \end{vmatrix} \tag{6.00}$$

$$= L_1 L_2 L_3 \ldots \left(\frac{1^2+\theta^2}{L_1} + \frac{3^2+\theta^2}{L_2} + \ldots \right) \left(\frac{2^2+\theta^2}{L_2} + \frac{4^2+\theta^2}{L_4} + \ldots \right). \tag{6.01}$$

The product form (6.01) of Δ_2 was obtained from (6.00) by direct expansion of Δ_2, but it might equally well have been obtained by performing the following series of operations on it: Change the signs of the first two columns. Next divide the third column by L_1, the fourth by L_2, the fifth by L_3, ..., etc., then to the first column add the third, fifth, seventh, ..., columns and to the second the fourth, sixth, eighth, ..., columns, so that

$$\underset{\mu \to 1}{Lt} \Delta_2 = \begin{vmatrix} \dfrac{1^2+\theta^2}{L_1} + \dfrac{3^2+\theta^2}{L_3} + \ldots & 0 & \dfrac{1^2+\theta^2}{L_1} & 0 & \dfrac{3^2+\theta^2}{L_3} & \ldots \\ 0 & \dfrac{2^2+\theta^2}{L_2} + \dfrac{4^2+\theta^2}{L_4} + \ldots & 0 & \dfrac{2^2+\theta^2}{L_2} & 0 & \ldots \\ 0 & 0 & 1 & 0 & 0 & \ldots \\ 0 & 0 & 0 & 1 & 0 & \ldots \\ \hdotsfor{6} \end{vmatrix}. \tag{6.02}$$

The effect of all the distant terms is now concentrated in the first two diagonal terms which are the same as those in (6.01).

The same operations may be performed on Δ_2 when $1-\mu$ is not small. In this case a determinant is obtained which does not reduce to the first two diagonal terms, but, on the other hand, all the other terms contain factors of the form $(m^2-n^2)^{-2}$. For this reason the determinant derived in this way, which will be called Δ_3, converges very much more rapidly than Δ_2.

$$
\Delta_3 = \begin{vmatrix}
\dfrac{1^2+\theta^2}{L_1}+\dfrac{3^2+\theta^2}{L_3}+\cdots & 0 & \dfrac{1^2+\theta^2}{L_1}, & 0 & \dfrac{3^2+\theta^2}{L_3} & \cdots \\[2ex]
0 & \dfrac{2^2+\theta^2}{L_2}+\dfrac{4^2+\theta^2}{L_4}+\cdots & 0 & \dfrac{2^2+\theta^2}{L_2} & 0 & \cdots \\[2ex]
\epsilon\!\left(\dfrac{1}{L_3(1^2-3^2)^2}+\dfrac{1}{L_5(1^2-5^2)^2}+\cdots\right) & \dfrac{1}{L_2(1^2-2^2)^2}+\dfrac{1}{L_4(1^2-4^2)^2}+\cdots & 1 & \dfrac{1}{L_2(1^2-2^2)^2} & \dfrac{\epsilon}{L_3(1^2-3^2)^2} & \cdots \\[2ex]
\dfrac{1}{L_1(2^2-1^2)^2}+\dfrac{1}{L_3(2^2-3^2)^2}+\cdots & \epsilon\!\left(\dfrac{1}{L_4(2^2-4^2)^2}+\dfrac{1}{L_6(2^2-6^2)^2}+\cdots\right) & \dfrac{1}{L_1(2^2-1^2)^2}, & 1 & \dfrac{1}{L_3(2^2-3^2)^2} & \cdots \\[2ex]
\epsilon\!\left(\dfrac{1}{L_1(3^2-1^2)^2}+\dfrac{1}{L_5(3^2-5^2)^2}+\cdots\right) & \dfrac{1}{L_2(3^2-2^2)^2}+\dfrac{1}{L_4(3^2-4^2)^2}+\cdots & \dfrac{1}{L_1(3^2-1^2)^2}, & \dfrac{1}{L_2(3^2-2^2)^2} & 1 & \cdots \\[2ex]
\cdots & \cdots & \cdots & \cdots & \cdots & \cdots
\end{vmatrix}
\qquad (6\cdot03)
$$

where ϵ is written for $3d/2R_1$.

Determination of Roots of $\Delta_3 = 0$

To determine the roots of the equation $\Delta_3 = 0$ it is necessary to assume a value for θ and a value for P, and to calculate the values of the determinant formed by taking the first 1, 2, 3, 4, ..., rows and columns of Δ_3. Owing to the fact that all the diagonal terms after the first two are equal to 1 the actual numerical value of the determinant converges to a definite limit. Taking a value of θ and a series of suitable values of P, the value of P for which Δ_3 changes sign is found by interpolation. This is the root of $\Delta_3 = 0$ which corresponds with the particular value of θ chosen. By taking a series of suitable values of θ the maximum root of $\Delta_3 = 0$ is found, and also the corresponding value of θ.

In evaluating Δ_3 for any values of P and θ the method adopted was first to find the numerical values of the terms, then to eliminate successively the third, fourth, fifth, etc, rows and columns. The effect of this procedure was to alter the values of the first four terms in the top left-hand corner of Δ_3. It was found, however, that after this operation had been repeated a few times no further alteration occurred, the effect of the distant terms being too small to be appreciated. By treating the determinant in this way it became obvious how many rows and columns should be taken in order to evaluate the root to the order of approximation which was desired.

7. EVALUATION OF P AND θ FOR THE CASE WHEN μ LIES BETWEEN 0 AND 1

The case first solved was that for which d is negligible compared with R_1—that is to say, the space between the cylinders is very small compared with their radii. Taking $\mu = 0$ and $\theta = 1$ it was found that if the first term only of Δ_3 is taken, the root is $P = 0.0571$. This result has already been given (see p. 55). On taking four rows and columns the root is $P = 0.0577$, an increase of 1%. On taking six rows and columns the further change in the root is of order 0.1%. It appears, therefore, that if an accuracy of 1% is desired it is unnecessary to take more than four rows and columns of Δ_3. Moreover, it was found that practically the whole change from 0.0571 to 0.0577 is due to the terms involving the factor $(2^2 - 1^2)^{-2}$.

Under these circumstances it appeared probable that the root of $\Delta_3 = 0$ could be obtained by adding a small correction of about 1% to the highest root of $f_1 = 0$, where f_1 is written for the first term of Δ_3, namely,

$$f_1 = \frac{1^2 + \theta^2}{L_1} + \frac{3^2 + \theta^2}{L_3} + \dots.$$

It has already been pointed out that as $\mu \to 1$ the root of $\Delta_3 = 0$ approaches that of $f_1 = 0$. It is clear that for all values of μ between 0 and 1 a small correction to the root of $f_1 = 0$ can be found which will give the root of $\Delta_3 = 0$.

The value of this correction which will be called $\delta_1 P$ may be found as follows: Taking four rows and columns of Δ, the part of Λ_3 due the extra terms containing $(2^2 - 1^2)^{-2}$ is found to be

$$-\left[\left(f_1 - \frac{1^2 + \theta^2}{L_1} \right) \left(f_2 - \frac{2^2 + \theta^2}{L_2} \right) \right] \bigg/ (2^2 - 1^2)^4 L_1 L_2$$

where f_2 is the second diagonal term in Δ_3, namely,

$$\frac{2^2+\theta^2}{L_2}+\frac{4^2+\theta^2}{L_4}+\dots$$

If P_1 is the root of $f_1=0$, $P_1+\delta_1 P$ is the root of $\Delta_3=0$, if

$$\frac{\partial\Delta_3}{\partial P}\delta_1 P-\left[\left(f_1-\frac{1^2+\theta^2}{L_1}\right)\left(f_2-\frac{2^2+\theta^2}{L_2}\right)\right]\Big/ L_1 L_2(2^2-1^2)^4=0. \tag{7.00}$$

The approximate value of Δ_3 is $f_1 f_2$ so that

$$\frac{\partial\Delta_3}{\partial P}=f_1\frac{\partial f_2}{\partial P}+f_2\frac{\partial f_1}{\partial P}, \tag{7.01}$$

and since $f_1=0$,
$$\frac{\partial\Delta_3}{\partial P}=f_2\frac{\partial f_1}{\partial P}. \tag{7.02}$$

Differentiating f_1 it is found that

$$\frac{\partial f_1}{\partial P}=-\left(\frac{1^2+\theta^2}{L_1^2}\frac{\partial L_1}{\partial P}+\frac{3^2+\theta^2}{L_3^2}\frac{\partial L_3}{\partial P}+\dots\right). \tag{7.03}$$

And, since in this case $L_m=\dfrac{\pi^2}{16m^2}\left(\dfrac{1+\mu}{1-\mu}-\dfrac{P}{\theta^2}(u^2+\theta^2)^3\right)$,

$$\frac{\partial L_m}{\partial P}=-\frac{\pi^2(m^2+\theta^2)^3}{16m^2\theta^2}. \tag{7.04}$$

Hence
$$\frac{\partial f_1}{\partial P}=\frac{\pi^2}{16\theta^2}\sum_{m\,\text{odd}}\frac{(m^2+\theta^2)^4}{m^2 L_m^2}. \tag{7.05}$$

Hence combining (7.00) and (7.05)

$$\delta_1 P=\left(f_1-\frac{1^2+\theta^2}{L_1}\right)\left(f_2-\frac{2^2+\theta^2}{L_2}\right)\Big/(2^2-1^2)^4 L_1 L_2 f_2\frac{\pi^2}{16\theta^2}\sum_{m\,\text{odd}}\frac{(m^2+\theta^2)^4}{m^2 L_m^2}. \tag{7.06}$$

In this expression $f_1(1^2+\theta^2)/L_1$, $f_2-(2^2+\theta^2)/L_2$, L_2 and f_2 are negative while L_1 is positive. Hence $\delta_1 P$ is positive.

Greatest root of $f_1=0$

The root, P', of the equation

$$\sum_{m\,\text{odd}}\frac{m^2+\theta^2}{m^2\{1-P'(m^2+\theta^2)^3/\theta^2\}}=0 \tag{7.07}$$

has already been evaluated for a certain range of value of θ.

It is evident that the root of

$$\sum_{m\,\text{odd}}\frac{m^2+\theta^2}{m^2\left\{\dfrac{1+\mu}{1-\mu}-P'(m^2+\theta^2)^3/\theta^2\right\}}=0$$

is
$$P=\left(\frac{1+\mu}{1-\mu}\right)P'.$$

In the case when $\theta = 1$, the root of $f_1 = 0$, is, therefore,

$$P = 0.0571 \left(\frac{1+\mu}{1-\mu}\right). \tag{7.08}$$

Evaluation of Correction to Root of $f_1 = 0$ and to θ

Using this value (7.08) in evaluating the various constitutents of (7.06) it is found that

$$\delta_1 P = 0.00056 \left(\frac{1-\mu}{1+\mu}\right). \tag{7.09}$$

This correction tends to zero when $\mu \to 1$ as was to be expected.

The next step is to find out whether this correction varies sufficiently with θ to alter the value of P which corresponds with the maximum root of $\Delta_3 = 0$. On inserting the values $\theta^2 = 1.2$ and $\theta^2 = 0.8$ in (7.06) it was found that $\delta_1 P$ increases with increasing values of θ, but that the increase is not sufficient to alter materially the maximum value of P. It is found that there is a slight increase in the value of θ which corresponds with the maximum value of P, but in the range of μ from 0 to 1 it is too slight to be worth discussing.

For the case when d is negligible compared with R_1 the greatest root of Δ_3, therefore, occurs when $\theta = 1$ and it is

$$P = 0.0571 \left(\frac{1+\mu}{1-\mu}\right) + 0.00056 \left(\frac{1-\mu}{1+\mu}\right). \tag{7.10}$$

Root of $\Delta_3 = 0$ when d/R_1 is small, but is not neglected

Returning now to the expression (6.03) for Δ_3, it will be seen that the ratio d/R_1 occurs in every term, either in the factor ϵ or in L_m. The terms containing ϵ are small compared with the terms which give rise to the terms $0.00056(1-\mu)/(1+\mu)$ in (7.10) and as we are only considering at present the range of values of μ for which this is small compared with $0.0571(1+\mu)/(1-\mu)$, it follows that all the terms in ϵ can be neglected. The correction to the expression (7.10) due to the fact that d/R_1 is not indefinitely small therefore appears in the analysis only as a change in the values of the terms L_m; and in these terms it always appears as a correction to be subtracted from the factor $(1+\mu)/(1-\mu)$. This correction may be divided into two parts.

(a) The part $d/2R_1$ which is the same for all values of m, and (b) the part $3/d2m^2\pi^2$ which becomes very small when m is large, but amounts to $\frac{1}{4}$ of $d/2R_1$ for $m = 1$. If the second part (b) did not exist, then evidently the approximate root of $\Delta_3 = 0$ given by (7.10) would still apply if $[(1+\mu)/(1-\mu)] - (d/2R_1)$ were substituted for $(1+\mu)/(1-\mu)$.

On looking at the expression (5.42) it will be noticed that owing to the factor $(m^2 + \theta^2)^3$ which occurs associated with P in the expression for L_m, the part contributed by the whole of the factor

$$\frac{1+\mu}{1-\mu} - \frac{d}{R_1}\left(\frac{1}{2} + \frac{3}{2m^2\pi^2}\right)$$

becomes very small compared with $(\theta^2 + m^2)^3 P/\theta^2$ as m increases. Hence it appears that if the part (b) were taken as constant and equal to $3d/2\pi^2 R_1$ for all values of m, very little error would be caused. To estimate its magnitude, the errors in L_1, L_2, L_3 and L_4 due to this erroneous approximation have been calculated for the most unfavourable case which will be required, namely, $P = 0.05, \theta = 1, \mu = 0, d/R_1 = \frac{1}{3}$. The errors are: in L_1, 0; in L_2, 0.7 %; in L_3, 0.1 %; in L_4, 0.02 %. The errors in $L_1 L_2 \ldots$ are therefore never so great as 1 % if this approximation is used.

The object with which these approximate calculations were undertaken was to provide a basis for comparison with experiments. As measurements of the speed at which instability sets in can hardly be expected to attain an accuracy greater than 1 % it does not seem worth while to attempt to attain greater precision than this. We shall therefore substitute

$$\left(\frac{1}{2} + \frac{3}{2\pi^2}\right)\frac{d}{R_1} \quad \text{or} \quad 0.652\frac{d}{R_1} \quad \text{for} \quad \left(\frac{1}{2} + \frac{3}{2\pi^2 m^2}\right)\frac{d}{R_1} \quad \text{in (5.42)}.$$

The value of P is therefore to this order of approximation given by the expression

$$P = 0.0571\left(\frac{1+\mu}{1-\mu} - 0.652\frac{d}{R_1}\right) + 0.00056\left(\frac{1+\mu}{1-\mu} - 0.652\frac{d}{R_1}\right)^{-1}. \tag{7.11}$$

This, together with the definition of P, (5.43), forms the criterion for stability. The expression (7.11) may be expected to hold for positive values of μ from 0 to 1, but it holds over a greater range than this. It holds in fact till the second term ceases to be a small correction. In calculating numerical values for P it was found that this occurred, in the cases considered, at about the value $\mu = -0.5$.

Evaluation of the Root of $\Delta_3 = 0$ when μ is Negative

In the case when μ is negative, that is when the cylinders rotate in opposite directions it is necessary to take account of several rows and columns of Δ_3. Being unable to discover any approximate formula of the type given in (7.11), it was decided to select particular values for μ, R_1 and R_2, substitute in equation (6.03), and deter-mine the maximum value of P and the corresponding value of θ by arithmetical exploration. It was expected that the results so obtained would bear a qualitative resemblance to the results obtained with any other negative value of μ.

The particular values chosen were $\mu = -1.5$, $R_1 = 3.80$ cm., $R_2 = 4.035$ cm. These values were chosen because, at the time this part of the work was begun, some of the measurements to be described in the second part of this paper had already been carried out by means of an apparatus which consisted of two cylinders of these two radii.

A certain amount of preliminary exploration was first undertaken. Assuming the value $\theta = 1$ the values of the determinants formed by taking the first 1, 2, 3, 4, ... etc., rows and columns of Δ_3 were found. Calling these $\Delta_{11}, \Delta_{22}, \Delta_{33}, \ldots$ it was found that they formed a series which appeared to converge rapidly to a definite limit after the fourth or fifth terms; it was found also that the limit towards which the series appeared to converge changed sign as P passed through a value in the neighbourhood of $0 \cdot 001$.

Further exploration seemed to show that the root increased as θ increased; accordingly after a number of trials to determine more precisely the range within which the root lay, the values of $\Delta_{11}, \Delta_{22}, \ldots, \Delta_{88}$ were calculated for values of P which appeared to lie on opposite sides of the root. These calculations were performed for the following values of θ^2, $1 \cdot 5$, $2 \cdot 0$, $2 \cdot 25$, $3 \cdot 0$, $4 \cdot 0$, $5 \cdot 0$. In this way the table (2) was constructed. In this table the values of θ and θ^2 are given in the first two columns. The third column contains assumed values of P. The fourth to the eighth columns contain the values of $\Delta_{44}, \Delta_{55}, \Delta_{66}, \Delta_{77}$ and Δ_{88}. The last column contains the value of P obtained by assuming that Δ_{88} varies uniformly with P in the small range between the two calculated values on either side of the root.

On inspecting the table it will be seen that the convergence of the determinant is very rapid after the fourth row and column, and that very little advantage is gained by using eight rows and columns instead of six or seven. On the other hand it was necessary to carry the calculations as far as Δ_{88} in order to be certain that this was the case.

Table 2. *Values of determinants used in calculating roots of* $\Delta_3 - 0$ *when*
$$\mu = -1 \cdot 5$$

θ	θ^2	P	$\Delta_{44} \times 10^{-4}$	$\Delta_{55} \times 10^{-4}$	$\Delta_{66} \times 10^{-4}$	$\Delta_{77} \times 10^{-4}$	$\Delta_{88} \times 10^{-4}$	Calculated Root
$1 \cdot 225$	$1 \cdot 5$	$0 \cdot 0012$	$-0 \cdot 55$	$-1 \cdot 33$	$-0 \cdot 86$	—	—	$0 \cdot 00124$
		$0 \cdot 0013$	$+1 \cdot 08$	$+0 \cdot 66$	$+1 \cdot 06$	—	—	
$1 \cdot 414$	$2 \cdot 0$	$0 \cdot 0012$	$-0 \cdot 08$	$-2 \cdot 12$	$-1 \cdot 30$	$-1 \cdot 19$	$-1 \cdot 18$	$0 \cdot 001286$
		$0 \cdot 0013$	$+0 \cdot 88$	$-0 \cdot 58$	$+0 \cdot 10$	$+0 \cdot 17$	$+0 \cdot 20$	
$1 \cdot 50$	$2 \cdot 25$	$0 \cdot 0013$	$+0 \cdot 96$	$-1 \cdot 07$	$-0 \cdot 19$	$-0 \cdot 14$	$-0 \cdot 13$	$0 \cdot 00131$
		$0 \cdot 0014$	$+2 \cdot 07$	$+0 \cdot 57$	$+1 \cdot 03$	$+1 \cdot 19$	$+1 \cdot 22$	
$1 \cdot 73$	$3 \cdot 0$	$0 \cdot 0013$	$+2 \cdot 38$	$-2 \cdot 36$	$-1 \cdot 06$	$-0 \cdot 90$	$-0 \cdot 86$	$0 \cdot 00134$
		$0 \cdot 0014$	$+3 \cdot 50$	$+0 \cdot 09$	$+1 \cdot 09$	$+1 \cdot 23$	$+1 \cdot 20$	
$2 \cdot 0$	$4 \cdot 0$	$0 \cdot 0012$	$+4 \cdot 61$	$-7 \cdot 30$	$-6 \cdot 18$	$-4 \cdot 57$	$-4 \cdot 63$	$0 \cdot 00130$
		$0 \cdot 0013$	$+6 \cdot 81$	$-0 \cdot 67$	$-0 \cdot 35$	$-0 \cdot 12$	$-0 \cdot 07$	
		$0 \cdot 0014$	$+4 \cdot 40$	$+2 \cdot 48$	$+3 \cdot 37$	—	—	
$2 \cdot 236$	$5 \cdot 0$	$0 \cdot 0012$	$+13 \cdot 91$	$-3 \cdot 61$	$-0 \cdot 83$	$-0 \cdot 41$	$-0 \cdot 31$	$0 \cdot 00121$
		$0 \cdot 0013$	$+16 \cdot 1$	$+2 \cdot 8$	$+4 \cdot 8$	$+5 \cdot 2$	$+5 \cdot 3$	
		$0 \cdot 0014$	$+16 \cdot 9$	$+6 \cdot 5$	$+8 \cdot 3$	$+8 \cdot 4$	$+8 \cdot 5$	
		$0 \cdot 0015$	$+17 \cdot 5$	$+9 \cdot 65$	$+10 \cdot 9$	$+11 \cdot 0$	$+11 \cdot 1$	

To find the maximum value of P the roots given in the last column of table 2 were plotted on a diagram, the ordinates being the corresponding values of θ. This diagram is shown in fig. 3. It will be seen that a smooth curve can be drawn through all the points, and that the maximum height of this curve occurs when $\theta = 1 \cdot 73$, and that at this point

$$P = 0 \cdot 00134. \tag{7.12}$$

In calculating the determinants given in table 2 it was found that the effect of the correction due to all the terms containing ϵ in Δ_3 was small compared with the correction $0 \cdot 652 \, d/R_1$ which is subtracted from $(1 + \mu)/(1 - \mu)$. If the effect of the terms containing ϵ be neglected, the work just described is applicable to other values of d and

R_1, but the value of μ must be altered so that the value of $[(1+\mu)/(1-\mu)] - 0.652\, d/R_1$ is the same as it was in the case which has been calculated.

In this way, for instance, in the case when $R_1 = 3.55$, $R_2 = 4.035$ it is found that the values $\theta = 1.73$, $P = 0.00134$ apply when

$$\mu = -1.347. \tag{7.13}$$

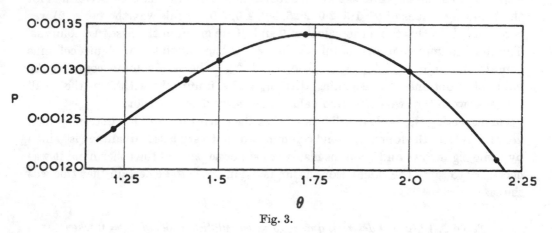

Fig. 3.

Stream Lines when Instability sets in

The results which were obtained in the preceding section will be used later in comparing the actual disturbances which arise in unstable fluid flow with those predicted theoretically. In the meantime, it is of interest to construct some diagrams showing the stream lines which are to be expected when instability sets in. These diagrams are useful in designing apparatus for testing the mathematical predictions, because the selections of the most suitable experimental method for demonstrating the instability of the flow, if it exists, will depend on the particular type of instability which is expected.

The particles of water flow in complicated three-dimensional curves. On the other hand the component of velocity in any meridian plane through the axis can evidently be represented by the Stokes Stream Function ψ. In the general case ψ is related to μ by the relation $u = \partial\psi/r\,\partial r$, so that

$$\psi = \frac{r}{\lambda} e^{\sigma t} \cos \lambda z \sum_{m=1}^{\infty} a_m B_1(\kappa_m r). \tag{7.20}$$

Dropping the factor $e^{\sigma t}/\lambda$ which does not affect the forms of the stream lines, in the approximate case when the asymptotic expression (5.21) is used for $B_1(\kappa_m r)$, this becomes

$$\psi = (R_1 + y)^{\frac{1}{2}} \cos \theta \kappa z \, \Sigma\, a_m \sin m\kappa y. \tag{7.21}$$

To construct the stream lines it is necessary therefore to calculate the constants a_m. Two cases will be considered: (a) the case where μ is nearly equal to 1, and (b) the case where $\mu = -1.5$.

(a) *Stream Lines when* $\mu = 1$, $\theta = 1$. In this case the values of a_m can be obtained directly from an inspection of equation (6·00). Retracing the operations by which (6·00) was derived from (5·43) and leaving out an arbitrary constant which determines the magnitude of the disturbance, it will be found that when m is odd

$$a_m = \frac{\theta^2 + 1}{mL_m},$$

where $\qquad L_m = \frac{\pi^2}{16m^2}\{1 - 0\cdot0571\,(m^2+1)^3\},$ $\qquad$ (7·22)

and when m is even $\qquad a_m = 0.$

The values of a_m obtained from (7·22) are given below (table 3).

Table 3

$a_1 = +0\cdot210$	$a_5 = -0\cdot0070$	$a_9 = -0\cdot0013$	$a_{13} = -0\cdot00045$
$a_3 = -0\cdot0306$	$a_7 = -0\cdot0028$	$a_{11} = -0\cdot00074$	—

Using these values of a_m the values of $\Sigma a_m \sin m\kappa y$ were calculated for values of y ranging from 0 to d, or π/κ. These are given in table 4, and are plotted in the curve

Table 4

$18y/d$	0 and 18	0·5 and 17·5	1 and 17	2 and 16	3 and 15	4 and 14
$\Sigma a_m \sin m\kappa y$	0	0·0038	0·0109	0·0408	0·0733	0·1138
$18y/d$	5 and 13	6 and 12	7 and 11	8 and 10	9	—
$\Sigma a_m \sin m\kappa y$	0·1519	0·1855	0·2136	0·2300	0·2347	—

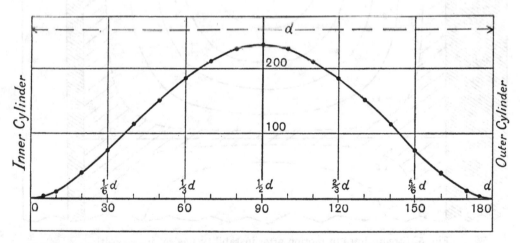

Fig. 4. Radial velocity u_1 on an arbitrary scale. Case when cylinders rotate in same direction, μ positive. Figures on under side of base line are values of $\pi y/d$ in degrees.

of fig. 4. It will be seen that the curve touches the axis at $y = 0$ and $y = d$, as was to be expected.

The value of ψ was next calculated for the case when d is small compared with R_1, so that the factor $\sqrt{(R_1 + y)}$ in (7·21) can be regarded as a constant. Curves were then drawn for various equidistant values of ψ, the numbers given in table 4 being multiplied by a factor so as to make $\psi = \pm 1$ at the centres of the pattern and $\psi = 0$ at the boundary. These curves are shown in fig. 5. Their spacing gives an idea of the

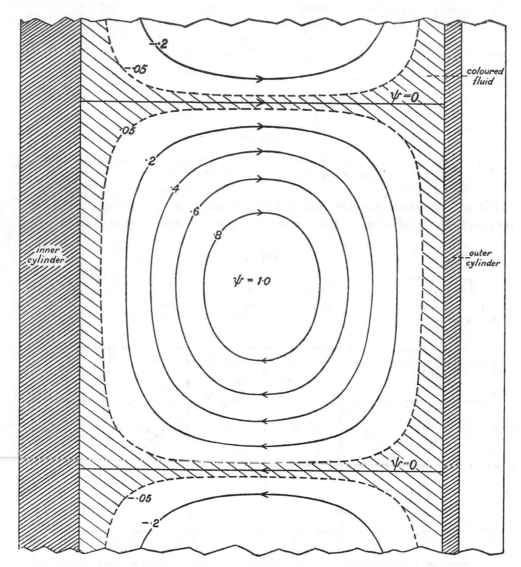

Fig. 5. Stream lines of motion after instability has set in, μ positive.

velocity of the flow at any point. It will be seen that the circulation in a section of the fluid by an axial plane consists of a series of vortices which fill square compartments extending from the inner to the outer cylinder. Alternate vortices rotate in opposite directions as though they were geared together.

(b) *Stream Lines when* $\mu = -1\cdot5$, $\theta = 1\cdot73$. In the case when $\mu = -1\cdot5$ it was necessary first to calculate the minors of Δ_3. Using the values $\mu = -1\cdot5$, $\theta = 1\cdot73$, $P = 0\cdot00134$, $R_1 = 3\cdot80$, $R_2 = 4\cdot035$, the values of L_m were first calculated, and the complete expression for Δ_{88} given below (7·23) was written out.

$$\Delta_{88} = \begin{vmatrix} 559\cdot2 & 0 & 24\cdot0 & 0 & 174 & 0 & 113 & 0 \\ 0 & 601\cdot3 & 0 & 117\cdot2 & 0 & 149 & 0 & 85 \\ 0\cdot21 & 1\cdot899 & -1 & 1\cdot862 & 0\cdot21 & 0\cdot35 & 0 & 0\cdot002 \\ 1\cdot258 & 0\cdot005 & 0\cdot669 & -1 & 0\cdot580 & 0\cdot005 & 0\cdot009 & 0 \\ 0\cdot008 & 0\cdot832 & 0\cdot008 & 0\cdot670 & -1 & 0\cdot160 & 0 & 0\cdot002 \\ 0\cdot374 & 0\cdot011 & 0\cdot027 & 0\cdot011 & 0\cdot296 & -1 & 0\cdot050 & 0 \\ 0\cdot005 & 0\cdot153 & 0 & 0\cdot038 & 0\cdot005 & 0\cdot097 & -1 & 0\cdot018 \\ 0\cdot065 & 0\cdot001 & 0\cdot005 & 0 & 0\cdot020 & 0\cdot001 & 0\cdot033 & -1 \end{vmatrix}.$$

$$(7\cdot23)$$

Using the first seven rows and all the eight columns, the eight minors formed by leaving out successively the 1st, 2nd, 3rd, ... 8th columns were then found. Denoting them by $M_1, M_2, M_3 \ldots M_8$ their values are given below (table 5).

Table 5

$M_1 = +1\cdot60 \times 10^4$	$M_3 = -9\cdot80 \times 10^4$	$M_5 = -3\cdot47 \times 10^4$	$M_7 = -0\cdot09 \times 10^4$
$M_2 = -1\cdot60 \times 10^4$	$M_4 = +6\cdot75 \times 10^4$	$M_6 = +0\cdot82 \times 10^4$	$M_8 = 0\cdot06 \times 10^{4*}$

To calculate $a_1, a_2, \ldots$ from these minors it was necessary to take account of the operations which were performed on the determinants Δ_1, Δ_2 and Δ_3, after the constants a_m had been eliminated. Retracing these operations it was found that when m is odd

$$a_m = (M_1 + M_{m+2})(\theta^2 + m^2)^2 (mL_m)^{-1}, \qquad (7\cdot24)$$

and when m is even

$$a_m = -(M_2 + M_{m+1})(\theta^2 + m^2)(mL_m)^{-1}. \qquad (7\cdot25)$$

Since all the terms can be divided by any factor without altering their relative values, the first factors of (7·24) and (7·25) were divided by 1·60 which is the numerical value of M_1 or M_2. For large values of m the first factor in (7·24) then becomes equal† to 1 and

$$a_m = (\theta^2 + m^2)(mL_m)^{-1}. \qquad (7\cdot26)$$

Using the formulae (7·24) and (7·25) for the first six terms and (7·26) for the higher terms, the following series of values were found for a_m (see table 6).

Table 6

$a_1 = -123\cdot5$	$a_3 = -67\cdot8$	$a_5 = 21\cdot6$	$a_7 = 9\cdot4$	$a_9 = 4\cdot6$
$a_2 = -189$	$a_4 = 18\cdot2$	$a_6 = 13\cdot8$	$a_8 = 6\cdot4$	$a_{10} = 3\cdot4$

* M_8 is probably slightly inaccurate owing to the method of reduction, but as will be seen later such an inaccuracy would have no appreciable effect on the result.

† The part due to M_{m+2} is small compared with the part due to M_1, so that errors in a_m due to errors in M_{m+2} become unimportant as m increases.

Using these numbers for a_m the values given in table 7 for $\Sigma a_m \sin m\kappa y$ were found. In table 7, d, the space between the cylinders, is divided into eighteen equal parts corresponding with changes of $10°$ or $\pi/18$ in κy.

Table 7

$18y/d$	0 and 18	1	2	3	4	5
$\Sigma a_m \sin m\kappa y$	0	55	164	284	348	341
$18y/d$	6	7	8	9	10	11
$\Sigma a_m \sin m\kappa y$	288	213	121	39	$-14\cdot6$	$-47\cdot8$
$18y/d$	12	13	14	15	16	17
$\Sigma a_m \sin m\kappa y$	$-59\cdot0$	$-51\cdot0$	$-38\cdot8$	$-28\cdot1$	$-16\cdot8$	$-5\cdot0$

From these numbers the value of u_1 is found by dividing by $(R_1+y)^{-\frac{1}{2}}$.

The curve given in fig. 6 shows the relation between u_1 and y. It will be noticed that as was to be expected the curve touches the axis at either end. The interesting thing about it, however, is that it crosses the axis at a point roughly half-way between the two cylinders. At this point the radial component of velocity is zero. This means that there is a certain cylindrical surface between the two rotating cylinders which divides the flow. The instability therefore produces a flow which is divided into two separate regions.

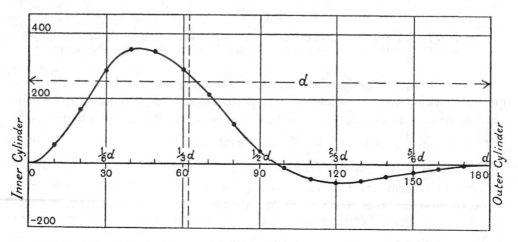

Fig. 6. Radial velocity u_1 on an arbitrary scale; case when cylinders rotate in opposite directions, $\mu = -1\cdot5$.

The stream lines of this flow were next calculated in the same way as in the previous case. They are shown in fig. 7, which is printed on the same scale as fig. 5 to facilitate comparison between them. It will be seen that the circulation now consists of two types of vortices. An inner region which extends out from the inner cylinder, about half-way to the outer one, is filled with vortices rotating alternately in opposite directions. These are very similar in character to the vortices found in the case when $\mu = 1$. They still fill rectangular compartments, and these compartments are still nearly square, though not so accurately square as in the case when $\mu = 1$. An effect of restricting the inner circulation to a region which is only about half the thickness of the total space between the cylinders appears to be to reduce also the spacing of the other sides of the rectangular boundaries of the vortices so that the compartments are still nearly square.

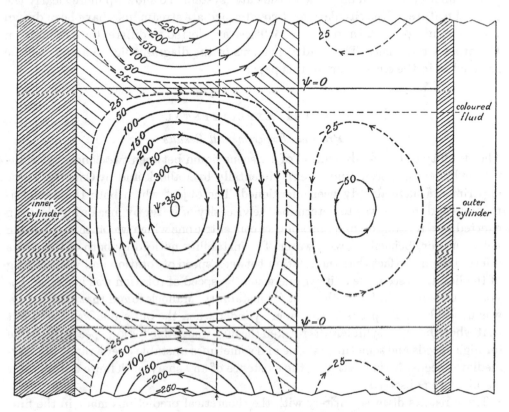

Fig. 7. Stream lines of motion after instability has set in, $\mu = -1.5$.

In the space outside the inner system of vortices is an outer system which is very much less vigorous than the inner system. These outer vortices rotate in the opposite directions to the inner ones with which they are in contact.

It seems probable that the physical explanation of this phenomenon is that the surface where the velocity is zero in the steady motion divides the space between the two cylinders into two regions. In the inner region the square of the circulation decreases outwards, so that centrigual force tends to make the flow unstable. In the outer region the square of the circulation increases so that centrifugal force tends to make the flow stable. The surface where the fluid is at rest in the steady motion is not coincident with the surface separating the two systems of vortices in the disturbed motion. In fig. 7 the section of the former surface is shown as a dotted line and it will be seen that the inner system of vortices extends outside the region where centrifugal force tends to produce instability. That this would be the case might have been anticipated on general grounds.

A remarkable feature of the vortex systems shown in fig. 7 is the great difference which exists between the vigour of the inner and outer systems. The stream lines are drawn for values of ψ differing by 50 units on an arbitrary scale. There are six of these in the inner system and only one in the outer system. To show up more clearly the general features of the circulations, two intermediate stream lines have been drawn for the values $\psi = +25$ and $\psi = -25$. These are dotted to differentiate them from the other stream lines. The shaded portions of the diagrams figs. 5 and 7, will be referred to in the second part of this paper.

Part II. Experimental

Previous Experimental Results

The stability of the steady motion of a viscous liquid between two concentric rotating cylinders has been studied experimentally by Mallock and by Couette. These experiments have already been mentioned. The object which both these experimenters had in view was to determine the viscosity of water by measuring the drag exerted by a rotating cylinder on another concentric one which was at rest, the space between them being filled with water. The instability noticed by both of them was inferred from the fact that the relation between speed of rotation and viscous drag of the liquid ceased to be a linear one when the speed of rotation was increased beyond a certain limit. Using this test for instability, Mallock found that steady flow was unstable at all speeds of the inner cylinder when the outer one was fixed, but that when the outer cylinder was rotated the flow was stable for low speeds, unstable for high speeds and sometimes stable and sometimes unstable over a range of intermediate speeds. No indication of the existence of any sharp or definite criterion for stability was observed.

These results disagree entirely with the theoretical predictions made in the first part of this paper. According to the foregoing theory the motion should be stable even at high speeds when the inner cylinder is at rest. When the outer cylinder is at rest the flow should be stable at low speeds of the inner cylinder, and there should be a definite speed at which instability should suddenly make its appearance as the speed is increased.

This disagreement between theory and experiment may be due to a variety of causes. It may be that the types of disturbance which actually arise are not symmetrical about the axis. On the other hand there are other possible causes besides instability which could give rise to a non-linear relation between speed of rotation and viscous drag in Mallock's experiments. In the first place the lengths of Mallock's cylinders were very little greater than their diameters. His outer cylinder for instance was 7·8 in. diameter while the depth of water used was only 8·5 in., and the thickness of the layer of water between the cylinders was 0·915 and 0·42 in., in his two sets of experiments. If the cylinders were infinitely long or if the thickness of the layer of fluid were very small, the steady two-dimensional flow contemplated in Mallock's experiments would no doubt occur, and a linear relation might be expected between speed and viscous drag. On the other hand in the neighbourhood of the bottom of the liquid the flow cannot be two-dimensional, and unfortunately the cylinder on which Mallock measured the viscous drag extended practically down to the bottom of the liquid. It therefore certainly penetrated into the region where the linear law does not hold. Mallock recognized this, for in the course of his experiments he substituted an ingenious mercury bottom for the rigid bottom with which he began. By this means he hoped to eliminate, partially at any rate, the effect of the bottom. The very large effect which this device had on his results showed that in his original experiments, at any rate, a large part of the drag which he observed might be attributed directly to the effect of the bottom. On the other hand, there is little evidence to show that it succeeded in eliminating this effect completely, or even that the bottom effect was not still large in the case when the outer cylinder was at rest.

It appears therefore that Mallock's experiments do not afford conclusive evidence of the existence of instability in the case when the outer cylinder is at rest, at any rate at slow speeds of the inner cylinder.

In the case when the inner cylinder is at rest Mallock's experiments appear more conclusive, because he observed sudden and violent changes in the drag on the inner cylinder. On the other hand it is by no means certain that this instability would have occurred if the inner cylinder had been supported differently. It has been shown by Von Hopf that instability may arise from flexibility in the bounding walls of a fluid in steady motion. In Mallock's experiments one of the cylinders had to be supported so that it could turn without resistance. This condition must, I think, have prevented this cylinder from being held so rigidly that small lateral movements were impossible.

Design of Apparatus

In designing apparatus for testing the conclusions reached in the first part of this paper, care was taken to eliminate as far as possible the disadvantages from which, in the author's opinion, Mallock's apparatus suffered. In the first place the cylinders were made as long as possible so as to eliminate end effects. They were 90 cm. long and the outer one was 4·035 cm. radius. In most of the experiments the thickness of the layer of liquid between the cylinders was less than 1 cm. In the second place both cylinders were held in heavy plane bearings at each end by heavy iron supports fixed to a stone floor and to the walls of the Cavendish Laboratory.

The general arrangement of the cylinders is shown in fig. 8. In that diagram the various parts of the apparatus are indicated by letters. The weight of each cylinder was taken by a steel ball, B, resting on a flat plate, C, below, and fitting into a conical centre in the end of the shaft which it supported. The bearings, J, were long and of exceptionally good fit, so that lateral motion could only occur by bending of the whole apparatus. In order that this might be minimized as far as possible the inner cylinder was fixed on a mild-steel shaft, A, of large diameter ($\frac{3}{4}$ inch).

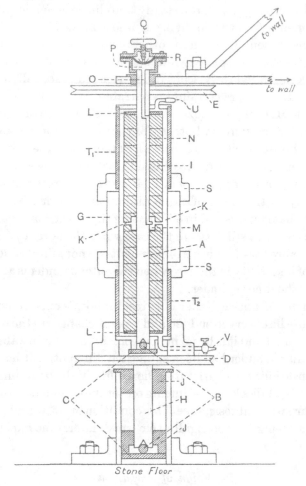

Fig. 8.

It is clear for the reasons given above that the methods of previous observers would not give the information required. It was necessary to devise some method which would show not only the exact speed at which instability occurs, but the type of motion which ensues.

The method employed by Osborne Reynolds in the case of flow through a circular pipe was to inject a thin continuous stream of colouring matter into the centre of the stream. When the breakdown of steady flow occurred the motion of the water could then be studied in some detail. In the present case a similar method was used, but certain modifications were necessary in view of the special type of motion which was expected. The actual stream lines in the cases calculated in Part I are complicated spiral curves which are not symmetrical about the axis. A method designed to show up or mark a single stream line would therefore yield results which would be difficult to interpret. On the other hand if colouring matter could be emitted simultaneously at all points of a circle placed symmetrically in the fluid, the motion of the sheets of colouring matter so produced would, if the motion were symmetrical, give exactly the information required, namely, the component of motion in an axial plane.

This condition was attained by emitting coloured liquid from six small holes placed on a symmetrical circle near the middle of the inner cylinder. The inner cylinder, I, was made by threading a large number of turned and bored sections made of paraffin wax on to the central steel shaft. These were held and pressed together by brass discs, L, of the same diameter at each end of the cylinder. In one of the paraffin sections were six very small symmetrical holes, K, which were connected together by means of a small groove, M, turned on the inner curved surface of the section. The coloured liquid was supplied to this groove by means of a small brass tube, N, which was let into a slot cut in one side of the central steel shaft. This duct led finally through a small central hole, O, in the upper end of the central shaft to a brass box, P, which was filled with the coloured liquid. To force the coloured liquid down the duct the milled head, Q, was turned. This pressed on a diaphragm of thick rubber, R, and so forced the colour out through the six small holes, K.

In order to see the colouring matter it was necessary to make the outer cylinder of glass. This requirement gave rise to considerable trouble on account of the difficulty of producing an accurately turned, bored, ground and polished glass tube 90 cm. long. The difficulty was surmounted by Messrs Powell, of the Whitefriars Glass Works, who succeeded in producing a satisfactory glass tube 8 in. long, 8·07 cm. bore and 10·5 cm. external diameter. This was turned, bored, roughly polished, and the ends faced square. It was then mounted in iron castings, S, which were fitted accurately on to the upper and lower sections, T_1 and T_2 of the outer rotating cylinder. These castings will be seen at the top and bottom of the photographs, figs. 9–16, pls. 1 and 2. The inside bore of the outer cylinder did not vary by as much as $\frac{1}{10}$ mm. in its whole length.

The whole apparatus was driven by an electric motor fitted with a governor so that the speed could be kept constant. The ratio of the speeds of the cylinders could be varied by means of a continuously variable speed gear.

Method of Performing Experiments

In performing an experiment the box, P, was filled with a solution of fluorescene, which was usually made of the same density as water by mixing with ammonia or alcohol, though in some experiments it was made slightly heavier. This fluorescent solution was found to be very good for eye observations, but it was useless when photographs were to be taken. A solution of eosin made up to the same density as water by mixture with alcohol was found to be the best for photographic purposes.

The space between the two cylinders was filled with water from which air had been expelled by boiling. In cases when the fluorescene solution was slightly heavier than water the liquid coming out of the six small holes fell down in six streams which kept fairly close to the inner cylinder. In cases where the fluorescene solution was of the same density as water some of the water was run out from the bottom of the apparatus, and at the same time fluorescene or eosin was forced out through the six holes. The downward movement of the water drew the fluorescene out into six thin vertical streams, which were found to be extremely close to the surface of the inner cylinder. The apparatus was then immediately started rotating at a slow speed and the shearing motion of the liquid in the annulus between the cylinders caused the six vertical lines of colour to broaden laterally till after a short time all the coloured liquid formed a uniform thin sheet on the surface of the inner cylinder.

The ratio of the speeds of the two cylinders was fixed during each experiment by the setting of the variable speed gear. The speed of the motor was then gradually increased till instability occurred.

Case when Cylinders are Rotating in the Same Direction

In this case we have seen that the type of motion to be expected when instability sets in is the same for all positive values of μ less than R_1^2/R_2^2. The flow in meridian planes consists of vortices contained in square partitions and rotating alternately in opposite directions. The effect which this system of vortices might be expected to have on the film of coloured fluid close to the inner cylinder can be seen by referring to fig. 5, Part I. Since the motion is evidently to the first order of small quantities a steady motion, the coloured liquid which lies close to the surface $\psi = 0$ will remain close to that surface. The surface $\psi = 0$ consists of the square partitions within which the vortices are contained. The coloured liquid will therefore mark out the edges of these square partitions. In fig. 5 the shaded portion represents a possible form of the coloured region after instability has set in.

General Result

On observing the apparatus from the side it was found that provided the experiment was carefully carried out, the instability made its appearance at a certain speed in every case when it was expected and in no case when it was not. This speed was quite definite, and the measurement could be repeated on different occasions with an accuracy of about 1 or 2 %.

The phenomenon which was observed was the same in each case. The layer of coloured liquid suddenly gathered itself into a series of equidistant films whose planes were perpendicular to the axis of rotation. These films were in each case spaced at a distance apart nearly equal to *twice* the thickness between the cylinders. The films seemed to spread out till they reached the inner surface of the outer cylinder. They then spread upwards and downwards close to that surface till they covered it with a thin film of coloured liquid. This film was almost invisible because it could not be seen edge-on. On the other hand, when the upward and downward flowing sheets met at the points half-way between the out-flowing films they formed inward-flowing films of the same type as the outward-flowing ones. The resulting appearance after the motion had been going on for about two or three seconds was that of a series of thin equidistant planes of coloured fluid spaced at a distance equal to the thickness of the space between the cylinders. In fact, after the first few seconds the motion appeared to get to a steady state in which it was impossible to distinguish the outward-flowing films from the inward-flowing ones, though all of them were extremely sharply defined.

Photographs of the Stream Lines

Considerable difficulty was experienced in obtaining satisfactory photographs of the phenomenon because when eosin was used instead of fluorescene a more concentrated solution was necessary; and it was found difficult to make up this solution so that its density remained the same as that of water when it was surrounded by water. It frequently happened in fact that the coloured liquid formed two columns one going up and the other down, when strong eosin solution mixed with alcohol was used. In spite of this and other difficulties some fairly good photographs were obtained.

Fig. 9, pl. 1, shows the appearance of the films shortly after their formation. This photograph shows a motion which is not so regular as most of those observed, but it has the advantage that one can see some of the intermediate stream lines marked out by some colouring matter which had got away from the surface of the inner cylinder before the instability set in. A particularly noticeable one occurs in the third partition from the bottom on the left-hand side. The photographs were taken with an ordinary magnesium flashlight apparatus.

Verification of Predicted Spacing of the Vortices

It will be noticed that the partitions shown in figs. 9, 10 and 11, pl. 1, appear square. This square appearance, however, is deceptive. The refraction of the glass and water magnify the horizonal dimensions without altering the vertical dimensions of objects in the water. On the other hand, the outer edge of the pattern is cut off altogether by refraction. These two effects neutralise one another so that the general appearance of the partitions in the photographs is square.

The photograph (fig. 9, pl. 1), was taken when the radius of the inner cylinder, R_1, was 2·93 cm.; R_2, the radius of the outer cylinder, was the same in all cases, namely, 4·035 cm. The distance apart of the partitions as measured on the original photograph was 0·47 cm., the external diameter of the glass cylinder on the photograph was 4·7 cm. and its true diameter was 10·5 cm. The true distance apart of the partitions was, therefore, $10·5 \times 0·47/4·7 = 1·05$ cm. The difference between the radii of the two cylinders was $4·035 - 2·93 = 1·105$ cm. Hence we have our first numerical verification of the theory of Part I.

$$\left.\begin{array}{ll} \text{Predicted spacing of vortices} & 1·105\,\text{cm} \\ \text{Observed spacing of vortices} & 1·05\,\text{cm.} \end{array}\right\} \text{Error } 5\,\%.$$

To show the effect of change in thickness of the layer of fluid a photograph of the bands or partitions, taken when $R_1 = 3.25$ cm., is shown in fig. 10, pl. 1. On measuring this spacing on the original photograph it was found that twelve of them occupied 3.95 cm. The magnification was 0.4095. Hence

True spacing of partitions was   0.804 cm.⎫
                                                                         ⎬ Error $3\frac{1}{2}$ %.
Predicted spacing was $4.035 - 3.25$ 0.785 cm.⎭

In order to show the accuracy with which these bars of coloured liquid space themselves when the experiments are carefully performed the photograph (fig. 11, pl. 1) is shown. The fineness of the partitions shown in this photograph approximates to the fineness which can easily be obtained with the fluorescene used for eye observations, but it is not actually quite so good.

Case when the Cylinders Rotate in Opposite Directions

When μ is negative—that is, when the cylinders rotate in opposite directions, only one case has been worked out completely, namely, that in which $\mu = -1.5$, $R_1 = 3.80$, $R_2 = 4.035$. The characteristic differences revealed by the analysis between the motion in this case, and that in the case when μ is positive, are:

(a) The spacing of the vortices is reduced in the ratio $1.73:1$. The predicted spacing of the vortices in this case is in fact $(4.035 - 3.80)/1.73 = 0.136$ cm; and

(b) The vortices in contact with the inner cylinder only extend out about halfway to the outer cylinder instead of extending right across the fluid annulus.

(a) *Spacing of Vortices.* To verify the conclusions reached in regard to the spacing of the vortices a number of measurements were taken when the radius of the inner cylinder was 3.80 cm., the values of μ ranging from $+0.65$ to -1.78.

Table 8. *Giving the observed spacing of the vortices for various values of μ in the case when $R_1 = 3.80$ cm. and $R_2 = 4.035$, so that $d = 0.235$ cm.*

μ	Observed spacing of vortices
0.65	0.241, 0.245
0.596	0.25, 0.24, 0.238, 0.25
0.40	0.244, 0.250, 0.244
0	0.236, 0.235
−0.388	0.230, 0.24, 0.236
−0.429	0.237, 0.238, 0.235
−0.640	0.232, 0.228
−0.716	0.201, 0.203, 0.198
−1.00	0.150, 0.165, 0.160, 0.157
−1.20	0.156, 0.165
−1.37	0.143, 0.146
−1.78	0.09, 0.105, 0.115

The results are given in table 8 and they are shown in the form of a curve in fig. 12. In this curve the abscissae represent μ and the ordinates represent the corresponding spacing of the vortices in centimetres. The predicted points, i.e. theoretical values of d/θ, are shown by means of circles and the observed points by means of dots.

The curve is drawn roughly through the dots. Unfortunately, owing to an oversight, no observation was taken for the value $\mu = -1.5$, but it will be seen that the calculated point lies almost exactly on the observed curve.

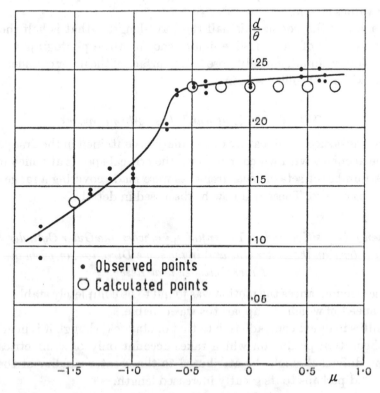

Fig. 12. Comparison between observed and predicted spacing of vortices for various values of μ; case when $R_1 = 3.80$ cm., $R_2 = 4.035$ cm.

It will be noticed that the predicted result that the spacing of the vortices should be the same for all positive values of μ and equal to the thickness of the layer of fluid is also strikingly verified. It will be observed that the spacing of the vortices does not begin to decrease much till μ has a considerable negative value, about -0.5.

(b) In the case when $R_1 = 3.80$ it was found difficult to verify the prediction that the vortices in contact with the inner cylinder should extend no further than about half-way out to the outer cylinder because the refraction of the glass cylinder prevented the extreme edge of this region from being seen. A simple calculation showed that this cause would make it impossible to see the outside edge of the inner circulation if it extended to within 0·37 cm. of the outer cylinder. For this reason, therefore, the inner cylinder was reduced to a diameter of 4 cm. and the photographs shown in figs. 13, 14, 15 and 16 (pls. 1 and 2) were taken. The values of μ were not measured very accurately but in figs. 13 and 14 it was about -1.05; in fig. 15 $\mu = -2.0$; and in fig. 16 $\mu = -2.3$.

On looking at the photographs it will be seen that the results predicted by theory are completely verified. The inner vortices do not penetrate to the outer parts of the fluid, the spacing of the vortices decreases as $-\mu$ increases, and the inner partitions remain of the same shape, approximately square, while they decrease in size with increasing values of $-\mu$.

The 'spacing of the vortices' is half the wave-length—that is half the distance apart of the centres of the ring-like figures shown in the photographs 13–16 into which the coloured liquid, initially close to the surface of the inner cylinder, suddenly forms itself when instability sets in.

Critical Speeds at which Instability appears

Perhaps the most successful feature of the analysis contained in the first part of this paper is the accuracy with which it predicts the critical speeds at which instability appears. A number of sets of measurements were made covering a range of values of μ from $-\infty$ to $+\infty$. These will now be discussed in detail.

Case when μ is positive and > 1 or infinite, i.e. when the Outer Cylinder Rotates Faster than the Inner One and in the same Direction, or when the Inner Cylinder is at Rest

Under these circumstances the motion was found to be completely stable even at the maximum speed of which the apparatus was capable.*

This result is in direct contradiction to that of Mallock, though it is in accordance with the theoretical prediction which takes account only of symmetrical disturbances. The difference might be attributed to the greater rigidity of the present apparatus, and perhaps to its greatly increased length.

Critical speeds when μ is less than R_1^2/R_2^2

Three complete sets of observations were taken: (1) with $R_1 = 3\cdot00$, (2) with $R_1 = 3\cdot80$, (3) with $R_1 = 3\cdot55$. Each observation consisted in observing the speed of rotation of the cylinders at which the vortices appeared, i.e. the speed when the partitions between the vortices suddenly spread out from the inner cylinder. In each case one or two rough readings were taken to find the approximate speed at which instability appeared. The governor was then set for this speed so that large changes in current produced only small changes in the speed. The final reading was then made by increasing the speed fairly rapidly† till the instability was on the point of occurring, then increasing the speed very gradually. In this way it was found that readings could be repeated with an average error of about 2 %.

When it was found that this order of accuracy could be obtained it became clear, that the temperature would have to be read with an error of $0\cdot2°$ C., or less, in order

---

* Five revolutions of the outer cylinder per second.
† The governor did not begin to act till a certain speed had been attained.

that viscosity might be known accurately enough to make full use of the accuracy of the stability measurements.

The speed of one of the cylinders was measured both just before and just after the instability occurred, and the observation was rejected if it was found that too great a jump in speed had been made. With the governor employed on the motor it was found that the variations in speed with a given setting of the apparatus were less than $\frac{1}{2}\%$. The ratio, μ, of the speeds of the cylinders was measured by timing them over a period of two or three minutes.

Table 9. *Showing observed and calculated speeds at which instability first appears when* $R_1 = 3 \cdot 00$, $R_2 = 4 \cdot 035\,cm$.

μ	Ω_2/ν observed	Ω_1/ν observed	Ω_1/ν calculated	μ	Ω_2/ν observed	Ω_1/ν observed
0·552	83·7	152·0	∞	−1·48	−130·0	87·9
0·530	51·6	97·0	105·2	−1·67	−163·0	97·0
0·520	40·1	77·2	84·9	−2·0	−236·0	118·0
0·455	28·8	63·5	50·5	−2·37	−326·0	137·0
0·423	21·6	51·1	44·5			
0·415	18·6	44·9	43·5			
0·410	19·3	47·1	42·6			
0·359	14·4	40·2	37·5			
0·245	8·6	35·3	31·6			
0	0	30·3	27·6			
−0·33	−10·6	32·1	30·0			
−0·33	−11·1	33·5	30·0			
−0·565	−21·0	37·2				
−0·60	−24·8	41·4				
−0·703	−29·8	42·3				
−0·905	−47·1	52·0				
−1·073	−67·0	62·8				
−1·285	−102·6	79·6				

In order to make the results comparable with one another it is necessary to divide the speed in each case by the coefficient of kinematical viscosity.* These coefficients were taken from Kaye and Laby's physical tables. The results are given in tables 9, 10 and 11. In column 1 of each table is given the value of μ. In columns 2 and 3 the observed values of Ω_1/ν and Ω_2/ν, where Ω_1 and Ω_2 are the angular velocities of the inner and outer cylinders, and ν represents the coefficient of kinematical viscosity which is equal to (the coefficient of viscosity) ÷ (density). In column 4 in each table is given the theoretical value of Ω_1 calculated for the corresponding value of μ from the criterion given in (7·11), Part I. On comparing columns 3 and 4 it will be seen that the agreement between theory and observation is extremely good in the cases where $R_1 = 3 \cdot 80$ and $R_1 = 3 \cdot 55$. It is not quite so good in the case where $R_1 = 3 \cdot 0$, but as the observations in this case were made before it was realised how high a degree of accuracy could be obtained in stability measurements, the temperature was only observed roughly once or twice during the experiments. Some uncertainty, therefore, exists as to the exact value of ν in this series of measurements.

* Two geometrically similar motions are also dynamically similar if the speed, divided by the coefficient of kinematic viscosity, is the same in the two cases.

Table 10. *Observed and calculated speeds at which instability first appears when* $R_1 = 3\cdot80$, $R_2 = 4\cdot035$ cm.

μ	Ω_2/ν	Ω_1/ν observed	Ω_1/ν calculated	μ	Ω_2/ν	Ω_1/ν observed	Ω_1/ν calculated
0·864	790·0	914·0	860·0	−0·553	−121·6	219·1	222·0
0·846	530·0	626·0	669·0	−0·621	−141·0	227·0	230·0
0·810	362·2	447·0	447·0	−1·0	−312·0	312·0	
0·788	340·0	431·5	424·0	−1·0	−320·0	320·0	
0·764	298·5	390·5	383·0	−1·0	−313·0	313·0	
0·788	278·0	353·0	424·0	−1·16	−400·7	345·3	
0·741	245·3	330·8	354·0	−1·26	−462·0	367·0	
0·666	196·3	294·0	294·0	−1·36	−539·2	396·6	
0·666	190·2	284·3	294·0	−1·428	−592·0	415·5	
0·631	172·8	273·8	276·0	−1·605	−718·0	447·5	
0·554	136·2	246·0	248·0	−1·714	−845·0	493·0	
0·450	99·1	220·1	225·0	−1·766	−876·0	496·0	
0·422	90·7	217·0	220·0	−1·916	−1005·0	524·0	
0·397	83·0	209·0	216·0	−1·953	−1056·0	540·4	
0·274	54·9	200·1	203·0	−1·996	−1104·0	553·0	
0·160	30·4	190·2	196·0	−2·24	−1362·0	608·0	
0	0	190·8	191·5	−2·51	−1672·0	666·0	
0	0	189·2	191·5	−2·865	−2113·0	737·0	
0	0	193·1	191·5	−2·891	−2120·0	733·0	
−0·082	−15·7	190·8	191·5				
−0·145	−27·8	192·0	192·3		calculated		calculated
−0·164	−31·0	189·5	193·0	−1·50	−712·0		475·0
−0·214	−41·5	192·0	194·5				
−0·378	−80·4	212·5	204·0				
−0·46	−101·5	219·0	209·0				

Table 11. *Observed and calculated speeds at which instability first appears when* $R_1 = 3\cdot55$, $R_2 = 4\cdot035$ cm.

μ	Ω_2/ν	Ω_1/ν observed	Ω_1/ν calculated	μ	Ω_2/ν observed	Ω_1/ν observed
0·765	303·0	396·0	470·0	−0·689	−66·2	96·2
0·7535	245·5	326·0	313·0	−0·793	−84·4	106·5
0·755	242·3	321·0	325·0	−0·800	−84·4	105·6
0·748	202·7	271·0	278·0	−0·843	−92·2	109·4
0·745	182·9	245·5	264·0	−1·00	−128·9	128·9
0·718	135·7	189·1	191·0	−1·00	−125·8	125·8
0·664	93·9	141·5	139·1	−1·129	−153·4	135·9
0·639	80·2	125·6	126·5	−1·244	−184·1	148·0
0·643	84·2	131·0	130·8	−1·302	−209·6	161·1
0·569	60·3	106·0	105·3	−1·489	−264·0	177·3
0·542	55·3	102·1	100·1	−1·63	−299·0	183·7
0·476	44·5	93·5	91·2	−1·795	−376·0	209·4
0·419	36·5	87·2	84·5	−1·925	−419·0	215·0
0·376	32·6	86·8	81·4	−2·00	−475·0	237·3
0·322	26·0	80·8	78·1	−2·17	−511·6	235·9
0·276	21·5	77·8	75·9	−2·32	−579·5	249·8
0·213	16·1	75·6	73·5	−2·53	−709·0	280·2
0	0	70·7	69·8	−2·68	−820·0	306·0
−0·144	−10·24	71·1	70·1	−2·84	−903·5	318·0
−0·236	−17·2	72·9	71·4	−3·25	−1278·0	393·0
−0·349	−26·9	75·6	74·1			
−0·479	−38·6	80·7	79·0*		Calculated	
−0·585	−52·6	89·9	84·8†			
−0·591	−53·5	90·5	84·0‡	−1·347	−232·3	172·8
−0·591	−53·8	91·0				

* Part due to second term of (7·11) 14 % of whole.
† Part due to second term of (7·11) 33 % of whole.
‡ Part due to second term of (7·11) 35 % of whole.

PLATE 1

Fig. 9. Vortices when $R_1 = 2 \cdot 93$ cm., $R_2 = 4 \cdot 035$ cm., μ positive.

Fig. 10. Vortices when $R_1 = 3 \cdot 25$ cm., $R_2 = 4 \cdot 035$ cm., μ positive.

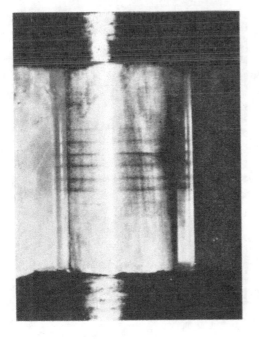

Fig. 11. Vortices when $\mu = 0$.

Fig. 13. Vortices when $R_1 = 2 \cdot 0$ cm., $R_2 = 4 \cdot 035$ cm., $\mu = -1 \cdot 05$ approximately.

PLATE 2

Fig 14. Vortices when $R_1 = 2\cdot0$ cm.,
$R_2 = 4\cdot035$ cm., $\mu = -1\cdot05$ approximately.

Fig. 15. Vortices when $R_1 = 2\cdot0$ cm.,
$R_2 = 4\cdot035$ cm., $\mu = -2\cdot0$ approximately.

Fig. 16. Vortices when $R_1 = 2\cdot0$ cm.,
$R_2 = 4\cdot035$ cm., $\mu = -2\cdot3$ approximately.

Fig. 20. Spiral form of instability.

In spite of this uncertainty there seems to be some evidence in the figures of column 4, table 9, to show that the mathematical approximation on which (6·03) is based is getting appreciably inaccurate when d/R_1 is as large as $\frac{1}{8}$, for the numbers in column 3 are systematically greater than those in column 4 from $\mu = +0·5$ to $\mu = -0·3$.

In working out the calculated values of Ω_1/ν for negative values of μ by the formula 7·11, Part I, it is assumed that the formula ceases to be applicable when the 'correction' term is more than 20 % of the main term. In table 11 it will be seen that when the correction is 33 % the value of Ω_1/ν is too low.

The calculated values of Ω_1/ν and Ω_2/ν for $\mu = -1·5$ in the case where $R_1 = 3·80$, and for $\mu = -1·347$ in the case where $R_1 = 3·55$ are given at the end of tables 10 and 11 (see (7·12) and (7.13)).

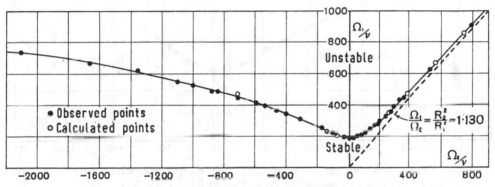

Fig. 17. Comparison between observed and calculated speeds at which instability first appears; case when $R_1 = 3·80$ cm., $R_2 = 4·035$ cm.

In order to give an idea of the uniformity of the experimental results and the accuracy of the theoretical predictions, two diagrams, figs. 17 and 18, have been prepared. In these diagrams, which may be called stability diagrams, the abscissae represent Ω_2/ν while the ordinates are Ω_1/ν. Every point in the diagram therefore represents a possible state of motion of the cylinders. The speeds at which instability sets in as the speed of rotation is slowly increased are represented by points on a curve. The observed points are shown as dots while the calculated ones are shown as circles centred at the points to which they refer. All points above the curve represent states of the apparatus in which the flow is unstable while those below it represent stable states.

The accuracy with which the observed and calculated sets of points fall on the same curve is quite remarkable. Attention is specially directed to the points corresponding with $\mu = -1·5$, $\Omega_1/\nu = 475$, $\Omega_2/\nu = -712$, in the case when $R_1 = 3·80$, and $\mu = -1·347$, $\Omega_1/\nu = 172·8$, $\Omega_2/\nu = -232·3$, in the case when $R_1 = 3·55$. These were calculated from (7·13) and (7·14). The accuracy with which these points fall on the curves appears remarkable when it is remembered how complicated was the analysis employed in obtaining them.

The curve, fig. 17, shows the relationship between Ω_1/ν and Ω_2/ν when $R_1 = 3\cdot80$ for the whole range over which measurements were taken. In the curve, fig. 18, which is the stability curve when $R_1 = 3\cdot55$, the extreme measurements have been left out in order that the curve might be drawn on a scale large enough to give an idea of the accuracy of the results.

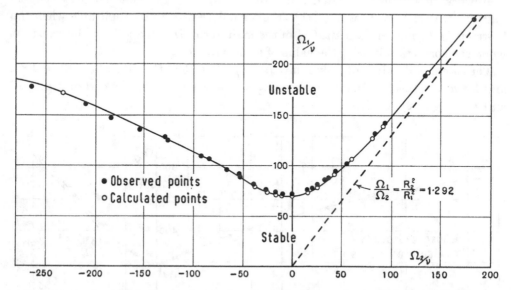

Fig. 18. Comparison between observed and calculated speeds at which instability first appears; case when $R_1 = 3\cdot55$ cm., $R_2 = 4\cdot035$ cm.

A noticeable feature of the results is the way in which the curves, figs. 17 and 18, approach asymptotically the lines $\Omega_1/\Omega_2 = R_2^2/R_1^2$. These lines are marked as dotted straight lines. The prediction of the late Lord Rayleigh that an inviscid fluid contained between two concentric cylinders would be stable if $\Omega_2/\Omega_1 > R_1^2/R_2^2$ is therefore true, and is applicable to viscous liquids.

The conclusion deduced from his theory that an inviscid liquid would be unstable if the cylinders rotated in opposite directions is not applicable to viscous fluids. In fact it is a remarkable feature of the curves that if the outer cylinder is rotating in the opposite direction of the inner one, the speed which it is necessary for the inner cylinder to attain in order that instability may arise is greater than it would be if the outer cylinder were at rest.

Spiral Form of Instability

In many cases a spiral form of instability was observed. In cases when the space between the cylinders was small compared with the radius, this form was very similar to the symmetrical type, except that each vortex in its square-sectioned partition was wrapped as a spiral round the inner cylinder. In this way a double-threaded screw or spiral was formed, the two 'threads' being vortices in the cross-sections of which the fluid rotated in opposite directions. It was noticed, however, that one of the vortices was usually wider than the other. The larger one was always the one for which the component of vorticity in the direction of the axis, was the same as that of the steady motion. For instance, in the case when the outer cylinder was at rest the appearance of the spiral would be that shown in fig. 19. The direction of rotation in the cross-sections of the spiral vortices in an axial plane is shown by means of curved arrows.

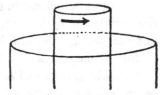

In the case where μ was < -1 the two vortices become so different in size that one of them almost disappeared altogether. The appearance of the coloured fluid was then that of a vortex rolled in a single-thread spiral on the inner cylinder. It was difficult to obtain photographs of this type of instability because there was no point from which the apparatus could be viewed so that the sheets of coloured fluid could be seen edgewise. One fairly good photograph was obtained; it is shown in fig. 20, pl. 2. It will be seen that the spiral form is a very definite form of instability.

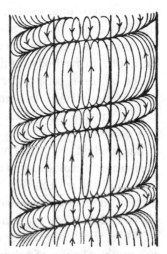

It was found that the formation of spiral instability was always connected with a circulation in the axial planes during the steady motion before the instability appeared. In order therefore to avoid the formation of spiral instability it was necessary to avoid such a circulation in the part of the apparatus where the observations were being made. Various methods were discovered for producing this effect, but it seems hardly necessary to go into such details here.

Fig. 19. Spiral form of instability which appears when steady motion is not strictly limited to two dimensions before instability sets in.

Since a very small component of velocity in the axial plane during steady motion was found to produce spiral instability, the formation of the symmetrical type of instability is, in itself, a good test for knowing whether the steady motion which exists in the apparatus before the instability sets in is a good approximation to the ideal two-dimensional motion which would exist if the cylinders were infinitely long.

Subsequent Motion of the Fluid

Though no attempt has been made to calculate the subsequent motion of the fluid certain observations were made. In all cases where R_1 was greater than 3·0 cm., it was found when μ was positive that if the speed of the apparatus was kept constant and very slightly higher than the speed at which the vortices formed, the vortices were permanent. They remained in perfectly steady motion so that the partitions marked by the coloured fluid were fixed. The photograph shown in fig. 11, pl. 1, is one of a steady motion which had been going for eight minutes when the photograph was taken. I do not remember to have heard of any other case in which two different steady motions are possible with the same boundary conditions. In this case evidently one of them, the two-dimensional one, is unstable; while the symmetrical three-dimensional one is stable.

A moderate increase in the speed of the apparatus merely increased the vigour of the circulation in the vortices without altering appreciably their spacing or position, but a large increase caused the symmetrical motion to break down into some kind of turbulent motion, which it was impossible to follow by eye.

The calculations in Part I indicate that at the exact speed at which instability begins the vortices form themselves infinitely slowly. In other words the calculated three-dimensional motion is steady, to the first order of small quantities, and is, to the first order, in neutral equilibrium. To determine whether it is really steady or whether it is unstable one would have to go to the second order, a matter of extreme difficulty in hydrodynamics. The experiments described above indicate that the effect of the second-order terms is to prevent the vortices from increasing indefinitely in activity. In some such way one might explain the formation of the true steady motion, consisting of alternate vortices, which is observed in the case when μ is positive.

Even when μ is negative the vortices formed when instability occurs appear to be permanent, provided μ is numerically less than a certain number which appears to vary slightly with R_1/R_2. In all the cases when $R_1 > 3 \cdot 55$ cm. it was found that the motion in alternate vortices was stable provided $-\mu < 1$, i.e. when the speed of the outer cylinder was numerically less than that of the inner cylinder.

When the speed of the outer cylinder increased above this value, however, the symmetrical rings of coloured fluid which invariably appeared in the first instance if the experiment was carefully performed, were found to break up shortly afterwards. In order to find out if possible how the fluid moves during the breakdown of the first symmetrical motion a careful examination was made into the nature of the flow when μ was nearly equal to -1. With a value of μ very slightly greater than this it was found that the breakdown occurred sufficiently slowly to enable the process to be observed by eye. Unfortunately attempts to photograph it failed, but it was sufficiently definite to be described.

Shortly after the symmetrical vortex system had formed itself, it was seen that every alternate vortex began to expand on one side and to contract on the opposite side of the cylinder. On the other hand the intermediate vortices began to expand to fill the spaces from which the first set had contracted and to contract in the parts where the first set had expanded. The effect is represented in sketch, fig. 21.

As seen in side elevation the effect was curious; it looked as though each vortex was pulsating so that its cross-section varied periodically, though with an oscillation of increasing amplitude. After a time it became impossible to follow the motion, owing partly, no doubt, to the fact that the system adopted for marking the liquid was not really suitable for observing any but symmetrical motions.

When experiments were tried with slightly greater values of $-\mu$ it was found that the breakdown occurred in a very similar manner, but that in this case each vortex

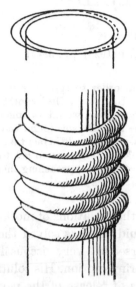

Fig. 21. Sketch illustrating appearance of vortices when they begin to break up; case when $\mu = -1$ approximately.

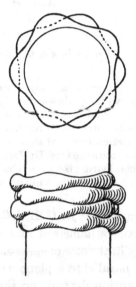

Fig. 22. Appearance of vortices when they begin to break up immediately after their formation: case when μ is less than -1.

seemed to expand in several points, equally spaced, round the cylinder. The appearance of the colouring matter was then similar to that shown in the sketch, fig. 22. The general impression gained by observing the phenomenon was that each vortex grew into the shape of a regular polygon, that these polygons were threaded on the inner cylinder and rotated in the same direction, and that the corners of each polygon were placed over the sides of the one below.

6

THE MOTION OF ELLIPSOIDAL PARTICLES IN A VISCOUS FLUID

REPRINTED FROM

Proceedings of the Royal Society, A, vol. CIII (1923), pp. 58–61

The experiments here described were undertaken in order to test experimentally an unproved hypothesis recently introduced by Dr G. B. Jeffery. The hypothesis was that ellipsoidal particles immersed in a moving viscous fluid would assume certain definite orientations in relation to the motion of the fluid. It was found that ellipsoidal particles made of aluminium and immersed in water-glass do in fact take up the positions indicated by Dr Jeffery, but that they take a long time to get to those positions. During the time which the particles are gradually approaching their final positions they oscillate in the way indicated in Dr Jeffery's analysis.

In a recent paper* Dr G. B. Jeffery has discussed the equations of motion of ellipsoidal particles immersed in a moving viscous fluid. He has solved the problem completely in the case of spheroidal particles immersed in a very viscous fluid which is moving parallel to a plane with a uniform shearing motion. His solution shows that the motion depends on the initial conditions of release of the particle. The motion is periodic, and there appears to be no tendency for a particle to set itself so that its axis lies in any particular direction. The particle, in fact, takes up the rotation of the fluid, and its axis of symmetry describes a kind of elliptic cone round the direction of the vortex filaments, that is, round the direction which is perpendicular to the plane in which the motion of the fluid takes place.

Though the analysis, which neglects the inertia terms in the equations of motion, gives no indication of any tendency for the axis to set itself in any particular direction, Dr Jeffery considers that ultimately the axis would probably adopt some special position, and he puts forward a 'minimum energy' hypothesis, which leads to the following definite, though unproved and unverified, results:

(1) A prolate spheroid, subject to the restriction imposed by this hypothesis, would set itself so that its long axis was parallel to the vortex lines, and therefore perpendicular to the plane in which the undisturbed motion of the fluid takes place. It would then rotate with the fluid, which would move in steady motion relative to it.

(2) An oblate spheroid would set itself so that an equatorial diameter was perpendicular to the plane of undisturbed motion of the fluid. It would then rotate about that diameter with a variable angular velocity, and the motion of the fluid would not be steady, but would be periodic.

* 'The Motion of Ellipsoidal Particles Immersed in a Viscous Fluid', *Proc. Roy. Soc.* A, CII (1922), 161.

At the end of this paper Dr Jeffery suggested that an experimental investigation of the matter might be valuable, and it struck me that some apparatus which I was using for another experiment would be suitable to use for such a purpose. Accordingly, I performed the experiments which are recorded below.

A glass tube of circular cross-section, 24·5 cm. long × 5·40 cm. diameter, was fixed to a brass base, placed on a table capable of rotating about a vertical axis and adjusted till it was concentric with the axis of rotation. An aluminium tube, 3·50 cm. diameter, was then fixed in such a position that it was concentric with the glass tube, and its lower end was close to the brass base to which the glass tube was fixed. The space between them was filled with a fluid known as 'water-glass'. This fluid, which is used for preserving eggs, consists of a highly concentrated solution of sodium silicate, and it can be obtained in a form in which it has a very high degree of viscosity.

On rotating the outer glass cylinder, a shearing motion was set up in the water-glass, which, though not identical with the laminar motion contemplated in Dr Jeffery's analysis, is yet sufficiently similar to form a good approximation to it. The plane of motion of the fluid is perpendicular to the common axis of the cylinders, and the direction of the vortex filaments is parallel to this axis.

Some ellipsoidal particles were made on a lathe from aluminium wire. Their dimensions were:

Prolate spheroids   0·100 × 0·230, 0·126 × 0·249, 0·106 × 0·270 cm.

Oblate spheroids   0·167 × 0·100, 0·265 × 0·120 cm.

These were dropped on the surface of the water-glass and pushed down into it by means of a rod. The viscosity of the water-glass was so great that it took over 2 hours for the aluminium particles to fall 1 cm. through the fluid. Under these circumstances it was to be expected that the particles would remain midway between the inner and outer cylinders for some time if originally placed in that position, and it was found that they did in fact remain there for about an hour when the apparatus was in action.

RESULTS

The results predicted by Dr Jeffery's analysis were verified, so far as qualitative experiments can verify quantitative mathematical results. It was found that the motion of the particles was periodic, that it depended on the initial position of the particle, and that the axes of the particles appeared to describe elliptical cones round the vertical, i.e. round the direction of the vortex filaments of the undisturbed motion of the fluid.

It was noticed that, in the case of the prolate spheroids, i.e. the long particles, the amplitudes of oscillation of their axes were greater in the vertical plane tangential to the cylinders than in the vertical plane through their axis of rotation. In the case of the oblate spheroids, or discs, the reverse was the case. Though it was not quite so obvious in this case that the axis of the spheroid moved on a surface like an elliptic cone, yet it could easily be seen that the equatorial plane of the disc was more nearly vertical when its axis (i.e. axis perpendicular to the equatorial plane) was in an axial plane of the cylinders, than when the axis was in the vertical plane tangential to the cylinders.

Both these observations are in accordance with the predictions of Dr Jeffery. If θ is the angle between the axis of symmetry of the particle and the vertical, and if ϕ is the angle between the vertical plane through the axis of the particle and the vertical plane through the axis of the cylinders, he obtains the equation*

$$\tan^2\theta = \frac{a^2b^2}{\kappa^2(a^2\cos^2\phi + b^2\sin^2\phi)},\tag{1}$$

for the cone on which the axis of a particle moves. In this formula κ is an arbitrary constant depending on the conditions of projection, a is the length of the axis of symmetry, and b is the diameter of the equatorial plane.

Putting $\phi = 0$ and $\phi = \frac{1}{2}\pi$ in equation (1), it will be seen that when $\phi = 0$, $\tan\theta = b/\kappa$; and when $\phi = \frac{1}{2}\pi$, $\tan\theta = a/\kappa$.

In the case of the prolate spheroid $a > b$, so that θ is greater when $\phi = \frac{1}{2}\pi$ than when $\phi = 0$. In the case of the oblate spheroid, on the other hand, $a < b$, so that θ is greatest when $\phi = 0$ and least when $\phi = \frac{1}{2}\pi$. The equatorial plane is most nearly vertical when θ is greatest, that is, when $\phi = 0$. These predictions are in agreement with the experimental observations described above.

VERIFICATION OF DR JEFFERY'S MINIMUM ENERGY HYPOTHESIS

It was found that it was possible to start the motion so that the angle θ, between the axis of revolution of the spheroid and the vertical, had any given value; but if the motion were continued for a long time this angle gradually changed till the spheroid assumed a definite position in relation to the cylinders. In the case of the oblate spheroid an equatorial diameter became vertical and the particle then rotated about this diameter. In the case of the prolate spheroid the long axis became vertical and the particle rotated about it.

The times taken by the particles to assume these final positions were surprisingly long. In the case when the outer cylinder was rotating at a rate of one revolution in 8 sec. it took about $4\frac{1}{2}$ min. for the oblate cylinder to get nearly to its final position, while the prolate spheroid took about 20 min.

In each case the particle executed about two and a half periods during one revolution of the outer cylinder, so that it took roughly 85 periods for the oblate spheroid to settle down, and 370 periods in the case of the prolate spheroid.

It appears, therefore, that these experiments confirm Dr Jeffery's minimum energy hypothesis, so far at any rate as its application to the motion of spheroids is concerned. It is not easy to understand how the forces are brought into play which

* *Loc. cit.*, equation (49), p. 171.

cause the axes of the particles to set themselves in definite directions. It seems clear that they must, in some way, depend on terms which are neglected in Dr Jeffery's approximate equations. These equations become more and more accurate the greater the viscosity of the fluid, but I do not know of any other case of viscous motion in which oscillations die down more and more slowly the greater the viscosity of the fluid.

7

ON THE DECAY OF VORTICES
IN A VISCOUS FLUID

REPRINTED FROM

Philosophical Magazine, vol. XLVI (1923), pp. 671–4

A number of problems have been solved in which the rate of decay of small oscillations or waves in a viscous fluid has been found, but the simplification brought about by considering only small motions excludes many of the most important problems of fluid motion. On the other hand, when the complete equations of motion involving terms containing the square of the velocity have been used very few solutions have been obtained. Certain problems in steady motion,* problems concerning the two-dimensional motion of a viscous liquid when it is symmetrical about an axis,† and problems concerning laminar motion parallel to a plane‡ probably complete the list.

The object of the present paper is to draw attention to a class of cases in which solutions of the complete equations of motion, including the 'inertia' terms, may be obtained for two-dimensional viscous flow.

If ψ represents the stream function so that the components of velocity are

$$u = -\frac{\partial \psi}{\partial y}, \quad v = \frac{\partial \psi}{\partial x},$$

the vorticity at any point is $\qquad \zeta = \nabla^2 \psi.$

The equation of motion may be written

$$\left(\frac{\partial}{\partial t} - \frac{\partial \psi}{\partial y} \frac{\partial}{\partial x} + \frac{\partial \psi}{\partial x} \frac{\partial}{\partial y} - \nu \nabla^2 \right) \nabla^2 \psi = 0, \tag{1}$$

where ν is the coefficient of kinematic viscosity. Now consider functions which satisfy the equation
$$\nabla^2 \psi = K\psi,$$

where K is a constant. These functions will also satisfy (1) if

$$\frac{\partial \psi}{\partial t} - \nu K \psi = 0, \tag{2}$$

the 'inertia' terms $\qquad -\dfrac{\partial \psi}{\partial y} \dfrac{\partial (\nabla^2 \psi)}{\partial x} + \dfrac{\partial \psi}{\partial x} \dfrac{\partial}{\partial y} (\nabla^2 \psi)$

---

* Flow through a pipe or between concentric cylinders.

† See Taylor, paper 9, vol. II, also Terazawa, *Report of Aeronautical Research Institute*, Tokyo, 1922.

‡ See Lamb's *Hydrodynamics*, ch. XI.

vanishing because the stream lines are also lines of constant vorticity. Equation (2) is satisfied if

$$\psi = \psi_1 e^{\nu K t},$$

where ψ_1 is a function of x and y only. Hence, if ψ_1 is a solution of

$$\nabla^2 \psi_1 = K \psi_1, \tag{3}$$

$$\psi = \psi_1 e^{\nu K t}$$

is a solution of (1).

ANALOGY WITH THE THEORY OF VIBRATING MEMBRANES

The equation of motion for a vibrating membrane is

$$\nabla^2 z = \frac{1}{c^2} \frac{\partial^2 z}{\partial t^2}, \tag{4}$$

where z is the displacement at any point from the position of equilibrium and c depends only on the tension and mass of the membrane. If the membrane vibrates in simple harmonic motion of period T, (4) becomes

$$\nabla^2 z + \frac{4\pi^2 z}{c^2 T^2} = 0. \tag{5}$$

This equation is of the same form as (3), and the amplitude of vibration of the membrane may be taken to represent ψ if

$$K = -4\pi^2 / c^2 T^2.$$

It appears therefore that if a solution of the problem of a vibrating membrane has been obtained so that the period and contours, or curves of equal displacement of the membrane, have been determined, a problem of viscous motion has also been solved in which the stream lines are the same as the contours of the vibrating membrane. The velocity of the flow dies down exponentially so that it is reduced in the ratio $e:1$ in time $c^2 T^2 / 4\pi^2$.

Though the analogy is exact, so far as it goes, it should be borne in mind that in general the boundary conditions in the two cases are different. In the case of a membrane held round its edges, for instance, the boundary condition is $z = 0$, while for the viscous fluid both ψ and $\partial \psi / \partial n$ must be zero at a solid wall. Another point of difference is that if two different solutions have been obtained in the membrane problem, they can be superposed. This is not true in the case of the viscous motion problem because the 'inertia' terms in (1), namely,

$$-\frac{\partial \psi}{\partial y} \frac{\partial}{\partial x} (\nabla^2 \psi) + \frac{\partial \psi}{\partial x} \frac{\partial}{\partial y} (\nabla^2 \psi),$$

vanish when ψ is a solution of

$$\nabla^2 \psi = K_1 \psi, \tag{6}$$

or when it is a solution of

$$\nabla^2 \psi = K_2 \psi, \tag{7}$$

but they do not vanish when ψ is a sum of solutions of (6) and (7).

Decay of a system of eddies rotating alternately in opposite directions and arranged in a rectangular array

As an example of the use of this type of solution of the equations of motion of a viscous fluid the function

$$\psi_1 = A \cos \frac{\pi x}{d} \cos \frac{\pi y}{d},$$

which is a solution of (3), may be considered. In this case K is $2\pi^2/d^2$, so that the corresponding solution of (1) is

$$\psi = A \cos \frac{\pi x}{d} \cos \frac{\pi y}{d} \exp - \frac{2\pi^2 \nu t}{d^2}.$$

This represents a system of eddies arranged in a square pattern, each rotating in the opposite direction to that of its four neighbours. d is the length of the side of a square containing one complete eddy, and such a system is reduced in intensity in the ratio $e:1$ in time $d^2/2\pi^2\nu$. The intensity of such a system of eddies, each 1 cm. in diameter, for instance, would be reduced to $1/e$th in $\frac{1}{3}$ sec. owing to the action of viscosity if the fluid were air, or $4\frac{1}{2}$ sec. in the case of water. The stream lines for this system of eddies are shown in the accompanying figure.

Fig. 1. Stream lines for system of eddies dying down under the action of viscosity.

8

EXPERIMENTS ON THE MOTION OF SOLID BODIES IN ROTATING FLUIDS

REPRINTED FROM

Proceedings of the Royal Society, A, vol. CIV (1923), pp. 213–18

Some years ago it was pointed out by Professor Proudman* that all slow steady motions of a rotating liquid must be two-dimensional. If the motion is produced by moving a cylindrical object slowly through the liquid in such a way that its axis remains parallel to the axis of rotation, or if a two-dimensional motion is conceived as already existing, it seems clear that it will remain two-dimensional. If a slow three-dimensional motion is produced, then it cannot be a steady one. On the other hand, if an attempt is made to produce a slow steady motion by moving a three-dimensional body† with a small uniform velocity (relative to axes which rotate with the fluid) three possibilities present themselves:

(*a*) The motion in the liquid may never become steady, however long the body goes on moving.

(*b*) The motion may be steady but it may not be small in the neighbourhood of the body.

(*c*) The motion may be steady and two-dimensional.

In considering these three possibilities it seems very unlikely that (*a*) will be the true one. In an infinite rotating fluid the disturbance produced by starting the motion of the body might go on spreading out for ever and steady motion might never be attained, but if the body were moved steadily in a direction at right angles to the axis of rotation, and if the fluid were contained between parallel planes also perpendicular to the axis of rotation, it seems very improbable that no steady motion satisfying the equations of motion could be attained. There is more chance that (*b*) may be true. A class of mathematical expressions representing the steady motion of a sphere along the axis of a rotating liquid has been obtained.‡ This solution of the problem breaks down when the velocity of the sphere becomes indefinitely small, in the sense that it represents a motion which does not decrease as the velocity of the sphere decreases. It seems unlikely that such a motion would be produced under experimental conditions.

There remains the third possibility (*c*). In this case the motion would be a very remarkable one. If the liquid were contained between parallel planes perpendicular to the axis of rotation, the only possible two-dimensional motion satisfying the

---

* *Proc. Roy. Soc.* A, XCII (1916), 408.

† E.g. a sphere or any body except an infinite cylindrical body with its axis parallel to the axis of rotation.

‡ Paper 4 above.

required conditions is one in which a cylinder of fluid moves as if fixed to the body. The boundary of such a cylinder would act as a solid body, and the liquid outside would behave as though a solid cylindrical body were being moved through it. No fluid would cross this boundary, and the liquid inside it would, in general, be at rest relative to the solid body. This idea appears fantastic, but the experiments now to be described show that the true motion does, in fact, approximate to this curious type.

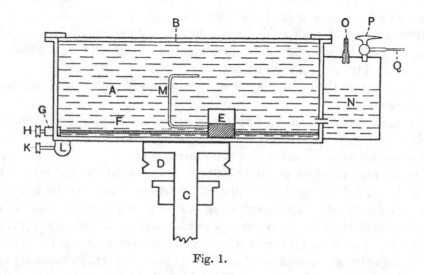

Fig. 1.

In these experiments, bodies were moved slowly through water contained in a rectangular tank which was rotating at a considerable speed. The streamlines relative to the rotating system were made visible by means of coloured fluid, and this was photographed by a camera placed on the axis of rotation (which was vertical) and aiming downwards through the plate-glass top of the tank. The arrangement is shown in fig. 1. In that diagram A is the tank, B is the removable plate-glass top which was screwed down with a watertight joint capable of standing considerable pressure. The tank was 9 in. wide by 12 in. long and 4 in. deep. C is the vertical shaft about which the tank rotates, and D is the driving pulley. The apparatus was rotated at a uniform speed by an electric motor fitted with a governor, but this is not shown in the diagram.

Since it was necessary to give the whole system, including the water, a uniform rotation before starting the experiment, the mechanism necessary for moving the body through the water and for operating the apparatus used for making the stream lines visible had to be fixed to the tank so as to rotate with it, but at the same time to be capable of being actuated from outside. This gave rise to the chief difficulty of the experiment. The body E (fig. 1) was moved slowly along a groove across the middle of the tank by means of a screw F of fine pitch, cut on a small steel shaft which passed through a stuffing box G. This shaft was driven by a small motor L through two pulley wheels H and K connected by a fine endless silk thread. A small electric motor L was fixed to the underside of the tank, and was connected with a battery and switch through a wire, which dipped into a fixed annular trough containing mercury, and placed concentric with the axis of rotation.

On operating the switch while the tank was in motion, the body E could thus be made to move slowly across the middle of the tank.

To make the streamline visible it was necessary to have a source of coloured fluid moving with the body E. This end was attained by using the body itself as a reservoir for the coloured fluid whence it was led by a very fine metal tube M to the point at which it was desired to start the stream line. In order to control the discharge of coloured fluid the upper part of this reservoir was filled with air, and the pressure of the water in the tank was gradually reduced while the experiment was proceeding. The reduction in pressure in the tank caused the air imprisoned in the upper part of the body E to expand and expel the coloured liquid through the tube M.

In order to get a steady stream of coloured fluid through M it was necessary to reduce the pressure in the tank at a uniform rate. This was accomplished by fixing a box N to the outside of the main tank, and the two were connected so that water could flow between them. The upper part of the box N was an air reservoir. The pressure in the tank could thus be raised by pumping air into N through a bicycle valve O which was soldered to the top of it.

In order to reduce the pressure at a uniform rate a fine capillary tube Q was fitted which allowed air to escape slowly out of N; and in order to keep the pressure up till the coloured liquid was wanted a stopcock P was inserted between N and Q. This stopcock was operated by a spring and trigger which could be released while the apparatus was in motion.

To photograph the streamlines several mercury vapour spark gaps of the type used by Mr C. T. R. Wilson were arranged in series round the apparatus, and to make a good background for the coloured streaks the bottom of the tank was silvered but not polished. Eosin was used as a colouring matter, and the eosin solution was made up to the same density as water by adding alcohol.

To perform an experiment the tap P was turned off and the trigger for releasing it was set. Air was then pumped into the reservoir N till the pressure was nearly, but not quite, sufficient to burst the glass top of the tank. The screw F was then turned till the body E was at the beginning of its path. The apparatus was then set rotating and left for some minutes till it was certain that all the water had attained a uniform rotation. When everything was ready to take a photograph the spring which turned the tap P was released. Directly the coloured liquid began to appear at the end of the tube M the switch operating the motor L was closed and the body E began to move. When the body got near the middle of the tank a spark was passed through the illuminating apparatus and an exposure made. Some practice was necessary before these operations could be performed in the correct order. The essential point aimed at in designing the apparatus was attained, for the speed of rotation was high while the speed of the body through the liquid was slow.

Results of the Experiments

In the first experiments the moving body was a cylinder 1·5 cm. diameter which extended from bottom to top of the tank. The tube M (fig. 1) was arranged so that the coloured stream emerging from it struck the cylinder centrally and divided into two. For reasons explained in a previous work,* the colouring matter remains in thin sheets, which appear as thin lines when seen edgewise from a point on the axis of rotation. These lines which were very visible in the photographs showed the well-known alternate vortices which are formed behind a cylinder moving in a fluid, but it must be remembered that these lines are not stream lines in the mathematical sense of the words when the motion is not steady.

The next experiment was made with a sphere, and the stream of coloured fluid was discharged from a point in front of its centre. The coloured streaks were very similar to those obtained with a cylinder, except that the line of the wake of alternate vortices did not lie directly behind the sphere; on the other hand, increasing speed of rotation made the wake appear more and more like that of a cylinder. This experiment seems to suggest that the motion is something like that outlined in the third alternative (c) (p. 93), a cylinder of fluid of the same diameter as the sphere moving with it and acting towards the rest of the fluid as if it were a solid cylinder.

To test this the apparatus was arranged so that a stream line at some height above the body E could be examined. The body E consisted in these experiments of a short cylinder, about 1 in. high by $1\frac{1}{4}$ in. diameter. This rested on the bottom of the tank, and there was thus about 3 in. of water between the top of it and the top of the tank. The coloured liquid was led by the pipe M to a level $1\frac{1}{2}$ in. above the top of the moving body. Under these circumstances the coloured liquid would, if there were no rotation, pass over the middle of the top of the body. The arrangement is that shown in fig. 1.

In the first of the experiments made with this body the coloured liquid was discharged from a point 1 in. in front of the imaginary vertical cylinder enclosing the body, as shown in fig. 1. It was found that the coloured stream flowed straight towards this imaginary cylinder to a point vertically above the foremost point of the body. At that point the stream divided as though it had struck a solid obstacle. The fact that this virtual 'solid obstacle' coincided with the imaginary cylinder

* Paper 3 above.

PLATE 1

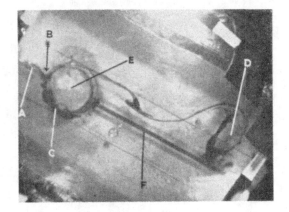

Fig. 2.

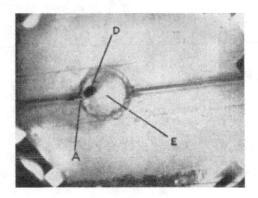

Fig. 3.

enclosing the body can be seen in the photograph (fig. 2, pl. 1). The point from which the coloured stream emerges is shown at A. The point at which the stream divides is marked with an arrow B. At this point part of the coloured stream passes to the right and collects in a sheath C which seems to remain close to the surface of the imaginary cylinder. The rest of the coloured liquid flows round the imaginary cylinder to the left and breaks away from it, forming large eddies D. The top view of the body E can be seen in the photograph, and it will be seen that the water contained in the cylinder vertically over the body E is quite clear. The broad black line which passes under it is the driving screw F (figs. 1 and 2). This line accordingly shows the direction of motion of the body.

In order to make certain that the liquid inside the imaginary vertical cylinder enclosing the body does in fact move with it, an experiment was next made in which the coloured liquid was discharged from a point inside this cylinder. A photograph taken under these conditions is shown in fig. 3 (pl. 1). In that photograph the end of the tube from which the coloured liquid issued is at A. It will be seen that it is almost exactly over the foremost point of the edge of the moving body E. The coloured liquid remained in a small compact mass (D) which travelled with the body. In this experiment the discharge of coloured fluid and the motion of the body were started simultaneously at the end of the tank, which is just outside the right-hand end of the photograph. Though the point from which the coloured liquid was issuing had travelled more than half the length of the tank, none of this fluid had escaped from the imaginary vertical cylinder containing the moving body E, although it was $1\frac{1}{2}$ in. above the level of the top of E.

This result confirms and supplements the observation previously recorded[*] that in the case when a sphere is moved slowly *along* the axis of rotation the motion tends to become two-dimensional, owing to the formation of a cylindrical dead-water region extending above the body and moving with it. On the other hand, no theoretical work has so far given any indication as to how such a motion could be established. The calculations of Mr S. F. Grace[†] on the motion of a solid sphere projected slowly in a liquid of the same density as itself, show that at one stage, at any rate, the disturbance in the surrounding fluid is greater in the region which lies above and below the sphere than it is in other directions. This may have some bearing on the subject, and it is to be hoped that some further light will be thrown on it when Mr Grace applies his method of analysis to the case of a sphere which is constrained to move uniformly along, or perpendicular to, the axis of rotation.

These experiments were carried out in the Cavendish Laboratory through the kindness of Sir Ernest Rutherford to whom the writer wishes to express his thanks.

[*] Paper 4 above.
[†] *Proc. Roy. Soc.* A, cii (1922), 89.

9

THE BURSTING OF SOAP-BUBBLES
IN A UNIFORM ELECTRIC FIELD*

REPRINTED FROM

Proceedings of the Cambridge Philosophical Society, vol. XXII (1925), pp. 728–30

The stability of a charged raindrop has been discussed mathematically by Lord Rayleigh. The case of an uncharged drop in a uniform electric field is perhaps of more meteorological importance† but a mathematical discussion of the conditions for stability turns out to be very much more difficult in this case, owing to the fact that the drop ceases to be spherical before it bursts. Moreover it does not seem possible to express its geometrical shape by means of any simple mathematical expressions. On the other hand, by using a soap bubble instead of a water drop it was found possible to carry out experiments under well-defined conditions in this case, whereas experiments with Rayleigh's charged drop would be difficult.

The method adopted in the present experiments was to place a soap bubble of measured volume on a horizontal wet aluminium plate. The half bubble thus formed was subjected to the influence of a uniform electric field by placing a second parallel plate above it and maintaining a known potential difference between the two. This is evidently equivalent to experimenting with a complete bubble of twice the volume freely floating in a uniform field. In the former case the film is at right angles to the wet plate along the circle of contact, and the plate, with the half bubble, forms an equipotential surface corresponding with the complete bubble and the equatorial equipotential plane.

The two circular metal plates were mounted horizontal and parallel to one another on insulating supports so that their distance apart could be varied. The lower one was earthed and the upper one was connected to the positive pole of an influence machine with a battery of Leyden jars in parallel. It was found that the potential of the upper plate could be adjusted with considerable delicacy by varying the distance from the knob of one of the jars of a pointed wire connected to earth.

A bubble of measured volume was placed on the middle of the lower plate by means of a pipette. The influence machine was set going and the field was gradually increased. The bubble was seen to elongate as the field increased. When the field had become sufficiently intense the bubble assumed a form roughly resembling the small end of an egg. Photographs of the bubble in this condition were taken by means of a spark from a set of Leyden jars which were independent of those connected to the plate. Fig. 1 (pl. 1) is a photograph taken at this stage.

* With C. T. R. Wilson.

† See 'Investigation of Lightning Discharges' by C. T. R. Wilson, *Phil. Trans. Roy. Soc.* A, CCXXI (1921), 104.

Shortly after the stage represented in fig. 1 had been reached it was found that quite a small increase in the field produced large changes in the shape of the bubble. While remaining egg-shaped the end rapidly became narrower and then pointed. At this stage it ceased to remain stationary. The end vibrated with great rapidity and as photographs showed, filaments or drops were being thrown off from the end.

The formation of the filaments of this kind has been studied by Zeleny[*] in the case of drops at the end of a tube, but it is not possible to deduce from his measurements information about isolated water drops.

Fig. 2 (pl. 1) shows the bubble in the vibrating stage and apparently just before the filament has been formed. Figs. 3 and 4 (pl. 1) show the bubble just after the formation of the filament. In fig. 3 the drops into which the filament breaks may be seen. At this stage the bubble is flattening very rapidly. The initial stages of the flattening can be seen in fig. 4, but the flat stage represented in fig. 5 (pl. 2) is rapidly reached and most of the photographs taken in the vibrating stage with exposures timed at random are of this type. The drops into which the filaments has broken can be seen in fig. 5.

In order to follow the rapidly changing shape a rapid succession of exposures was made on the same plate, by allowing a battery of large Leyden jars to discharge through a water resistance and into a smaller single jar with short illuminating spark gap in parallel. Two photographs obtained in this way are shown in figs. 6 and 7 (pl. 2). Fig. 8 (pl. 2) is a tracing taken from a photograph which was too faint for reproduction. It is interesting because it shows the almost perfect cone into which the top of the bubble is drawn at the moment when the filament is first formed. In some of Zeleny's experiments the electrified liquid surfaces assumed similar conical forms.

The equilibrium of the bubble requires that

$$T\left(\frac{1}{\rho_1}+\frac{1}{\rho_2}\right)-2\pi\sigma^2 = P,$$

where P is the excess of air pressure inside the bubble, σ is the surface density of charge, ρ_1 and ρ_2 the principal radii of curvature at any point and T is the effective surface tension (in this case twice the two surface tensions of soap solution).

For bubbles of given shape but different sizes the surface density of charge at corresponding points is proportional to the field but independent of the linear dimensions. The field therefore which produces a given shape should be proportional to

$$(T/\text{linear dimensions})^{\frac{1}{2}}$$

and in particular the field which causes the bubble, in the present experiments, to become just unstable should be inversely proportional to the square root of the radius of the undistorted bubble.

A number of measurements were made of the field required to burst bubbles of different volumes. The potential difference between the plates was measured by means of a micrometer spark gap using Kaye and Lahy's tables for sparking potentials between balls of 5 cm. diameter.

[*] *Proc. Camb. phil. Soc.* xviii (1914), 71; *Phys. Rev.* x (1917), 1.

The results are given in the following table:

Volume of half drop (cm.³)	Radius, r cm.	F, Field necessary for bursting (V./cm.)	$F \sqrt{r}$
0·028	0·251	7520	3770
0·056	0·299	7000	3830
0·112	0·377	6060	3720
0·224	0·475	5150	3550
0·56	0·645	4410	3540
0·70	0·694	4410	3670
1·00	0·782	4130	3650
2·50	1·062	3550	3660

As will be seen from the table the product $F \times \sqrt{r}$ is nearly constant. This agrees with the theory outlined above.

It is of interest to find the size of the largest stable water drop in the greatest field that can exist without sparking at normal atmospheric pressure.

Taking this to be 30,000 V./cm. the radius of the largest soap bubble would be $(3700)^2/(30,000)^2 = 0·015$ cm. If the surface tension of soap solution is taken as 29 c.g.s. units (so that the effective surface tension of the film is 58) and that of water as 75, the largest water drop is $(0·015 \times 75)/58 = 0·02$ cm.

PLATE 1

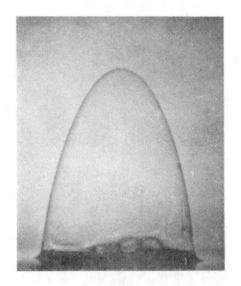

Fig. 1.

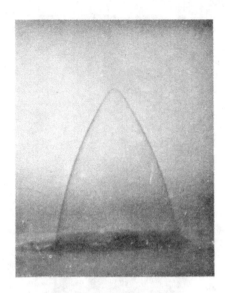

Fig. 2.

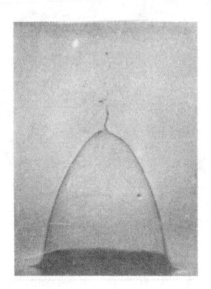

Fig. 3.

Fig. 4.

PLATE 2

Fig. 5.

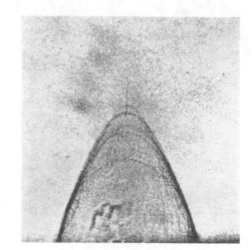

Fig. 6.

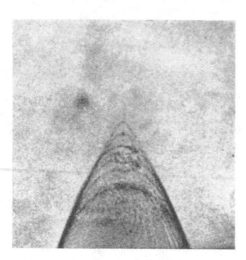

Fig. 7.

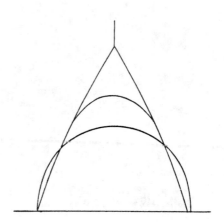

Fig. 8.

10

THE VISCOSITY OF A FLUID CONTAINING SMALL DROPS OF ANOTHER FLUID

REPRINTED FROM

Proceedings of the Royal Society, A, vol. CXXXVIII (1932), pp. 41–8

Einstein's expression for the viscosity of a fluid containing solid spheres in suspension is extended so as to include the case when the spheres are liquid. The expression obtained is valid provided the surface tension is great enough to keep the drops nearly spherical. When the rate of distortion of the fluid or the radius of the drop is great enough, the drops tend to break up, and an approximate expression is given for determining the size of the largest drop that can exist in a fluid which is undergoing distortion at any given rate.

The viscosity of a fluid in which small solid spheres are suspended has been studied by Einstein as a problem in theoretical hydrodynamics.[*] Einstein's paper gave rise to many experimental researches on the viscosity of fluids containing solid particles, and it soon became clear that though complete agreement with the theory might be expected when the particles are true spheres, some modification is necessary when the particles are flattened or elongated. The theory of such systems was developed by G. B. Jeffery,[†] who calculated the motion of ellipsoidal particles in a viscous fluid and their effect on the mean viscosity. Some of his conclusions have been verified by observation.[‡]

So far no one seems to have extended Einstein's work to liquids containing small drops of another liquid in suspension. The difficulties in the way of a complete theory when solid particles are replaced by fluid drops are almost insuperable, partly because the correct boundary conditions are not known, and partly because a fluid drop would deform under the combined action of viscous forces and surface tension. Even if the boundary conditions were known to be those commonly used in hydrodynamical theory, the calculation of the shape of the deformed drop would be exceedingly difficult. When the radius of the suspended drops or the velocity of distortion of the fluid are small, surface tension may be expected to keep them nearly spherical, and in that case Einstein's analysis may be extended so as to include the case of liquid drops. For this purpose the following assumptions will be made:

(1) The drops are so small that they remain nearly spherical.

(2) There is no slipping at the surface of the drop.

---

[*] *Annln. Phys.* XIX (1906), 289, and a correction to that paper, *Annln. Phys.* XXXIV (1911), 591.

[†] *Proc. Roy. Soc.* A, CII (1922), 161.

[‡] Paper 6 above.

(3) The tangential stress parallel to the surface is continuous at the surface of the drop, so that any film which may exist between the two liquids merely transmits tangential stress from one fluid to the other.

A general method for analysing the slow motion of viscous fluids has been given by Lamb,* according to which the components of velocity may be regarded as containing three types of terms. The complete expression for one component of velocity is

$$u = \left[\frac{1}{\mu} \Sigma \frac{r^2}{2(2n+1)} \frac{\partial p_n}{\partial x} + \frac{nr^{2n+3}}{(n+1)(2n+1)(2n+3)} \frac{\partial}{\partial x} \left(\frac{p_n}{r^{2n+1}} \right) \right]$$

$$+ \left[\Sigma \frac{\partial \phi_n}{\partial x} \right] + \left[\Sigma \left(z \frac{\partial \chi_n}{\partial y} - y \frac{\partial \chi_n}{\partial z} \right) \right]. \quad (1)$$

The terms in the first square bracket are connected with the pressure distribution, which is represented by $p = \Sigma p_n$, p_n being a solid harmonic function of degree n. The terms in the second bracket represent an irrotational motion which can exist in a field of uniform pressure. The terms in the third bracket represent vortex motion which can exist in a field of uniform pressure. χ_n is an arbitrary function of degree n.

In Einstein's equations the co-ordinate axes were chosen parallel to the principal axes of distortion, and with this choice of axes terms containing χ_n disappear. Einstein used the most general type of flow near a solid sphere, but little loss of generality is suffered by taking the special case when the mean motion of the whole system is two-dimensional. In this case, if the origin of co-ordinates is taken at the centre of a drop, the flow at great distances from it may be represented by the irrotational flow $\phi_2 = \frac{1}{4}\alpha(x^2 - y^2)$, the constant $\frac{1}{4}\alpha$ being chosen so that the flow is identical, except for a rotation of the whole field, with the familiar case of uniformly shearing laminar flow which is commonly represented by taking the axis of x in the direction of flow when the system is $(u = \alpha y, v = 0)$.

Choice of Functions ϕ_n and p_n

Outside the drop the appropriate functions are

$$\phi_2 = \tfrac{1}{4}\alpha(x^2 - y^2), \quad \phi_{-3} = B_{-3} a^5 \frac{x^2 - y^2}{r^5}, \quad p_{-3} = \mu A_{-3} a^3 \frac{x^2 - y^2}{r^5} \quad (2)$$

while inside the drop

$$\phi_2' = B_2(x^2 - y^2), \quad p_2 = \mu' A_2 a^{-2}(x^2 - y^2) \quad (3)$$

where B_{-3}, A_{-3}, B_2, A_2, are constants to be determined by the boundary conditions, μ and μ' are the viscosities of the main body of fluid and the drop, $r^2 = x^2 + y^2 + z^2$, and a is the radius of the drop.

---

* *Hydrodynamics*, ch. XI.

Substituting these expressions in (1), the components of velocity outside the drop are

$$u = \tfrac{1}{2}A_{-3}\,a^3 x\,\frac{x^2-y^2}{r^5} + B_{-3}\,a^5\left[-\frac{5x(x^2-y^2)}{r^7} + \frac{2x}{r^5}\right] + \tfrac{1}{2}\alpha x,$$

$$v = \tfrac{1}{2}A_{-3}\,a^3 y\,\frac{x^2-y^2}{r^5} + B_{-3}\,a^5\left[-\frac{5y(x^2-y^2)}{r^7} - \frac{2y}{r^5}\right] - \tfrac{1}{2}\alpha y, \tag{4}$$

$$w = \tfrac{1}{2}A_{-3}\,a^3 z\,\frac{x^2-y^2}{r^5} + B_{-3}\,a^5\left[-\frac{5z(x^2-y^2)}{r^7}\right].$$

Similarly inside the drop

$$u' = A_2 a^{-2}\left[-\tfrac{5}{21}xr^2 - \tfrac{2}{21}x(x^2-y^2)\right] + 2B_2 x,$$

$$v' = A_2 a^{-2}\left[-\tfrac{5}{21}yr^2 - \tfrac{2}{21}y(x^2-y^2)\right] - 2B_2 y, \tag{5}$$

$$w' = A_2 a^{-2}\left[-\tfrac{2}{21}z(x^2-y^2)\right].$$

Boundary Conditions at $r = a$

Continuity of velocity requires

$$u = u', \quad v = v', \quad w = w' \tag{6}$$

and the drop remains spherical if

$$ux + vy + wz = 0. \tag{7}$$

At $r = a$,

$$u = \tfrac{1}{2}A_{-3}a^{-2}x(x^2-y^2) + B_{-3}a^{-2}[-5x(x^2-y^2) + 2a^2 x] + \tfrac{1}{2}\alpha x$$

and

$$u' = \tfrac{5}{21}A_2 x - \tfrac{2}{21}A_2 a^{-2}x(x^2-y^2) + 2B_2 x$$

so that $u = u'$ if

$$\tfrac{1}{2}A_{-3} - 5B_{-3} = -\tfrac{2}{21}A_2 \tag{8}$$

and

$$2B_{-3} + \tfrac{1}{2}\alpha = \tfrac{5}{21}A_2 + 2B_2; \tag{9}$$

when (8) and (9) are satisfied it will be found that $v = v'$ and $w = w'$, so that all three conditions (6) are satisfied.

To satisfy (7) a formula given by Lamb may be used, namely,

$$xu + yv + zw = \frac{1}{\mu}\Sigma\,\frac{nr^2}{2(2n+3)}\,p_n + \Sigma u\phi_n$$

This provides a third equation between the four undetermined constants, namely,

$$\tfrac{1}{2}A_{-3} - 3B_{-3} + \tfrac{1}{2}\alpha = 0 \tag{10}$$

and the problem will be soluble in this form if only one further equation is necessary in order to satisfy both conditions for continuity of tangential stress.

The components of stress acting across unit area of a spherical surface are p_{rx}, p_{ry}, p_{rz}. The general expression* for p_{rx} is

$$rp_{rx} = \Sigma \left\{ \frac{n-1}{2n+1} r^2 \frac{\partial p_n}{\partial x} + \frac{2n^2 + 4n + 3}{(n+1)(2n+1)(2n+3)} r^{2n+3} \frac{\partial}{\partial x} \left(\frac{p_n}{r^{2n+1}} \right) \right\}$$

$$+ 2\mu \Sigma (n-1) \frac{\partial \phi_n}{\partial x} + \mu \Sigma (n-1) \left(y \frac{\partial \chi_n}{\partial z} - z \frac{\partial \chi_n}{\partial y} \right). \quad (11)$$

In the present case this reduces outside the drop to

$$\frac{rp_{rx}}{\mu} = A_{-3} a^3 \left[\frac{x}{r^3} - \frac{4x}{r^5} (x^2 - y^2) \right] - 8B_{-3} a^5 \left[\frac{2x}{r^5} - \frac{5x}{r^7} (x^2 - y^2) \right] + \alpha x \quad (12)$$

and inside the drop to

$$\frac{rp_{rx}}{\mu} = A_2 a^{-2} \left[\tfrac{16}{21} r^2 x - \tfrac{19}{21} x (x^2 - y^2) \right] + 4B_2 x, \quad (13)$$

with somewhat similar expressions for p_{ry} and p_{rz}.

Writing

$$A_{-3} - 16B_{-3} + \alpha = \gamma, \quad \tfrac{16}{21} A_2 + 4B_2 = \gamma', $$
$$4A_{-3} - 40B_{-3} = \beta, \quad \tfrac{19}{21} A_2 = \beta', \quad (14)$$

the stress components on the outside of the surface $r = a$ are

$$\frac{ap_{rx}}{\mu} = \gamma x - \beta a^{-2} x (x^2 - y^2),$$
$$\frac{ap_{ry}}{\mu} = -\gamma y - \beta a^{-2} y (x^2 - y^2), \quad (15)$$
$$\frac{ap_{rz}}{\mu} = -\beta a^{-2} z (x^2 - y^2),$$

while identical expressions serve to express the stress components inside, provided μ, β, γ are replaced by μ', β', γ'. It will be noticed that it is not possible to ensure continuity of all three components of stress. To do so would require both $\mu\beta = \mu'\beta'$ and $\mu\gamma = \mu'\gamma'$, and these equations cannot both be satisfied as well as (8), (9) and (10).

To apply the condition of continuity of tangential stress it is necessary to transform the stress components p_{rx}, p_{ry}, p_{rz} into components p_{rr} acting normal to the surface of the sphere, $p_{r\theta}$ parallel to the surface of the sphere and in a plane passing through the axis of x, $p_{r\phi}$ parallel to the surface of the sphere and perpendicular to the axis of x. The scheme of direction cosines for this transformation is

	p_{rx}	p_{ry}	$p_{r\theta}$
p_{rr}	x/r	y/r	z/r
$p_{r\theta}$	$-(1 - x^2/a^2)^{\frac{1}{2}}$	$xya^{-2}/(1 - x^2/a^2)^{\frac{1}{2}}$	$xza^{-2}/(1 - x^2/a^2)^{\frac{1}{2}}$
$p_{r\phi}$	0	$-za^{-1}/(1 - x^2/a^2)^{\frac{1}{2}}$	$ya^{-1}/(1 - x^2/a^2)^{\frac{1}{2}}$

* See Lamb's *Hydrodynamics*, ch. XI.

Applying this transformation, the stresses at $r = a$ in the outer fluid are

$$\left.\begin{aligned}
p_{rr} &= \mu a^{-2}(x^2 - y^2)\,(\gamma - \beta), \\
p_{r\theta} &= \frac{\mu\gamma x(x^2 - y^2 - a^2)}{a^2\sqrt{(a^2 - x^2)}}, \\
p_{r\phi} &= \frac{\mu\gamma yz}{a\sqrt{(a^2 - x^2)}}.
\end{aligned}\right\} \tag{16}$$

Identical expressions serve to represent the stress at $r = a$ inside the drop if β, γ, μ are replaced by β', γ', μ'. It will be seen from (16) that continuity of both $p_{r\theta}$ and $p_{r\phi}$ is ensured if

$$\mu\gamma - \mu'\gamma' \text{ or } A_{-3} - 16B_{-3} \mid \alpha = \frac{\mu'}{\mu}(\tfrac{16}{21}A_2 + 4B_2), \tag{17}$$

but when (17) is satisfied p_{rr} is discontinuous at $r = a$.

DETERMINATION OF A_{-3}, B_{-3}, A_2, B_2

The four equations (8), (9), (10), (17) can now be used to determine the constants. The solution is

$$A_{-3} = -\frac{5\alpha}{2}\left(\frac{\mu' + \tfrac{2}{5}\mu}{\mu' + \mu}\right), \quad B_{-3} = -\frac{\alpha}{4}\frac{\mu'}{\mu' + \mu}, \quad A_2 = \frac{21\alpha\mu}{4(\mu' + \mu)},$$

$$B_2 = -\frac{3\alpha\mu}{8(\mu' + \mu)}. \tag{18}$$

VISCOSITY OF THE SUSPENSION

The effect of the presence of solid spheres in suspension on the viscosity of a fluid was shown by Einstein to depend only on p_{-3}, and the same reasoning is still true when the spheres are liquid. In the case of a solid sphere $A_{-3} = -5\alpha/2$; thus it will be seen from (18) that Einstein's expression*

$$\mu^{\times} = \mu(1 + 2{\cdot}5\Phi) \tag{19}$$

for the viscosity of a fluid containing solid spheres must be replaced by

$$\mu^{\times} = \mu\left\{1 + 2{\cdot}5\Phi\left(\frac{\mu' + \tfrac{2}{5}\mu}{\mu' + \mu}\right)\right\} \tag{20}$$

when the spheres are fluid. In these formulae $\mu^{\times}$ is the mean viscosity and Φ is the small proportion of the whole volume occupied by the spheres. The two formulae are identical, as would be expected when μ' becomes infinite.

The factor $(\mu' + \tfrac{2}{5}\mu)/(\mu' + \mu)$ by which Einstein's term must be multiplied in order to take account of the currents set up inside the drop, may be compared with the factor† $(\mu' + \tfrac{2}{5}\mu)/(\mu' + \mu)$ by which Stokes' expression for the resistance of a

---

* *Annln. Phys.*, xxxiv (1911), 592.

† Hadamard, *C.R. Acad. Sci.*, Paris, CLII (1911), 1735.

solid sphere in a viscous fluid must be multiplied in order to take account of the internal currents in a liquid drop falling through another liquid.

LIMITS TO THE SIZE OF DROPS CONTAINED IN A SHEARING FLUID

In order that the drops may be nearly spherical the pressure difference due to viscous forces must be small compared with that due to surface tension, namely, $2T/a$, where T is the surface tension.

The difference in pressure between the inside and outside of the drop is

$$[p_{rr}]_{\text{inside}} - [p_{rr}]_{\text{outside}},$$

and from (16) this is $\quad P = [\mu'(\gamma' - \beta') - \mu(\gamma - \beta)]\dfrac{x^2 - y^2}{a^2},$

hence substituting from (14) and (18)

$$P = \frac{\alpha\mu}{\mu' + \mu}(\tfrac{19}{4}\mu' + 4\mu)\frac{x^2 - y^2}{a^2}. \tag{21}$$

The maximum and minimum values of $(x^2 - y^2)/a^2$ are ± 1, so that the drop will be nearly spherical so long as α is small compared with

$$\frac{\mu' + \mu}{\mu\left(\dfrac{19\mu'}{4} + 4\mu\right)}\left(\frac{2T}{a}\right). \tag{22}$$

On the other hand, the drop whose radius is

$$a = \frac{2T(\mu' + \mu)}{\alpha\mu(\tfrac{19}{4}\mu' + 4\mu)}, \tag{23}$$

is of such a size that the disruptive forces due to viscosity tending to burst the drop are about equal to the force due to surface tension which tends to hold it together. An approximate expression of this kind might have some interest in connection with the mechanical formation of emulsions, but since the fluid would certainly be turbulent in any emulsifying machine, the appropriate value to be taken for α would need much consideration.

11

THE FORMATION OF EMULSIONS IN DEFINABLE FIELDS OF FLOW

REPRINTED FROM

Proceedings of the Royal Society, A, vol. CXLVI (1934), pp. 501–23

The distortion of a drop of one fluid by the viscous forces associated with certain mathematically definable fields of flow of another fluid which surrounds it is discussed. An expression is found for small distortions from the spherical form which occur at slow speeds. If L is the greatest, and B the least diameter, $(L-B)/(L+B) = F$ approximately, where F is a non-dimensional quantity, proportional to the speed of flow, which involves the surface tension, viscosity, and the radius of the drop.

Apparatus for producing in golden syrup two definable fields of flow were constructed and their effects on drops of various oily liquids were observed and registered photographically.

Agreement with theory was found in the range of low speeds where agreement might be expected. At higher speeds the effect produced by the flow varies greatly with μ'/μ, μ and μ' being the viscosities of the syrup and drop.

For $\mu'/\mu = 0.0003$ the drop elongates almost indefinitely, but does not burst at the highest speeds attainable. For $\mu'/\mu = 0.5$ the drop burst at $F = 1.4$. For $\mu = \mu'$ the drop burst at about $F = 0.5$. For $\mu'/\mu = 20$ the drop burst at $F = 0.3$ in one type of field, but in the other the drop did not burst even at the highest speed attainable.

The difficulty experienced in bursting drops of viscous fluid by a disruptive field of flow in a surrounding fluid of considerably less viscosity is shown to be qualitatively in accordance with a theory of drops in a laminar shearing field of flow when F and μ'/μ are both large. According to this theory $(L-B)/(L+B) = 5\mu/4\mu'$.

The physical and chemical condition of emulsions of two fluids which do not mix has been the subject of many studies, but very little seems to be known about the mechanics of the stirring processes which are used in making them. The conditions which govern the breaking up of a jet of one fluid projected into another have been studied by Rayleigh* and others, but most of these studies have been concerned with the effect of surface tension or dynamical forces in making a cylindrical thread unstable so that it breaks into drops. The mode of formation of the cylindrical thread has not been discussed. As a rule in experimental work it has been formed by projecting one liquid into the other under pressure through a hole. It seems that studies of this kind which neglect the disruptive effect of the viscous drag of one fluid on the other, though interesting in themselves, tell us very little about the manner in which two liquids can be stirred together to form an emulsion.

When one liquid is at rest in another liquid of the same density it assumes the form of a spherical drop. Any movement of the outer fluid (apart from pure rotation or translation) will distort the drop owing to the dynamical and viscous forces which then act on its surface. Surface tension, however, will tend to keep the drop spherical.

* *Proc. Roy. Soc.* XXIX (1879), 71.

When the drop is very small, or the liquid very viscous, the stresses due to inertia will be small compared with those due to viscosity.

Recently the present writer made a rough theoretical estimate*, based on the hydrodynamical equations of a spherical drop in a shearing fluid, of the maximum size of drop which surface tension might be expected to hold together against the disruptive forces due to the viscous drags of the shearing fluid. Since the drop must depart very markedly from the spherical form before it bursts, this theoretical estimate is unlikely to be of much value except as an indication of the conditions under which marked deviations from the spherical form began to occur. It seemed worth while therefore to make some experiments on the deformation and bursting of a drop of one fluid in another under controlled conditions measuring the interfacial tension of the two liquids their viscosities and the rate of deformation of the outer fluid.

Among the infinite variety of possible fields of flow two have been chosen which can easily be produced in an actual fluid and at the same time can be represented by simple mathematical equations. The first is that represented by

$$u = Cx, \quad v = -Cy, \tag{1}$$

the stream lines of which are rectangular hyperbolas. The second is that represented by

$$u' = \alpha y', \quad v' = 0. \tag{2}$$

'FOUR ROLLER' APPARATUS

To produce approximately the field of flow represented by (1) the apparatus represented in the sketch of fig. 1 was constructed. Four brass cylinders 3·81 cm. × 2·39 cm. diameter were mounted at the corners of a square the sides of which were 3·18 cm. Their axles ran in brass bearings fixed in two glass plates which formed the sides of a box the internal dimensions of which were 7·6 × 7·6 × 3·9 cm. The remaining sides were brass and one of them was pierced by a large hole through which the apparatus could be filled with liquid. The cylinders were driven in the directions indicated by arrows in fig. 1 by means of two vertical shafts and bevel wheels. These two shafts were rotated at the same speed but in opposite directions by a motor through reduction gears. The box was filled with golden syrup, illuminated by an electric lamp, and observed by means of a long focus camera (magnification 2·5) set with its axis on the centre line in a direction parallel to the axes of the cylinders.

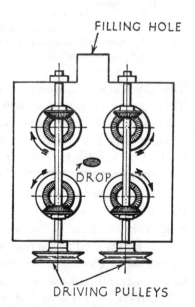

Fig. 1. 'Four roller' apparatus.

* See paper 10 above.

Slight variations in the water content of the golden syrup caused streaks to appear in the image on the camera screen as soon as the apparatus was set in motion. These streaks mark out stream lines and the photograph, fig. 2, pl. 1. shows that at any rate in the centre of the field they are very like the rectangular hyperbolas which are stream lines of the flow represented by (1).

To compare the velocity in the field of flow actually produced with equation (1) a series of lines were ruled horizontally and vertically on the screen of the camera at distances $\pm 0{\cdot}5$, $\pm 0{\cdot}1$, $\pm 1{\cdot}5$, $\pm 2{\cdot}0$, and $\pm 2{\cdot}5$ cm. from the centre. The times at which images of small particles in the golden syrup passed successive lines were observed and also the time τ of one revolution of the vertical shafts. In one set of such measurements τ was 48·7 sec. and the corresponding times for covering the 0·5 cm. distances between successive lines are given in table 1.

Table 1

Distances, in cm., from centre	Time, in sec.	Value of C from formula (3)
0·5–1·0	7	0·099
1·0–1·5	3·9	0·104
1·5–2·0	2·7	0·106
2·0–2·5	2·0	0·111

If the field of flow is represented by (1) the velocity of a particle is $dx/dt = Cx$ so that the time of passage of a particle from x_1 to x_2 is $C^{-1}(\log x_2 - \log x_1)$. Hence

$$C = (\log x_2 - \log x_1) \div (\text{time from } x_1 \text{ to } x_2). \qquad (3)$$

The values of C corresponding with this formula are given in table 1. It will be seen that C is nearly constant over the range from 0·5 to 2·5 cm. from the centre of the field in the apparatus itself.

The values of C in any experiment are proportional to the speed of rotation of the cylinder, i.e. to $1/\tau$. Taking the average value of C in this set of measurements as 0·105 corresponding with $\tau = 48 \cdot 7$ it will be seen that in any other experiment for which τ is measured the flow will be represented by (1) provided

$$C = (48 \cdot 7)\,(0 \cdot 105)\,\tau^{-1} = 5 \cdot 1/\tau. \qquad (4)$$

The number 5·1 is a dimensionless constant of the apparatus.

'PARALLEL BAND' APPARATUS

To produce the flow represented by (2) two endless celluloid bands of cinema film 35 mm. wide were stretched between rollers, one of which in each case was fitted with pins to engage in the regularly spaced holes at the edge of the film. The two bands could be driven at any speed in either direction, the ratio of their speeds being controlled by a continuously variable gear. The band and rollers were contained in a glass sided box the width of which was 3 mm. greater than that of the film. The apparatus is illustrated in the sketch fig. 3.

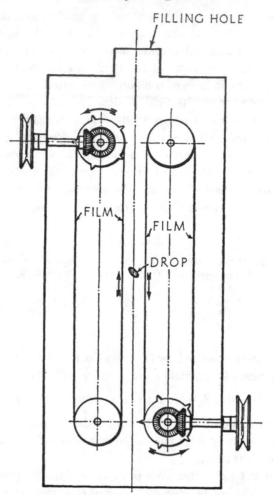

Fig. 3. 'Parallel band' apparatus.

With this construction the speed of the band was definitely related to the speed of the rollers so that each revolution of a driving pulley moved the corresponding endless band through 7·10 cm.

The distance apart of the bands was 1·36 cm. so that if the shearing motion of the liquid between them is represented by $u' = \alpha y'$, y' being measured perpendicular to the bands,

$$\alpha = \frac{7·10}{1·36}\left(\frac{1}{\tau_1}+\frac{1}{\tau_2}\right) = 5·22\left(\frac{1}{\tau_1}+\frac{1}{\tau_2}\right), \tag{5}$$

where τ_1 and τ_2 are the times of revolution of the two driving pulleys. If the constant C in (1) is chosen to be equal to $\frac{1}{2}\alpha$ and if the whole system (2), i.e. the axes of co-ordinates (x', y') is given a rotation with angular velocity $\frac{1}{2}\alpha$ and if the axes (x', y') are instantaneously at 45° to the axes (x, y) then at that instant the two fields of flow are identical. Effects therefore which depend only on the instantaneous distribution of velocity and are unaffected by a rotation of the whole system will be identical in the two pieces of apparatus when they are operated at corresponding speeds, i.e. so that $C = \frac{1}{2}\alpha$. On the other hand, effects which do not depend only on the instantaneous distribution of velocity but are dependent on a sequence of such distributions are very different in the two.

To illustrate the significance of these remarks consider the effects of the two kinds of flow of a very viscous fluid on an elongated symmetrical solid body, say a prolate spheroid, placed with its long axis in the plane xy. If this body is placed in the field of flow (1) with its long axis at angle θ to the axis of x its surface will be subjected to exactly the same stresses that would act on it in the field of flow (2) if its long axis were placed at $45° + \theta$ to the axis of x'. The resultant effects of the two fields of flow on the motion of the body over a period of time are, however, very different. In the field of flow (1) the body will set itself permanently with its long axis parallel to the axis of x, whereas in the flow (2) the body will continually roll over and over rotating, at a variable speed, about an axis perpendicular to the plane $x'y'$.

CALCULATION OF SMALL DEFORMATIONS

Before describing the experimental results obtained with the two forms of apparatus we may see how far theory can predict the effects of the two fields of fluid flow (1) and (2) on drops of another fluid immersed in them.

Suppose that a spherical drop of a fluid of viscosity μ' is immersed in a fluid of viscosity μ and that the latter is caused to flow in the velocity distribution represented by $(u = Cx, v = -Cy)$. If the flow is very slow the viscous drag will deform the drop only slightly from the spherical form. This small deformation will cause only a small change in the distribution of stresses in either fluid, accordingly in the stress conditions which must be satisfied at the surface of the drop, namely, (a) continuity of tangential stress, and (b)

$$T(r_1^{-1} + r_2^{-1}) = \text{constant} + p_i - p_0, \tag{6}$$

the stresses may be reckoned as those which occur when the drop is held spherical by a distribution of normal force at the surface. In equation (6) T is the interfacial surface tension between the two liquids, p_i and p_0 are the normal pressures inside and outside the drop, r_1 and r_2 are the principal radii of curvature of the deformed drop.

It follows that the analysis previously given by the present writer* for the flow in the neighbourhood of a spherical drop can be applied directly in the present work. The value of $p_i - p_0$ is therefore

$$p_i - p_0 = \frac{1}{2}C\mu \frac{19\mu' + 16\mu}{\mu' + \mu}\left(\frac{x^2 - y^2}{a}\right) + \text{constant.} \tag{7}$$

---

* Paper 10 above. The expression here given in (7) is identical with equation (21) of that paper except that $2C$ has been substituted for α.

Comparing (6) and (7) it will be seen that it is necessary to find the shape of the nearly spherical drop for which the variation in $(r_1^{-1}+r_2^{-1})$ is proportional to $(x^2-y^2)\,a^{-2}$. It can be verified that for the surface whose equation is

$$r = a+b(x^2-y^2)\,a^{-2} \tag{8}$$

$$r_1^{-1}+r_2^{-1} = 2a^{-1}+4b(x^2-y^2)\,a^{-4}. \tag{9}$$

Combining (6), (7) and (9) and equating coefficients of the variable part of the pressure it will be seen that (8) represents the deformed drop provided

$$\tfrac{1}{2}C\mu\,\frac{19\mu'+16\mu}{\mu'+\mu} = \frac{4Tb}{a^2}. \tag{10}$$

A convenient method for expressing the results of experiment is to measure L, the length of the drop in the direction of the x axis, and B, the breadth in the direction of the y axis. These measurements are connected with the constant b of equation (8) by the formula

$$\frac{L-B}{L+B} = \frac{b}{a}, \tag{11}$$

so that (10) becomes

$$\frac{L-B}{L+B} = F\,\frac{19\mu'+16\mu}{16\mu'+16\mu}, \tag{12}$$

where

$$F = 2C\mu a/T. \tag{13}$$

F is non-dimensional.

It will be noticed that over the whole range of ratios μ'/μ from 0 to α

$$(19\mu'+16\mu)/(16\mu'+16\mu)$$

varies only from 1·0 to 1·187, so that $(L-B)/(L+B)$ is nearly equal to F.

For the flow $(u'=\alpha y',\ v'=0)$ the deformation of the drop at slow rates of flow should also be represented by (12), but in that case

$$F = \mu a\alpha/T, \tag{14}$$

and the long axis should lie in the direction making 45° with the axis of x' (i.e. at 45° to the celluloid film as shown in fig. 3).

METHOD OF EXPERIMENT

In all the experiments to be described the apparatus was filled with golden syrup (which is a concentrated sugar solution) diluted with a small quantity of water till the viscosity was between 50 and 150 c.g.s. A number of liquids which do not mix with water were used for the drop so as to cover a large range of values of μ'/μ. For low values of μ'/μ a mixture of carbon tetrachloride with the paraffin oil sold as 'Nujol' was made up to be of the same density as the syrup and was coloured purple with dissolved iodine. For values of μ' rather smaller than μ a lubricating oil sold as 'BB' was used. For values of μ' equal to μ a black lubricating oil was used. For values of μ' considerably greater than μ a mixture of coal tar and pitch was made up so that its viscosity was about 2000 c.g.s. These values together with the interfacial surface tensions are given in table 2.

Table 2

	μ' (c.g.s.)	T (c.g.s.)
CCl$_4$ and paraffin	0·034	23
'BB' oil	60	17
Black lubricating oil	100	8
Tar-pitch mixture	2000	(23)

Before setting the apparatus in motion the drop was introduced into the syrup by means of a pipette which was lowered through the hole in the top of the apparatus. The drop became spherical under the influence of surface tension and was photographed in that condition in order to measure its radius. The apparatus was then set in motion at a slow speed and adjusted till the drop was steady and stationary. A photograph was then taken. The speed was then increased and another photograph taken, these operations being repeated till some limiting conditions such as the bursting of the drop was reached.

EXPERIMENTS WITH THE 'FOUR ROLLER' APPARATUS

After introducing the drop at the top of the apparatus the latter was slowly set in motion so as to carry the drop towards the middle of the field. After prolonged running the drop gradually took up a position in the central horizontal plane. It was however, highly unstable so far as horizontal motion was concerned tending to move off to the right or left if it got displaced from the centre. This instability was controlled by varying the speed of the right- or left-hand pair of rollers. It has already been mentioned that the two rollers on the right were driven by one shaft and the two on the left by another. Each pair was driven by a belt which could be made to slip. When uncontrolled neither of the belts slipped so that all four rollers rotated at the same speed. If the drop was observed to get slightly off centre to the right, the right-hand pair of rollers was retarded and the drop moved back towards the centre. After a little practice it was possible to keep the drop very close to the centre of the field.

The photograph of fig. 4, pl. 1, shows a drop which had been maintained in this way symmetrically in the middle of the field for a time which was long enough to ensure that the drop and surrounding fluid was in a steady state. One-quarter of each of the four rollers is shown in the corners of the photograph and the direction of rotation of each is shown by means of an arrow. Each of the photographs taken with this apparatus was similar to that of fig. 4, but in order to economize space only the part of the field immediately surrounding the drop is shown in the remaining photographs.

$\mu'/\mu = 0.0003$. The sequence of photographs of fig. 5, pl. 1, shows the effect of the field of flow represented by equation (1) on drops of the CCl$_4$ and paraffin mixture ($\mu' = 0.034$). At each setting of the apparatus the time of revolution τ of the rollers was observed, the apparatus being uncontrolled during the exposure, and for

8

sufficient time before it, to ensure steady conditions. Using equation (4), the constant C was found. The diameter $2a$ of the undistorted drop was measured on the photograph and T were measured independently by the methods described in the Appendix so that the non-dimensional quantity $F = 2c\mu a/T$ could be calculated.

The analysis of the slightly distorted drop shows that the shape of the drop should depend on the two non-dimensional quantities μ'/μ and F. The shape of the drop can be defined by the ratio $(L-B)/(L+B)$ where L is the greatest length and B the breadth. For any given value of μ'/μ the experimental results can, therefore be expressed by means of a single curve giving $(L-B)/(L+B)$ for all values of F. In each case the maximum and minimum diameters of the photographs of distorted drops were measured and $(L-B)/(L+B)$ was found. Under each photograph of fig. 5 the measured values of F and $(L-B)/(L+B)$ are given and over each photograph the radius a of the drop in centimetres.

In the first series of photographs (fig. 5) it will be seen that at the slowest speed, $F = 0.18$, the drop is only slightly distorted from the spherical form, and the distortion, measured by $(L-B)/(L+B)$, is 0.15. In this drop $\mu'\mu = 0.034/100 = 0.00034$ so that the predicted relationship (12) is nearly exactly $F = (L-B)/(L+B)$. The experiment therefore is in good agreement with the theory.

In the second photograph of fig. 5 $F = 0.28$ and $(L-B)/(L+B) = 0.26$, so that the theory still appears to represent the experimental conditions. In the third photograph the drop has developed into a form which is very far from spherical, indeed the ends have become pointed, so that the theory can no longer be applied. It will be noticed, however, that $F = 0.41$, $(L-B)/(L+B) = 0.44$, so that the theoretical relationship for small deformations is still nearly correct. The drop shown in photograph 3, fig. 5, was not in a truly steady state, for after developing the point shown in the photograph a thin skin appeared to slip off its surface and the ends of the drop again became rounded. This condition persisted, as shown in the fourth photograph, till $F = 0.54$. At $F = 0.65$ (fifth photograph) the ends of the drop again became pointed and these points remained as F increased up to the highest speed at which the apparatus could be operated.

The sixth, seventh, and eighth photographs, fig. 5, show a pointed drop which increases in length and decreases in thickness with increasing speed. Thus as F increases from 0.65 to 2.45 $(L-B)/(L+B)$ increases from 0.51 to 0.87. It was not possible to realize values of F higher than 2.45 because the ends of the drop got into the region where the field of flow ceased to be even approximately that represented by (1). There was no sign, however, that the drop would have burst even if considerably higher speeds had been attained.

The results described above, together with some others for which the drop is not here reproduced, are shown graphically in fig. 6 where values of F and $(L-B)/(L+B)$ are plotted in a diagram. The theoretical relationship $F = (L-B)/(L+B)$ applicable to small values of F is shown by means of a broken line.

$\mu'/\mu = 0.9$. Using black lubricating oil for which $\mu' = 100$ c.g.s. in syrup for which $\mu = 110$ c.g.s. the value of μ'/μ was 0.91. The interfacial surface tension T was found to be 8·0 c.g.s. A drop of oil 0·144 cm. diameter was used and the series of photographs shown in fig. 7, pl. 2, were obtained. In the first of these the time of revolution of the rollers was 97 sec. so that from (4) and (13)

$$F = \frac{10·2(0·144)(110)}{(8·0)(97)} = 0·21.$$

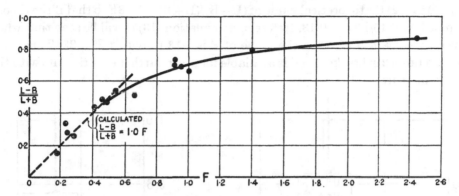

Fig. 6. CCl$_4$-paraffin. 'Four roller' apparatus, $\mu'/\mu = 0·0003$.

The value of $(L-B)/(L+B)$ obtained by measuring the photograph was 0·19 and its theoretical value from (12) is 1·09F = 0·23. It will be seen, therefore, that the observed value 0·19 is in fair agreement with its theoretical value 0·23. The error is certainly not greater than that which arises from the uncertainties involving in measuring T.

The second photograph shows the drop at $F = 0·30$ when the theoretical value of $(L-B)/(L+B)$ was 1·09(0·30) = 0·33. The observed value was 0·29 so that again the theory is confirmed. In the third photograph $F = 0·37$ and the drop has become much elongated. The observed value of $(L-B)/(L+B) = 0·54$ is now considerably greater than the value (1·09) (0·37) = 0·40 obtained by extrapolating the theory.

When the speed reached the point at which $F = 0·39$ the drop began to pull out into a thread-like form. This is shown in the fourth and fifth photographs of fig. 7 which were taken while the drop was bursting in this way. Shortly after taking the fifth photograph the apparatus was stopped. The thread of oil which had seemed quite stable while the apparatus was in motion then gradually broke up into a number of small drops. The final appearance of the oil in this condition is shown in the sixth photograph of fig. 7.

These results are shown graphically in fig. 8. It will be seen that when $\mu'/\mu = 0·9$ the drop remains coherent so long as $F < 0·39$ but burst as soon as F reaches this value. This is indicated by the broken line continuation of the curve.

$\mu'/\mu = 20$. To investigate the effect of high viscosity of the drop a mixture of tar and pitch was made with viscosity 2000 c.g.s. This had a density not far from that of the syrup, namely, 1·40. In these experiments the viscosity of the syrup was 99 c.g.s. No way was found for making an independent and reasonably accurate measurement of the interfacial surface tension. On the other hand, if the truth of the theory of slightly deformed drops is assumed equations (12) and (13) can be applied to find T.

A drop 0·16 cm. radius was used. The first photograph, fig. 9, pl. 1, shows this drop in its undeformed condition. The apparatus was set in motion so that the time of revolution τ of the rollers was 54 sec. and the second photograph of fig. 9 was taken. By measuring this photograph it was found that $(L-B)/(L+B) = 0.15$. Taking μ'/μ as 20 in (12) the theoretical value of $(L-B)/(L+B)$ is $1.18F$ so that if the theory is correct $F = 0.15/1.18 = 0.13$. Referring to equation (13) it will be seen that when $a = 0.16$, $\tau = 54$, $\mu = 99$, then the value of F is 0·13 provided $T = 23$. This value is given in table 2 in brackets and is available for use in further experiments with the tar-pitch mixture.

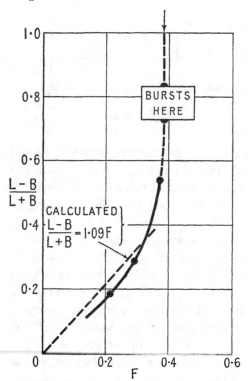

Fig. 8. 'Four roller' apparatus. Black lubricating oil, $\mu'/\mu = 0.9$.

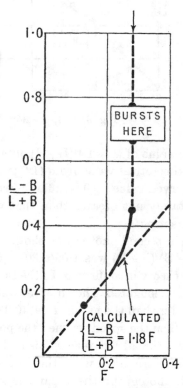

Fig. 10. 'Four roller' apparatus, tar-pitch mixture, $\mu'/\mu = 20$.

On increasing the speed of the rollers till $\tau = 24$ sec. the drop was found to be bursting. The third photograph of fig. 9, pl. 1, shows the drop in the act of bursting. Using $T = 23$ c.g.s. the corresponding value of F was 0·28. It appears, therefore that when $\mu'/\mu = 20$ the drop bursts for some value of F rather less than 0·28.

These results are indicated in the diagram, fig. 10.

EXPERIMENTS WITH THE 'PARALLEL BAND' APPARATUS

This apparatus was driven by a motor through various reduction gears and a variable gear so that the speed could be varied and also, independently, the ratio of the speeds of the two bands. A drop of fluid was placed in the apparatus and photographed. The apparatus was then set in motion. If the drop happened to be placed exactly midway between the bands it would remain at rest when they both moved at the same speed. If the drop was not quite central then it could be brought to rest by adjusting sightly the ratio of the speeds of the two bands..

When the conditions had become steady a photograph was taken and speeds of both bands measured by timing the revolutions of the driving rollers.

$\mu'/\mu = 0.0003$. A drop of the same mixture of CCl_4 and paraffin as that used in the 'four roller' apparatus was introduced. The first photograph of fig. 11, pl. 2, shows this drop before starting the apparatus. The dark horizontal lines at the top and bottom are the celluloid bands which in the apparatus itself were vertical. The arrow above the whole series of pictures shows the direction of motion of the band which is at the top of the photograph. The other band moves in the opposite direction. Measurement of the first photograph showed that the radius of the drop was 0.157 cm.

The second photograph of fig. 11 shows the drop distorted by viscous drag, the two periods of revolution of the driving rollers being $\tau_1 = 192$ sec., $\tau_2 = 233$ sec. It will be seen that the long axis of the distorted drop lie at about 45° to the bands, i.e. in the direction in which lines of particles are elongating at the greatest rate. In this experiment it was found that $\mu = 123$ c.g.s., so that from (5) and (14)

$$\mathbf{F} = (5.22)\ (0.157)\ (123)\ T^{-1}\ (\tau_1^{-1} + \tau_2^{-1}).$$

If T is taken as 23 c.g.s. this becomes $\mathbf{F} = 4.4\ (\tau_1^{-1} + \tau_2^{-1})$ so that $\mathbf{F} = 0.04$. Measurement of the photograph gives $(L-B)/(L+B) = 0.08$ so that the theoretical relationship $\mathbf{F} = (L-B)/(L+B)$ is not fulfilled.

The third photograph taken when $\mathbf{F} = 0.10$ gives $(L-B)/(L+B) = 0.22$ so that again the observed value of $(L-B)/(L+B)$ is about twice as great as the prediction. The consistency of the error makes it seem probable that the surface tension of the drop in the syrup was considerably less during this series of experiments than it had been previously, probably owing to impurities in the syrup. The sequence of drops shown in the eight photographs of fig. 11 shows that as the speed of the apparatus is increased the drop elongates but does not burst even at the highest speed obtainable. The highest value of F shown in fig. 11 is 2.30, but this photograph was obtained with a different filling of syrup from the others. One experiment, however, was tried with a very large drop ($a = 0.54$) and it was found that this drop was still coherent at $\mathbf{F} = 5.3$ if $T = 23$ (or $\mathbf{F} = 11$ if $T = 11$).

It will be noticed that the fifth photograph of fig. 11 shows a thin streak coming off each end of the drop. This was a transient phenomenon and is evidently the same as that previously described in connection with the third photograph of fig. 5, pl. 1.

The results obtained with two fillings of syrup are set out in fig. 12 as 'First series' and 'Second series.'

$\mu'/\mu = 0.5$. Using fresh syrup and a drop of 'BB' oil the results set forth in table 3 were obtained. T was measured independently and found to be 17 c.g.s. In the first

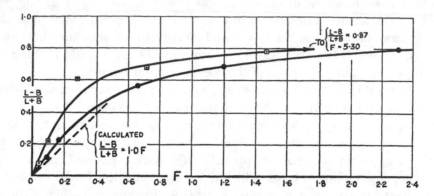

Fig. 12. 'Parallel band' apparatus. CCl_4-paraffin. $\mu'/\mu = 0.0003$.
● First series; ⊡ Second series.

set of observations μ was 135 c.g.s. while in the second it was 110 c.g.s. The second set is shown in the photograph, fig. 13, pl. 3. It will be seen that for low values of F the drop is, as predicted, an ellipse with its long axis at 45° to the celluloid bands. As the speed of the apparatus is increased the drop becomes elongated till at $F = 1.43$ a steady motion ceases to be possible and the drop elongates into a thead-like form.

Table 3. *BB oil drop in 'parallel band' apparatus.* $\mu' = 60, T = 17$

a (cm.)	μ	τ_1 (sec.)	τ_2 (sec.)	F	$\dfrac{L-B}{L+B}$
0·140	110	62	51	0·17	0·17
0·167	135	157	828	0·05	0·07
0·123	135	99	87	0·11	0·10
0·123	135	83	85	0·12	0·19
0·123	135	24	25	0·42	0·45
0·101	135	9·5	10·5	0·84	0·00
0·100	135	7·0	8·0	1·12	0·81
0·100	135	5·5	6·2	1·43	burst

These results are set out graphically in fig. 14. For $\mu'/\mu = 0.5$ the theoretical relationship (12) is $(L-B)/(L+B) = 1.06F$. This is represented in fig. 14 by a broken line. It will be seen that the agreement is good.

$\mu'/\mu = 0.9$. With the same black lubricating oil which had previously been used in the 'four roller' apparatus, the results given in table 4 were obtained. Some of the photographs are shown in fig. 15, pl. 3. It will be noticed that when $F = 0.28$ the drop is not symmetrical, one end being more pointed than the other. This asymmetry is still more pronounced when $F = 0.55$. It always occurred when there was a large difference in density between the drop and the syrup.

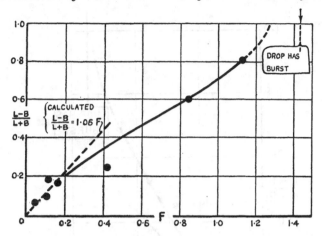

Fig. 14. 'Parallel band' apparatus. 'BB' oil. $\mu'/\mu = 0.5$.

Table 4. *Black lubricating oil in syrup, 'parallel band' apparatus.*
$$\mu = 110,\ \mu' = 100,\ T = 8{\cdot}0$$

a	τ_1 (sec.)	τ_2 (sec.)	F	$\dfrac{L-B}{L+B}$
0·097	262	207	0·06	0·07
0·097	141	111	0·11	0·15
0·097	92	71	0·17	0·22
0·097	57	44	0·28	0·45
0·086	40	31	0·36	0·53
0·086	22	23	0·55	burst

The relationship between $(L-B)/(L+B)$ and F is shown in fig. 16 and the theoretical relationship for $\mu'/\mu = 0{\cdot}9$, namely, $(L-B)/(L+B) = 1{\cdot}09F$, is there shown by a broken line.

$\mu'/\mu = 20$. Using the pitch and tar mixture for which μ' was about 2000 c.g.s. the drop again became elliptical with its long axis at 45° to the celluloid bands when the apparatus was run at very slow speeds. This is shown in the second photograph of fig. 17, pl. 3, taken when $F = 0{\cdot}08^*$. As the speed increased to $F = 0{\cdot}27$, third photograph, fig. 17, the drop became slightly more elongated, but its long axis became more nearly parallel to the celluloid bands. At $F = 1{\cdot}13$, fourth photograph, fig. 17, its deformation was only slightly greater than at $F = 0{\cdot}27$ and its long axis was still more nearly parallel to the bands. At $F = 1{\cdot}69$, fifth photograph, fig. 17, its shape and orientation were almost identical with those at $F = 1{\cdot}13$.

In comparing these results with those shown in fig. 9, pl. 1, which were obtained with the 'four roller' apparatus a very striking difference will be noticed. In the 'four roller' apparatus the drop burst at a very low speed represented by $F = 0{\cdot}28$. In the 'parallel band' apparatus the drop did not burst however fast the apparatus was run. On the contrary, it attained at high speeds a constant condition in which $(L-B)/(L+B)$ was $0{\cdot}26$ and the long axis was parallel to the bands.

* The values of F are found by assuming $T = 23$ c.g.s., see table 2.

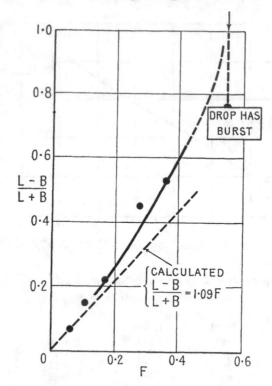

Fig. 16. 'Parallel band' apparatus. Black lubricating oil. $\mu'/\mu = 0.9$.

The explanation of this phenomenon is simple. In the 'four roller' apparatus, which produces the field of flow (1), the lines of particles which are extending at the greatest rate, namely, those parallel to the axis of x, remain in the direction of maximum rate of elongation as long as the flow continues. The disruptive stress due to the viscous drag of the syrup is therefore always tending to extend the drop in the same direction. As soon as this stress is able to overcome the cohesive effect of surface tension the drop bursts.

In the 'parallel band' apparatus the lines of particles which lie in the direction of maximum rate of elongation, namely, at 45° to the bands, are continually being rotated away from that position towards the line parallel to the bands which is neither elongating nor contracting. After attaining this position further rotation brings the line of particles into an orientation where they are contracting. When the drop is very viscous compared with the surrounding medium it rotates almost like a slightly plastic solid body, lines of particles in it elongating slowly while they are within 45° of the direction of maximum rate of elongation and contracting slowly while they are within 45° of the direction of maximum rate of contraction. In this way the surface of the drop attains a permanent position with its long axis parallel to the bands. The tar-pitch mixture rotates round inside this fixed envelope; in fact, accidental small unevennesses could be seen to move round the contours of the

PLATE 1

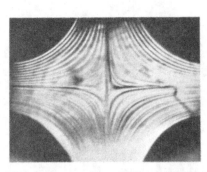

Fig. 2. Stream lines in 'four roller' apparatus.

Fig. 4. Large drop in 'four roller' apparatus.

$a =$	0·20	0·20	0·25	0·16	0·12

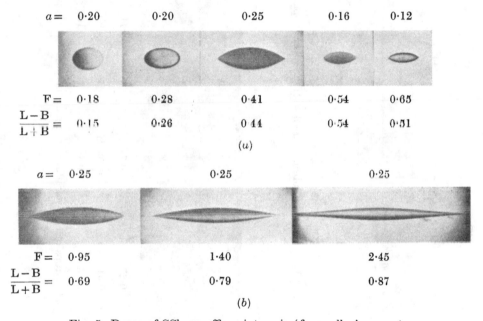

	0·20	0·20	0·25	0·16	0·12
$F =$	0·18	0·28	0·41	0·54	0·65
$\frac{L-B}{L+B} =$	0·15	0·26	0·44	0·54	0·51

(a)

$a =$	0·25	0·25	0·25
$F =$	0·95	1·40	2·45
$\frac{L-B}{L+B} =$	0·69	0·79	0·87

(b)

Fig. 5. Drops of CCl_4-paraffin mixture in 'four roller' apparatus.

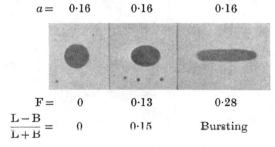

$a =$	0·16	0·16	0·16
$F =$	0	0·13	0·28
$\frac{L-B}{L+B} =$	0	0·15	Bursting

Fig. 9. Drop of tar-pitch mixture in 'four roller' apparatus.

PLATE 2

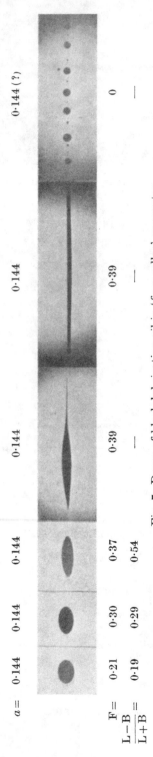

$a=$	0·144	0·144	0·144	0·144	0·144	0·144	0·144 (?)
$F=$	0·21	0·30	0·37	0·39	0·39		0
$\dfrac{L-B}{L+B}=$	0·19	0·29	0·54	—	—		—

Fig. 7. Drops of black lubricating oil in 'four roller' apparatus.

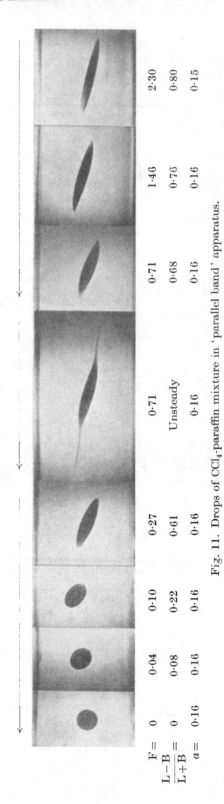

$F=$	0	0·04	0·10	0·27	0·71	0·71	1·46	2·30
$\dfrac{L-B}{L+B}=$	0	0·08	0·22	0·61	Unsteady	0·68	0·76	0·80
$a=$	0·16	0·16	0·16	0·16	0·16	0·16	0·16	0·15

Fig. 11. Drops of CCl$_4$-paraffin mixture in 'parallel band' apparatus.

PLATE 3

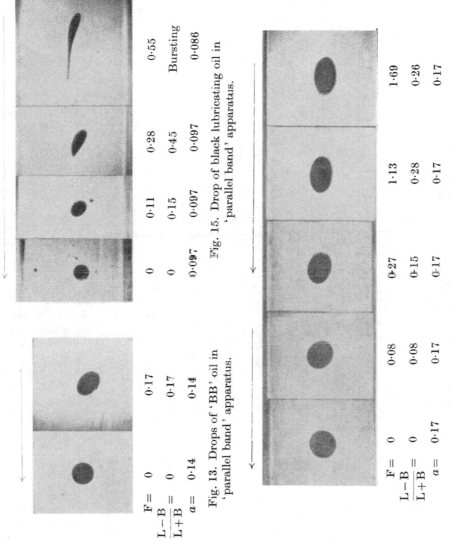

F = 0   0·17
$\frac{L-B}{L+B}$ = 0   0·17
a = 0·14   0·14

Fig. 13. Drops of 'BB' oil in 'parallel band' apparatus.

F = 0   0   0·11   0·28   0·55
$\frac{L-B}{L+B}$ = 0   0·15   0·45   Bursting
a = 0·097   0·097   0·097   0·097   0·086

Fig. 15. Drop of black lubricating oil in 'parallel band' apparatus.

F = 0   0·08   0·27   1·13   1·69
$\frac{L-B}{L+B}$ = 0   0·08   0·15   0·28   0·26
a = 0·17   0·17   0·17   0·17   0·17

Fig. 17. Drop of tar-pitch mixture in 'parallel band' apparatus.

drops. As the speed increases the viscous stresses in the surrounding medium and in the drop itself increase in the same ratio. The stresses due to surface tension do not increase so that ultimately at comparatively high speeds the drop assumes a form which is quite independent of surface tension.

ANALYTICAL TREATMENT OF VERY VISCOUS DROP

The idea just put forward can be analysed mathematically in the case when μ' is large compared with μ and the speed is so great that the forces due to surface tension can be neglected compared with those due to viscosity.

Referring to the treatment* previously given for finding the distribution of flow inside and outside a drop held spherical by surface tension the conditions of continuity of velocity at the surface of the drop must be satisfied so that equations (8) and (9) of the previous paper still hold. These are

$$\tfrac{1}{2}A_{-3} - 5B_{-3} = -\tfrac{2}{21}A_2 \tag{8a}$$

$$2B_{-3} + \tfrac{1}{2}\alpha = \tfrac{5}{21}A_2 + 2B_2. \tag{9a}$$

The condition of continuity of tangential stress is

$$A_{-3} - 16B_{-3} + \alpha = \frac{\mu'}{\mu}(\tfrac{16}{21}A_2 + 4B_2). \tag{17a}$$

Neglecting the effect of surface tension the condition of continuity of normal stress p_{rr} is

$$\mu(\gamma - \beta) = \mu'(\gamma' - \beta')$$

or

$$\mu(-3A_{-3} + 24B_{-3} + \alpha) = \mu'(-\tfrac{3}{21}A_2 + 4B_2). \tag{17b}$$

The solution of equations $(8a)$, $(9a)$, $(17a)$, $(17b)$ is

$$\left.\begin{aligned} A_{-3} &= 10B_{-3} = -5\alpha\left(\frac{\mu' - \mu}{2\mu' + 3\mu}\right) \\[2mm] A_2 &= 0, \quad B_2 = \frac{5\alpha\mu}{4(2\mu' + 3\mu)}. \end{aligned}\right\} \tag{15}$$

Referring to equation (5) of paper 10, the components of velocity at the surface are

$$u = 2B_2 a\cos\phi\sin\theta, \quad v = -2Ba\sin\phi\sin\theta, \quad w = 0, \tag{16}$$

where

$$x = a\cos\phi\sin\theta, \quad y = a\sin\phi\sin\theta.$$

When μ'/μ is large these are

$$u = \frac{5}{4}\frac{\mu}{\mu'}a\alpha\cos\phi\sin\theta, \quad v = -\frac{5}{4}\frac{\mu}{\mu'}a\alpha\sin\phi\sin\theta. \tag{17}$$

This expression gives the instantaneous rate of deformation of the drop referred to polar coordinates (θ, ϕ) at 45° to the celluloid bands of the apparatus.

***

* Paper 10 above.

In the central plane $\theta = \frac{1}{2}\pi$ the radial component of velocity is

$$\tfrac{5}{4}a\alpha\frac{\mu}{\mu'}\cos 2\phi. \tag{18}$$

It has already been pointed out that the condition of a drop in the shearing field $(u' = \alpha y, v' = 0)$ is identical with that in the field $(u = \frac{1}{2}\alpha x, v = -\frac{1}{2}\alpha y)$ when $(x'y')$ is at $45°$ to (xy) when the latter is rotated with angular velocity $\frac{1}{2}\alpha$.

If the result of this rotation and this rate of deformation is to keep the outer surface of the drop in a fixed position in the apparatus though the particles move round on it, and if this surface is nearly spherical its central section may be represented by the equation
$$r = a\{1 + bf(\phi)\}. \tag{19}$$

The condition that this surface may remain fixed in space is that the component of velocity to the surface due to rotating it at angular velocity $\frac{1}{2}\alpha$ shall be equal to the radial component of velocity of deformation. Thus

$$\tfrac{1}{2}\alpha ab\frac{d}{d\phi}f(\phi) = \tfrac{5}{4}a\alpha\frac{\mu}{\mu'}\cos 2\phi. \tag{20}$$

The solution of (20) is
$$bf(\phi) = \frac{5}{4}\frac{\mu}{\mu'}\sin 2\phi,$$

and the equation to the surface is therefore

$$r = a\left(1 + \frac{5}{4}\frac{\mu}{\mu'}\sin 2\phi\right). \tag{21}$$

Remembering that $\phi = 0$ is the line at $45°$ to the celluloid bands of the apparatus it will be seen that (21) indicates:

(a) that the long axis of the drop is parallel to the bands
(b) the ratio $(L-B)/(L+B)$ is equal to $\frac{5}{4}(\mu/\mu')$, $\qquad\qquad$ (22)

and this ratio is independent of the size of drop and speed of the apparatus.

Photographs were taken with a view to verifying the formula (22). In each case the value of $(L-B)/(L+B)$ was about 0.25, while the value of $\frac{5}{4}\mu/\mu'$ was about 0.07, so that (22) was not verified quantitatively, though the prediction that the drop will assume a permanent shape with its long axis parallel to the bands was verified (see fig. 18).

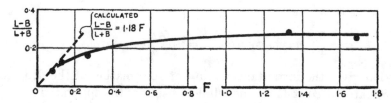

Fig. 18. 'Parallel band' apparatus, tar-pitch mixture, $\mu'/\mu = 20$.

Conclusions

It has been shown that when a drop is slightly distorted from the spherical form by the stresses in a viscous fluid which is in motion round it the deformation depends only on the instantaneous conditions. The drop becomes elongated in the direction along which lines of particles are elongating at the greatest rate. Thus in the 'four roller' apparatus the long axis of the drop is horizontal while in the 'parallel band' apparatus it is at 45° to the bands. The shape of a slightly deformed drop is in complete agreement with a theory which shows that

$$\frac{L - B}{L + B} = \frac{19\mu' + 16\mu}{16\mu' + 16\mu} F.$$

As the viscous drag on the surface of the drop increases the drop elongates, but its shape no longer depends on the instantaneous conditions of flow. Thus the two types of apparatus produce different effects. The ultimate fate of the drop as the speed of distortion of the outer fluid increases depends also very much on the ratio μ'/μ. For very small values of μ'/μ, e.g. 0·0003, the drop remains coherent in spite of the fact that it gets very long and narrow in both types of apparatus.

As the viscosity of the drop increases the speed necessary to burst it gets less with both kinds of flow: thus, with the 'parallel band' apparatus, it bursts when $F = 1·2$ approximately, for $\mu'/\mu = 0·5$. For $\mu' = \mu$ it bursts at $F = 0·5$ with the parallel band apparatus, and $F = 0·4$ with the 'four roller' apparatus.

When μ'/μ has increased to 20 the two kinds of disrupting field have very different effects. In the 'four roller' apparatus the drop bursts at $F = 0·28$, whereas the 'parallel band' apparatus seems incapable of bursting the drop even at the highest speeds attainable. At high speeds in the latter apparatus the drop, in fact, attains constant shape which depends only on μ/μ'. In this condition theoretical considerations lead to the prediction that

$$(L - B)/(L + B) = 5\mu/4\mu',$$

but this formula is not confirmed by these experiments except that the drop does assume a constant shape.

It is remarkable that in the 'parallel band' apparatus the ease with which the drop can be burst increases as the viscosity of the drop rises from small values to be equal to that of the surrounding medium, but when the viscosity of drop rises to be several times as great as that of the outer fluid the viscous drag of the latter is incapable of bursting it however big the viscous stresses may be. The drop merely rotates remaining nearly spherical.

Finally the manner in which the drops burst is of interest. The act of bursting is always an elongation to a threadlike form. When this thread breaks up it degenerates into drops which are of the order of 1/100th of the size of the original drop. This seems to be related to the known fact that when an emulsion is formed mechanically it contains drops which cover a very large range of sizes.

The experiments here described were carried out in the Cavendish Laboratory through the kindness of Lord Rutherford, to whom the author wishes to express his thanks.

APPENDIX

Measurement of Viscosity. For the oils, and the syrup viscosity was measured by weighing the amount of fluid flowing in a given time through a glass tube. This method was unsuitable for measuring the viscosity of the highly viscous mixture of tar and pitch, accordingly the apparatus shown in the sketch of fig. 19 was constructed.

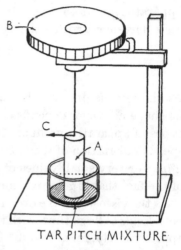

A brass cylinder A 1·27 cm. diameter × 3.9 cm. long was hung by a fine steel wire from the torsion head B. A small metal dish containing a washer 1·2 mm. thick was placed below it and the torsion head gently lowered till the bottom of the cylinder was just in contact with the washer. The washer was then removed and some of the tar-pitch mixture was dropped into the dish so that the depth was slightly more than 1·2 mm. The brass cylinder was then carefully dropped on to the surface of the mixture so that it was separated from the dish by a disc of the mixture 1·2 mm. thick × 1·27 cm.

Fig. 19. Viscometer for very viscous liquids.

diameter. A light pointer C served to mark the position of the cylinder as it rotated about its vertical axis.

If the torsion head was rotated suddenly through an angle θ_0 the cylinder would begin to rotate in the same direction, the rate of rotation being proportional to $\theta_0 - \theta$ where θ is the angle of rotation of the cylinder from its initial position. If the torsional couple of the wire is $K(\theta_0 - \theta)$ and this is equal to the couple due to action of viscosity μ' on the base of the cylinder, then the equation of motion is

$$K(\theta_0 - \theta) = \mu' \frac{d\theta}{dt} \int_0^a \frac{2\pi r^3 \, dv}{d}, \qquad (23)$$

where d is the thickness of the disc of fluid which in this case was 0·12 cm. and $2a = 1·27$ cm. The right-hand side of (23) is in c.g.s. units $2·12 \, \mu' \, d\theta/dt$. To find K the cylinder was allowed to oscillate freely and its period timed. In this way it was found that $K = 2·11 \times 10^2$ in c.g.s. units. Hence (23) becomes on integration

$$\log\left(\frac{\theta_0}{\theta_0 - \theta}\right) = \frac{211}{2·12} \frac{t}{\mu'} = 100 t/\mu'. \qquad (24)$$

In making measurements it was found convenient to turn the torsion head suddenly through half a turn so that $\theta_0 = \pi$. The time taken for the cylinder to turn through $\frac{1}{2}\pi$ and $3\pi/4$ was observed and it was found, as would be expected, that the time for $3\pi/4$ was exactly twice that for $\frac{1}{2}\pi$.

Using formula (24) it will be seen that the time corresponding with $\theta = \frac{1}{2}\pi$ is $\mu' \log 2/100$, while that corresponding with $\theta = 3\pi/4$ is $\mu' \log 4/100$.

This apparatus was used in making a mixture of tar and pitch of the desired viscosity. In the mixture chosen for some of the experiments the time for $\theta = \frac{1}{2}\pi$ was found to be 14 sec. while that for $\theta = 3\pi/4$ was 28 sec., so that

$$\mu' = (100)\,(14)/\log 2 = 2 \times 10^3 \,\text{c.g.s.}$$

Measurement of Surface Tension. In measuring the interfacial surface tension between two liquids one or both of which are very viscous it is advisable to avoid the use of any apparatus containing an appreciable length of capillary tube, for this may unduly prolong the time necessary for the system to attain equilibrium. In the present work the surface tension was measured by means of the apparatus shown in fig. 20. This sketch is self-explanatory. The syrup–oil interface was at the end of a very short length of capillary tube whose internal diameter was 0·37 mm. The syrup was in an open glass-walled box and the interface could be observed through a microscope. The oil was in a U-tube and the pressure of the air over the air–oil surface, which was of large area, could be varied and measured by a manometer.

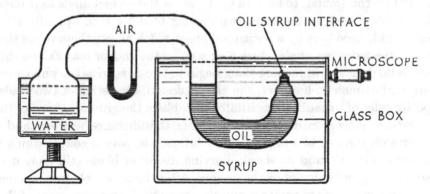

Fig. 20. Apparatus used for measuring surface tension between oil and syrup.

The air pressure was first adjusted till the syrup–oil interface appeared flat in the microscope (the oil wetted the glass) and then measured. The air pressure was then increased till it attained its maximum steady value when the interface was hemispherical. The difference between the air pressure when the interface is flat and the air pressure when it is hemispherical with radius a is $2T/a$.

With this apparatus the values given in table 2 were measured.

12

THE HOLDING POWER OF ANCHORS

REPRINTED FROM

The Yachting Monthly and Motor Boating Magazine, April 1934

The essential principle in the action of all anchors is that a surface set at an acute angle to the ground will dig in if pulled horizontally. In order that an anchor may function properly it must satisfy two conditions. The first is that in whatever position it may fall as it strikes the sea floor, it must begin to dig in as soon as the pull comes on the chain. The second is that it shall remain in the 'digging in' position while it is dragged into the ground, or, in other words, it shall be stable in the ground. Up to the present only two essentially different designs have been used, or, if the grapnel is included, three.

In the traditional form, with two flukes and shank and a stock which is longer than the flukes, the stock serves a double purpose. It ensures that the anchor shall fall on the ground in such a position that one of the digging planes or palms will function as soon as the chain begins to drag along the ground; and it also serves, by lying flat on the ground, to keep the palm set at the correct angle as it buries itself.

In stockless anchors there are two digging blades set on opposite sides of the shank, and hinged to it by a horizontal hinge which allows them to set themselves at the correct digging angle which ever way up the anchor may fall on the ground There is no difficulty in making them begin to bite as soon as the pull comes on the chain. Unfortunately, however, the simple design, consisting of two blades set on opposite sides of a shank, is essentially unstable in the ground. If one of the blades penetrates slightly more deeply than the other, the difference between the downward pressures on the two blades necessarily brings into play a couple which tends to rotate the anchor round its shank, burying the lower blade still more deeply and raising the upper blade out of the ground. After the upper blade has come out the lower blade continues to rotate round the shank till it also comes out of the ground on the opposite side to that on which it went in.

In order to counteract this tendency to rotate round the shank, the upper parts of the blades of stockless anchors are usually made to flare out, so that only the lower part of the blade can bury itself. This method of stabilizing is often not completely successful. Experimenting with one of the best known makes, I found it nearly always dragged along with one blade lower than the other, but even when it is successful the weight of the part which must then remain above the ground causes the anchor to be very inefficient for its weight.

The efficiency of an anchor may be expressed as the ratio (holding force ÷ weight of anchor). In dry sand, recent experiments with models of varying linear dimensions but the same shape have proved that this ratio is constant for any particular design, though linear dimensions varied in the ratio 5 : 1 and the weights in the ratio 125 : 1. In different types of holding ground the relative efficiency of different designs varies, but this variation is never great enough to mask the very great superiority of the traditional pattern over all existing stockless varieties. The reason why stockless anchors are frequently preferred in large sizes appears to be entirely a question of ease in handling. The stock is, of course, a very awkward thing to deal with while bringing the anchor on board, and when the anchor is on board the stock makes stowage difficult unless it is unshipped. The manner in which a stockless anchor houses itself into a specially prepared hawse hole is its chief recommendation.

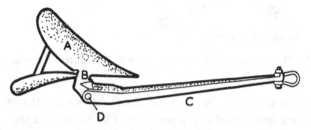

Fig. 1. The 'C.Q.R.' anchor as it takes the ground.

In considering how the efficiency of anchors might be improved so that lighter ones could be used than are needed at present, the designer naturally fixes his attention on the parts which are not essential for digging in. In the traditional pattern the stock is only needed for stability, and in action, only one fluke buries itself. One fluke and the shank seem to be essential, so that the problem which naturally suggests itself is to devise an anchor with a shank and one fluke which shall yet satisfy the requirements of taking hold of the ground as soon as the chain becomes taut and being stable when it has buried itself. The design which will now be described seems to be a solution of this problem.

The figures show the anchor in two positions: (1) The horizontal position which it assumes as it falls on the ground, and (2) the vertical position which it takes up when it is dragged through the ground. The fluke (marked A in figures 1 and 2) resembles a double-bladed ploughshare, which is welded on to a steel arm (marked B) of triangular section. The end of this arm is bent so that it comes into line with the shank C. A steel pin D passes with a driving fit through the end of the fluke member. The end of the shank is expanded into the form of jaws, through which holes are bored to take the pin D, leaving considerable clearance. The opening of the jaws is considerably larger than would be necessary to take the end of the fluke member, so that the joint between the two members is very loose in every direction.

The pin lies in the plane of symmetry of the anchor so that it is horizontal in position 1 and nearly vertical in position 2.

When the anchor falls on the ground in position 1 it will be seen from the photograph that the point of the fluke is aiming obliquely into the ground. Thus it is originally in a position suitable for digging in. As the chain begins to pull the anchor along the ground the point begins to penetrate, and since the centre line of the pin lies a short distance behind the point of the fluke, the downward earth pressure on the point begins to turn the fluke still more down into the ground.

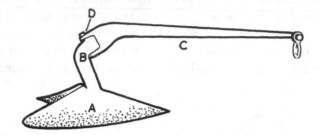

Fig. 2. The 'C.Q.R.' anchor in the position it naturally takes up.

As the penetration proceeds the parts of the fluke behind the centre-line of the pin begin to receive a downward earth pressure. The fluke then ceases to turn down and begins to set itself in line with the shaft. Finally, when the fluke has penetrated so that the whole of the double blade is buried, the anchor has turned over in to the vertical position 2 (fig. 2), the shank lying along the surface of the ground. In this position the anchor is very stable, so that if during dragging it strikes a stone or other irregularity, which pushes it out of the vertical position, it very quickly recovers itself.

The stability in the ground was strikingly verified by an experiment which I made with a 1¼lb. model about 1 ft. long. After dragging it into a sandy beach, I pulled it round in a circle. The anchor, while remaining under the sand, dragged round in a circle 5 ft. diameter. On digging round the anchor at one stage of the test, to see how it was lying, I found that it had banked itself in order to do the turn, just like an aeroplane.

The anchor which has just been described has been called the 'C.Q.R.' patent anchor (provisional application No. 8455/33). It was designed entirely by means of models weighing from 1 to 1½lb. The correct angles and position of the hinge were found by trial and error and guessing. After completing the main elements of the design in this way, a large model, weight 42¼lb., was made and sent for test to the Seaplane Experimental Station at Felixstowe. We had concluded, as a result of our model tests that in good holding ground the holding power would be something like 1200 lb., and we designed the anchor to take that load. When the tests were carried out, however, the force on the cable rose steadily as the anchor dragged into the ground to the huge figure of 25 cwt., which was the maximum force measurable by the spring balance which was used. On raising the anchor the shank was found, to be bent. A heavier shank was then made, which raised the total weight to 52½lb., and the tests repeated with a more powerful spring balance. It was then found that the holding power was 29 cwt., which is sixty-two times the weight of the anchor.

This ratio 62 is about three times as great as that obtained in the same ground with the best anchor of the traditional type, and six or more times as great as the best result which would be obtained with any of the standard stockless anchors.

The results seemed so encouraging that we made some specimens weighing 23 lb. on the same lines as the first 42½ lb. anchor, in order that comparisons might be made with the best obtainable specimens of the same weight but of the traditional and stockless types.

Before starting our tests we made enquiries to find out whether any data was available showing how the traditional compares with the stockless anchor. We were unable to find any, except some recent work published by the Air Ministry on tests with some special seaplane anchors. Our results may therefore be of interest to yachtsmen independently of the data concerning our new anchor.

In comparing two anchors it is essential to make the comparison under identical conditions, and in as many different kinds of ground as possible. It is also useful to measure, if possible, the holding power with different lengths of cable, so that the pull on the anchor is at various angles to the horizontal.

Some of the tests were carried out by dragging the anchors horizontally along the foreshore just after the tide had left it, while others were made from a boat anchored by means of a very large anchor in the River Crouch. In the latter case the anchors to be tested were dropped from a dinghy at a distance of 35 fathoms from the moored boat, and pulled towards it by a winch. The cable was manila, ¾ in. diameter. The tension in the cable was measured by means of a spring balance operating at the end of a lever, which reduced the tension in the ratio 9½ : 1.

Four anchors were treated:

(*A*) The new 'C.Q.R.' anchor, 23 lb.

(*B*) An anchor of the traditional pattern, with convex spear-shaped palms, 25 lb.

(*C*) An anchor of the traditional pattern with flat heart-shaped palms, 20 lbs.

(*D*) A 20 lb. stockless anchor.

As may be seen from the table of results the traditional type is always more efficient than the stockless, and it seems that the new type is, in all kinds of ground, more efficient than the traditional type. Indeed, the superiority of the new type is very marked, for it gives a rule about two-and-a-half times as big a drag as the best anchor of the same weight obtainable at the present time. The new anchor was designed primarily for possible use in seaplanes, where reduction in weight is so important that it overrides many other considerations which may affect the yachtsman. The cost of construction, for instance, is considerably greater than that of any other anchor of the same weight. It must be remembered however, that this anchor should be as good as one of the traditional type weighing twice as much. This saving of weight in short-handed yachts may be worth much. Whether the new type will prove easy to handle and stow is a question which I hope to be able to answer at the end of the coming season, for I propose to substitute a 60 lb. specimen for my present very excellent 120 lb. Nicholson in my 19 ton cutter Frolic. One thing is certain, however, the new anchor has no stock to foul the jib sheets when coming about, so that one continuously acting source of blasphemy will be removed.

For the benefit of people who like to try new things, it is proposed to put this anchor on the market at the beginning of the coming season in three sizes, namely 20, 35 and 60 lb. These are intended to replace ordinary anchors up to double their weight.

Table 1. *Average values of efficiency ratio,* $\dfrac{holding\ power\ or\ drag}{weight\ of\ anchor}$

	New Type Anchor A	Traditional		Stockless Anchor D
		Anchor B	Anchor C	
Soft sand just covered with water	22·7	9·9	—	0*
Foreshore, consisting of mud and shingle	(25 fms.) 35·0	13·0	15·0	1·4
At anchor in 3½ fathoms, mud bottom	(25 fms.) 31·8 (20 fms.) 28·9 (15 fms.) 22·7	(20 fms.) 5·7 (15 fms.) 6·6	(20 fms.) 5·0 (10 fms.) 5·0	(25 fms.) 2·3
At anchor in 2¼ fathoms, holding ground, consisting of a thin layer of soft mud over a bed of stiff clay	(15 fms.) 7·0 (8 fms.) 8·6 (5 fms.) 0	(20 fms.) 4·1 (15 fms.) 0	(25 fms.) 0	

The figures in brackets show the length of cable in fathoms.

* The figure 0 indicates that the anchor did not take hold. When the anchor D was pressed into the sand by hand it gave a drag equal to 5·5 times its weight.

13

MEASUREMENTS WITH A HALF-PITOT TUBE

REPRINTED FROM

Proceedings of the Royal Society, A, vol. CLXVI (1938), pp. 476–81

The use of a half-pitot tube for measuring the surface friction of a fluid flowing past a smooth surface was introduced by the late Sir Thomas Stanton.* The sketch, fig. 1, shows how the method works. In fig. 1, A is the half-pitot. This is a pitot tube set facing the fluid, one wall of which could, in Stanton's apparatus, be raised or lowered. The other wall is the smooth surface over which friction measurements are to be made. The difference in pressure between the fluid in the half-pitot tube and fluid in a tube connected to a hole (C, fig. 1) in the surface is measured by means of a manometer M. To calibrate the instrument it was set up in a pipe through which fluid was flowing at a speed below the critical velocity. Under these conditions the distribution of velocity is parabolic across the section of the tube, and the tangential stress at the fluid surface is known.

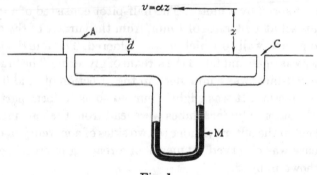

Fig. 1.

Stanton expressed his results in terms of an 'effective distance' d'. This is the distance from the wall at which the velocity, v, is such that the pressure in the half-pitot, p, is equal to $\frac{1}{2}\rho v^2$. When d, the opening of the pitot, was large it was found that d'/d was approximately $\frac{1}{4}$. On the other hand, it was found that when d is very small d'/d increases. Stanton's measurements of corresponding values of d' and d seem to indicate that d' tends to a finite limit as d tends to zero, so that d'/d would, in that case, tend to become infinite when d is very small. This conclusion cannot, however, be regarded as an experimentally established fact because the highest value of d'/d obtained in Stanton's experiments was only 2·2.

The idea that the pressure, p, in a half-pitot is likely to tend to a limiting finite value as d tends to zero was put forward on theoretical grounds by the present writer

---

* Stanton, *Proc. Roy. Soc.* A, XCVII (1920), 413.

during a discussion of Stanton's paper. If α represents the rate of shear of the fluid at the solid surface through which the half-pitot projects, the tangential stress $f = \mu\alpha$, μ being the viscosity. On general dynamical grounds it was thought that if the Reynolds number $d^2\alpha/\nu$ is very small the flow in the neighbourhood of the half-pitot must be determined entirely by the balance of the viscous forces, the inertia forces being negligible. In this case the pressure in the half-pitot depends only on μ and α and is independent of d. The limiting pressure, p_1, in a very small half-pitot tube must in fact be proportional to the tangential stress $\mu\alpha$.

Let
$$p_1 = kf = k\mu\alpha. \tag{1}$$

The constant k might be determined theoretically. In the work described below k was measured experimentally and found to be approximately equal to 1·2. It is hoped that the problem of determining k theoretically will attract the attention of some mathematician.

Apparatus

In order that a low value of $d^2\alpha/\nu$ might be attained the working fluid was glycerine. The viscosity of the glycerine used was measured and found to be $\mu = 8\cdot2$ c.g.s. at 18° C. Its density was 1·20 so that $\nu = 6\cdot83$. The annular space between two concentric cylinders 2 and 3 in. diameter was filled with glycerine. The inner cylinder was rotated at constant speed by a motor. The half-pitot consisted of a very thin rectangular brass plate set at a distance of 1 mm. from the surface of the outer cylinder by two sides and a back wall to which it was soldered. The fourth side of the box formed in this way was open and faced the stream of glycerine. The brass plate forming the top of the half-pitot was 5 mm. deep in the direction of fluid flow by 10 mm. in the transverse direction. It was slightly curved so as to form part of a cylinder concentric with the outer cylinder. Tubes were lead from the half-pitot and from a small round hole cut in the outer cylinder to two sides of a mercury manometer. The level of the mercury was observed by means of a reading microscope. The general arrangement is shown in fig. 2.

The manometer gives the difference in pressure between the fluid in the half-pitot and that in a hole in the solid surface through which the half-pitot projects. If the pressure in this hole is assumed to be the same as the static pressure in the fluid, the manometer gives p, the pressure in the half-pitot.

In making observations the rotational speed of the inner cylinder was maintained constant long enough for the reading of the manometer to become steady. The number n of revolutions per second and the reading x of the microscope were observed.

Results

The results are given in table 1 and are shown in fig. 3, where x and n are plotted in a diagram. It will be seen that in the range covered x is proportional to n. The straight line shown in fig. 3 represents the relation
$$x/n = 0\cdot72. \tag{2}$$

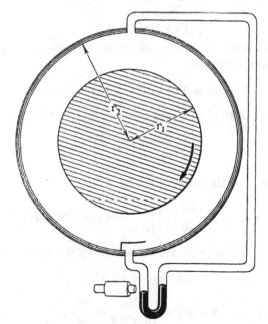

Fig. 2.

Table 1

n rev./sec.	x divisions	$d^2\alpha/\nu$	d'/d
2·35	1·6	0·035	8·3
3·25	2·4	0·048	7·1
3·5	2·4	0·051	6·9
3·7	2·7	0·055	6·7
4·15	3·0	0·061	6·3
5·15	3·7	0·076	5·7
5·15	3·7	0·076	5·7
6·30	4·5	0·093	5·1

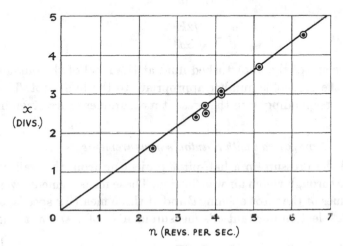

Fig. 3.

Calculation of k

Calibration of the manometer and reading microscope gave $p = 140x$ dynes so that the experimental result (2) is

$$p/n = 140 \times 0\cdot72 = 101 \text{ c.g.s.} \tag{3}$$

If r_1, r_2 are the radii of the inner and outer cylinders the velocity of the fluid at radius r is

$$v = \frac{2\pi n r_1^2}{r_2^2 - r_1^2}\left(r - \frac{r_2^2}{r}\right).$$

The tangential stress, f, at the outer cylinder is

$$f = \mu\alpha = \mu r_2\left\{\frac{d}{dr}\left(\frac{v}{r}\right)\right\}_{r=r_2} = 4\pi\mu n r_1^2 (r_2^2 - r_1^2)^{-1}.$$

In the present case $r_1 = 1\cdot0$ in., $r_2 = 1\cdot5$ in., so that $r_1^2(r_2^2 - r_1^2)^{-1} = 4/5$ and

$$f = 3\cdot2\pi n\mu = 10\cdot06 n\mu. \tag{4}$$

Hence from (3) and (4)
$$k = p/f = \frac{101 n}{10\cdot06 n\mu} = \frac{10\cdot0}{\mu}. \tag{5}$$

The measured value of μ at $18°\,\mathrm{C}$ was $8\cdot2$ so that the final result of these measurements is

$$p/f = k = 1\cdot2. \tag{6}$$

'Effective' opening of half-pitot

The effective opening of the half-pitot tube is the distance d' at which

$$\tfrac{1}{2}\rho v^2 = p = k\mu,$$

and since $v = \alpha d'$
$$d' = \sqrt{\frac{2kv}{\alpha}}, \tag{7}$$

or
$$\frac{d'}{d} = \sqrt{\frac{2kv}{\alpha d^2}}. \tag{8}$$

Since αd is the velocity of the undisturbed fluid at the level of the outside of the half-pitot $\alpha d^2/\nu$ is the Reynolds number appropriate to the half-pitot. The values of $\alpha d^2/\nu$ and d'/d corresponding with the present measurements are given in cols. 3 and 4 of table 1.

Comparison with Stanton's measurements

Stanton measured the pressure in a half-pitot projecting from the wall of a pipe $0\cdot269$ cm. diameter through which air was flowing. These measurements were made with a series of values of the pitot opening d and at three mean air speeds, 955, 570 and 370 cm./sec. A selection of Stanton's measurements* of d and d' are given in table 2.

* Stanton, *Proc. Roy. Soc.* A, xcvii (1920), 413, table vi.

Table 2. *Stanton's measurements in* 0·269 *cm. pipe.*
Mean speed 955 cm./sec., $\alpha = 2·84 \times 10^4$

d (mm.)	0·0254	0·0508	0·0762	0·1016	0·1524	0·2032	0·3556
d' (mm.)	0·0559	0·0737	0·0813	0·0940	0·1143	0·1372	0·2006
d'/d	2·20	1·45	1·06	0·925	0·760	0·59	0·564
$d^2\alpha/\nu$	1·31	5·2	10·8	20·9	47·0	83·5	256·0

To compare Stanton's results among themselves and also with the present results it is necessary to reduce them to a non-dimensional form taking account of dynamical similarity. For this reason the corresponding values of $d^2\alpha/\nu$ and d'/d have been calculated. One set of these is given in table 2. This set and also the other two sets of measurements are shown on a logarithmic scale in fig. 4. Since the flow was laminar α, the shear at the surface of the tube, was given by the expression

$$\alpha = 8 \left(\frac{\text{mean velocity}}{\text{diameter of tube}} \right).$$

It will be seen in fig. 4 that Stanton's observations are fairly consistent among themselves when plotted on a non-dimensional basis. They are also consistent with the present observations in the sense that the curve through Stanton's points can be drawn tangential to the straight line through the points representing the present results.

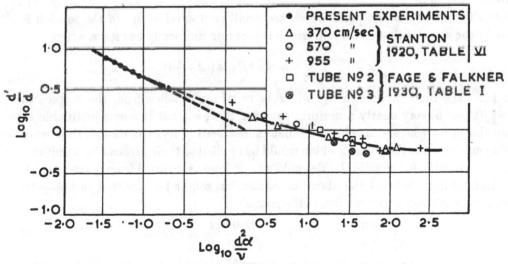

Fig. 4.

Some observations by Fage and Falkner[*] are also plotted on the same diagram. These observations were made with a flat channel instead of a round pipe. In this case the formula for finding α is

$$\alpha = 6 \left(\frac{\text{mean velocity}}{\text{distance between walls}} \right).$$

[*] Fage & Falkner, *Rep. Memo. aeronaut. Res. Comm., Lond.*, no. 1315 (1930).

14

NOTE ON R. A. BAGNOLD'S EMPIRICAL FORMULA FOR THE CRITICAL WATER MOTION CORRESPONDING WITH THE FIRST DISTURBANCE OF GRAINS ON A FLAT SURFACE

REPRINTED FROM

Proceedings of the Royal Society, A, vol. CLXXXVII (1946), pp. 16–18

In §11 of his paper Bagnold* calls attention to the fact that at the stage when oscillating water is just beginning to move sand grains no turbulence is produced even at a distance from the surface comparable with one grain diameter. The motion of water near a smooth plate which is oscillating horizontally has been calculated by Lamb†. In Bagnold's notation the horizontal velocity u of the water at height y above the plane is

$$u = \omega R\, e^{-\beta y} \cos(\omega t - \beta y), \qquad (1)$$

where $\beta = \sqrt{(\omega/2\nu)}$, ν is the kinematic viscosity, and R is the amplitude of oscillation of the plate.

At distances from the plate which are small compared with $1/\beta$ the motion is nearly the same as that of a liquid which is shearing uniformly at rate α, where

$$\alpha = \left[\frac{\partial u}{\partial y}\right]_{y=0} = \nu^{-\frac{1}{2}} \omega^{\frac{3}{2}} R \cos(\omega t + \tfrac{1}{4}\pi).$$

In fact, the motion through the whole layer which extends from the surface to $y = \sqrt{(2\nu/\omega)}$ is very nearly a uniform shearing flow, so that it seems justifiable to consider cases where grains have diameters less than $\sqrt{(2\nu/\omega)}$ as though they were being subjected to the stresses which would be applied to their surfaces by a uniform shearing motion in the fluid. In Bagnold's experiments ω varied from 0·5 to 6 sec.$^{-1}$, so that taking $\nu = 0\cdot011$ the above consideration might be expected to apply to grains whose diameters were below the range

$$\sqrt{\frac{2(0\cdot011)}{0\cdot5}} = 0\cdot2 \quad \text{to} \quad \sqrt{\frac{2(0\cdot011)}{6}} = 0\cdot06\,\text{cm}. \qquad (2)$$

Three grades of quartz sands used by Bagnold satisfied this condition, namely, those of mean diameter 0·0016, 0·016 and 0·036 cm.

The system of flow patterns produced in a uniformly shearing fluid by a grain of given shape when the diameter, d, the rate of shear α or the kinematic viscosity ν vary depends on a single variable only, namely the Reynolds number

$$z = \alpha d^2/\nu, \qquad (3)$$

---

* *Proc. Roy. Soc.* A, CLXXXVII (1946), 1.

† Lamb, *Hydrodynamics* (6th ed.), p. 620.

and the force which fluid of density ρ exerts on the grain must be of the form

$$F = d^2 \rho (\alpha d)^2 f(z),\qquad(4)$$

where f is a function which depends on the grain shape only and not on its dimensions. The condition that a grain of density σ will move must therefore be of the form

$$\rho \alpha^2 d^4 f(z) = A(\sigma - \rho) g d^3,\qquad(5)$$

where A is a constant depending on the grain shape only. (5) may be rewritten in the form

$$\frac{\gamma g d^3}{\nu^2} = \frac{1}{A} z^2 f(z),\qquad(6)$$

where

$$\gamma = \frac{\sigma - \rho}{\rho}.\qquad(7)$$

(6) may now be regarded as an equation for (z) and can be solved to give

$$\frac{\alpha d^2}{\nu} = z = \phi(\zeta),\qquad(8)$$

where

$$\zeta = \frac{\gamma g d^3}{\nu^2}.\qquad(9)$$

In this form (8) may be compared with Bagnold's empirical equation (4) §11 which gives the way in which ω depends on R, d and γ for substituting the maximum value of α, namely $\alpha = \nu^{-\frac{1}{3}} \omega^{\frac{2}{3}} R$, (8) becomes

$$\omega = R^{-\frac{3}{2}} \nu d^{-\frac{3}{2}} \chi \left(\frac{\nu g d^3}{\nu^2} \right),\qquad(10)$$

where $\chi(\zeta) = [\phi(\zeta)]^{\frac{3}{2}}$.

It will be seen that this analysis leads to the expectation that the exponent of R will be -0.66 instead of the empirical value -0.75.

The particular form of χ which most nearly coincides with Bagnold's empirical relation is $\phi(\zeta) = B\zeta^{\frac{2}{3}}$. This gives

$$\omega = (\text{constant}) R^{-\frac{2}{3}} (g\gamma)^{\frac{3}{5}} \gamma^{-\frac{1}{4}} d^{\frac{1}{3}},\qquad(11)$$

making the exponent of $g\gamma$, 0.45 instead of the empirical 0.5; and that of d, $\frac{1}{3}$ instead of 0.325.

It is of interest to note the two limiting forms which might be expected when $\alpha d^2/\nu$ is very small or very large. In the former case (Stokes' law), $f(\alpha d^2/\nu)$ is proportional to $\nu/\alpha d^2$ and (10) reduces to

$$\omega = (\text{constant}) R^{-\frac{2}{3}} (g\gamma)^{\frac{2}{3}} \nu^{-\frac{1}{3}} d^{\frac{2}{3}} \text{ (Stokes' law, small values of } \alpha d^2/\nu),\qquad(12)$$

while in the latter

$$\omega = (\text{constant}) R^{-\frac{2}{3}} (g\gamma)^{-\frac{1}{3}} \nu^{\frac{1}{3}} d^{-\frac{1}{3}} \text{ (constant drag coefficient, large values of } \alpha d^2/\nu).\qquad(13)$$

It seems from (11) that the empirical formula gives exponents for γ and d which are intermediate between those of (12) and (13) but, so far as the variation of ω with d is concerned, the empirical formula is much closer to the Stokes' law limit (equation (12)).

It is of interest to make a rough approximate calculation of the critical value of α at which grains might be expected to begin to roll. Taking the case of a cubical grain of side d, if the tangential traction on the upper face due to viscous drag is the same as that on the surface on which the grain rests, it will give rise to a moment about the down-stream lower edge of the cube of amount $\mu\alpha d^3$. The face of the cube which faces the stream might be expected to experience a pressure of approximately the same magnitude as that in a half-pitot tube placed facing the stream. This has been measured* and found to be approximately $1\cdot2\,\mu\alpha$ when the opening of the tube is so small that Stokes' law is obeyed. If the rear face of the cube has a suction of this magnitude, the moment of the normal components on the front and rear faces about the lower edge is $1\cdot2\,\mu\alpha d^3$. It seems likely that the normal component on the top face and the tangential components on the front and back faces will contribute less than the components already mentioned. The tangential drag on the side faces will make some positive contribution towards upsetting the cube, but it is difficult to make an estimate of its magnitude. These considerations lead to the conclusion that a cubical grain might be expected to roll when

$$2\cdot2\,\mu\alpha d^3 > \tfrac{1}{2}(\sigma-\rho)\,gd^4, \tag{14}$$

or expressed in terms of ω and R if

$$\omega > (4\cdot4)^{-\frac{2}{3}}(\gamma g)^{\frac{1}{3}}\nu^{-\frac{1}{3}}R^{-\frac{2}{3}}d^{\frac{2}{3}}. \tag{15}$$

If this calculated value of ω is represented by the symbol ω_{cube}, the following table gives a comparison between ω_{cube} and the observed value of ω at which grains begin to move ($\omega_{\text{obs.}}$) taken from Bagnold's figure 4.

d (cm.)	R (cm.)	$\omega_{\text{obs.}}$ (sec.$^{-1}$)	ω_{cube} (sec.$^{-1}$)
0·009	11	1·0	0·2
0·016	8	1·5	0·3
0·036	10	1·8	0·5

It will be seen that $\omega_{\text{obs.}}$ and ω_{cube} are of the same order of magnitude, but that $\omega_{\text{obs.}}$ is greater than ω_{cube}. If similar rough arguments had been applied to the case of a hexagonal grain of height d in which the moment of the gravity component about the rolling edge is volume of grain by $\tfrac{1}{2}(\sigma-\rho)\,gd/\sqrt{3}$ instead of volume by $\tfrac{1}{2}(\sigma-\rho)\,gd$, better agreement between the calculated and observed values of ω would have been obtained.

* Paper 13 above.

15

THE PATH OF A LIGHT FLUID
WHEN RELEASED IN A HEAVIER FLUID
WHICH IS ROTATING

Paper written for the Aeronautical Research Council (1950)

In a recent note* an attempt has been made to calculate how a fluid would move if released in a fluid of another density while the whole system was rotating at uniform angular velocity ω. In attempting to solve this problem a number of assumptions are made, some explicitly and some implicitly. To find out what implicit assumptions are made it is necessary to refer to the equation of motion which is deduced in order to guess the physical problem to which it applies. Fluid density $\alpha\rho$ is supposed to be emitted at radius R into a fluid of density ρ which is in uniform rotation with angular velocity ω. The source is supposed to rotate with the fluid. The equation of motion given by Messrs Alcock and Armstrong in their paper is

$$\tfrac{1}{2}\alpha\rho\left(\frac{dr}{dt}\right)^2 - \alpha\rho\,(R-r)\frac{d^2r}{dt^2} \quad \rho(R\quad r)\frac{d^2r}{dt^2} - \tfrac{1}{2}\rho\omega^2(R^2 - r^2)\,(1-\alpha) = 0. \tag{1}$$

where r is the distance from the centre of rotation of the inner end of the column of liquid $\alpha\rho$. This column is assumed to move radially inwards towards the centre of rotation.

In this equation the last term represents the difference in the pressure drop along columns of liquid of densities ρ and $\alpha\rho$, each of length $R-r$. It correctly represents therefore the hydrostatic pressure which drives the lighter liquid $\alpha\rho$ inwards. The first term represents the pressure required to accelerate the fluid $\alpha\rho$ up to a velocity dr/dt. The second term represents the pressure difference necessary to accelerate a column of length $R-r$ and density $\alpha\rho$ with radial acceleration relative to the rotating axes d^2r/dt^2. In order that this term may be a correct representation of a physical problem it is necessary that the cross-section of the stream of fluid $\alpha\rho$ shall be uniform and hence the velocity uniform at all points in the column. This is an implicit assumption and is inconsistent with another assumption, implied in the first term, that the liquid accelerates from rest relative to rotating axes. To invert a problem to which the first two terms apply one might imagine that the liquid $\alpha\rho$ flows through a radial tube which has a uniform cross-section that is small compared with the cross-section at the point where the liquid is supplied. Another implicit assumption is that the liquid $\alpha\rho$ is supplied at its source at a rate which is proportional to dr/dt.

* 'Thermo-centrifugal Convection in Combustion Chambers', by J. F. Alcock and W. D. Armstrong. Paper no. 13226 of the Aeronautical Research Council, January 1949.

Besides all these implicit assumptions it is explicitly assumed that the motion is entirely radial. This could only be assumed if it is a physical condition of the problem that the whole flow is contained within a material tube which prevents any but radial flow.

It seems to me that the limitation to motion in a radial tube of uniform cross-section must prevent any conclusions from being drawn about the motion which thermo-centrifugal convection would produce when the radial tube was not present.

The simplest ideal problem which could be posed to illustrate the dynamics of thermo-centrifugal convection would be to find the shape and velocity of a stream of a light fluid released at a constant rate from a point which is fixed with reference to axes which rotate with the main uniformly rotating heavier fluid. It is a legitimate assumption in such a case to take the stream as steady with respect to the rotating system. In this ideal problem we may assume that there is no tangential friction at the boundary between the two fluids and that the only condition that must be satisfied there is that the pressure shall be continuous. We have therefore first to find the analogue to Bernoulli's theorem applicable to a stream of fluid which is steady relative to a rotating system. If q is the fluid velocity the component of acceleration along a stream line which makes angle χ with the radius vector (see fig. 1) is

$$q\frac{dq}{ds} + \omega^2 r \cos\chi.$$

With the convention that χ is measured from the inward-pointing radius as shown in fig. 1, $\cos\chi\, ds = -dr$ so that the acceleration along a stream line is

$$q\frac{dq}{ds} - \omega^2 r \frac{dr}{ds}.$$

Bernoulli's equation in the fluid $\alpha\rho$ is therefore found by integrating along a stream line. It is

$$\frac{p}{\alpha\rho} = -\tfrac{1}{2}q^2 + \tfrac{1}{2}\omega^2 r^2 + \text{constant}. \tag{2}$$

The presure in the fluid ρ is

$$p = \tfrac{1}{2}\rho\omega^2 r^2 + \text{constant}.$$

Since pressure is assumed to be continuous at the interface between the two fluids the velocity q of the fluid $\alpha\rho$ relative to the rotating system is given by

$$q^2 = \text{constant} - \omega^2 r^2 \left(\frac{1}{\alpha} - 1\right). \tag{3}$$

If the fluid $\alpha\rho$ is emitted at rest relative to rotating axes at radius R, (3) becomes

$$q^2 = \omega^2(R^2 - r^2)\left(\frac{1}{\alpha} - 1\right). \tag{4}$$

Equation (4) can be satisfied if the fluid $\alpha\rho$ flows through a stream tube of any shape provided the cross-section of the tube is such that the equation of continuity is satisfied as well as (4), i.e. the cross-section of the tube must be proportional to $(R^2 - r^2)^{-\frac{1}{2}}$.

So far the motion is indeterminate; any path could be chosen arbitrarily, including a radial one as assumed by Messrs. Alcock and Armstrong and the equation for longitudinal acceleration along the path could be satisfied. The problem can only be made determinate when the effect of the acceleration perpendicular to the path is taken into consideration. This contains two terms. One is the well known

$$q^2/\text{(radius of curvature of the path)}.$$

The second is due to the rotation and is $2\omega q$. The equation which ensures that the transverse pressure gradient shall be the same inside the stream tube as it is outside it in the rotating fluid is

$$2\omega\rho\alpha q - \frac{q^2\alpha\rho}{\text{(radius of curvature of path)}} = \omega^2 r(1-\alpha)\rho\sin\chi \qquad (5)$$

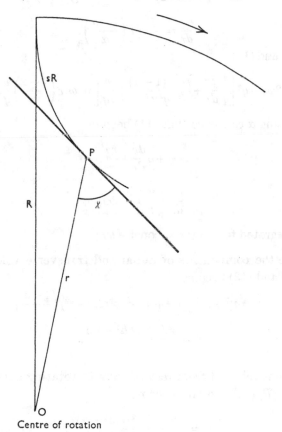

Centre of rotation

Fig. 1.

where χ is the angle shown in fig. 1. If the non-dimensional quantity s is defined as

$$s = \text{(length of path from some fixed point)}/R,$$

the radius of curvature of the path is given by

$$\frac{R}{\text{radius of curvature}} = \frac{d(\chi - \theta)}{ds} \qquad (6)$$

where r, θ are the polar co-ordinares of a point on the path (see fig. 1). With the convention for the sign of χ which is shown in fig. 1,

$$\frac{ds}{dr} = -\frac{1}{R\cos\chi}, \tag{7}$$

and if v is defined as the component of fluid velocity, relative to rotating axes, in the direction perpendicular to r,

$$\frac{v}{q} = \sin\chi = \frac{r}{R}\frac{d\theta}{ds}. \tag{8}$$

Thus from (6)

$$\frac{1}{\text{radius of curvature}} = -\frac{d}{dr}\left(\frac{v}{q}\right) - \frac{1}{r}\frac{v}{q}. \tag{9}$$

From (3)

$$q\frac{dq}{dr} = -\omega^2 r\left(\frac{1-\alpha}{\alpha}\right). \tag{10}$$

Combining (5), (9) and (10)

$$2\omega q + q^2\left[\frac{1}{q}\frac{dv}{dr} + \frac{v\omega^2 r(1-\alpha)}{q^3\alpha} + \frac{v}{rq}\right] = \omega^2 r\left(\frac{1-\alpha}{\alpha}\right)\frac{v}{q}. \tag{11}$$

The terms containing α cancel so that (11) becomes

$$2\omega q + q\frac{dv}{dr} + \frac{vq}{r} = 0$$

or

$$\frac{dv}{dr} + \frac{v}{r} = -2\omega, \tag{12}$$

which may be integrated to $\quad vr = \text{const} - \omega r^2.$ (13)

If U and V are the components of radial and transverse velocity at the point $(r = R, \theta = 0)$, (3) and (13) become

$$q^2 = u^2 + v^2 = U^2 + V^2 + \omega^2(R^2 - r^2)\left(\frac{1-\alpha}{\alpha}\right) \tag{14}$$

and

$$v = \frac{VR + \omega(R^2 - r^2)}{r}. \tag{15}$$

When the fluid $\alpha\rho$ is released from rest relative to rotating axes $U = V = 0$, and if x is written for r/R, (14) and (15) become

$$u = \frac{\omega R(1-x^2)^{\frac{1}{2}}(x^2 - \alpha)^{\frac{1}{2}}}{x\alpha^{\frac{1}{2}}} \tag{16}$$

$$v = \omega R\left(\frac{1-x^2}{x}\right). \tag{17}$$

The equation to the stream lines is then

$$\frac{v}{u} = -r\frac{d\theta}{dr} = -x\frac{d\theta}{dx}$$

so that (16) and (17) reduce to

$$d\theta = -\alpha^{\frac{1}{2}} \frac{dx}{x} \left(\frac{1-x^2}{x^2-\alpha}\right)^{\frac{1}{2}}.$$

(18)

Integrating (18) and taking the constant of integration so that $\theta = 0$ when $x = 1$

$$\theta = \tan^{-1}\left\{\frac{\alpha(1-x^2)}{x^2-\alpha}\right\}^{\frac{1}{2}} - \alpha^{\frac{1}{2}}\tan^{-1}\left\{\frac{1-x^2}{x^2-\alpha}\right\}^{\frac{1}{2}}.$$

(19)

At $r = R$, $x = 1$ and $d\theta/dx = 0$ so that the fluid begins to move inward towards the centre of rotation under the influence of centrifugal force. The stream however is rapidly turned from its radial course by the geostrophic force acting at right angles to its path till at the point $r = R\sqrt{\alpha}$, $\theta = \frac{1}{2}\pi(1-\sqrt{\alpha})$,

$$dr/d\theta = 0.$$

At this point the stream makes its nearest approach to the centre of rotation. The path then continues till it meets the circle $r = R$ at the point $(R, (1-\sqrt{\alpha})\pi)$. The maximum velocity is

$$\omega R \left(\frac{1-\alpha}{\sqrt{\alpha}}\right).$$

(20)

The stream lines for the case $\alpha = 0.5$ and $\alpha = 0.25$ are shown in fig. 2 where they are marked as $(\alpha = 0.5, \epsilon = 0)$ and $(\alpha = 0.25, \epsilon = 0)$ respectively.

To calculate the time corresponding with any given value of r/R the equation

$$-\frac{dr}{dt} = -R\frac{dx}{dt} = \frac{\omega R}{x\sqrt{\alpha}}(1-x^2)^{\frac{1}{2}}(x^2-\alpha)^{\frac{1}{2}}$$

may be integrated. The result after adding a constant of integration so that $t = 0$ when $x = 1$ is

$$t = \frac{\sqrt{\alpha}}{2\omega}\left\{\pi - \cos^{-1}\left(\frac{1+\alpha-2x^2}{1-\alpha}\right)\right\}.$$

(21)

The time taken by the fluid to reach the nearest point to the centre is therefore $\frac{1}{2}\pi\sqrt{\alpha}/\omega$, and to complete the path it is $\pi\sqrt{\alpha}/\omega$.

ANALOGY WITH PARTICLE DYNAMICS

Though the problem has so far been treated as one of flow of a perfect fluid, the same path and velocity would have been attained by a particle of unit volume, and mass $\rho\alpha$, moving freely under the influence of a central force $\rho\omega^2 r$ (i.e a particle of mass m under the influence of a force $\omega^2 rm/\alpha$). Relative to fixed axes such a particle moves in harmonic motion and describes an ellipse in a period $2\pi\sqrt{\alpha}/\omega$. The time taken by the particle to pass from the maximum to minimum distance from the centre is therefore $\frac{1}{2}\pi\sqrt{\alpha}/\omega$ which is identical with the time taken for the stream to reach the nearest point to the centre (see equation (21)). The paths shown in fig. 2 are simply the paths, relative to rotating axes, of points moving in elliptic paths relative to fixed axes.

The fact that in steady motion relative to rotating axes a stream of a perfect fluid follows the same path as a freely moving particle suggests that the path of a stream of fluid which is resisted or mixes with the surrounding fluid, could be calculated as a problem in particle dynamics. Suppose a volume W of fluid of density $\alpha\rho$ is released from relative rest at radius R in a rotating fluid of density ρ, and assume

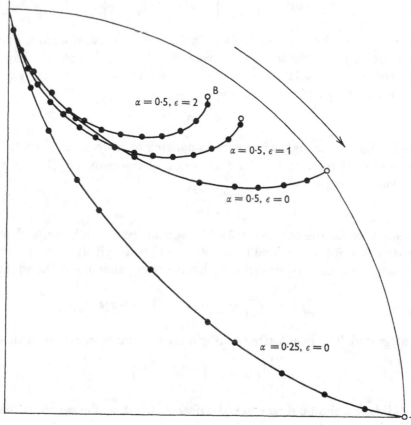

Fig. 2.

that is mixes with the fluid ρ as it flows along its path in such a way that a volume $\epsilon W/R$ enters the 'particle' while it travels over unit length. At distance Rs along the path the volume of the particle is therefore $W(1+\epsilon s)$ and its mass is $W\rho(\alpha+\epsilon s)$. To set up the equations of motion we may express the relevant quantities in the co-ordinates sR and χ (see fig. 1). The change in velocity along the path due to mixture in the path element Rds is

$$\delta_1 q = -q\frac{\epsilon ds}{\alpha+\epsilon s}, \tag{22}$$

The change in velocity due to the action of the hydrostatic centrifugal forces is

$$\delta_2 q = \frac{(1-\alpha)\,\omega^2 r \cos\chi ds}{q(\alpha+\epsilon s)}. \tag{23}$$

The equation of motion along the path is therefore

$$\frac{dq}{ds} = \frac{(1-\alpha)\,\omega^2 r \cos\chi}{q(\alpha+\epsilon s)} - \frac{\epsilon q}{\alpha+\epsilon s}. \tag{24}$$

The equation for acceleration perpendicular to the path is

$$-2\omega q(\alpha+s\epsilon) + q^2(\alpha+s\epsilon)\frac{d(\chi-\theta)}{ds} = -\omega^2 r(1-\alpha)\sin\chi. \tag{25}$$

Writing w for $q/\omega R$, $x = r/R$, (24) and (25) become

$$\frac{dw}{ds} = \frac{(1-\alpha)\,x\cos\chi}{(\alpha+\epsilon s)\,w} - \frac{\epsilon w}{\alpha+\epsilon s} \tag{26}$$

$$\frac{d\chi}{ds} = \frac{2}{w} - \frac{(1-\alpha)\,x\sin\chi}{(\alpha+\epsilon s)\,w^2} + \frac{d\theta}{ds}. \tag{27}$$

The transformation from co-ordinates (s, χ) to (r, θ) is expressed by

$$\frac{d\theta}{ds} = \frac{\sin\chi}{x} \tag{28}$$

$$\frac{dx}{ds} = -\cos\chi. \tag{29}$$

For any given values of α and ϵ the quantities w, x, χ and θ can be calculated step-by-step using small increments of s and applying (26) to (29). For this purpose the initial values $x = 1$, $\theta = \chi = w = s = 0$ are taken. It can be shown, as might be expected, that when $\epsilon = 0$ the equations (26) to (29) can be integrated to give the path represented by (19). In two cases, namely those corresponding with $\alpha = 0.5$, $\epsilon = 1$ and $\alpha = 0.5$, $\epsilon = 2$, equations (26) to (29) were integrated using increments in ds of 0.05. Since the equations are difficult to use in the early stage of the path approximate expressions were found in which change in mass due to mixture was neglected for the portion of the path between $r = R$ and $r = 0.9R$. The equations (26) to (29) were then applied. The results are given in table 1 and the corresponding paths are shown in fig. 2.

Comparing the cases $\epsilon = 0$, $\epsilon = 1$ and $\epsilon = 2$ it will be seen that mixture of the lighter with the heavier fluid does not enable the path to approach any nearer to the centre than it does when there is no mixture. The figures in table 1 show that the fluid is nearly brought to rest at the points shown in fig. 2 as the ends of the path but the equations would have to be examined in greater detail to enable us to know whether the fluid particle actually comes to rest or whether it makes a small loop. In either case the existence of a particle of fluid which is at rest or nearly so in a rotating fluid would give rise to a problem of exactly the same nature as that already solved. To find how the fluid particle would move after coming to rest at the point R (fig. 2) it would be necessary to start with radius $0.915R$ (see last entry in table 1) and take the density as $\rho(\alpha+\epsilon s)/(1+\epsilon s)$. For the case represented by B in fig. 2 this density is $\rho\{\frac{1}{2}+2(0.70)\}/\{1+2(0.70)\} = 0.79\rho$.

Comparing what happens when a particle density $\frac{1}{2}\rho$ starts from rest at radius R with and without mixing it seems that the second loop will not carry the 'particle' B (fig. 2) inside the radius $(0.915R)\sqrt{0.79} = 0.814R$.

Table 1. *Step-by-step solutions of equations* (26) *to* (29)

	$\alpha = 0.5, \epsilon = 1$				$\alpha = 0.5, \epsilon = 2$		
s	x	w	θ	s	x	w	θ
0	1·00	0	0	0	1·00	0	0
0·10	0·900	0·447	0·030	0·10	0·900	0·447	0·030
0·15	0·856	0·484	0·055	0·15	0·866	0·447	0·056
0·20	0·816	0·502	0·090	0·20	0·826	0·440	0·091
0·25	0·781	0·507	0·134	0·25	0·792	0·426	0·136
0·30	0·752	0·503	0·186	0·30	0·766	0·408	0·190
0·35	0·730	0·492	0·246	0·35	0·748	0·386	0·251
0·40	0·717	0·475	0·312	0·40	0·740	0·361	0·317
0·45	0·713	0·453	0·382	0·45	0·743	0·331	0·384
0·50	0·718	0·425	0·452	0·50	0·758	0·295	0·448
0·55	0·733	0·391	0·518	0·55	0·784	0·253	0·504
0·60	0·757	0·351	0·577	0·60	0·820	0·202	0·547
0·65	0·790	0·302	0·626	0·65	0·865	0·136	0·573
0·70	0·830	0·236	0·662	0·70	0·915	0·041	0·575
0·75	0·878	0·157	0·682				
0·80	0·928	0·040	0·682				

It seems from this analysis that thermo-centrifugal force is unlikely to bring a fluid of density $\alpha\rho$ nearer to the centre than $\sqrt{\alpha}$ times the radius at which it is released, but turbulent mixing will ultimately cause the area of mixture to spread through the whole volume. It may be pointed out that when a fluid rotates inside a fixed cylindrical boundary the turbulent mixing is very much increased by another effect of centrifugal force which has nothing to do with variations in density. Fluid which has lost its rotation by friction with the walls is driven radially inward by the radial pressure gradient due to the centrifugal force of the part of the fluid which has not lost its rotation by friction with the walls.

It might form the subject of an interesting experimental investigation to find out how rapid the mixing is when a fluid of the same density as the rotating fluid is introduced and to compare it with the case of a lighter fluid.

16

ANALYSIS OF THE SWIMMING OF MICROSCOPIC ORGANISMS

REPRINTED FROM

Proceedings of the Royal Society, A, vol. CCIX (1951), pp. 447–61

Large objects which propel themselves in air or water make use of inertia in the surrounding fluid. The propulsive organ pushes the fluid backwards, while the resistance of the body gives the fluid a forward momentum. The forward and backward momenta exactly balance, but the propulsive organ and the resistance can be thought about as acting separately. This conception cannot be transferred to problems of propulsion in microscopic bodies for which the stresses due to viscosity may be many thousands of times as great as those due to inertia. No case of self-propulsion in a viscous fluid due to purely viscous forces seems to have been discussed.

The motion of a fluid near a sheet down which waves of lateral displacement are propagated is described. It is found that the sheet moves forwards at a rate $2\pi^2b^2/\lambda^2$ times the velocity of propagation of the waves. Here b is the amplitude and λ the wave-length. This analysis seems to explain how a propulsive tail can move a body through a viscous fluid without relying on reaction due to inertia. The energy dissipation and stress in the tail are also calculated.

The work is extended to explore the reaction between the tails of two neighbouring small organisms with propulsive tails. It is found that if the waves down neighbouring tails are in phase very much less energy is dissipated in the fluid between them than when the waves are in opposite phase. It is also found that when the phase of the wave in one tail lags behind that in the other there is a strong reaction, due to the viscous stress in the fluid between them, which tends to force the two wave trains into phase. It is in fact observed that the tails of spermatozoa wave in unison when they are close to one another and pointing the same way.

INTRODUCTION

The manner in which a fish swims by causing a wave of lateral displacement to travel down its body from head to tail seems to be understood through the work of James Gray and his colleagues. This movement gives rise to circulations round the body which, in a fluid of small viscosity like water, are necessary to produce a forward force by dynamical reaction. In other words, the creature owes it ability to propel itself entirely to the inertial forces set up in the surrounding fluid by its muscular movements. Viscosity is important only in so far as it plays a part in the mechanics of the boundary layer, which in turn plays a part in determining the magnitude of the circulations with which the inertia reaction of the water is associated.

The propelling organs of some very small living bodies (spermatozoa for instance) bear a superficial resemblance to those of fish, in that propulsion is achieved by sending waves of lateral displacement down a thin tail or flagellum. The direction of movement of the organism is, like that of the fish, opposite to that of propagation of the waves of lateral displacement. The dynamics of a body as small as a spermatozoon—say 5×10^{-3} cm. long with a tail 10^{-5} cm. diameter—swimming in water must, however, be completely different from that of a fish. If L is some characteristic length defining the size of a body moving in water with velocity V, in a fluid of density ρ, and viscosity μ, the Reynolds number, $R = LV\rho/\mu$, expresses in numerical form the order of magnitude of the ratio

$$\frac{\text{stress in fluid due to inertia}}{\text{stress due to viscosity}}.$$

In most fishes R is of order many thousands, in a tadpole it is perhaps of order 10^2, and in bodies of the size of spermatozoa it is of order 10^{-3} or less. It will be seen, therefore, that the forces due to viscosity, which may legitimately be neglected in comparison with inertia forces, in studying the motions of fish, may be thousands of times as great as the inertia forces in the case of the smallest swimming bodies.

Reynolds number is usually defined in relation to a body moving steadily through a fluid with velocity V. In cases where the body is vibrating the inertia stresses arise from the reaction between the vibrating surface and the surrounding fluid. The number which corresponds with Reynolds number in describing the order of magnitude of the ratio

$$\frac{\text{stress due to inertia}}{\text{stress due to viscosity}} \quad \text{is} \quad nL^2\rho/\mu,$$

where n is the frequency of vibration. In the case of a spermatozoon n is of order 100 c./sec., and μ for water is 10^{-2}. The length which is of importance in considering the stress in the fluid is the diameter of the tail rather than its length, so that L is of order 10^{-5} cm. and $nL^2\rho/\mu$ is of order 10^{-6}. In considering the motions of spermatozoa therefore it is necessary only to take account of viscous forces. Inertia forces may legitimately be neglected.

These considerations naturally give rise to the following question. How can a body propel itself when the inertia forces, which are the essential element in self-propulsion of all large living or mechanical bodies, are small compared with the forces due to viscosity?

An attempt will be made to answer this question by showing that self-propulsion is possible in a viscous fluid when bodies immersed in it execute movements which bear a strong resemblance to those which spermatozoa are known to make.

SELF-PROPULSION IN A VISCOUS FLUID

The only problems concerning the motion of solids in viscous fluids which have so far been solved relate to bodies which are moved by the application of an external force like gravity. The motion of spheres and ellipsoids in an infinitely extended fluid under external forces or couples has been analysed. It has been found that such bodies tend to move along with them a very large volume of the surrounding fluid. Long cylindrical bodies move so much fluid that the whole volume, extending to infinity, moves with the body. The fact that a cylinder in steady motion gives rise to finite fluid velocity at an infinite distance was discussed by Stokes* who also obtained the solution to the problem presented by an oscillating cylinder in which inertia stresses are comparable with those due to viscosity. He pointed out that as the frequency of oscillation decreases the volume of fluid which moves with the cylinder increases till, as the frequency approaches zero, the disturbance tends to extend to infinity.

When large bodies like ships or aeroplanes are propelled by some internal mechanism through a fluid, the mechanics of their motion is always analysed by considering separately (*a*) a propelling mechanism like a paddle wheel or airscrew which develops a forward force by pushing fluid backwards, and (*b*) resistance which arises because the body entrains some of the surrounding fluid and thus gives it a forward momentum. When the self-propelled body is moving at a steady pace it is clear that the backward momentum considered under (*a*) exactly balances the forward momentum of (*b*).

When a body propels itself in a viscous fluid it is still true that the total rate of production of momentum is zero. In other words the resultant force which the fluid exerts on the body must be zero. On the other hand, it is clear that when inertia stresses are negligible compared with those due to viscosity it is no longer possible to use the conception of propulsion as being due to the separable effects of a propulsive unit and fluid resistance. The truth of this statement is at once obvious if the body considered is two-dimensional, in the form of an infinitely long cylinder, for the effect on the fluid of moving the cylinder—considered independently of the propulsive unit—would, as Stokes showed, be to move the whole fluid in which it was immersed. There is no reason to suppose that a self-propelling body would move a great volume of the fluid surrounding it, in fact in the particular problem the solution of which fills most of this paper, the influence of a self-propelling body extends only a very short distance from it.

Provided that no attempt is made to separate propulsion from resistance, but the motions of the whole fluid and the body are considered as inseparable, Stokes' difficulty disappears. Though microscopic swimming creatures are certainly three-dimensional, yet the great simplicity of two-dimensional, compared with three-dimensional analysis, makes it worth while to discuss the problem of self-propulsion in a viscous fluid in two dimensions.

The propelling organ of a spermatozoon is a thin tail down which the organism sends waves of lateral displacement. Whether this tail moves in two or three dimensions is not clear.† The analogous two-dimensional problem is that of a sheet down which waves of lateral displacement are propagated. This problem will be investigated with a view to finding out whether such waves can give rise to viscous stresses which drive the sheet forwards.

* G. G. Stokes, *Trans. Camb. phil. Soc.* IX (1851), Pt. 2, 8.
† Lord Rothschild, *Biol. Rev.* XXVI (1951), 1.

Waves of small amplitude in sheet immersed in a viscous fluid

Take axes which are fixed relative to the mean position of the particles of the sheet. The waving surface will be taken represented by

$$y_0 = b \sin (kx - \sigma t). \tag{1}$$

The velocity of the wave is σ/k and it moves in the direction x positive. The wavelength is $2\pi/k = \lambda$. t represents time. b, the amplitude, will be assumed small compared with λ. If the sheet is inextensible and the amplitude of the wave small, material particles will oscillate in a path which is nearly parallel to the axis of y, though, as will be seen later, their actual paths are narrow figures of 8. The components of velocity of a particle of the sheet are u_0, v_0 where

$$u_0 = 0, \quad v_0 = \frac{\partial y_0}{\partial t} = -b\sigma \cos (kx - \sigma t). \tag{2}$$

The problem is therefore to find a motion in a viscous fluid which satisfies (2) as a boundary condition on $y_0 = b \sin (kx - \sigma t)$. The field equation which viscous flow in two dimensions satisfies when inertia is neglected is

$$\nabla^4 \psi = 0, \tag{3}$$

where ψ is a stream function and the components of velocity are

$$u = -\frac{\partial \psi}{\partial y}, \quad v = \frac{\partial \psi}{\partial x}. \tag{4}$$

As a first approximation when bk is small assume for ψ

$$\psi = (A_1 y + B_1) e^{-ky} \sin (kx - \sigma t) - Vy. \tag{5}$$

This function satisfies (3). The velocity of the fluid at infinity is V, so that if V has a finite value the particles of the waving surface will move relative to the main body of viscous fluid with velocity $-V$. The conditions to be satisfied as the surface $y_0 = b \sin (kx - \sigma t)$ are

$$\frac{\partial \psi}{\partial x} = v_0 = -b\sigma \cos (kx - \sigma t), \quad -\frac{\partial \psi}{\partial y} = u_0 = 0. \tag{6}$$

To the first order (when bk is small) the values of u and v at $y = 0$ will be the same as those at $y = b \sin (kx - \sigma t)$, so that the boundary conditions (6) are satisfied if

$$-V + (A_1 - B_1 k) \sin (kx - \sigma t) = -u_0 = 0 \tag{7}$$

and

$$B_1 k \cos (kx - \sigma t) = v_0 = -b\sigma \cos (kx - \sigma t). \tag{8}$$

(7) and (8) are satisfied if

$$V = 0, \quad A_1 = B_1 k = -b\sigma. \tag{9}$$

Inserting values of A_1 and B_1 in (5) it is found that

$$\psi = -\frac{b\sigma}{k}(1+ky)\,e^{-ky}\sin(kx-\sigma t),\qquad(10)$$

and ψ represents the flow near a sheet down which waves of small amplitude are travelling. It will be noticed that since $V=0$ the waves in the sheet do not propel it through the fluid. This conclusion, however, will be modified when the equations are treated, using a higher order of accuracy than that which led to (7) and (8).

The dissipation of energy can be found by calculating the work done per unit area of the sheet against viscous stress. Its mean value is

$$W = -\overline{\frac{dy_0}{dt}Y_y},\qquad(11)$$

where Y_y is the stress normal to the sheet and*

$$Y_y = -p+2\mu\frac{\partial v}{\partial y}.\qquad(12)$$

Here $-p$ is the mean value of the principal stress components. p is described as pressure. The pressure associated with the stream function (10) is

$$p = 2\sigma b k\mu\, e^{-ky}\cos(kx-\sigma t),\qquad(13)$$

and since at the surface $u=0$, $\partial u/\partial x=0$ so that $\partial v/\partial y=0$. Hence

$$W = 2b^2\sigma^2 k\mu\,\overline{\cos^2(kx-\sigma t)} = b^2\sigma^2 k\mu.\qquad(14)$$

Since the motion vanishes at infinity, it is clear from what has been said that the total force on the sheet must be zero. In fact the forward component of the force which the pressure exerts on the plate is

$$F_1 = \overline{p\frac{dy_0}{dx}} = -\mu\sigma b^2 k^2.\qquad(15)$$

This is negative so that the pressure tends to drive the sheet in the direction $-x$. The force, due to the tangential component of stress at any point, is $\mu(\partial u/\partial y)$. To the first order of small quantities the mean value F_2 of the tangential stress on the sheet is zero, but it must be remembered that the tangential stress actually acts over the wave surface $y_0 = b\sin(kx-\sigma t)$. Thus taking the variation in tangential stress due to this fact into account, the mean stress is the mean value of $\mu(\partial u/\partial y)$ over this surface is

$$F_2 = \mu\sigma bk\overline{(ky-1)\,e^{-ky}\sin(kx-\sigma t)};\qquad(16)$$

putting $y = b\sin(kx-\sigma t)$ in (16), and remembering that $\overline{\sin^2(kx-\sigma t)}=\tfrac{1}{2}$,

$$F_2 = \mu\sigma b^2 k^2.\qquad(17)$$

The total mean force per unit area exerted by the fluid on the surface is F_1+F_2. From (15) and (17) it is seen that $F_1+F_2=0$, a result which was anticipated on general principles.

* H. Lamb, *Hydrodynamics* (6th ed.), p. 574.

Propulsive effect of waves which are not small

It has been shown that waves of small amplitude travelling down a sheet do not give rise to propulsive stresses in the surrounding viscous fluid. It is now proposed to discuss the effect of waves whose amplitude is not so small that terms containing b^2k^2 can be neglected. It does not seem to be possible to discuss by analytical methods waves whose amplitude is unlimited, but it is possible to consider the effect of waves of finite amplitude by expanding the various terms in the mathematical expressions representing the disturbance produced by the waving sheet in powers of bk. This expansion will be carried to include terms containing powers of bk as high as $(bk)^4$. To simplify the analysis the equations will be written in non-dimensional form by taking $k = 1$. If z is written for $x - \sigma t$ the appropriate form to assume for ψ is

$$\frac{\psi}{\sigma} = \sum_{n \text{ odd}}^{\infty} (A_n y + B_n)\, e^{-ny} \sin nz + \sum_{n \text{ even}}^{\infty} (C_n y + D_n)\, e^{-ny} \cos nz - \frac{Vy}{\sigma}. \qquad (18)$$

This satisfies $\nabla^4 \psi = 0$, and the disturbance rapidly decreases with distance from the sheet.

The term Vy/σ is again inserted to allow for the possibility that the waving sheet may move relatively to the fluid far distant from it with velocity $-V$.

Boundary conditions

It will be assumed that the form of the sheet is

$$y_0 = b \sin z, \qquad (19)$$

even when b is not small. The boundary condition which must be satisfied by the fluid in contact with the sheet is that there is no slip at its surface. The fact that the sheet is in the form $y_0 = b \sin z$ controls the component of velocity normal to its surface, but some further physical assumption must be made about the sheet before the component parallel to its surface can be expressed in mathematical form. This assumption will be that the sheet is *inextensible*.

Velocity of particles in an inextensible sheet disturbed by transverse waves

The velocity of the waves is σ. Their external shape can be reduced to rest by imparting to the whole fluid a velocity $-\sigma$. The velocity of particles of an inextensible sheet moving along the fixed curve $y = b \sin z$ is

$$Q = \sigma \times \left(\frac{\text{length of a curve in one wave-length}}{\text{one wave-length}} \right). \qquad (20)$$

The ratio is
$$\frac{1}{2\pi} \int_0^{2\pi} (1 + b^2 \cos^2 z)^{\frac{1}{2}}\, dz. \qquad (21)$$

Expanding (21) in powers of b up to b^4

$$\frac{Q}{\sigma} = 1 + \tfrac{1}{4}b^2 - \tfrac{3}{64}b^4. \tag{22}$$

The velocity components of particles in the sheet relative to axes which travel with the waves are
$$u_1 = -Q \cos \theta, \quad v_1 = -Q \sin \theta, \tag{23}$$

where
$$\tan \theta = \frac{dy_0}{dz} = b \cos z. \tag{24}$$

After some reduction it is found from (22), (23) and (24), retaining all terms up to those containing b^4, that

$$\frac{u_1}{\sigma} + 1 = -\tfrac{1}{32}b^4 + (\tfrac{1}{4}b^2 - \tfrac{1}{8}b^4) \cos 2z - \tfrac{3}{64}b^4 \cos 4z, \tag{25}$$

$$\frac{v_1}{\sigma} = -(b - \tfrac{1}{8}b^3) \cos z - (\tfrac{1}{8}b^3) \cos 3z. \tag{26}$$

Since $u_1 + \sigma, v_1$ are the components of velocity of the particles of the sheet relative to the original axes the boundary conditions for ψ are

$$\frac{1}{\sigma} \left[\frac{\partial \psi}{\partial y} \right]_{y = b \sin z} = -\tfrac{1}{32}b^4 + (\tfrac{1}{4}b^2 - \tfrac{1}{8}b^4) \cos 2z - \tfrac{3}{64}b^4 \cos 4z, \tag{27}$$

$$\frac{1}{\sigma} \left[\frac{\partial \psi}{\partial z} \right]_{y = b \sin z} = -(b - \tfrac{1}{8}b^3) \cos z - \tfrac{1}{8}b^3 \cos 3z. \tag{28}$$

It will be noticed that if only terms containing b are retained the particles oscillate in the lines parallel to the axis y. If terms containing b^2 and b are retained particles of the sheet traverse paths in the form of figures of 8.

It remains to find the values of $\partial \psi / \partial z$ and $\partial \psi / \partial y$ on the boundary. For this purpose it is convenient to expand ψ near $y = 0$ in powers of y. Thus

$$(A_n y + B_n) e^{-ny} = B_n + (A_n - nB_n) y + \left(-nA_n + \frac{n^2}{2!} \right) y^2 + \left(\frac{n^2}{2!} A_n - \frac{n^3}{3!} B_n \right) y^3 + \dots$$

and

$$\frac{d}{dy}(A_n y + B_n) e^{-ny} = A_n - nB_n + 2 \left(-nA_n + \frac{n^2}{2!} B_n \right) y + 3 \left(\frac{n^2}{2!} A_n - \frac{n^3}{3!} B_n \right) y^2 + \dots.$$

At $y = y_0 = b \sin z$

$$\begin{aligned}
\frac{1}{\sigma} \frac{\partial \psi}{\partial y} &= \{A_1 - B_1 + y_0(-2A_1 + B_1) + y_0^2(\tfrac{3}{2}A_1 - \tfrac{1}{2}B_1) + y_0^3(-\tfrac{2}{3}A_1 + \tfrac{1}{6}B_1)\} \sin z \\
&\quad + \{C_2 - 2D_2 + y_0(-4C_2 + 4D_2) + y_0^2(6C_2 - 4D_2)\} \cos 2z \\
&\quad + \{A_3 - 3B_3 + y_0(-6A_3 + 9B_3)\} \sin 3z \\
&\quad + \{C_4 - 4D_4\} \cos 4z - V/\sigma,
\end{aligned} \tag{29}$$

$$\frac{1}{\sigma}\frac{\partial \psi}{\partial z} = [B_1 + y_0(A_1 - B_1) + y_0^2(-A_1 + \tfrac{1}{2}B_1) + y_0^3(\tfrac{1}{4}A_1 - \tfrac{1}{6}B_1)]\cos z$$
$$+ [D_2 + y_0(C_2 - 2D_2) + y_0^2(-2C_2 + 2D_2)](-2\sin 2z)$$
$$+ [B_3 + y_0(A_3 - 3B_3)](3\cos 3z)$$
$$+ D_4(-4\sin 4z). \tag{30}$$

In order that the boundary conditions may be satisfied for all values of z it is necessary to express terms like $y_0^n \begin{pmatrix}\cos\\\sin\end{pmatrix} mz$ in (29) and (30) in the form $\Sigma A_l \begin{pmatrix}\cos\\\sin\end{pmatrix} lz$, l being an integer. The coefficients of $\begin{pmatrix}\cos\\\sin\end{pmatrix} lz$ in the expressions for the boundary conditions (27) and (28) may then be equated. The expressions necessary for developing the expansions up to terms containing b^4 are given in table 1. Using this table, the coefficients given in the table 2 may be equated to zero.

Table 1. *Relations necessary for expanding boundary conditions in powers of b up to b^4*

$y_0 = b\sin z$

$y_0 \sin z = \tfrac{1}{2}b(1 - \cos 2z)$ $\quad y_0^2 \sin z = \tfrac{1}{4}b^2(3\sin z - \sin 3z)$ $\quad y_0^3 \sin z = \tfrac{1}{8}b^3$
$\times (3 - 4\cos 2z + \cos 4z)$

$y_0 \cos z = \tfrac{1}{2}b\sin 2z$ $\quad y_0^2 \cos z = \tfrac{1}{4}b^2(\cos z - \cos 3z)$ $\quad y_0^3 \cos z = \tfrac{1}{8}b^3$
$\times (2\sin 2z - \sin 4z)$

$y_0 \sin 2z = \tfrac{1}{2}b(\cos z - \cos 3z)$ $\quad y_0^2 \sin 2z = \tfrac{1}{4}b^2(2\sin 2z - \sin 4z)$

$y_0 \cos 2z = \tfrac{1}{2}b(\sin 3z - \sin z)$ $\quad y_0^2 \cos 2z = \tfrac{1}{4}b^2(-1 + 2\cos 2z - \cos 4z)$

$y_0 \sin 3z = \tfrac{1}{2}b(\cos 2z - \cos 4z)$

$y_0 \cos 3z = \tfrac{1}{2}b(\sin 4z - \sin 2z)$

Table 2. *Coefficients to be equated to zero in the development of (27) and (28)*

1	$(-2A_1 + B_1)\tfrac{1}{2}b + (-\tfrac{3}{4}A_1 + \tfrac{1}{6}B_1)\tfrac{3}{8}b^3 - \tfrac{1}{4}b^2(6C_2 - 4D_2) - \tfrac{1}{32}b^4 - V/\sigma$	(a)
$\sin z$	$(A_1 - B_1) + (\tfrac{3}{4}A_1 - \tfrac{1}{2}B_1)\tfrac{3}{4}b^2 - \tfrac{1}{2}b(-4C_2 + 4D_2)$	(b)
$\cos 2z$	$-(-2A_1 + B_1)\tfrac{1}{2}b - \tfrac{1}{2}b^3(-\tfrac{3}{4}A_1 + \tfrac{1}{6}B_1) + C_2 - 2D_2 + \tfrac{1}{2}b^2(6C_2 - 4D_2) + \tfrac{1}{2}b(-6A_3 + 9B_3) + \tfrac{1}{4}b_2 - \tfrac{1}{8}b^4$	(c)
$\sin 3z$	$-\tfrac{1}{4}b^2(\tfrac{3}{4}A_1 - \tfrac{1}{2}B_1) + (-4C_2 + 4D_2)\tfrac{1}{2}b + A_3 - 3B_3$	(d)
$\cos 4z$	$+\tfrac{1}{8}b^3(-\tfrac{3}{4}A_1 + \tfrac{1}{6}B_1) - \tfrac{1}{2}b^2(6C_2 - 4D_2) - \tfrac{1}{2}b(-6A_3 + 9B_3) + C_4 - 4D_4 - \tfrac{3}{64}b^4$	(e)
$\cos z$	$B_1 + \tfrac{1}{4}b^2(-A_1 + \tfrac{1}{2}B_1) - 2(C_2 - 2D_2)\tfrac{1}{2}b + b - \tfrac{1}{8}b^3$	(f)
$\sin 2z$	$\tfrac{1}{2}b(A_1 - B_1) + (\tfrac{1}{2}A_1 - \tfrac{1}{8}B_1)\tfrac{1}{4}b^3 - 2D_2 - b^2(-2C_2 + 2D_2) - \tfrac{3}{2}b(A_3 - 3B_3)$	(g)
$\cos 3z$	$-\tfrac{1}{4}b^2(-A_1 + \tfrac{1}{2}B_1) + b(C_2 - 2D_2) + 3B_3 + \tfrac{1}{8}b^3$	(h)
$\sin 4z$	$-\tfrac{1}{8}b^3(\tfrac{1}{2}A_1 - \tfrac{1}{8}B_1) + b^2(-C_2 + D_2) + \tfrac{3}{2}b(A_3 3 - B_3) - 4D_4$	(i)

It will be noticed that C_4 and D_4 occur only in (e) and (i). The constants A_1, B_1, C_2, D_2, A_3 and B_3 may be obtained by equating (b), (c), (d), (f), (g), (h) to zero. The equation (1) can then only be satisfied when V has a particular value. It will be noticed that in order that (b) and (f) may be satisfied is it necessary that A_1 and B_1 shall be of the form

$$A_1 = -b + \text{higher powers of } b$$

and

$$B_1 = -b + \text{higher powers of } b.$$

In fact, in order that the six equations may be satisfied for all values of b it is necessary that A_1, B_1, C_2, D_2, A_3 and B_3 shall be of the form

$$A_1 = -b(1+\alpha b^2), \quad B_1 = -b(1+\beta b^2), \quad C_2 = \gamma_1 b^2 + \gamma_2 b^4,$$

$$D_2 = \delta_1 b^2 + \delta_2 b^4, \quad A_3 = \epsilon b^3, B_3 = \eta b^3.$$

It remains to determine α, β, γ_1, γ_2, δ_1, δ_2, ϵ, η from the equations (b), (c), (d), (f), (g) and (h). It can be verified that the appropriate values are

$$\alpha = -\tfrac{1}{2}, \quad \beta = -\tfrac{1}{4}, \quad \gamma_1 = \tfrac{1}{4}, \quad \gamma_2 = -\tfrac{1}{8}, \quad \delta_1 = 0, \quad \delta_2 = \tfrac{1}{12}, \quad \epsilon = 0, \quad \eta = -\tfrac{1}{12}. \tag{31}$$

Inserting these in (e) and (i) it is found that

$$C_4 = \tfrac{29}{192} b^4, \quad D_4 = \tfrac{1}{24} b^4. \tag{32}$$

PROPULSIVE EFFECT OF PROPAGATING
TRANSVERSE WAVES IN A SHEET

(a) may be written in the form

$$\frac{V}{\sigma} = \tfrac{1}{2}b^2 + (\alpha - \tfrac{1}{2}\beta + \tfrac{3}{16} - \tfrac{3}{2}\gamma_1 - \tfrac{1}{32}) b^4. \tag{33}$$

Hence from (31)
$$\frac{V}{\sigma} = \tfrac{1}{2}b^2(1 - \tfrac{19}{16}b^2)$$

In the non-dimensional units the velocity of the wave relative to the particles of the sheet is σ. When dimensional units are used (33) is written

$$\frac{Vk}{\sigma} = \tfrac{1}{2}b^2k^2(1 - \tfrac{19}{16}b^2k^2),$$

or if the velocity of the waves of lateral displacement relative to the material of the sheet is V

$$\frac{V}{U} = \frac{2\pi^2 b^2}{\lambda^2}\left(1 - \frac{19}{4}\frac{\pi^2 b^2}{\lambda^2}\right). \tag{34}$$

V is the velocity of the fluid at infinity relative to the material of the sheet. Since V is positive the sheet moves with velocity $-V$ relative to the fluid at infinity when waves of lateral displacement travel with velocity $+U$ down the sheet.

VISCOUS FLUID ON BOTH SIDES OF THE SHEET

In the foregoing discussion the reaction of the viscous fluid on one side only of the waving sheet has been coinsidered. In applying the results to the swimming of microscopic organisms it is necessary to suppose that the sheet is in contact with the fluid on both surfaces. In that case for a wave of given amplitude the sheet will move relative to the fluid at infinity at the same speed V that has been calculated when fluid on one side only was contemplated. On the other hand, the rate of dissipation of energy is $2W$ instead of W where W has the same meaning as in (14),

It has been proved therefore that when small but not infinitesimal waves travel down a sheet immersed in a viscous fluid they propel the sheet at a rate which is $2\pi^2 b^2/\lambda^2$ times the wave velocity and in the opposite direction to that of propagation of the waves. It would have been less laborious to calculate only the first term of the expression for V/U. The second term containing the factor b^4/λ^4 was calculated in order to form some idea of how large the amplitude might be before a serious error might be expected in the analysis. The outside limit at which the formula might be expected to give reasonably accurate results would be when the second term was, say, one-quarter, as big as the first. That is when

$$\frac{b}{\lambda} = \sqrt{\frac{1}{19\pi^2}} = 0\cdot073.$$

In that case $\qquad\qquad V/U = \tfrac{3}{4}(\tfrac{2}{19}) = 0\cdot079.$

The shape of the tail in this case is shown in fig. 1. A tail of the shape shown in fig. 1 would have to oscillate $1/0\cdot079 = 12\cdot7$ times in order to progress 1 wave-length. It will be noticed that the wave shown in fig. 1 is not very large. It may

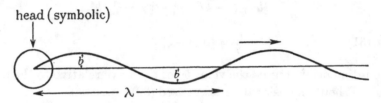

Fig. 1. Symbolic representation of microscopic organism swimming. Shape of waving tail when $kh = 0\cdot25$, $b/\lambda = 0\cdot073$; →, direction of propagation of waves in tail.

well be that waves of larger amplitude would propel the sheet more than $1/12\cdot7$ of a wave-length per oscillation, but the method of analysis here adopted could hardly be used in discussing such a case without great labour. The Southwell's relaxation technique might perhaps be employed.

STRESS IN THE TAIL

The internal mechanism necessary to produce lateral motion can only be due to tensions and compressions acting across each normal section of the tail so as to produce a couple M. This couple varies along the tail. Its magnitude can be calculated when the distribution of pressure along the tail is known. In the case of a waving sheet which has fluid on both sides the equilibrium equation is

$$\frac{dM}{dx} = F, \quad \frac{dF}{dx} = -P, \tag{35}$$

where P is the difference of pressure on the two sides of the sheet. The pressure variations are equal in magnitude but of opposite signs on the two sides of the sheet. Equation (13) therefore gives

$$P = 4\sigma bk\mu \cos(kx - \sigma t)$$

so that (35) becomes

$$\frac{d^2M}{dx^2} = -4\sigma bk\mu \cos(kx - \sigma t).$$

Hence

$$M = \frac{4\sigma b\mu}{k} \cos(kx - \sigma t).$$

The maximum value of M is $4\sigma b\mu/k$ or $4nb\mu\lambda$, where n is the frequency of vibration of the tail. The magnitude of the maximum stress can only be calculated if the thickness d of the tail is known. The minimum possible value of the maximum stress is then $4M/d^2$ or $16\mu bn\lambda/d^2$. Taking the case when $\lambda = 10^{-3}$ cm., $\mu = 10^{-2}$, $b = \frac{1}{4}\lambda$, $d = 10^{-5}$ cm., $n = 50$ c./sec., this stress is 2×10^4 or 20 g. weight/sq.cm.

MECHANICAL REACTION BETWEEN NEIGHBOURING WAVING TAILS

It has been observed that when two or more spermatozoa are close to one another there is a strong tendency for their tails to vibrate in unison. James Gray* writes: 'Numerous authors have observed that when the heads of individual spermatozoa are in intimate contact their tails beat synchronously and a very striking example of the phenomenon can be observed in *Spirochaeta balbianii*.' Fig. 2, which is reproduced from fig. 78, p. 119 of James Gray's book *Ciliary Movement*, shows his

Fig. 2. *Spirochaeta balbianii* forming aggregates, the individuals in which soon establish synchronous movements. (Reproduced from Gray's *Ciliary Movement*.)

idea of the way in which aggregates of these organisms which vibrate in unison are formed. Rothschild† attributes certain comparatively large-scale motions in dense suspensions of bull or ram spermatozoa to 'periodic aggregation of spermatozoa the tails of which probably beat synchronously in the aggregations'.

* J. Gray, *Ciliary Movement* (1928).
† Lord Rothschild, *Nature*, Lond. CLXIII (1949), 358.

Among the various possible explantions of this phenomenon it might be supposed that the stresses set up in the viscous fluid between neighbouring tails may have a component which would tend to force their waves into phase. It is of interest therefore to analyse the field of flow between two waving sheets when their waves are not in phase in order to find out whether the viscous stresses are of such a nature as to tend to force them into phase.

Taking axes of co-ordinates midway between the two sheets which are at $y = \pm h$ it will be assumed that waves of the same amplitude, b, travel down each sheet with the same velocity σ/k. It will be assumed also that the phase of the sheet at $y = +h$ lags behind that of the sheet at $y = -h$ by an angle 2ϕ. All cases will be covered if ϕ is taken to lie in the range $0 < \phi < \frac{1}{2}\pi$.

The equations to the two sheets are then

$$y = h + y_1 = h + b \sin(z + \phi),$$

and
$$y = -h + y_2 = -h + b \sin(z - \phi),$$

where
$$z = kx - \sigma t.$$

(36)

Fig. 3c shows the sheets when $\phi = 45°$ so that y_1 lags $90°$ behind y_2.

The stream function is assumed to be

$$\psi = (A_1 y \sinh ky + B_1 \cosh ky) \cos \phi \sin z + (A_2 y \cosh ky + B_2 \sinh ky) \sin \phi \cos z. \quad (37)$$

The condition $\partial \psi / \partial y = 0$ at $y = \pm h$ is satisfied provided

$$\frac{B_1 k}{A_1} = -(kh \coth kh + 1), \quad \frac{B_2 k}{A_2} = -(kh \tanh kh + 1); \quad (38)$$

the second condition to be satisfied at $y = +h$ is

$$\frac{\partial \psi}{\partial x} = \frac{\partial y_1}{\partial z} \frac{\partial z}{\partial t} = -\sigma \frac{\partial y_1}{\partial z},$$

and at $y = -h$ is
$$\frac{\partial \psi}{\partial x} = -\sigma \frac{\partial y_2}{\partial z}.$$

Both these are satisfied if

$$A_1 h \sinh kh + B_1 \cosh kh = -b\sigma/k,$$
$$A_2 h \cos kh + B_2 \sinh kh = -b\sigma/k.$$

(39)

From (38) and (39)

$$A_1 = \frac{b\sigma \sinh kh}{\sinh kh \cosh kh + kh}, \quad B_1 = -\frac{b\sigma}{k} \left(\frac{kh \cosh kh + \sinh kh}{\sinh kh \cosh kh + kh} \right),$$

$$A_2 = \frac{b\sigma \cosh kh}{\sinh kh \cosh kh - kh}, \quad B_2 = -\frac{b\sigma}{k} \left(\frac{kh \sinh kh + \cosh kh}{\sinh kh \cosh kh - kh} \right).$$

(40)

It is now possible to calculate the stress which the viscous fluid exerts on the sheets. The component perpendicular to the sheet is as in (12)

$$Y_y = -p + 2\mu \frac{\partial v}{\partial y} = -p. \quad (41)$$

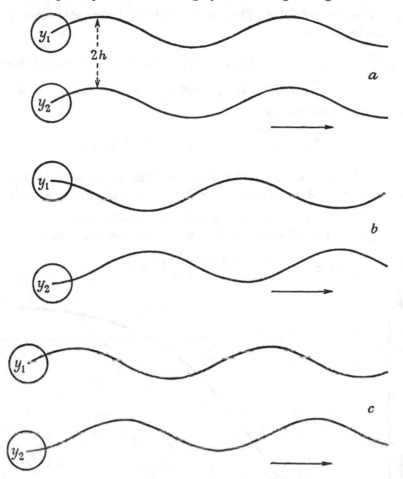

Fig. 3. *a*, waves in phase, $\phi = 0$, $b/\lambda = 0.073$, $2h/\lambda = \frac{3}{8}$. *b*, waves in opposite phase, $\phi = \frac{1}{2}\pi$, $b/\lambda = 0.073$, $2h/\lambda = \frac{3}{8}$. *c*, wave y_1 lags behind y_2, $\phi = \frac{1}{4}\pi$. The arrows indicate the direction of wave propagation.

The pressure p corresponding with the stream function (37) is given by

$$\frac{p}{2\mu k b\sigma} = \frac{\sinh ky \sinh kh \cos\phi \cos z}{\sinh kh \cosh kh + kh} - \frac{\cosh ky \cosh kh \sin\phi \sin z}{\sinh kh \cosh kh - kh}. \tag{42}$$

At $y = h$ the pressure is p_1, where

$$\frac{p_1}{2\mu k b_0} = \alpha \cos\phi \cos - \beta \sin\phi \sin z \tag{43}$$

and
$$\alpha - \frac{\sinh^2 kh}{\sinh kh \cosh kh + kh}, \quad \beta = \frac{\cosh^2 kh}{\sinh kh \cos kh - kh}. \tag{44}$$

The mean rate of dissipation of energy between the two sheets is equal to the mean rate at which the sheets do work. The rate at which unit length of the sheet $y = h + y_1$ does work on the fluid is

$$-p_1\frac{\partial y_1}{\partial t} = (\alpha\cos\phi\cos z - \beta\sin\phi\sin z)(\cos\phi\cos z - \sin\phi\sin z)(z\mu k b^2\sigma^2). \quad (45)$$

Since
$$\overline{\cos^2 z} = \overline{\sin^2 z} = \tfrac{1}{2} \quad \text{and} \quad \overline{\sin z\cos z} = 0,$$

the mean rate of doing work is

$$\bar{E} = -\overline{p_1\frac{\partial y_1}{\partial t}} = \mu k b^2\sigma^2(\alpha\cos^2\phi + \beta\sin^2\phi). \quad (46)$$

Since α is less than β for all values of kh the rate of dissipation is least when $\phi = 0$, so that the waves are in phase as in fig. 3a. $\bar{E}$ is greatest when $\phi = \tfrac{1}{2}\pi$ as in fig. 3b. The ratio

$$\frac{\bar{E}_1}{\bar{E}_2} = \frac{\text{rate of dissipation when waves are in phase}}{\text{rate of dissipation when waves are in opposite phase}}$$

$$= \tanh^2 kh\left(\frac{\sinh kh\cosh kh - kh}{\sinh kh\cosh kh + kh}\right). \quad (47)$$

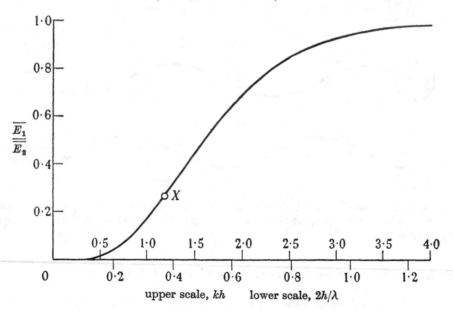

Fig. 4.

The way in which $\bar{E}_1/\bar{E}_2$ depends on kh is shown in fig. 4. It will be seen that E_1 is much less than E_2 when kh is small. The point X in fig. 4 refers to the sheets shown in fig. 3a and b. In fig. 3a, b and c the sheets are separated by a distance $\tfrac{3}{8}\lambda$. It will be seen that even at this distance there is a very large difference between $\bar{E}_1$ and $\bar{E}_2$. For $2h/\lambda = \tfrac{1}{10}$ which corresponds with the waves shown in Gray's drawing, fig. 2, $\bar{E}_1/\bar{E}_2 = 8\times 10^{-4}$. The effort required to make the tails wave in unison is in this case only one-thousandth of that necessary to make them wave if out of phase.

The reduction in the rate of dissipation of energy when the waves on the two sheets get into phase is very striking when kh is small, but without knowledge of the internal mechanism which moves the tail it is not possible to say with certainty that the tails will in fact get into the position where least energy is dissipated. On the other hand, it seems that whatever that mechanism may be, it is likely that a component of pressure which is in phase with the displacement of the sheet y_1 will tend to decrease the frequency of oscillation, while a component which is in the opposite phase will increase it. When the two sheets are so far away from one another that they do not influence one another the relationship between pressure and displacement of each sheet is that expressed by (1) and (13) so that the phases of p_1 and y_1 differ by $\frac{1}{2}\pi$. The reaction of the fluid therefore does not exert any *direct* force tending to increase or decrease the frequency. This must not be taken to imply that it has no effect. The work done by the sheet may have a great indirect effect on the frequency which the internal mechanism of the tail of a living object may be able to excite.

When kh is not large the sheets influence one another through the medium of the fluid. Comparing (43) with (36) it will be seen that it is only when $\alpha = \beta$ that y_1 is exactly out of phase with p_1. To find whether the *direct* effect of pressure is to increase or decrease the frequency of the sheet y_1 it is necessary to find the sign of $\overline{y_1 p_1}$. The direct effect of the viscous stress would be to increase or decrease the frequency according as $\overline{y_1 p_1}$ is negative or positive.

Writing (43) in the form

$$\frac{p_1}{2\mu k b\sigma} = C\cos{(z+\phi+\epsilon)}, \tag{48}$$

$$C^2 = \alpha^2+\beta^2 \quad \text{and} \quad \tan{(\phi+\epsilon)} = \frac{\beta}{\alpha}\tan\phi. \tag{49}$$

Since from (44) $\beta > \alpha$ and by definition $0 < \phi < \frac{1}{2}\pi$, (49) shows that ϵ is positive.

The mean value of $y_1 p_1$ is

$$\overline{y_1 p_1} = 2\mu k b^2 \sigma C(-\tfrac{1}{2}\sin\epsilon),$$

so that $\overline{y_1 p_1}$ is negative. The direct effect of pressure is therefore to tend to increase the frequency of the sheet y_1. At the sheet $y = -h+y_2$, the condition that the direct effect of pressure shall be to increase frequency is that $\overline{y_2 p_2}$ shall be positive (a positive pressure presses y_1 in the positive direction and y_2 in the negative direction).

Using (42) the pressure p_2 at the sheet y_2 is

$$\frac{p_2}{2\mu k b\sigma} = -\alpha\cos\phi\cos z - \beta\sin\phi\sin z$$

$$= -C\cos{(z-\phi-\epsilon)}, \tag{50}$$

where α, β, and c have the same meaning as before and

$$\tan{(\phi+\epsilon)} = \frac{\beta}{\alpha}\tan\phi,$$

so that ϵ also has the same meaning as before. From (36) and (50)

$$\frac{y_2\,p_2}{2\mu k b^2 \sigma} = -C\sin(z-\phi)\cos(z-\phi-\epsilon), \tag{51}$$

so that
$$\frac{\overline{y_2\,p_2}}{2\mu k b^2 \sigma} = -\tfrac{1}{2}C\sin\epsilon. \tag{52}$$

Since ϵ is positive, $\overline{y_2\,p_2}$ is negative. Thus the *direct* effect of pressure on the sheet y_2 is to decrease its frequency. Since the phase of y_1 lags behind that of y_2, the direct effect of the reaction between the two sheets is to increase the velocity of the waves in sheet y_1 and decrease that of the waves in y_2. In other words, the *direct* effect of the reaction of one sheet on the other through the viscous medium is to make the wave get into phase as illustrated in fig. 3 a.

In conclusion, I should like to express my thanks to Professor James Gray and Lord Rothschild for calling my attention to this problem.

17

THE ACTION OF WAVING CYLINDRICAL TAILS IN PROPELLING MICROSCOPIC ORGANISMS

REPRINTED FROM

Proceedings of the Royal Society A, vol. CCXI (1952), pp. 225–39

The action of the tail of a spermatozoon is discussed from the hydrodynamical point of view. The tail is assumed to be a flexible cylinder which is distorted by waves of lateral displacement propagated along its length. The resulting stress and motion in the surrounding fluid is analysed mathematically. Waves propagated backwards along the tail give rise to a forward motion with velocity proportional to the square of the ratio of the amplitude of the waves to their length. The rate at which energy must be supplied to maintain the waves against the reaction of the surrounding fluid is calculated.

Similar calculations for the case when waves of lateral displacement are propagated as spirals show that the body is propelled at twice the speed given it by waves of the same amplitude when the motion is confined to an axial plane. An externally applied torque is necessary to prevent the reaction of the fluid due to spiral waves from causing the cylinder to rotate. This is remarkable because the cylinder itself does not rotate.

A working model of a spermatozoon was made in which spiral waves could travel down a thin rubber tube without rotating it. The torque just referred to was observed and was balanced by an eccentric weight. The performance of the model while swimming freely in glycerine was compared with the calculations. The calculated speed of the model was higher than was observed, but this discrepancy could be accounted for by the fact that the model has a body containing its motive power while the calculations refer to a disembodied tail.

1. INTRODUCTION

The study of microscopic swimming creatures opens up a new field in hydrodynamics. The self-propulsion of aeroplanes, ships and large fishes depends entirely on the inertia of the surrounding fluid. A propelling unit generates backward momentum which is exactly balanced by the forward momentum associated with fluid resistance. When microscopic organisms swim in water the forces due to viscosity are so much greater than those due to inertia that the latter can be neglected. Self-propulsion of a flexible body is achieved by so distorting its surface that the body must move forward in order that the total force on it due to the viscous stress in the surrounding fluid may be zero. It is possible to imagine a simple example of self-propulsion. Consider an animal shaped like a anchor ring, or the rubber ring which is used to hold together the ends of the spokes of a folded umbrella. This shape is sketched in fig. 1*a*. It may be likened to a circular cylinder bent so that its axis forms a circle. If internal muscular contractions could cause the ring to rotate like a vortex ring as indicated by the curved arrows in fig. 1*a* its motion in space would depend on the reaction between its surface and the medium with which it was in contact. If, for instance, the ring were threaded on a solid cylinder like the rubber ring on the stick of an umbrella, it would move as indicated diagrammatically by the arrows in fig. 1 *b*. If it were inside a tube as shown in fig. 1 *c*, the bodily motion of the ring would be in the opposite direction. If it were placed in a viscous fluid, then the bodily motion would be in such a direction that no resultant force due to viscous stress would act on it. In fact it would move in a manner which is exactly analogous to that of a vortex ring and in the direction shown by the central arrow in fig. 1 *a*.

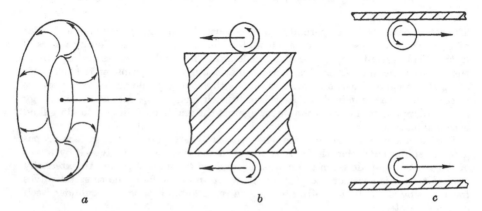

a   *b*   *c*

Fig. 1 *a*. Hypothetical ring-shaped animal capable of rotating its body in the direction indicated. *b*. Direction of motion when the ring rolls on the outside of a cylinder. *c*. Direction of motion when the ring rolls on the inside of a cylindrical tube.

In nature many microscopic bodies swim by causing waves of transverse displacement to travel down a thin tail. In a recent paper* the mechanics and propulsive effect of such a movement was studied in the mathematically simple case when the tail is assumed to be a thin sheet, and the results of calculation showed qualitative agreement with what is observed in nature. Quantitative results for detailed comparison with particular organisms could not be obtained because actual tails or flagellae resemble threads rather than sheets.

Results which might be expected to apply quantitatively would be obtained if a flexible cylinder waving in a viscous fluid were substituted for the flexible sheet of the previous work. In the following pages this problem is discussed making an assumption analogous to that made for the waving sheet, that the displacement of the axis of the flexible cylinder from a straight line is small. This limitation, however, may be far more restrictive in the case of the waving cylinder than it is for the waving sheet. In the latter case, the analytical expression of the effect of a wave of amplitude b and wave-length λ in a sheet was developed in powers of b/λ and b/λ was assumed to be small. In developing analogous expressions for the disturbance

* Paper 16 above.

due to a waving cylinder of radius u the Bessel functions used must be expanded in terms of b/a rather than b/λ. If b/a is to be regarded as small the limitation is serious when a/λ is small. In nature, even though b/λ may be rather small, b is usually greater than a, so that results obtained by analysis which assumes a/λ small require careful scrutiny before they can be accepted as applicable to organisms which occur in nature. This question will be referred to later.

2. FLAGELLAE WAVING IN A PLANE

Motion of a viscous fluid near a flexible circular cylinder which is distorted by small waves of lateral displacement

Small waves of lateral displacement unaccompanied by changes in cross-section distort a cylinder of radius a into a form which in cylindrical co-ordinates (r, θ, x) is represented by

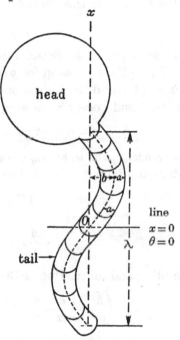

$$r = a + b \cos \theta \sin \kappa(x + Ut). \qquad (2\cdot1)$$

Here b is the amplitude, $\kappa = 2\pi/\lambda$, where λ is the wave-length, U is the velocity with which the waves are propagated in the direction in which x is decreasing. Fig. 2 shows the position of the flexible cylinder at time $t = 0$.

Neglecting the inertia of the fluid the equations of motion of a viscous fluid in cylindrical co-ordinates are

$$\frac{1}{\mu}\frac{\partial p}{\partial r} = \nabla^2 u - \frac{u}{r^2} - \frac{2}{r^2}\frac{\partial v}{\partial \theta}, \qquad (2\cdot2)$$

$$\frac{1}{\mu r}\frac{\partial p}{\partial \theta} = \nabla^2 v + \frac{2}{r^2}\frac{\partial u}{\partial \theta} - \frac{v}{r^2}, \qquad (2\cdot3)$$

$$\frac{1}{\mu}\frac{\partial p}{\partial x} = \nabla^2 w. \qquad (2\cdot4)$$

Fig. 2. Shape of waving tail at time $t = 0$.

Here u, v, w are the components of velocity parallel to r, θ, x respectively. μ is the viscosity, p is the pressure defined by

$$-p = \tfrac{1}{3}(\widehat{rr} + \widehat{\theta\theta} + \widehat{xx}), \qquad (2\cdot5)$$

and $\widehat{rr}, \widehat{\theta\theta}, \widehat{xx}$ are stress components in the radial, tangential and axial directions. ∇^2 is written for

$$\frac{\partial^2}{\partial r^2} + \frac{1}{r}\frac{\partial}{\partial r} + \frac{1}{r^2}\frac{\partial^2}{\partial \theta^2} + \frac{\partial^2}{\partial x^2}. \qquad (2\cdot6)$$

The equation of continuity is

$$\frac{\partial u}{\partial r} + \frac{u}{r} + \frac{1}{r}\frac{\partial v}{\partial \theta} + \frac{\partial w}{\partial x} = 0. \qquad (2\cdot7)$$

Eliminating u, v, w it is found* that

$$\nabla^2 p = 0. \tag{2.8}$$

Writing $s = \kappa(x + Ut)$, $\qquad\qquad z = \kappa r$, $\tag{2.9}$

a solution of (2.8) may be chosen in the form

$$\frac{p}{\mu\kappa} = F(z)\cos\theta\cos s. \tag{2.10}$$

Substituting this in (2.8)

$$\left(\frac{\partial^2}{\partial z^2} + \frac{1}{z}\frac{\partial}{\partial z} - \frac{1}{z^2} - 1\right)F(z) = 0.$$

The solution of this equation which vanishes when z is infinite is

$$F(z) = AK_1(z), \tag{2.11}$$

where $K_1(z)$ is the Bessel function of order 1 and imaginary argument.†

Using the expression for p obtained by eliminating $F(z)$ from (2.10) and (2.11) in (2.2), (2.3) and (2.4), equations for u, v and w are found. The form in which the variables θ and s occur in these equations suggests that solutions exist in the form

$$u = u_1\cos\theta\cos s, \quad v = v_1\sin\theta\cos s, \quad w = w_1\cos\theta\sin s, \tag{2.12}$$

where u_1, v_1 and w_1 are functions of z only. Substituting from (2.12) in the equations (2.2) and (2.3) it is found that

$$AK_1'(z) = \frac{\partial^2 u_1}{\partial z^2} + \frac{1}{z}\frac{\partial u_1}{\partial z} - \frac{2u_1}{z^2} - \frac{2v_1}{z^2} - u_1, \tag{2.13}$$

$$-\frac{A}{z}K_1(z) = \frac{\partial^2 v_1}{\partial z^2} + \frac{1}{z}\frac{\partial v_1}{\partial z} - \frac{2v_1}{z^2} - \frac{2u_1}{z^2} - v_1. \tag{2.14}$$

Adding and subtracting (2.13) and (2.14) the dependent variables can be separated

$$\left(\frac{\partial^2}{\partial z^2} + \frac{1}{z}\frac{\partial}{\partial z} - \frac{4}{z^2} - 1\right)(u_1 + v_1) = A\left\{K_1'(z) - \frac{1}{z}K_1(z)\right\} = -AK_2(z) \tag{2.15}$$

and

$$\left(\frac{\partial^2}{\partial z^2} + \frac{1}{z}\frac{\partial}{\partial z} - 1\right)(u_1 - v_1) = A\left\{K_1'(z) + \frac{1}{z}K_1(z)\right\} = -AK_0(z). \tag{2.16}$$

Using the recurrence formula for the Bessel functions (Watson, p. 79), it can be verified that particular integrals of (2.15) and (2.16) are

$$u_1 + v_1 = \tfrac{1}{2}AzK_1(z) \quad \text{and} \quad u_1 - v_1 = \tfrac{1}{2}AzK_1(z),$$

and since the complementary functions of (2.15) and (2.16) which do not become infinite as $z \to \infty$ are $K_2(z)$ and $K_0(z)$ respectively, the solutions of (2.15) and (2.16) are

$$2u_1 = BK_2(z) + CK_0(z) + AzK_1(z), \tag{2.17}$$

$$2v_1 = BK_2(z) - CK_0(z), \tag{2.18}$$

---

* Lamb, *Hydrodynamics*, (6th ed.), p. 595.
† G. N. Watson, *Bessel Functions* (1922).

where A, B and C are constants to be determined later. w_1 is found by substituting in the equation of continuity

$$-w_1 = \frac{du_1}{dz} + \frac{u_1}{z} + \frac{v_1}{z},\tag{2.19}$$

so that making use of the recurrence formulae

$$K_2'(z) + \frac{2}{z}K_2(z) = -K_1(z), \quad K_0'(z) = -K_1(z), \quad \frac{d}{dz}\{zK_1(z)\} = -zK_0(z),$$

$$2w_1 = BK_1(z) + CK_1(z) + A\{zK_0(z) - K_1(z)\}.\tag{2.20}$$

Boundary conditions

The cross-sections of the cylinder are assumed to remain circular, and to move as rigid laminae in planes perpendicular to the axis $r = 0$. The direction of motion is parallel to the plane $\theta = 0$. The conditions to be satisfied at the surface of the cylinder which in the first approximation is taken as $r = a$, or $z = \kappa a = z_1$, are

$$u_1 = \kappa Ub, \quad v_1 = -\kappa Ub, \quad w_1 = 0.\tag{2.21}$$

These conditions yield three equations for determining A, B, and C:

$$2\kappa Ub = BK_2(z_1) + CK_0(z_1) + Az_1K_1(z_1),\tag{2.22}$$

$$-2\kappa Ub = BK_0(z_1) - CK_0(z_1),\tag{2.23}$$

$$0 = B + C + A\left\{\frac{z_1K_0(z_1)}{K_1(z_1)} - 1\right\}.\tag{2.24}$$

Solving (2.22), (2.23), and (2.24)

$$\frac{A}{2\kappa Ub} = \frac{1}{\phi(z_1)},\tag{2.25}$$

$$\frac{B}{2\kappa Ub} = \frac{-z_1K_1(z_1)}{2K_2(z_1)\,\phi(z_1)},\tag{2.26}$$

$$\frac{C}{2\kappa Ub} = \frac{1}{K_0(z_1)}\left(1 - \frac{1}{2}\frac{z_1K_1(z_1)}{\phi(z_1)}\right),\tag{2.27}$$

where

$$\phi(z_1) = z_1K_1(z_1)\left\{\frac{1}{2} + \frac{1}{2}\frac{K_0(z_1)}{K_2(z_1)} - \left[\frac{K_0(z_1)}{K_1(z_1)}\right]^2\right\} + K_0(z_1).\tag{2.28}$$

Using tables of $K_0(z)$, $K_1(z)$, $K_2(z)$* the distribution of velocity can be obtained for any given value of $z_1 = \kappa a$.

Limiting cases

It is worth while giving the limiting values for A, B, C, (a) when κa is small and (b) when κa is large. In the latter case the wave-length is small compared with the radius of the cylinder, so that in the limiting case the disturbance is identical with that due to waves on a plane sheet.†

* G. N. Watson, *loc. cit.*
† Paper 16 above.

(a) *κa small*. In this case

$$\lim_{z_1 \to 0} K_0(z_1) = \frac{2}{\pi} \ln\left(\frac{2}{rz_1}\right) = 0.1159 - \ln(z_1),$$

$$\lim K_1(z_1) = \frac{1}{z_1}, \quad \lim K_2(z_1) = \frac{2}{z_1^2}.$$

Hence

$$\phi(z_1) \to \tfrac{1}{2} + K_0(z_1), \quad \frac{A}{2\kappa Ub} \to \frac{1}{\tfrac{1}{2} + K_0(z_1)},$$

$$\frac{B}{2\kappa Ub} \to -\tfrac{1}{4}z_1^2\left(\frac{1}{K_0(z_1) + \tfrac{1}{2}}\right), \quad \frac{C}{2\kappa Ub} \to \frac{1}{K_0(z_1) + \tfrac{1}{2}}. \tag{2.29}$$

(b) *κa large*. In this case it is found by substituting in (2·28) the asymptotic expansions of $K_0(z)$, $K_1(z)$, $K_2(z)$ valid when z is large, that

$$\lim_{z_1 \to \infty} \phi(z_1) \to K_0(z_1).$$

The asymptotic values of A, B and C are therefore

$$A = \frac{2\kappa Ub}{K_0(z_1)}, \quad B = -2\kappa Ub\left(\frac{z_1}{2K_0(z_1)}\right), \quad C = -2\kappa Ub\left(\frac{z_1}{2K_0(z_1)}\right). \tag{2.30}$$

Energy output required to maintain the flagellae in motion

The rate dQ, at which work is done by unit area of the surface of a cylinder, is found by taking the components of velocity of the surface, multiplying them by the corresponding stresses, and multiplying the result by -1. (This is because stress is called positive when it is a tension.) Hence

$$\frac{dQ}{Ub\kappa} = -\widehat{rr}\cos s\cos\theta + \widehat{r\theta}\cos s\sin\theta, \tag{2.31}$$

where $s = \kappa(x + Ut)$. The components of stress in a viscous fluid are

$$\widehat{rr} = -p + 2\mu\frac{\partial u}{\partial r} = \mu\kappa\left\{-AK_1(z) + 2\frac{\partial u_1}{\partial z}\right\}\cos\theta\cos s, \tag{2.32}$$

$$\widehat{r\theta} = \mu\left(\frac{\partial v}{\partial r} - \frac{v}{r} + \frac{1}{r}\frac{\partial u}{\partial\theta}\right) = \mu\kappa\left\{\frac{\partial v_1}{\partial z} - \frac{v_1}{z} - \frac{u_1}{z}\right\}\sin\theta\cos s, \tag{2.33}$$

since at the surface $u_1 = -v_1 = \kappa Ub$ and $w_1 = 0$, (2·19) shows that $du_1/dz = 0$.

Hence

$$\frac{dQ}{\mu\kappa^2 Ub} = \cos^2 s\left[AK_1(z_1)\cos^2\theta + \left\{\frac{dv_1}{dz}\right\}_{z_1}\sin^2\theta\right]. \tag{2.34}$$

The mean value of the contribution to the total work by unit length of the flagellum per second is therefore

$$\bar{Q} = \tfrac{1}{4}2\pi a(\mu\kappa^2 Ub)\left[AK_1(z_1) + \left(\frac{dv_1}{dz}\right)_{z_1}\right]. \tag{2.35}$$

Inserting the value of v_1 from (2·18) and using a recurrence formula this gives

$$\bar{Q} = \tfrac{1}{2}\mu\pi\kappa^2 ab U \left[AK_1(z_1) - \tfrac{1}{2}B\left\{ K_1(z_1) + \frac{2K_2(z_1)}{z_1} \right\} + \tfrac{1}{2}CK_1(z_1) \right]. \qquad (2\cdot36)$$

Limiting cases

(a) *κa small.* It is of interest to find the value of $\bar{Q}$ when the flagellum is of small diameter compared with the wave-length of the oscillations. Substituting from (2·29) in (2·36) it is found that when κa is small.

$$\bar{Q} = 2\mu\pi U^2 b^2 \kappa^2 \left(\frac{1}{K_0(\kappa a) + \tfrac{1}{2}} \right). \qquad (2\cdot37)$$

(b) *κa large.* In this case the diameter of the cylinder is large compared with the wave-length, so that the motion of the fluid in the neighbourhood of the surface is similar to that considered in the two-dimensional problem of waves in a plane sheet. The amplitude of the waves at any point is $b \cos s \cos \theta$, and the mean rate of doing work per unit area is $\lim\limits_{\kappa a \to \infty} dQ$, where dQ is given by (2·34). Using the asymptotic formulae for $K_0(z)$, $K_1(z)$ and $K_2(z)$ it is found that

$$\lim_{z_1 \to \infty} \left[\frac{dv}{dz} \right] = \kappa U b \quad \text{and} \quad \lim_{z_1 \to \infty} AK_1(z_1) = 2\kappa U b.$$

Defining the mean value of dQ as s varies, keeping θ constant, as $\overline{dQ}$,

$$\overline{dQ} = \mu\kappa^3 U^2 b^2 (\cos^2 \theta + \tfrac{1}{2} \sin^2 \theta). \qquad (2\cdot38)$$

In the plane case (paper 16 above) the rate of dissipation of energy per unit area due to waves of amplitude b and frequency $\sigma/2\pi$ was found to be

$$\overline{W} = \mu\kappa b^2 \sigma^2 = \mu\kappa^3 U^2 b^2,$$

so that the rate of dissipation at $\theta = 0$ in the cylindrical case is the same as in the plane case. This affords some verification of the accuracy of the analysis.

Propulsive effect of a waving cylindrical tail

In the preceding discussion it has been assumed that the velocity in the fluid at the surface of the tail is the same as though the tail were truly cylindrical and straight at the moment considered, but had a surface velocity equal to that which it has when waves of displacement travel down it. This may be described as a treatment which gives first-order effects, but it is useless for discussing second-order effects which depend on the product of the velocity and displacement of the surface. In the case of the waving sheet it was found that the velocity at which the sheet was propelled by the fluid reactions to waves of lateral displacement is a second-order effect. For this reason an attempt will be made to carry the discussion of waves in a cylinder to the second order with a view to calculating the propulsive effect of these waves.

The first step is to show that the solution represented by (2·12) can be generalized so as to include velocity distribution of the form

$$u = u_{mn} \cos m\theta \cos ns,$$
$$v = v_{mn} \sin m\theta \cos ns,$$
$$w = w_{mn} \cos m\theta \sin ns,$$

(2·39)

where u_{mn}, v_{mn}, w_{mn} are functions of $nz = n\kappa r$ only. It can be verified that a solution in terms of Bessel functions of type $K_n(nz)$ is

$$2u_{mn} = B_{mn} K_{m+1}(nz) + C_{mn} K_{m-1}(nz) + A_{mn} nz K_m(nz),$$

(2·40)

$$2v_{mn} = B_{mn} K_{m+1}(nz) - C_{mn} K_{m-1}(nz),$$

(2·41)

$$-nw_{mn} = \frac{d}{dz}(u_{mn}) + \frac{1}{z}(u_{mn} + mv_{mn}).$$

(2·42)

The pressure is

$$p = \mu\kappa p_{mn} \cos m\theta \cos ns,$$

(2.43)

where

$$p_{mn} = A_{mn} n K_m(nz).$$

(2·44)

Boundary conditions

It will be assumed that the cross-section of the tail remains circular and that the velocity of its centre is $b \sin s$.

In polar co-ordinates the surface is

$$r = a + b \cos\theta \sin s.$$

(2·45)

In applying the boundary conditions to the first order, u_1, v_1, w_1, are given their values at $z = z_1$, or $r = a$. In applying them so as to include second-order effects the values of u_1, v_1, w_1 are corrected so as to allow for the variation in r. The change in z is $b\kappa \cos\theta \sin s$, and the corresponding value of u_1 is taken as

$$[u_1]_{z=z_1} + b\kappa \cos\theta \sin s [u_1']_{z=z_1}.$$

(2·46)

The boundary conditions at $r = a + b\cos\theta \sin s$ are

$$u = bU\kappa \cos\theta \cos s, \quad v = -bU\kappa \sin\theta \cos s, \quad w = 0.$$

(2·47)

To the first order, these are satisfied by $u_1 = bU\kappa, v_1 = -bU\kappa, w_1 = 0$, but when second-order terms are taken into account and u_1 is given the value shown in (2·46), the boundary condition (2·47) can only be satisfied by adding further solutions of the equations of motion of the form (2·40) to (2·44). The necessary functions are shown in table 1, where five such solutions $(a), (b), (c), (d), (e)$ are shown in columns and the corresponding components of the boundary conditions in rows. In this table the values of all functions of z are taken at $z = z_1$.

Since

$$\cos^2\theta = \tfrac{1}{2}(1 + \cos 2\theta), \quad \sin\theta \cos\theta = \tfrac{1}{2}\sin 2\theta, \quad \sin^2 s = \tfrac{1}{2}(1 - \cos 2s)$$

the three boundary conditions can be satisfied provided

$$
\left.
\begin{aligned}
u_1 &= bU\kappa, \quad v_1 = -bU\kappa, \quad w_1 = 0, \\
u_{02} &= -b\kappa u_1', \quad v_{02} = -\tfrac{1}{4}b\kappa w_1', \quad -w_{02} = -\tfrac{1}{4}b\kappa w_1', \\
u_{22} &= -\tfrac{1}{4}b\kappa v_1', \quad v_{22} = -\tfrac{1}{4}b\kappa v_1', \quad w_{22} = \tfrac{1}{4}b\kappa w_1', \quad V = \tfrac{1}{4}b\kappa w_1'.
\end{aligned}
\right\}
\tag{2.48}
$$

Table 1. *Boundary conditions*

(a)	(b)	(c)	(d)	(e)
$bU\kappa\cos\theta\cos s =$				
$u_1\cos\theta\cos s + b\kappa u_1'\cos\theta\cos s\cos\theta\sin s$	$+u_{02}\sin 2s$	0	$+u_{22}\cos 2\theta\sin 2s$	0
$-bU\kappa\sin\theta\cos s =$				
$v_1\sin\theta\cos s + b\kappa v_1'\sin\theta\cos s\cos\theta\sin s$	0	0	$+v_{22}\sin 2\theta\sin 2s$	0
$0 =$				
$w_1\cos\theta\sin s + b\kappa w_1'\cos\theta\sin s\cos\theta\sin s$	$-w_{02}\cos 2s$	$+w_{20}\cos 2\theta$	$-w_{22}\cos 2\theta\cos 2s$	$-V$

It will be noticed that the solution (a) contains three constants A_1, B_1, C_1. Likewise the solution (d) contains three constants A_{22}, B_{22}, C_{22}. Solution (b) is of the degenerate class for which $m = 0$, so that $K_{m+1}(z) = K_{m-1}(z)$. When $v = 0$, (2.41) shows that $B_{02} = C_{02}$, so that solution (b) contains only two arbitrary constants. In solution (c) $u = v = 0$ and p is constant, so that (2.2) and (2.3) are satisfied. The equation for w is $\nabla^2 w = 0$, so that if $w = w_{02}\cos 2\theta$, where w_{02} is a function of r only, the equation for w_{02} is

$$
\left(\frac{\partial^2}{\partial r^2} + \frac{1}{r}\frac{\partial}{\partial r} - \frac{4}{r^2}\right)w_{02} = 0.
\tag{2.49}
$$

The complete solution of (2.49) is

$$
w_{02} = Ar^2 + Br^{-2},
\tag{2.50}
$$

which contains two constants A and B. Since w_{02} must be small when r is large, $A = 0$. Solution (c) therefore contains only one arbitrary constant. There are thus ten constants (including V) connected by the ten linear equations (2.48). The problem is, therefore, uniquely solved.

It will be seen, therefore, that if the disturbance due to the motion of the flexible cylinder is to vanish at infinity (a condition assumed by the use of the Bessel functions $K_n(z)$), it must move at velocity V relative to the fluid at infinity, where

$$
V = \tfrac{1}{4}b\kappa[w_1']_{z=\kappa a}.
\tag{2.51}
$$

w_1' can be expressed in terms of Bessel functions, differentiating (2.20)

$$
2w_1' = -(B+C-A)\left\{\frac{1}{z_1}K_1(z_1) + K_0(z_1)\right\} + A\{K_0(z_1) - z_1 K_1(z_1)\}.
\tag{2.52}
$$

Since $w_1 = 0$ at $z = z_1$

$$
(B+C-A)K_1(z_1) + Az_1 K_0(z_1) = 0,
$$

so that (2.52) becomes

$$
2w_1' = A\left\{2K_0(z_1) - z_1 K_1(z_1) + \frac{z_1 K_0^2(z_1)}{K_1(z_1)}\right\}.
\tag{2.53}
$$

Substituting the value of A from (2·25)

$$\frac{V}{U} = \frac{1}{2}\frac{b^2\kappa^2}{\phi(z_1)}\left\{K_0(z_1) - \tfrac{1}{2}z_1 K_1(z_1) + \frac{1}{2}\frac{z_1 K_0^2(z_1)}{K_1(z_1)}\right\}, \qquad (2·54)$$

where $\phi(z_1)$ is the function given in (2·28).

It will be noticed that V/U is positive. From the definitions of U and V it will be seen that this proves that the direction of propulsion of the waving tail is opposite to that in which the waves of lateral displacement are propagated.

Case when κa is small

The most interesting case is when κa is small so that the tail is of small diameter compared with the wave-length of the disturbances propagated down it. When $z_1 \to 0$, $z_1 K_1(z_1) \to 1$ and also $\phi(z_1) \to K_0(z_1) + \tfrac{1}{2}$, so that

$$\frac{z_1 K_0^2(z_1)}{K_1(z_1)} \to 0;$$

when κa is small
$$\frac{V}{U} = \tfrac{1}{2}b^2\kappa^2\left\{\frac{K_0(\kappa a) - \tfrac{1}{2}}{K_0(\kappa a) + \tfrac{1}{2}}\right\}. \qquad (2·55)$$

This result is similar to that obtained in the plane case where*

$$\frac{V}{U} = \tfrac{1}{2}b^2\kappa^2. \qquad (2·56)$$

3. FLAGELLAE WAVING IN SPIRALS

Analysis of flow produced by a spiral wave of lateral displacement

When looking through a microscope it is not possible to appreciate motion in the line of sight. When, therefore, an organism is seen to be swimming by causing waves of lateral displacement to travel down its tail, it cannot be stated whether the particles of the tail are moving in straight lines or circles or ellipses. It has been suggested as a result of inspecting certain photographs that the wave seen may really be a wave of lateral displacement in which the tail at any instant is in the form of a spiral. It is of interest, therefore, to calculate the flow when such waves travel down a flexible cylinder immersed in a viscous fluid. The calculation gains additional interest from the fact that it is possible to construct a working model which makes this movement, though it is very difficult to make a working model of a flexible cylinder vibrating in progressive waves whose motion is confined to a plane.

The equation representing the shape, at any time, of a cylinder of radius a, which has been slightly distorted into a spiral by progressive wave is

$$r = a + b\cos(\theta \mp s), \qquad (3·1)$$

where $s = \kappa(x + vt)$, and b/a is supposed small.

* Paper 16 above.

If the negative sign is taken, (3·1) represents the surface of a cylinder bent into the form of a right-handed spiral or screw when it rotates at angular velocity κU about its axis in the direction of increasing θ. This sense of rotation appears clockwise to an observer looking along the axis in the direction where x is positive. This statement does not imply that the particles on the surface of the material cylinder are moving like those of a rigid spiral rotating in the sense described. If the positive sign is taken (3·1) represents the surface of a left-handed cylinder rotating in the direction of decreasing θ. In the analysis which follows the negative sign will be taken in (3·1), but it is equally applicable to the motion which is the mirror image in an axial plane and corresponds with the positive sign in (3·1).

Since the particles of the cylinder do not rotate and are assumed to remain in a plane perpendicular to the axis of x, the components of velocity of the surface particles expressed in cylindrical co-ordinates are

$$u = Ub\kappa \sin(\theta - s), \quad v = Ub\kappa \cos(\theta - s), \quad w = 0. \tag{3·2}$$

To find the motion produced in the fluid by this distribution of velocity at the surface of the cylinder it is necessary to find solutions of the equation of motion ((2·3) to (2·5)) which satisfy (3·2) at the surface represented by (3·1).

First approximation

If only first-order terms are considered it can be taken that (3·2) is satisfied on the undistorted cylinder $r = a$. Since (3·2) can be written

$$
\begin{aligned}
u &= Ub\kappa(\sin\theta\cos s - \cos\theta\sin s), \\
v &= Ub\kappa(\cos\theta\cos s + \sin\theta\sin s), \\
w &= 0,
\end{aligned}
\tag{3·3}
$$

the motion can be regarded as the sum of the two motions obtained by altering the origins of θ or s by $\frac{1}{2}\pi$.

The motion in the fluid is therefore represented by the sum of two solutions α and β, namely,

$$
\left.
\begin{aligned}
u &= +_\alpha u_1 \sin\theta\cos s, \\
v &= -_\alpha v_1 \cos\theta\cos s, \\
w &= +_\alpha w_1 \sin\theta\sin s,
\end{aligned}
\right\}
\quad \text{and} \quad
\left.
\begin{aligned}
u &= -_\beta u_1 \cos\theta\sin s, \\
v &= -_\beta v_1 \sin\theta\sin s, \\
w &= +_\beta w_1 \cos\theta\cos s.
\end{aligned}
\right\}
\tag{3·4}
$$

Each of these solutions contain three constants, so that it is possible to satisfy the six linear equations which result from equating (3·3) to (3·4) at $z = \kappa a$.

Comparing (3·3) and (3·4) it will be seen that the signs adopted in (3·4) ensure that the boundary conditions at $z = z_1$ are

$$_\alpha u_1 = {}_\beta u_1 = Ub\kappa, \quad _\alpha v_1 = {}_\beta v_1 = -Ub\kappa,$$

so that the functions of z occurring in the two solutions (α) and (β) are identical with those which occur in the problem where the waves are limited to a plane and the amplitude is the same.

As might be expected the result is simply the sum of two first-order solutions of the type already discussed in which the motion takes place in perpendicular planes. The amplitude of the two is the same and their phases differ by $\frac{1}{2}\pi$.

Second approximation

The second approximation is obtained in the same way as before. The first-order solution which is the sum of (α) and (β) is

$$u = u_1 \sin(\theta - s), \quad v = -v_1 \cos(\theta - s), \quad w = w_1 \cos(\theta - s), \qquad (3\cdot5)$$

and u_1, v_1, w_1 are the same functions of z that were considered in the case of plane motion so that at $z = z_1$, $u_1 = Ub\kappa$, $v_1 = -Ub\kappa$, $w_1 = 0$. The radial displacement of a particle from the surface $z = \kappa a$ is $b\cos(\theta - s)$. The values of u_1, v_1 and w_1 at the surface of the deformed cylinder are therefore taken as

$$\left. \begin{aligned} u &= u_1 \sin(\theta - s) + b\kappa u_1' \sin(\theta - s) \cos(\theta - s), \\ v &= -v_1 \cos(\theta - s) - b\kappa v_1' \cos^2(\theta - s), \\ w &= w_1 \cos(\theta - s) + b\kappa w_1' \cos^2(\theta - s), \end{aligned} \right\} \qquad (3\cdot6)$$

the values for $u_1, v_1, w_1, u_1', v_1', w_1'$ being those $z = \kappa a$.

The boundary condition $(3\cdot2)$ may be satisfied so far as the variable parts of $(3\cdot6)$ are concerned by adding to $(3\cdot6)$ a solution of the equations of motion in the form

$$\left. \begin{aligned} u &= u_{22} \sin 2(\theta - s), \\ v &= -v_{22} \cos 2(\theta - s), \\ w &= w_{22} \cos 2(\theta - s). \end{aligned} \right\} \qquad (3\cdot7)$$

This solution has three arbitrary constants. The variable part of the boundary condition, obtained by equating $(3\cdot2)$ to the sum of $(3\cdot6)$ and $(3\cdot7)$, can be satisfied by this choice because the six equations

$$u_1 = b\kappa U, \qquad \tfrac{1}{2}b\kappa u_1' + u_{22} = 0,$$

$$-v_1 = b\kappa U, \qquad -\tfrac{1}{2}b\kappa v_1' - v_{22} = 0,$$

$$w_1 = 0, \qquad \tfrac{1}{2}b\kappa w_1' + w_{22} = 0,$$

can be satisfied by a proper choice of the six available arbitrary constants in (u_1, v_1, w_1) and (u_{22}, v_{22}, w_{22}). After satisfying the variable part of the boundary conditions in this way it is necessary to annul the constants left when

$$\tfrac{1}{2}\{1 + \cos 2(\theta - s)\}$$

is substituted for $\cos^2(\theta - s)$ in $(3\cdot6)$; these are $-\tfrac{1}{2}b\kappa v_1'$ in v and $\tfrac{1}{2}b\kappa w_1'$ in w.

The second of these can be annulled by adding the solution $(w = 0, v = 0, w = -V)$, and the w-component of the boundary condition is then completely satisfied provided

$$V = \tfrac{1}{2}b\kappa[w_1']_{z=z_1}. \qquad (3\cdot8)$$

The first can be annulled by adding the solution

$$\left(u = 0, \quad v = \frac{\Omega}{r} = \frac{\Omega\kappa}{z}, \quad w = 0\right).$$ (3·9)

The v-component of the boundary condition is then satisfied completely if

$$\frac{\Omega\kappa}{z_1} = -\tfrac{1}{2}b\kappa[v_1']_{z=z_1}.$$ (3·10)

Interpretation of (3·8)

The interpretation of (3·8) is similar to that of the analogous equation (2·51) in the case where the waving motion was confined to a plane.

The waving tail propels itself in the opposite direction to that of propagation of the spiral waves. Since the function w_1 has been chosen so that it has exactly the same form and is expressed by the same symbols as the corresponding function in the plane case, it can be seen by comparison of (3·8) and (2·51) that for a given amplitude the spiral wave propels the tail at twice the speed of the plane wave.

When κa is small the rate of progress V produced by the spiral waves is therefore given by

$$\frac{V}{U} = b^2\kappa^2 \left\{\frac{K_0(\kappa a) - \tfrac{1}{2}}{K_0(\kappa a) + \tfrac{1}{2}}\right\}.$$ (3·11)

Interpretation of (3·10)

The motion represented by (3·10) is a circulation round the axis of the cylinder. Since all the Bessel functions involved in the expressions for u, v and w ultimately die away exponentially with increasing z, the velocity component $v = \Omega\kappa/z$ is ultimately greater than all the other constituents, of v. The existence of this term implies that a constant couple must be applied to the tail about its length in order that it may not rotate. The magnitude of this couple can be found by calculating the couple exerted by the fluid at infinity over a cylinder of radius so large that only the disturbance represented by (3·9) is appreciable. The mean couple exerted per unit length of the waving cylinder is therefore

$$G = 2\pi\mu r^2 \left(\frac{dv}{dr} - \frac{v}{r}\right),$$

where v is given by (3·9). Using (3·9) and (3·10) it is found that

$$G = -2\pi\mu ab\kappa[v_1']_{z=z_1}.$$ (3·12)

When κa is small this becomes

$$G = -4\pi\mu b^2\kappa U/\{K_0(\kappa a) + \tfrac{1}{2}\}.$$ (3·13)

It appears therefore that when waves which bend a cylinder into a right-handed spiral travel in the direction of $-x$, a couple G must be applied in the direction of decreasing θ, i.e. counter-clockwise as seen by an eye looking in the direction $+x$. The torque is therefore in the same direction that would be needed to hold the cylinder if it rotated about its axis without distortion in the same direction as that which the centres of sections move round the axis $r = 0$ owing to the spiral waves.

It is of interest to notice that this torque occurs in the absence of any rotation of the material of the waving cylinder. That such a torque must exist was pointed out to me in conversation by Professor James Gray, who reached this conclusion from data derived from a general study of undulatory animals. In the experiments to be described later it was so noticeable that a large gravity-controlled torque had to be applied to counterbalance it.

4. MODEL EXPERIMENTS

Preliminary experiments with model spermatozoa

When spermatozoa are observed swimming in the field of a microscope the oscillations of their tails are too rapid to follow in detail. Some interesting cinema records have recently been taken which show that the amplitude of the oscillations is larger than could legitimately be represented by analysis which assumes that b/a is small. It would be of interest if a model could be made to represent on a very much larger scale the flagellum of a spermatozoon.

Such a working model would not be required to operate at exactly the same Reynolds number as the organism. All that would be necessary to attain dynamical similarity would be that this number was low enough to ensure that the stress in the fluid due to inertia was small compared with that due to viscosity. If such a model could be produced it would not only throw light on zoological problems but it could be used for comparing its performance with that deduced from calculations in cases where the amplitude was small. If approximate agreement were found in such cases the performance of the model would be of great interest when the amplitude was so large that the calculations could not legitimately be used.

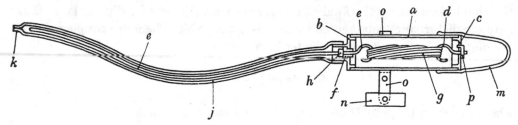

Fig. 3. Working model of swimming spermatozoon.

Unfortunately, it is a very difficult matter to design any mechanism which would cause waves of lateral displacement in a plane to travel down a cylinder or sheet, at any rate under conditions which would permit the model to be used in a very viscous fluid. The difficulty may not be insuperable, but before embarking on an attempt to make such a model, a simpler though possibly less realistic model has been made. This is shown in the drawing of fig. 3.

The model consists of an aluminium tube a fitted with end-pieces b and c. Wires d and e pass through these ends, and the bearing of e is bushed with a plastic at f to reduce the friction due to end-thrust of the collar h. The wire e is bent into the form of a long spiral. Rubber rings g are threaded on the hooks at the inboard ends of d and e; d is prevented from turning in the case a by the collar p, but e can be rotated so that the rubber rings are twisted. When the wire e is released the rubber rings untwist and cause it to rotate. In this condition the model obviously does not represent any living creature because the wire e rotates as a whole relative to the body a, a type of distortion which is impossible in a living organism. To get over this objection the end b was made so that a short projecting tube surrounded the wire e without touching it. A rubber tube j was threaded over the spiral wire e and forced over the projecting portion of the end b. After lubricating the surface of contact between the wire e and the rubber tube j with water or glycerine the end k of the rubber tube was closed. A rubber thimble m was forced over the end c to make the apparatus watertight. The model was then ballasted by means of a lead weight n attached to a ring-shaped clamp o. The weight was adjusted so that the model was as nearly as possible of the same density as the surrounding medium and at the same time would lie horizontally.

When the model was released after being wound up, the spiral wire caused a spiral wave to travel down the rubber tube without rotating it relative to the body. This deformation is a possible one for a living organism to make, and the calculations necessary for comparison with observation have been described in §3.

Experimental procedure

Experiments were made in a tank of glycerine. In the first form of model the weight n (fig. 3) was close to the surface of the cylindrical body a. On releasing the mechanism the body with its accompanying weight n rotated rapidly about its axis and the material rubber surface rotated with it.

The wave of lateral deformation travelled slowly down the tail. This effect was clearly due to the existence of the couple G (see equation (3·13)). To counteract this couple the clamp holding the weight n was extended till it was removed from the body to a distance of twice the diameter of the body from the axis as shown in fig. 3. In this condition the body of the model no longer rotated, and it was observed to move forward rather slowly. The experiments were performed by the author using rubber gloves, and the technique of keeping the glycerine pure and keeping it away from the hands and clothes are learned in the course of the work.

Good agreement between the theory which assumes that the tail operates without having to drive a body along, and experiments with the model shown in fig. 3, is not to be expected, but it is worth making the comparison even if only to help in forming an opinion whether it would be worth while to make a more realistic experiment.

In one set of experiments the tail was 5 in. long and the pitch of the helix 8 in.

The radius of the cylinder on which the spiral was wound was $b = \frac{5}{8}$ in. The diameter of the rubber tube was $\frac{1}{4}$ in. Thus

$$b\kappa = \frac{5}{8} \frac{2\pi}{8} = 0\cdot50, \quad \kappa a = \frac{2\pi}{8} \frac{1}{8} = 0\cdot1 \quad \text{and} \quad K_0(0\cdot1) = 2\cdot4$$

(*Bessel Functions*, p. 737), so that the predicted value of V/U is, according to (3·10),

$$\left(\frac{V}{U}\right)_{\text{calc}} = (0\cdot5)^2 \left(\frac{2\cdot4 - 0\cdot5}{2\cdot4 + 0\cdot5}\right). \tag{4·1}$$

$$= 0\cdot19.$$

Using this model it was observed that the model propelled itself 4 in. while the tail made 20 oscillations. Since the wave-length was 8 in.

$$\left(\frac{V}{U}\right)_{\text{expt}} = \frac{4 \text{ in.}}{8 \text{ in.} \times 20} = 0\cdot025. \tag{4·2}$$

Comparing (4·2) with (4·1) it will be seen that the observed rate of progress is smaller than that calculated. This discrepancy is of the order of magnitude which would be expected in view of the fact that in the experiment the tail has to push a rather large body.

It will be noticed that the condition assumed in the analysis that b/a shall be small is very far from being true, for $b/a = \frac{5}{8}/\frac{1}{8} = 5$. It might be expected that first-order effects like the rate of dissipation of energy which depends on the velocity of the surface of the helical cylinder and not critically on the amount of the displacement of its axis from a straight line, would be correctly predicted by theory. No such expectation would be justified *a priori* in calculations depending on second-order effects such as that of the velocity of propulsion. For this reason alone it seems that it will be worth while to construct a more realistic model of a cylinder capable of being deformed by helicoidal waves of displacement.

18

ANALYSIS OF THE SWIMMING OF LONG AND NARROW ANIMALS

REPRINTED FROM

Proceedings of the Royal Society, A, vol. CCXIV (1952), pp. 158–83

The swimming of long animals like snakes, eels and marine worms is idealised by considering the equilibrium of a flexible cylinder immersed in water when waves of bending of constant amplitude travel down it at constant speed. The force of each element of the cylinder is assumed to be the same as that which would act on a corresponding element of a long straight cylinder moving at the same speed and inclination to the direction of motion. Relevant aerodynamic data for smooth cylinders are first generalised to make them applicable over a wide range of speed and cylinder diameter. The formulae so obtained are applied to the idealised animal and a connection established between B/λ, V/U and R_1. Here B and λ are the amplitude and wave-length, V the velocity attained when the wave is propagated with velocity U, R_1 is the Reynolds number $Ud\rho/\mu$, where d is the diameter of the cylinder, ρ and μ are the density and viscosity of water.

The results of calculation are compared with James Gray's photographs of a swimming snake and a leech.

The amplitude of the waves which produce the greatest forward speed for a given output of energy is calculated and found, in the case of the snake, to be very close to that revealed by photographs.

Similar calculations using force formulae applicable to rough cylinders yield results which differ from those for smooth ones in that when the roughness is sufficiently great and has a certain directional character propulsion can be achieved by a wave of bending which is propagated forward instead of backward. Gray's photographs of a marine worm show that this remarkable method of propulsion does in fact occur in the animal world.

1. INTRODUCTION

The motions which fishes, snakes and other animals make when they swim have been studied photographically by Gray*. The way in which their muscles produce the observed movements of their flexible bodies seems to be understood. The external mechanics of the locomotion of snakes on land has also been discussed,† but attempts to analyse swimming from the point of view of hydrodynamics have failed because, in general, there is no way in which experiments made with rigid bodies can be used to predict the forces acting on flexible bodies. In the special case when the flexible body is very long compared with its lateral dimensions it may be legitimate to assume that the reaction of the surrounding water on any section of it is the same as though that section were part of a long cylinder moving at the same speed and in the same direction as itself. This assumption has been used successfully in calculating the curves made by the cable of a captive balloon in a

---

* J. Gray, *Proc. Roy. Soc.* B, CXXVIII (1939), 28.

† J. Gray, *J. exp. Biol.* XXIII (1946), 101 and *J. exp. Biol.* XXVI (1949), 354.

wind or by the underwater part of a fishing line when trolling for mackerel. The data on which these calculations were based were obtained by setting up a long straight cylinder or wire at various angles of incidence to the air current in a wind-tunnel and measuring the force acting on it.

Attempts to apply similar methods to the mechanics of swimming might be successful if the swimmer were sufficiently long in comparison with its thickness. For this reason the swimming of snakes, leeches and certain marine worms have been studied. Since wind-tunnel measurements give the force on a cylinder set at angle of incidence i in a wind of velocity Q, the most direct method of applying the basic assumptions would be to measure the velocity and direction of motion of each element of the body of a swimming animal at successive intervals of time. This could be done by making measurements on successive frames of a cinematograph record, but the work would be very laborious, and it might be impossible to make measurements sufficiently accurate to determine reliable values of Q and i. Even if the work could be carried out and the basic assumption used in conjunction with wind-tunnel measurements to calculate the force on each element of the body, the only possible final result would consist in verifying (or not verifying) that the integrated resultant force and couple acting on the whole animal are both zero. Such a result would contribute little to our understanding of the general principles of the mechanics of swimming.

For this reason a less direct method has been adopted. A study of successive frames in some of Gray's photographs of long animals swimming has revealed two main features which seem likely to be significant in the mechanics of swimming: (1) the animal sends waves of lateral displacement down its body and (2) in some cases, particularly in the case of a snake, these waves increase in amplitudes as they pass down the body.

In the analysis which forms the subject of this work the first of these features is studied as a problem in the mechanics of an idealized or 'mathematical' animal which consists of a flexible cylinder of uniform section. It will be assumed to swim by sending waves of uniform length and amplitude at a uniform speed down its body.

The work is divided into nine sections. In §2 the relevant aerodynamic data are examined and formulae are given for the lateral and longitudinal components of force acting on a cylinder set obliquely to a stream of fluid. Two sets of formulae are given. The first refers to smooth and the second to rough cylinders. In §3 equations are given which represent the geometry of a flexible cylinder down which waves of bending of constant amplitude are being propagated. In §4 the swimming charac-teristics of a smooth flexible cylinder are calculated and the results shown in a 'swimming diagram'. In §5 measurements of Gray's photographs of smooth animals swimming are compared with the calculations of §4. In §6 the energy required for propulsion of smooth animals is calculated and the amplitudes of the waves which drive them fastest for a given output of energy is found. In §7 calculations analogous to those of §4 are made for animals with a rough surface. It is found that when the surface is sufficiently rough and the roughness has certain directional characteristics, an animal could swim forwards by sending waves *forwards* along its body, and in §8 the performance of a marine worm which actually swims in this way is compared with the calculations. In §9 some limitations to the application of the analysis are mentioned and the swimming characteristics of a very small animal discussed.

2. THE FORCE ON A CIRCULAR CYLINDER SET OBLIQUELY TO A STREAM OF FLUID

Very few experimental data have been published on this subject. Relf & Powell* gave measurements of the force on a smooth cylinder $\frac{3}{8}$ in. in diameter set at angles varying by 10° intervals from 0 to 90° to the wind direction. The transverse force, F_N, and the longitudinal force, F_L, expressed as pounds weight per foot run of the cylinder in a wind of 40 ft./sec., is given in table 1 taken from their paper.

Table 1. *Force on an inclined cylinder $\frac{3}{8}$ in. diameter in a wind 40 ft./sec.*

1	2	3	4	5	6	7
	F_N	F_L				F_L
$i°$	(lb./ft. run)	(lb./ft. run)	$0.07 \sin^2 i$	F_N (calc.)	$\dfrac{F_L}{\cos i\,(\sin i)^{\frac{1}{2}}}$	$\dfrac{F_L}{\cos i}$
0	0	0·0016	0	0	—	0·0016
10	0·0025	0·0016	0·0021	0·0021	0·0039	0·0016
20	0·0090	0·0021	0·0081	0·0081	0·0038	0·0022
30	0·0191	0·0024	0·0175	0·0172	0·0039	0·0029
40	0·0297	0·0023	0·0290	0·0282	0·0037	0·0030
50	0·0415	0·0019	0·0410	0·0400	0·0034	0·0029
60	0·0525	0·0015	0·0525	0·0509	0·0032	0·0030
70	0·0606	0·0012	0·0618	0·0598	0·0036	0·0035
80	0·0657	0·0003	0·0680	0·0657	[0·0017]	0·0017
90	0·0672	0	0·0700	0·0677	—	—

$$F_N \text{ (calc.)} = 5.91 \times 10^{-2} \{1.1 \sin^2 i + 0.045 (\sin i)^{\frac{1}{2}}\}$$

It was pointed out by Relf & Powell that F_N is nearly proportional to $\sin^2 i$, where i is the angle between the axis of the cylinder and the wind direction. The values of $0.07/\sin^2 i$ are given in column 4 of table 1 for comparison with values of F_N in column 2. It will be seen that the agreement is fairly good. If Q is the wind velocity $Q \sin i$ is the component of velocity at right angles to the cylinder, and since the drag on a cylinder placed at right angles to the wind is very nearly proportional to Q^2, the interpretation of their measurements, which Relf & Powell gave, is that the normal component of velocity determines the normal force independently of the longitudinal component $Q \cos i$. This result was to be expected on theoretical grounds because at the Reynolds number of Relf & Powell's experiments (7.0×10^3) the boundary layer is laminar. The field of flow is therefore one in which the three

* E. F. Relf & C. H. Powell, *Rep. Memo. aeronaut. Res. Comm. Lond.* no. 307 (1917).

components of velocity u, v, w and also the pressure, p, are functions of two variables x and y only. Under these circumstances u, v and p are independent of w in the sense that their values are unaltered by any change in w, though w is dependent on u and v.

Relf & Powell's measurements were successfully used by McLeod* to calculate the shape of a flexible cable used for towing weights under an aeroplane. For this purpose McLeod found that sufficiently accurate results could be obtained if F_L were neglected altogether. There is, however, an intrinsic interest in applying the principle that the transverse components of velocity are independent of the longitudinal components, to make predictions about the longitudinal force.† In the present investigation the longitudinal component of force turns out to be of paramount importance. It is not possible to apply Relf & Powell's data directly to cases in which the Reynolds number differs greatly from that at which their experiments were made. For this reason, as well as for the intrinsic interest of the subject, a theoretical prediction about the effect of Reynolds number on the force acting on a cylinder placed obliquely in a stream of fluid is needed.

Normal component of force

The component of force acting on unit length of a cylinder at right angles to its axis when placed obliquely in a fluid stream will be represented by N, and N depends only on $Q \sin i$ so far as variations in Q and i are concerned.

The experimental results on smooth cylinders set at right angles to a fluid stream of velocity Q are represented in fig.1.‡ In this figure the drag coefficient C_D is plotted against $R = dQ\rho/\mu$; C_D is defined by the equation

$$N = \tfrac{1}{2}\rho Q^2 d C_D, \tag{2·1}$$

and $d = 2a$ is the diameter of the cylinder, ρ is the density of fluid and μ the viscosity. Curve a, fig. 1, shows the value of C_D. Fig. 1 also shows in b the part, $[C_D]_p$, of C_D which is due to the component of stress normal to the surface of the cylinder and in c the part $[C_D]_f$ due to the tangential component. Evidently

$$[C_D]_p + [C_D]_f = C_D. \tag{2·2}$$

It will be seen in fig. 1 that $[C_D]_p$ is nearly constant in the range $20 < R < 10^5$, where it varies only between $0·9$ and $1·1$. On the other hand, $[C_D]_f$ is found to be nearly equal to $4R^{-\frac{1}{2}}$ in this range.§ Using this value for $[C_D]_f$ the principle that lateral components of velocity are independent of longitudinal components yields the following expression for N:

$$N = \tfrac{1}{2}\rho d Q^2 \{[C_D]_p \sin^2 i + 4R^{-\frac{1}{2}} \sin^{\frac{3}{2}} i\}. \tag{2·3}$$

* A. R. McLeod, *Rep. Memo. aeronaut. Res. Comm., Lond.* no. 554 (1918).

† W. B. Sears, *J. aeronaut. Sci.* xv (1948), 49 and J. M. Wild, *J. aeronaut. Sci.* xvi (1949), 41.

‡ S. Goldstein, *Modern Developments in Fluid Dynamics* (1938), 425.

§ A. Thom, *Rep. Memo. aeronaut. Res. Comm. Lond.* nos. 1176 and 1194 (1928).

In Relf & Powell's experiments $R = 7.9 \times 10^3$, so that $4R^{\frac{1}{2}} = 0.045$. Their measurements of F_N are expressed in lb./ft. at 40 ft./sec., so that the factor of C_D in (2·1) is

$$\frac{30.48}{981 \times 453.6} \left(\tfrac{1}{2}\right) (0.00122) \left(\tfrac{3}{8} \times 2.54\right) (40 \times 30.48)^2 = 5.91 \times 10^{-2}$$

and (2·3) gives $\qquad F_N = 5.91 \times 10^{-2}\{[C_D]_p \sin^2 i + 0.045 \sin^{\frac{3}{2}} i\}.$ \hfill (2·4)

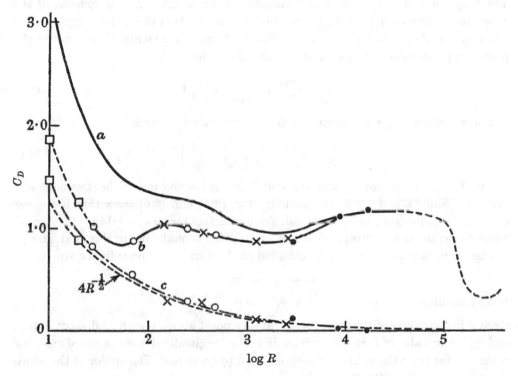

Fig. 1. Drag coefficients for smooth circular cylinders.

The value of $[C_D]_p$ which, when $i = 90°$, gives the best agreement with the measurements of column 2, table 1, is $[C_D]_p = 1.1$. Using this value in (2·4) the figures given in column 5, table 1, were calculated. Comparing columns 2 and 5 it will be seen that (2·3) is a good representation of Relf & Powell's measurements.

Table 2. *Values of C_D for a circular cylinder*

R	Tomotika's calculation	$1.0 + 4R^{-\frac{1}{2}}$
1	5·93	5·0
2	4·04	3·8
3	3·39	3·3
4	2·92	3·0

It is of interest to note that when $i = 90°$ (2·3) applies approximately to very low Reynolds numbers. The value of C_D for values of R from 0·4 to 4·0 has recently been calculated by Tomotika & Aoi.* The corresponding values of $C_D = 1·0 + 4R^{-\frac{1}{2}}$ which result from taking $[C_D]_p = 1$ in (2·3) are given in table 2 for comparison. It will be seen that when $i = 90°$ no large error will result from applying (2·3) down to $R = 2$.

Longitudinal component of force L

The longitudinal component is due entirely to the longitudinal component of the tangential stress which depends only on the distribution of w, the component of velocity parallel to the axis of the cylinder. If the components of velocity in the plane perpendicular to this axis are u and v the equation for w is

$$u\frac{\partial w}{\partial x} + v\frac{\partial w}{\partial y} = \frac{\mu}{\rho}\left(\frac{\partial^2 w}{\partial x^2} + \frac{\partial^2 w}{\partial y^2}\right). \tag{2·5}$$

The equation to the conduction of heat in two dimensions is

$$u\frac{\partial \vartheta}{\partial x} + v\frac{\partial \vartheta}{\partial y} = \frac{\kappa}{\rho\sigma}\left(\frac{\partial^2 \vartheta}{\partial x^2} + \frac{\partial^2 \vartheta}{\partial y^2}\right), \tag{2·6}$$

where ϑ is the temperature, κ the conductivity and σ the specific heat at constant pressure. Since u and v are independent of w, (2·5) and (2·6) show that if $\kappa = \mu\sigma$ the equations for ϑ and w are identical. Though in fact for air $\kappa = 1·14\mu\sigma$, it is worth while to make the assumption that $\kappa = \mu\sigma$ in order to make use of this analogue. The boundary conditions to be satisfied at the surface of the cylinder are

$$\vartheta = \vartheta_0, \quad w = 0,$$

and at infinity $\qquad\qquad \vartheta = 0, \quad w = W.$

Here ϑ is the excess of the temperature at any point above the air at distant points and ϑ_0 is the value of ϑ at the surface. W is the longitudinal component of velocity of the air far from the cylinder which is taken to be at rest. The analogue therefore shows that at all points of the field

$$\frac{W - w}{W} = \frac{\vartheta}{\vartheta_0}. \tag{2·7}$$

The rate of loss of heat from unit length of the cylinder is $H = \kappa\int\frac{\partial\vartheta}{\partial n}\,ds$, and the longitudinal force is $L = \mu\int\frac{\partial w}{\partial n}\,ds$, the integrations being taken round the perimeter of a cross-section. Hence

$$\frac{L}{H} = \frac{\mu}{\kappa}\frac{W}{\vartheta_0}. \tag{2·8}$$

It is therefore possible to use measurements of H to determine L.

As a result of a large number of measurements of the rate of loss of heat from wires stretched at right angles to a wind stream, King† concluded that the rate of loss

---

* S. Tomotika & T. Aoi, *Quart. J. Mech. appl. Math.* IV (1951), 401.

† L. V. King, *Phil. Trans. Roy. Soc.* A, CCXIV (1914), 373.

of energy per unit length of wire of radius u is $1 \cdot 432 \times 10^{-3} \vartheta_0 \sqrt{(aQ)}$ watts. Dividing by $4 \cdot 18$ King's experimental result is therefore

$$\frac{H}{\vartheta_0} = 3 \cdot 417 \times 10^{-4} \sqrt{(aQ)} \text{ cal/sec.} \tag{2.9}$$

Hence if $\kappa = \mu\sigma$,

$$\frac{L}{W} = 3 \cdot 14 \times 10^{-4} \frac{\mu}{\kappa} \sqrt{(aQ)}. \tag{2.10}$$

Assuming that air has no viscosity but has conductivity, Boussinesq* obtained for heat loss from a cylinder

$$\frac{H}{\vartheta_0 \sqrt{(aQ)}} = \frac{8}{\sqrt{\pi}} \sqrt{(\rho\kappa\sigma)} = 4 \cdot 51 \sqrt{(\rho\kappa\sigma)}. \tag{2.11}$$

King, on the other hand, made the same assumption about viscosity but assumed a different surface condition of heat transfer. His theoretical result was

$$\frac{H}{\vartheta_0 \sqrt{(aQ)}} = 2\sqrt{\pi} \sqrt{(\rho\kappa\sigma)} = 3 \cdot 55 \sqrt{(\rho\kappa\sigma)}. \tag{2.12}$$

The value of L appropriate to a flat plate of breadth b placed edgewise in a wind is†

$$L = 1 \cdot 328 \rho^{\frac{1}{2}} \mu^{\frac{1}{2}} Q^{\frac{3}{2}} b^{\frac{1}{2}}. \tag{2.13}$$

If therefore $\mu = \kappa/\sigma$ (2·8) gives

$$\frac{[H]_{\mu=\kappa/\sigma}}{\vartheta_0} = \frac{\kappa}{\mu} \frac{L}{Q} = 1 \cdot 328 \sqrt{(\kappa\rho\sigma Qb)}. \tag{2.14}$$

Making the assumption of Boussinesq and King that $\mu = 0$ it is found that for a flat plate

$$\frac{[H]_{\mu=0}}{\vartheta_0} = \frac{4}{\sqrt{\pi}} \sqrt{(\kappa\rho\sigma Qb)}. \tag{2.15}$$

Hence for a flat plate

$$\frac{[H]_{\mu=\kappa/\sigma}}{[H]_{\mu=0}} = \frac{1 \cdot 328 \sqrt{\pi}}{4} = 0 \cdot 589. \tag{2.16}$$

The effect of taking account of the motion in the viscous boundary layer is therefore to reduce the estimate made using Boussinesq's assumption by a factor $0 \cdot 589$. If the same factor applies to the cooling of cylinders Boussinesq's boundary condition, namely, $\vartheta = $ constant at $r = a$, would give

$$\frac{H}{\vartheta_0 \sqrt{(aQ)}} = 4 \cdot 51 \times 0 \cdot 589 \sqrt{(\rho\kappa\sigma)} = 2 \cdot 65 \sqrt{(\rho\kappa\sigma)}, \tag{2.17}$$

and King's boundary condition $(\partial\vartheta/\partial n = $ constant at $r = a)$ would give

$$\frac{H}{\vartheta_0 \sqrt{(aQ)}} = 3 \cdot 55 \times 1 \cdot 589 \sqrt{(\rho\kappa\sigma)}. \tag{2.18}$$

---

* J. Boussinesq, *J. Math. pures appl.* (1905), 285.
† H. Blasius, *Z. Math. Phys.* LVI (1908), 285.

It will be noticed that all these formulae are of the form

$$\frac{H}{\vartheta_0 \sqrt{(aQ)}} = A \sqrt{(\rho \kappa \sigma)}, \qquad (2\cdot19)$$

A being a constant. Using the values $\rho = 0\cdot00122$, $\kappa = 5\cdot0 \times 10^{-5}$, $\sigma = 0\cdot2417$, in King's experimental result $(2\cdot9)$, $(2\cdot19)$ is satisfied if $A = 2\cdot85$, which is not far from the value $2\cdot65$ predicted by taking the temperature of the air at the surface of the cylinder as the same as that of the cylinder and assuming that the reduction in heat transfer due to viscosity is the same as for a flat strip.

If it were true that $\kappa = \mu\sigma$ it would be possible to predict L directly using $(2\cdot8)$ and $(2\cdot19)$ with $A = 2\cdot85$, but since one set of measurements of L is available it is better to use the form suggested by $(2\cdot19)$ and determine the value of A which best fits the observations when $(2\cdot19)$ is used in conjunction with $(2\cdot8)$. In the case of a cylinder set obliquely Q must be replaced by $Q \sin i$. Assume therefore

$$\frac{L}{W} = A \frac{\mu}{\kappa} \sqrt{(\rho \kappa \sigma)} \sqrt{(aQ \sin i)}. \qquad (2\cdot20)$$

The kinetic theory of gases gives the relationship $\kappa = 1\cdot603\mu$, $C_v = 1\cdot14\mu\sigma$ for air, and since $W = Q \cos i$, $(2\cdot20)$ can be written in the form

$$L = \tfrac{1}{2} d\rho Q^2 \sqrt{\left(\frac{2}{1\cdot14}\right)} A \left(\frac{1}{R}\right)^{\tfrac{1}{2}} \cos i \sin^{\tfrac{3}{2}} i \qquad (2\cdot21)$$

where $R = dQ\rho/\mu$.

The first step in comparing this formula with Relf & Powell's observation is to divide the value of F_L given in column 3 of table 1 by $\cos i(\sin i)^{\tfrac{1}{2}}$. The results are given in column 6. It will be seen that except for the value at $i = 90°$, which is indeterminate and at $i = 80°$ where the observation is probably inaccurate and at $i = 0$ where they have little meaning, the values are very constant. The values of $F_L/\cos i$ are given in column 7 for comparison with those in column 6. It will be seen that those in column 6 are more constant than those in column 7. The mean value of $F_L(\cos i)^{-1} (\sin i)^{-\tfrac{1}{2}}$ in column 6 is $0\cdot0036$ lb./ft., so that the value of A which fits the observations is found by inserting in $(2\cdot21)$

$$0\cdot0036 \cos i \sin^{\tfrac{3}{2}} i \times \frac{453\cdot6 \times 981}{30\cdot48}$$

instead of L, and

$$d = \tfrac{3}{8} \times 2\cdot54, \quad Q = 40 \times 30\cdot48, \quad \rho = 0\cdot00122,$$

and

$$R = \frac{\tfrac{3}{8} \times 2\cdot54 \times 40 \times 30\cdot48 \times 0\cdot00122}{1\cdot8 \times 10^{-4}} = 7\cdot9 \times 10^3.$$

It is found in this way that Relf & Powell's measurements of F_L are closely predicted by $(2\cdot21)$ if A is taken as $4\cdot1$ or

$$A \sqrt{\left(\frac{2}{1\cdot14}\right)} = 5\cdot4,$$

so that (2·21) becomes

$$L = \tfrac{1}{2}\rho dQ^2(5\cdot4R^{-\frac{1}{2}})\cos i \sin^{\frac{1}{2}}i. \tag{2·22}$$

If A had the value 2·85, deduced by assuming $\kappa = \mu\sigma$ and using King's heat-transfer measurements, the numerical factor in (2·22) would have been 3·78 instead of 5·4. Though for air $a = 1\cdot14\mu\sigma$, it seems hardly likely that this error in the assumptions could account for the difference between 5·4 and 3·78.

It would be of interest to test (2·22) experimentally, particularly at low values of R.

Rough cylinders

If the cylinder is so rough that the boundary layer is not laminar the force cannot be analysed by the method used for smooth cylinders. In general, it is not possible to make any theory of the aerodynamics of rough cylinders because the force would depend on the exact nature of the roughness. If the roughness consisted of a number of long projections pointing equally in all directions, it is likely that the force on them would be in the direction opposite to that of their motion. The normal component of force N might be divided into portions due to the pressure and to the skin friction, the friction being the resultant force on the projections. In that case the force component formulae might be,

$$N = \tfrac{1}{2}\rho dQ^2\{[C_D]_p \sin^2 i + C_f \sin i\}, \\ L = \tfrac{1}{2}\rho dQ^2 C_f \cos i. \tag{2·23}$$

This case is illustrated as b in fig. 2. In the limiting case when the diameter of the cylinder was so small that C_D is negligible compared with C_f the 'cylinder' would look like a hairy string. The force components might then be taken as

$$N = \tfrac{1}{2}\rho dQ^2 C_f \sin i, \\ L = \tfrac{1}{2}\rho dQ^2 C_f \cos i. \tag{2·24}$$

(2·24) might also be expected to apply to a body in the form of a fine thread on which a number of equally spaced spherical beads were threaded. This case is illustrated as c in fig. 2.

Another possible form of roughness might consist of thin discs or plates set at right angles to a cylinder. In that case the roughness would make a much greater contribution to L than to N, and the appropriate formulae might be

$$N = \tfrac{1}{2}\rho dQ^2[C_D]_p \sin^2 i, \\ L = \tfrac{1}{2}\rho dQ^2 C_f \cos i. \tag{2·25}$$

This case is illustrated as d in fig. 2. All these formulae, except those for a smooth cylinder shown as a in fig. 2, are entirely speculative; they are set down because by using them in an analysis of swimming it might be possible to derive qualitative ideas as to how the nature of the surface of an animal affects its quality as a swimmer.

$$N/\tfrac{1}{2}\rho dQ^2 \qquad\qquad L/\tfrac{1}{2}\rho dQ^2$$

(a) $[C_D]_p \sin^2 i + 4R^{-\frac{1}{2}}\sin^{\frac{3}{2}} i$ $5\cdot4R^{-\frac{1}{2}}\cos i \sin^{\frac{1}{2}} i$

(b) $[C_D]_p \sin^2 i + C_f \sin i$ $C_f \cos i$

(c) $C_f \sin i$ $C_f \cos i$

(d) $[C_D]_p \sin^2 i$ $C_f \cos i$

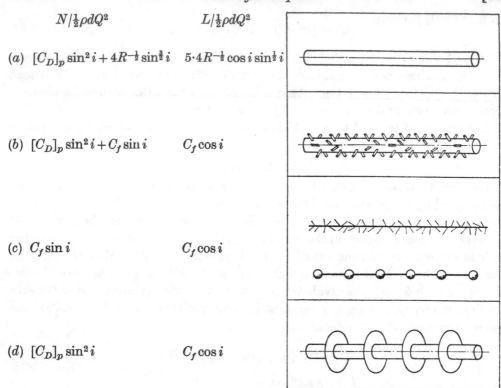

Fig. 2. Types of roughness with corresponding force formulae.

3. GEOMETRICAL AND KINEMATICAL CONSIDERATIONS AND ASSUMPTIONS

If the backward velocity of the waves relative to the mean position of any material element of the cylinder is U and the velocity with which these waves drive it forwards V, and if the centre line of the cylinder is deformed into a sine curve of amplitude B and wave-length λ, the equation which represents the centre line at time t is

$$y = B\sin\frac{2\pi}{\lambda}\{x+(U-V)t\}. \tag{3.1}$$

Here x is the co-ordinate representing distance relative to fixed axes in the direction along which the animal is swimming, y is at right angles to x.

The analysis is simplified by giving the whole field a velocity $U-V$ in the direction $+x$. This reduces the centre line to rest, but each element of the flexible cylinder is now travelling parallel to the centre line with velocity q which is constant if the centre line is assumed to be inextensible. Though, in fact, real animals are by no means inextensible it does not appear that in swimming, as distinct from progressing over solid ground, the centre lines suffer appreciable extension or contraction. The self-propelling property of an inextensible cylinder will be investigated and q will therefore be taken as constant. Fig. 3 shows the geometry of the field. Since the fluid is now moving parallel to the axes of x with velocity $U-V$, the angle of incidence of the stream on a *fixed* sine curve would be θ, the angle between the tangent to the sine curve and the axis of x (see fig. 3). Since the particles of the 'snake'

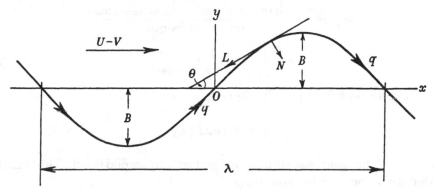

Fig. 3. Definition of axes.

are moving along the curve with velocity q, the angle of incidence i, which determines the mechanical reaction of the fluid on an element of the cylinder, is given by

$$\tan i = \frac{(U-V)\sin\theta}{q-(U-V)\cos\theta},\qquad(3\cdot2)$$

and the relative velocity Q is given by

$$Q^2 = q^2 - 2q(U-V)\cos\theta + (U-V)^2.\qquad(3\cdot3)$$

It is convenient to express these equations in non-dimensional form. Writing

$$V = nU,\quad q = \gamma U,\quad z = \frac{2\pi}{\lambda}\{x+(U-V)t\},\quad \frac{2\pi B}{\lambda} = \tan\alpha,\qquad(3\cdot4)$$

(3·1) becomes

$$\frac{2\pi y}{\lambda} = \tan\alpha\sin z,\qquad(3\cdot5)$$

so that

$$\tan\theta\,\frac{dy}{dx} = \tan\alpha\cos z,\qquad(3\cdot6)$$

and $$\cos\theta = (1+\tan^2\alpha\cos^2 z)^{-\frac12},\quad \sin\theta = \tan\alpha\cos z(1+\tan^2\alpha\cos^2 z)^{-\frac12};\qquad(3\cdot7)$$

(3·3) becomes

$$\frac{Q^2}{U^2} = \gamma^2 + (1-n)^2 - 2\gamma(1-n)\cos\theta;\qquad(3\cdot8)$$

(3·2) with (3·8) gives

$$\frac{Q}{U}\sin i = (1-n)\sin\theta,\qquad(3\cdot9)$$

and

$$\frac{Q}{U}\cos i = \gamma - (1-n)\cos\theta.\qquad(3\cdot10)$$

Before going further it may be remarked that γ is a function of α only. Since the velocity U of the waves is defined as the velocity relative to the mean position of the particles of the animal, the time taken for a wave travelling with velocity U to go one wave-length is equal to the time taken for a particle moving with velocity q along the centre line of the body to transverse one wave-length, so that

$$\gamma = \frac{q}{U} = \frac{\text{length of one wave-length of a sine curve}}{\text{one wave-length}}.$$

Hence

$$\gamma = \int_0^\lambda \frac{ds}{\lambda} = \frac{1}{2\pi} \int_0^{2\pi} \frac{dz}{\cos\theta}. \tag{3.11}$$

Substituting for $\cos\theta$ from (3·7) it will be found that

$$\gamma = \frac{2}{\pi} \sec\alpha E(\alpha), \tag{3.12}$$

where $E(\alpha)$ is the complete elliptic integral of the second kind. This has been tabulated for values of α between 0 and $\frac{1}{2}\pi$.

Equilibrium conditions

It will be assumed that the animal forms itself into an integral number of wave-lengths. The equilibrium condition is then that the resultant force on one wave-length is zero. The component of force in the direction y is certainly zero because the symmetry of the assumed shape ensures that this shall be the case. The equilibrium condition is therefore given by

$$\int_0^\lambda N \sin\theta \, ds = \int_0^\lambda L \cos\theta \, ds. \tag{3.13}$$

Further progress can only be made by substituting expressions like those given in §2 to represent N and L in terms of Q and i. If one of these be selected and Q and i given their values in terms of z using (3·3), (3·7), (3·9) and (3·10), the equilibrium equation (3·13) is then expressed as a relation between definite integrals. For any given value of α and n these can be integrated numerically, and a relation between α, n and the experimentally determined constants contained in the expressions for N and L can be found. The two types of expression (2·3) and (2·22) appropriate to smooth animals and (2·23), (2·24) or (2·25) for rough animals will next be discussed separately.

4. SMOOTH ANIMALS

The expressions (2·3) and (2·22) developed in §2 for the components of force on a smooth cylinder contain $R = Qd\rho/\mu$. In discussing the swimming characteristics of any waving movement the animal may make, it is more convenient to define the Reynolds number in terms of velocity U which is the same for all parts of the body. Thus if

$$R_1 = Ud\rho/\mu, \tag{4.1}$$

from (3·8)
$$R = R_1\{\gamma^2 + (1-n)^2 - 2\gamma(1-n)\cos\theta\}^{\frac{1}{2}}; \qquad (4·2)$$

substituting for R, Q and i in (2·3) and (2·22)

$$N = \tfrac{1}{2}\rho dU^2\{[C_D]_p(1-n)^2\sin^2\theta + 4R_1^{-\frac{1}{2}}(1-n)^{\frac{3}{2}}\sin^{\frac{3}{2}}\theta\} \qquad (4·3)$$

and
$$L = \tfrac{1}{2}\rho dU^2\{5·4(1-n)^{\frac{1}{2}}R_1^{-\frac{1}{2}}\sin^{\frac{1}{2}}\theta[\gamma - (1-n)\cos\theta]\}. \qquad (4·4)$$

At this stage it is necessary to point out that θ may be positive or negative. The first term in the expression (4·3) for N contains $\sin^2\theta$ as a factor, but the direction of the normal force is reversed when θ changes sign. For this reason when the expression for N is inserted in (4·3), $(\sin\theta)|\sin\theta|$ should be written instead of $\sin^2\theta$, but if the limits in the integrals (3·13) are taken as 0 and $\tfrac{1}{4}\lambda$ instead of 0 and λ, θ does not change sign and symmetry ensures that the resulting equation is true if (4·3) is true. After inserting (4·3) and (4·4) in (3·13), the equilibrium is found to be

$$[C_D]_p R_1^{\frac{1}{2}}C = 5·4\gamma(1-n)^{-\frac{1}{2}}\frac{2}{\pi}\int_0^{\frac{1}{2}\pi}\sin^{\frac{1}{2}}\theta\,dz$$

$$-(1-n)^{-\frac{1}{2}}\frac{2}{\pi}\left\{5·4\int_0^{\frac{1}{2}\pi}\sin^{\frac{1}{2}}\theta\cos\theta\,dz + 4\int_0^{\frac{1}{2}\pi}\sin^{\frac{5}{2}}\theta\sec\theta\,dz\right\}, \qquad (4·5)$$

where
$$C = \frac{2}{\pi}\int_0^{\frac{1}{2}\pi}\frac{\sin^3\theta}{\cos\theta}\,dz. \qquad (4·6)$$

When the expressions (3·7) for $\sin\theta$ and $\cos\theta$ are inserted in (4·5) the integrals are intractable but they can be integrated numerically. For this purpose a value of α was first fixed. (3·7) was then used to find the values of $\cos\theta$ and $\sin\theta$ corresponding with the ten values of z which divide 90° into nine equal parts. When integrating, the values of the integrands at $z = 0$ and $z = \tfrac{1}{2}\pi$ were halved and added to the sum of the other eight values. The result was divided by 9. This gives an approximation to the value of $\dfrac{2}{\pi}\displaystyle\int_0^{\frac{1}{2}\pi}$ (integrand) dz which is accurate enough for the present work.

Values of
$$I_1 = \frac{2}{\pi}\int_0^{\frac{1}{2}\pi}\sin^{\frac{1}{2}}\theta\,dz, \quad I_2 = \frac{2}{\pi}\int_0^{\frac{1}{2}\pi}\sin^{\frac{1}{2}}\theta\cos\theta\,dz,$$

$$I_3 = \frac{2}{\pi}\int_0^{\frac{1}{2}\pi}\sin^{\frac{5}{2}}\theta\sec\theta\,dz, \quad I_4 = \frac{2}{\pi}\int_0^{\frac{1}{2}\pi}\sin^{\frac{1}{2}}\theta\sec\theta\,dz$$

obtained in this way are given in table 3.

Table 3

α	I_1	I_2	I_3	I_4
0	0	0	0	0
10	0·315	0·312	0·006	0·318
20	0·446	0·429	0·034	0·463
30	0·547	0·501	0·098	0·599
40	0·633	0·537	0·213	0·750
50	0·711	0·542	0·413	0·954
60	0·782	0·505	0·762	1·267
70	0·849	0·424	1·450	1·873
80	0·910	0·276	3·388	3·603
90	1·000	0	∞	∞

When the expressions (3·7) for $\cos\theta$ are substituted in (4·6) the integral C can be expressed explicitly as a function of α. In fact

$$C = \frac{2}{\pi}\left[\tan\alpha - \cos\alpha\log\tan\left(\tfrac{1}{4}\pi + \tfrac{1}{2}\alpha\right)\right]. \tag{4·7}$$

Values of C and γ and $B/\lambda = (1/2\pi)\tan\alpha$ are given in columns 2, 3 and 4 of table 4. Values of $[C_D]_p R_1^{\frac{1}{2}}$ obtained by inserting these values in (4·5) are given in table 5.

Table 4

$\alpha°$	B/λ	γ	C	A_1	A_2	A_3	A_4	A_5	A_6
0	0	1·0000	0	1·0000	1·0000	1·0000	1·0000	1·0000	1·0000
5	0·0139	1·0019	0·0004	0·9981	0·9962	0·9943	0·9924	0·9905	0·9887
10	0·0281	1·0077	0·0022	0·9923	0·9848	0·9773	0·9699	0·8626	0·9560
15	0·0426	1·0178	0·0077	0·9827	0·9659	0·9518	0·9335	0·9216	0·9028
20	0·0579	1·0324	0·0185	0·9691	0·9397	0·9116	0·8847	0·8591	0·8346
25	0·0742	1·0523	0·0367	0·9514	0·9063	0·8644	0·8256	0·7894	0·7553
30	0·0919	1·0788	0·0647	0·9294	0·8660	0·8071	0·7577	0·7097	0·6698
35	0·1114	1·1132	0·1053	0·9029	0·8192	0·7471	0·7254	0·6329	0·5829
40	0·1335	1·1577	0·1558	0·8713	0·7660	0·6793	0·6077	0·5524	0·4985
45	0·1592	1·2160	0·2398	0·8347	0·7071	0·6080	0·5303	0·4689	0·4198
50	0·1897	1·2930	0·3452	0·7921	0·6428	0·5239	0·4520	0·3839	0·3486
55	0·2273	1·3970	0·4877	0·7430	0·5736	0·4598	0·3812	0·3277	0·2856
60	0·2757	1·5420	0·6835	0·6864	0·5000	0·3856	0·3125	0·2641	0·2305
65	0·3413	1·7531	0·9600	0·6211	0·4226	0·3132	0·2490	0·2092	0·1824
70	0·4373	2·0817	1·3713	0·5453	0·3420	0·2437	0·1910	0·1602	0·1400
75	0·5940	2·6476	2·0418	0·4561	0·2588	0·1826	0·1385	0·1197	0·1018
80	0·9026	3·8133	3·3410	0·3485	0·1736	0·1152	0·0894	0·0756	0·0664
85	1·8191	7·3732	7·1028	0·2127	0·0872	0·0567	0·0439	0·0377	0·0329
90	—	—	—	0	0	0	0	0	0

Table 5. *Smooth animals: values of* $[C_D]_p R_1^{\frac{1}{2}}$

n	α ... 20°	20°	30°	40°	50°	60°	70°	80°
0·95	—	—	—	—	—	—	—	482
0·9	$2 \cdot 2 \times 10^4$	$3 \cdot 8 \times 10^3$	$1 \cdot 4 \times 10^3$	713	413	274	201	163
0·8	$6 \cdot 9 \times 10^3$	1200	444	226	131	87·6	64·6	52·6
0·7	$3 \cdot 3 \times 10^3$	603	213	109	63·5	42·7	31·6	25·9
0·6	1820	321	119	61·1	35·9	24·3	18·2	15·0
0·5	1083	192	71·6	37·0	21·0	15·0	11·4	9·50
0·3	655	118	44·2	23·1	13·80	9·60	7·37	6·26
0·4	384	71·0	26·9	14·3	8·70	6·17	4·84	4·19
0·2	204	39·5	15·3	8·38	5·26	3·87	3·13	2·80
0·1	78·4	17·6	7·25	4·27	2·86	2·26	1·94	1·82
0	—	1·82	1·41	1·28	1·12	1·08	1·06	1·11
−0·1	—	—	—	—	—	0·20	0·41	0·56
−0·2	—	—	—	—	—	—	—	0·16

The results of these calculations can conveniently be displayed in a diagram which may be called a 'swimming diagram' in which contours of equal values of $[C_D]_p R_1^{\frac{1}{2}}$ are shown, the ordinates representing $n = V/U$ and the abscissae, α. To produce such a diagram it is necessary to interpolate between the calculated values given in table 5. The resulting diagram is shown in fig. 4. Since the angle α is not directly measurable on photographs of swimming animals, the scale for $B/\lambda = (1/2\pi)\tan\alpha$ is marked on the top of the diagram. Limitations to the application of this analysis are discussed in §9.

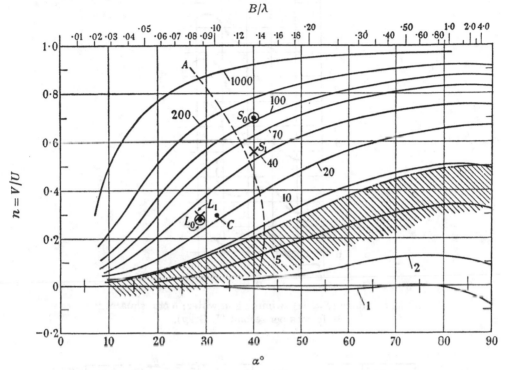

Fig. 4. Swimming diagram for long, smooth animals showing contours for constant values of $\lfloor C_D \rfloor_p R_l^{\frac{1}{4}}$. Line A represents conditions for maximum speed with given energy output.

5. COMPARISON WITH PHOTOGRAPHS OF SWIMMING
SNAKE AND LEECH

Snake

Fig. 5 shows the tracings of photographs of a snake taken at intervals of $\frac{1}{16}$ sec. by Professor James Gray. The snake is swimming in a shallow trough over a grid of 5 cm. squares. Each photograph is displaced two squares downwards from its predecessor. It will be seen that the waves increase as they pass from head to tail. To measure V and U the centre lines of the snake in each position were traced and the results superposed to form the diagram of fig. 6. The position of the head in each case is shown as a black dot; the broken line AB in fig. 6 is drawn so that it passes closely along the path pursued by the head. It will be seen that the head only deviates slightly from this line, but that the tail moves violently. Using fig. 6 Gray found that the head moves 10 cm. in $\frac{5}{16}$ sec. so that $V = 32$ cm./sec. To measure U is not easy, partly because the amplitude of the motion is not constant. The velocities of the maxima (i.e. the points on the curves furthest from the broken line

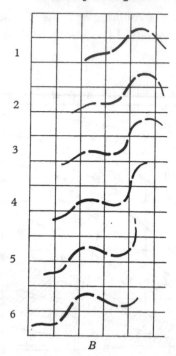

Fig. 5. Snake (*Natrix*) swimming in water; 5 cm. squares,
16 frames per second (J. Gray).

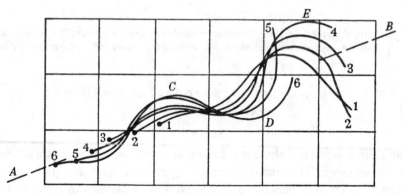

Fig. 6. Centre lines of snake superposed. AB, mean line of motion of head;
C, D, E, maximum distances from AB.

AB in fig. 6) were measured. The maximum marked D in fig. 6 appears in all the
frames. During the $\frac{5}{16}$ sec. interval between frame 1 and frame 6 it moved back
4·5 cm. The maximum marked C is pronounced only between frames 3 and 6.
During the corresponding interval of $\frac{3}{16}$ sec. it moved back 1·8 cm. The maximum
marked E exists between frames 1 and 4. During the corresponding $\frac{3}{16}$ sec. it moves
3·4 cm. The wave velocities corresponding with the maxima C, D and E respectively

are therefore 9·6, 14·5 and 18·0 cm./sec. These velocities are equal to $U - V$. Their mean is 14 cm./sec., so that if $V = 32$ cm./sec.

$$n = \frac{V}{U} = \frac{32}{32+14} = 0·7. \tag{5·1}$$

A rough estimate of the mean amplitude of the waves can be obtained by finding the distance between lines parallel to the broken line AB which touch the maxima at D and E. In the case of frame 6 this is $2B = 4·7$ cm, and the corresponding wavelength is 17·6 cm. So that

$$B/\lambda = 0·134. \tag{5·2}$$

The point corresponding with $n = 0·7$, $B/\lambda = 0·134$ is marked in fig. 4 at S_0. It will be seen that it falls near the line $[C_D]_p R_1^{\frac{1}{2}} = 100$. The diameter of cross-section of the snake as measured on the photographs was 0·6 cm. The velocity U was $32 + 14 = 46$ cm./sec. and for water $\mu = 0·011$, so that

$$R_1 = \frac{46 \times 0·6}{0·011} = 2500.$$

It has been seen in §2 that $[C_D]_p$ is approximately 1·0, so that

$$[C_D]_p R_1^{\frac{1}{2}} = (2500)^{\frac{1}{2}} = 50. \tag{5·3}$$

The point where the line $[C_D]_p R_1^{\frac{1}{2}} = 50$ cuts the abscissa $B/\lambda = 0·134$ is shown as a cross at S_1 in fig. 4. This corresponds to $V/U = 0·55$. The swimming efficiency which may be judged by the value of U/V is therefore rather larger than that predicted assuming a wave of constant amplitude, but the fact that the measurements of B/λ and of V/U vary over such a wide range would make accurate prediction impossible.

Leech

Fig. 7 shows eight successive frames in a sequence of photographs of a large leech swimming from right to left over a grid of 2 cm. squares. The frequency of the photographs is 15/sec. It was found that the average speed of the head forward was 2·0 cm. in seven frames so that $V = 2 \times \frac{15}{7} = 4·3$ cm./sec. The velocity backwards of the crests and troughs of the waves were obtained by measuring the slopes of the broken lines drawn through them in fig. 7. The velocity backwards of the first wave corresponding with the crest which extends from frame 1 to frame 6 is $U - V = 12$ cm./sec., that corresponding with the first trough extending from frame 3 to frame 8 is also 12 cm./sec. The velocity of the second crest extending from frame 5 to frame 8 is 7·5 cm./sec., so that if the first crest and first trough which each appear on six frames are used $U = 12 + 4·3 = 16·3$ cm./sec., while the last crest which appears only in the four frames 5 to 8 give $U = 7·5 + 4·3 = 11·8$ cm./sec. Perhaps the best way to weight these observations is to take

$$U = \frac{5 \times 16·3 + 5 \times 16·3 + 3 \times 11·8}{13} = 15·3 \text{ cm./sec.,}$$

so that
$$n = \frac{V}{U} = \frac{4 \cdot 3}{15 \cdot 3} = 0 \cdot 28. \qquad (5 \cdot 4)$$

To obtain a mean value for B/λ a tangent line was drawn wherever possible to touch the curved profile at two points, and the maximum distance of the body of the leech from this line was measured in each frame. The distance was taken as $2B$ and the distance between the points of contact as λ; in this way the following values of B/λ were obtained: $0 \cdot 06$, $0 \cdot 07$, $0 \cdot 09$, $0 \cdot 09$, $0 \cdot 08$, $0 \cdot 16$, $0 \cdot 07$. The mean value is

$$B/\lambda = 0 \cdot 089. \qquad (5 \cdot 5)$$

The point $V/U = 0 \cdot 28$, $B/\lambda = 0 \cdot 089$ is marked as L_0 in fig. 4. It will be seen that it lies between the contours

$$[C_D]_p R_1^{\frac{1}{2}} = 20 \quad \text{and} \quad [C_D]_p R_1^{\frac{1}{2}} = 40.$$

Unfortunately, the leech is not circular in section but approximately elliptical; in fact, the dimensions of the axes or the ellipse as seen in fig. 7 are $0 \cdot 2$ and $0 \cdot 9$ cm. Taking d as the mean of these

$$d = 0 \cdot 55 \text{ cm.},$$

so that
$$R_1 = \frac{0 \cdot 55 \times 15 \cdot 3}{0 \cdot 011} = 770.$$

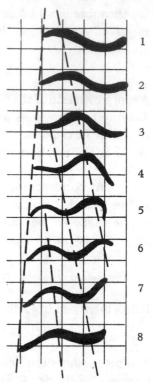

Fig. 7. Leech swimming in water; 2 cm. squares, 15 frames per second (J. Gray).

Taking $[C_D]_p$ as $1 \cdot 0$ this gives $\qquad [C_D]_p R_1^{\frac{1}{2}} = 28.$ $\qquad (5 \cdot 6)$

The point L_1 in fig. 4 corresponds with the value of V/U predicted by the diagram.

It will be seen therefore that the swimming performance of the leech is close to what would be expected if it were a smooth cylinder.

6. ENERGY REQUIRED FOR PROPULSION OF SMOOTH ANIMALS

The rate, W, at which the animal does work on the surrounding fluid per unit length of its body is the mean value of

$$N(U - V) \sin \theta + L\{q - (U - V) \cos \theta\}$$
or
$$U(1 - n)\{N \sin \theta - L \cos \theta\} + qL. \qquad (6 \cdot 1)$$

The first term in $(6 \cdot 1)$ vanishes owing to the equilibrium condition $(3 \cdot 13)$ so that from $(4 \cdot 4)$

$$W = q \times (\text{mean value of } L)$$
$$= \tfrac{1}{2}\rho d(5 \cdot 4\gamma U^3 R_1^{-\frac{1}{2}})(1 - n)^{-\frac{1}{2}} \frac{2}{\pi} \int_0^{\frac{1}{2}\pi} \{\gamma - (1 - n) \cos \theta\} \sin^{\frac{1}{2}} \theta \sec \theta \, dz. \qquad (6 \cdot 2)$$

It is of interest to know the amplitude of wave which would propel the animal at a given speed with the least output of energy. For this reason it is useful to express (6·2) in terms of V rather than U. Remembering that R_1 contains U as a factor, the required expression is

$$W = \frac{5\cdot4}{2}\rho d V^3 \left(\frac{\mu}{\rho d V}\right)^{\frac{1}{2}} G(n,\alpha),\tag{6·3}$$

where

$$G(n,\alpha) = (1-n)^{\frac{1}{2}} n^{-\frac{3}{2}}\{(\gamma^2-\gamma)I_4 + n\gamma I_1\}\tag{6·4}$$

and

$$I_4 = \frac{2}{\pi}\int_0^{\frac{1}{2}\pi}\sin^{\frac{1}{2}}\theta\sec\theta\,dz,\quad I_1 = \frac{2}{\pi}\int_0^{\frac{1}{2}\pi}\sin^{\frac{1}{2}}\theta\,dz.\tag{6·5}$$

Values of I_1 and I_4 are given in table 3 and values of $G(n\alpha)$ are given in table 6. Using auxiliary diagrams it is possible to construct from the figures of table 6 a diagram analogous to fig. 4 which displays lines of constant $G(n,\alpha)$ on a diagram (fig. 8) whose ordinates are n and abscissae α.

Table 6. *Values of $G(n,\alpha)$*

$\alpha°$ $n \ldots$ 0·1	0·2	0·3	0·4	0·5	0·6	0·7	0·8	0·9
0	—	—	—	—	—	—	—	—
—	—							
10	11·2	3·44	1·80	1·01	0·66	0·44	0·30	0·20	0·12
20	23·8	6·27	2·91	1·66	1·05	0·70	0·47	0·31	0·18
30	49·6	11·2	4·81	2·62	1·61	1·04	0·69	0·45	0·26
40	104	20·1	8·37	4·33	2·56	1·62	1·05	0·67	0·38
50	212	43·0	16·1	7·98	4·54	2·78	1 76	1·10	0·62
60	578	102	36·8	17·5	9·64	5·74	3·54	2·16	1·19
70	1957	335	117	54·0	28·0	16·8	10·1	6·06	3·27
80	14780	2481	848	385	203	116	68·6	40·4	21·4

Wave for minimum output of energy

The lines displayed in fig. 8 correspond with conditions under which constant energy output is required to propel an animal at a given speed. By superposing the diagram on that of fig. 4 which displays lines of constant $[C_D]_p\, R_1^{\frac{1}{4}}$ a set of possible values of R_1 are obtained, but since for a constant V and d, R_1 is not constant, this superposition has little meaning. To find the minimum value of W corresponding with any given value of d and V it is necessary to construct a diagram similar to fig. 4 but displaying contours of constant $[C_D]_p\left(\dfrac{dV\rho}{\mu}\right)^{\frac{1}{4}}$. That is to say, the values of $[C_D]_p\, R_1^{\frac{1}{4}}$ given in table 5 must be multiplied by $n^{\frac{1}{4}}$ to give the figures of table 7 and a new set of contours drawn. The diagram obtained in this way is shown in fig. 9. It will be seen that its curves are similar in general appearance to those of fig. 4.

The set of values of α and n which characterize the propulsion of the animal at a given value of V are found by moving along the appropriate contour in fig. 9. To find the particular value of α which corresponds with the least energy output,

$G(n, \alpha)$

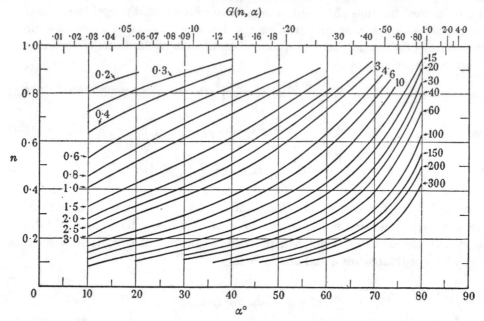

Fig. 8. $G(n, \alpha)$. Shows in non-dimensional form the rate of dissipation of energy.

Table 7. *Values of* $[C_D]_p \left(\dfrac{dV\rho}{\mu}\right)^{\frac{1}{2}}$

$\alpha°$ n ... 0·1	0·2	0·3	0·4	0·5	0·6	0·7	0·8	0·9
10 24·8	91·1	210	414	766	1415	2757	6203	20968
20 5·57	17·7	38·9	74·6	136	249	504	1079	3632
30 2·29	6·85	14·7	28·0	50·6	92·2	178	397	1280
40 1·35	3·75	7·83	14·6	21·2	47·3	90·8	202	676
50 0·90	2·35	4·77	8·75	15·5	27·8	53·1	117	392
60 0·71	1·73	3·38	6·07	10·6	18·8	35·7	78·3	260
70 0·61	1·40	2·65	4·66	8·03	14·1	26·4	57·8	191
80 0·58	1·25	2·30	3·96	6·71	11·7	21·7	47·1	154

fig. 8 may be superposed on fig. 9. The required value of α will be that corresponding with the point on the contour of $[C_D]_p \left(\dfrac{dV\rho}{\mu}\right)^{\frac{1}{2}}$, where it touches one of the super-posed contours of $G(n, \alpha)$. In this way the points on the broken line shown as A in fig. 4 were found. Though there is no physical reason to suppose that animals do in fact swim by forming the particular type of wave which involves least output of energy, it is worth noticing that the point in fig. 4, which corresponds to a swimming snake is close to this line. The point corresponding with a swimming leech is well to the left of the line, showing that this specimen bent its body rather less than would be expected if the most efficient movement were used.

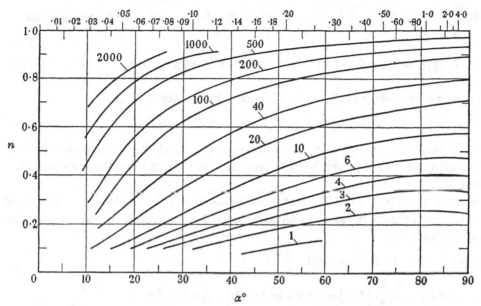

Fig. 9. Contours of $[C_D]_n \left(\dfrac{dV\rho}{\mu}\right)^{\frac{1}{2}}$.

7. Swimming of a rough flexible cylinder

Similar analysis to that applied to a smooth swimmer can be applied to one with a rough surface, using any of the force-component formulae discussed in §2. When the geometrical relationships (3·2), (3·9), (3·10) giving i and Q as functions of U, θ, n and α are inserted, the force components appropriate to the cases b, c, d of fig. 2 are

(b) $\dfrac{N}{\frac{1}{2}\rho dU^2} = [C_D]_p (1-n^2)\sin^2\theta + C_f(1-n)\sin\theta\{\gamma^2+(1-n)^2-2\gamma(1-n)\cos\theta\}^{\frac{1}{2}}$ (7·1)

$\dfrac{L}{\frac{1}{2}\rho dU^2} = C_f\{\gamma-(1-n)\cos\theta\}\{\gamma^2+(1-n)^2-2\gamma(1-n)\cos\theta\}^{\frac{1}{2}};$ (7·2)

(c) is obtained from (b) by setting $[C_D]_p = 0$ and for (d)

$\dfrac{N}{\frac{1}{2}\rho dU^2} = [C_D]_p (1-n)^2\sin^2\theta,$ (7·3)

$\dfrac{L}{\frac{1}{2}\rho dU^2} = C_f\{\gamma-(1-n)\cos\theta\}\{\gamma^2+(1-n)^2-2\gamma(1-n)\cos\theta\}^{\frac{1}{2}}.$ (7·4)

Substituting for L and N in the equation of equilibrium (3·13) a relationship between $[C_D]_p/C_f$ and n, γ and α is obtained in each case. These are

(b) $\dfrac{[C_D]_p}{C_f}(1-n)^2 C = \dfrac{2}{\pi}\displaystyle\int_0^{\frac{1}{2}\pi}\{\gamma-(1-n)\sec\theta\}\{\gamma^2+(1-n)^2-2\gamma(1-n)\cos\theta\}^{\frac{1}{2}}\,dz,$

(7·5)

(d) $\dfrac{[C_D]_p}{C_f}(1-n)^2 C = \dfrac{2}{\pi}\displaystyle\int_0^{\frac{1}{2}\pi} \{\gamma-(1-n)\cos\theta\}\{\gamma^2+(1-n)^2-2\gamma(1-n)\cos\theta\}^{\frac{1}{2}}\,dz,$

$$(7\cdot6)$$

(c) $\displaystyle\int_0^{\frac{1}{2}\pi} \{\gamma(1-n)\sec\theta\}\{\gamma^2+(1-n^2)-2\gamma(1-n)\cos\theta\}^{\frac{1}{2}}\,dz = 0.$ $\qquad(7\cdot7)$

The occurrence of $\sec\theta$ in $(7\cdot5)$ arises because the last term

$$C_f(1-n)\sin\theta\{\gamma^2+(1-n)^2-2\gamma(1-n)\cos\theta\}^{\frac{1}{2}}$$

in $(7\cdot1)$ is multiplied by $\sin\theta\,dz/\cos\theta$ and transferred to the right-hand side of $(7\cdot5)$, where $\sin^2\theta/\cos\theta$ combines with $\cos\theta$ to produce $\sec\theta$. Here C has the same meaning as in $(4\cdot6)$.

To integrate $(7\cdot5)$ and $(7\cdot6)$ numerically the method adopted in the case of smooth cylinders may be used. As an alternative the integrands can be expanded in powers of $\cos\theta$:

$$\{\gamma^2+(1-n)^2-2\gamma(1-n)\cos\theta\}^{\frac{1}{2}}$$

$$= \{\gamma^2+(1-n)^2\}^{\frac{1}{2}}\Big\{1-\tfrac{1}{2}m\cos\theta-\tfrac{1}{8}m^2\cos^2\theta-\tfrac{1}{16}m^3\cos^3\theta$$

$$-\frac{5}{2^7}m^4\cos^4\theta-\frac{7}{2^8}m^5\cos^5\theta-\frac{21}{2^{10}}m^6\cos^6\theta\ldots\Big\}, \quad (7\cdot8)$$

where $\qquad\qquad\qquad m = \dfrac{2\gamma(1-n)}{\gamma^2+(1-n)^2}.$ $\qquad\qquad(7\cdot9)$

Writing $\qquad A_n = \dfrac{2}{\pi}\displaystyle\int_0^{\frac{1}{2}\pi}\cos^n\theta\,dz = \dfrac{2}{\pi}\displaystyle\int_0^{\frac{1}{2}\pi}(1+\tan^2\alpha\cos^2 z)^{-\frac{1}{2}n}\,dz,$ $\qquad(7\cdot10)$

$$\frac{2}{\pi}\int_0^{\frac{1}{2}\pi}\{\gamma-(1-n)\sec\theta\}\{\gamma^2+(1-n)^2-2\gamma(1-n)\cos\theta\}^{\frac{1}{2}}\,dz$$

$$= \{\gamma^2+(1-n)^2\}^{\frac{1}{2}}\Big\{-(1-n)A_{-1}+[\tfrac{1}{2}m(1-n)+\gamma]A_0-[\tfrac{1}{2}m\gamma-\tfrac{1}{8}m^2(1-n)]A_1$$

$$-[\tfrac{1}{8}m^2\gamma-\tfrac{1}{16}m^3(1-n)]A_2-\Big[\tfrac{1}{16}m^3\gamma-\frac{5}{2^7}m^4(1-n)\Big]A_3$$

$$-\Big[\frac{5}{2^7}m^4\gamma-\frac{7}{2^8}m^5(1-n)\Big]A_4-\Big[\frac{7}{2^8}m^5\gamma-\frac{21}{2^{10}}m^6(1-n)\Big]A_5$$

$$-\Big[\frac{21}{2^{10}}m^6\gamma-\frac{33}{2^{11}}m^7(1-n)\Big]A_6-\ldots\Big\} \quad (7\cdot11)$$

and

$$\frac{2}{\pi}\int_0^{\frac{1}{2}\pi}\{\gamma-(1-n)\cos\theta\}\{\gamma^2+(1-n)^2-2\gamma(1-n)\cos\theta\}^{\frac{1}{2}}\,dz$$

$$= \{\gamma^2+(1-n)^2\}^{\frac{1}{2}}\Big\{\gamma-[\tfrac{1}{2}m\gamma+1-n]A_1+[\tfrac{1}{2}m(1-n)-\tfrac{1}{8}\gamma m^2]A_2$$

$$+[\tfrac{1}{8}m^2(1-n)-\tfrac{1}{16}m^3\gamma]A_3+\Big[\tfrac{1}{16}m^3(1-n)-\frac{5}{2^7}m^4\gamma\Big]A_4+\ldots\Big\}. \quad (7\cdot12)$$

Expressions for the integrals A_n can be derived from recurrence formula

$$\frac{\mathrm{d}}{\mathrm{d}\alpha}(A_n) = \frac{2}{\pi}\int_0^{\frac{1}{2}\pi}\frac{\mathrm{d}}{\mathrm{d}\alpha}(1+\tan^2\alpha\cos^2 z)^{-\frac{1}{2}n}\,\mathrm{d}z$$

$$= -n\cot\alpha\sec^2\alpha(A_n - A_{n+2}),$$

so that
$$A_{n+2} = A_n - \frac{1}{n}\cos\alpha\sin\alpha A_n', \tag{7.13}$$

where A_n' is written for $\mathrm{d}A_n/\mathrm{d}\alpha$. The expressions for A_3, A_4, A_5 and A_6 given in table 8 were obtained from A_1 and A_2, using (7.13). Here K is the complete elliptic integral of the first kind, namely

$$K(\alpha) = \int_0^{\frac{1}{2}\pi}\frac{\mathrm{d}z}{(1-\sin^2\alpha\sin^2 z)^{\frac{1}{2}}}, \tag{7.14}$$

and K' and K'' are $\mathrm{d}K/\mathrm{d}\alpha$ and $\mathrm{d}^2K/\mathrm{d}\alpha^2$.

Table 8

$$A_{-1} = \frac{2}{\pi}\int_0^{\frac{1}{2}\pi}\sec\theta\,\mathrm{d}z = \frac{2}{\pi}\sec\alpha E(\alpha) = \gamma$$

$$A_0 = 1$$

$$A_1 = \frac{2}{\pi}\cos\alpha\,K(\alpha)$$

$$A_2 = \frac{2}{\pi}\int_0^{\frac{1}{2}\pi}(1+\tan^2\alpha\cos^2 z)^{-1}\,\mathrm{d}z = \cos\alpha$$

$$A_3 = \frac{2}{\pi}[K(\alpha)\cos^3\alpha + K'(\alpha)\cos^2\alpha\sin\alpha]$$

$$A_4 = \tfrac{7}{8}\cos\alpha + \tfrac{1}{8}\cos 3\alpha$$

$$A_5 = A_3 - \frac{2}{\pi}\{\tfrac{1}{8}[K(\alpha)-\tfrac{1}{3}K''(\alpha)](\cos\alpha-\tfrac{1}{2}\cos 3\alpha-\tfrac{1}{2}\cos 5\alpha)-\tfrac{1}{15}K'(\alpha)(\sin 3\alpha+\sin 5\alpha)\}$$

$$A_6 = A_4 - \tfrac{5}{64}\cos\alpha + \tfrac{7}{128}\cos 3\alpha + \tfrac{3}{128}\cos 5\alpha$$

Table 4 gives values of A_1 to A_6; B/λ, γ and C are also given.

The results of calculations can be displayed in a swimming diagram like that of fig. 4, but the quantity $[C_D]_p R_1^{\frac{3}{4}}$, the values of which were shown in fig. 4, is replaced by $[C_D]_p/C_f$. The results for the type of roughness shown diagrammatically as (d) in fig. 2 are shown in fig. 10. The contours of equal values of $[C_D]_p/C_f$ shown in fig. 10 differ from those for smooth cylinders, fig. 4, in that negative values of V/U occur in a part of the diagram which might correspond with physically possible circumstances. This cannot be said of fig. 4.

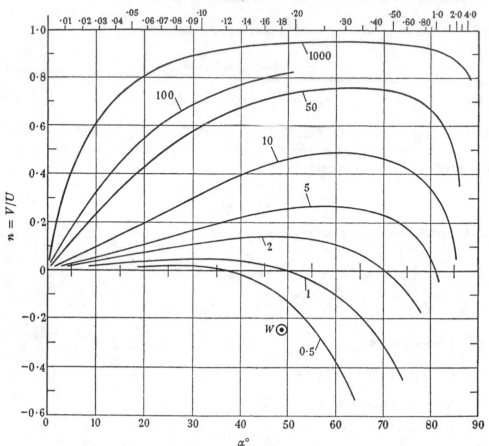

Fig. 10. Swimming diagram for rough cylinder. Values of $[C_D]_p/C_f$.

8. THE SWIMMING OF THE MARINE WORM *NEREIS DIVERSICOLOR*

In preparing the diagram shown in fig. 10 the parts of the curves corresponding with negative values of V/U were calculated merely for the sake of mathematical completeness. On showing the diagram to Professor Gray and pointing out that a negative value of V/U would correspond with a case in which a very rough animal could swim by sending waves of displacement in the same direction as that in which it wished to go, Gray called my attention to a set of photographs he had taken of a marine worm *Nereis diversicolor* which does in fact swim in this way. This set of photographs is reproduced from his paper* in fig. 11. The worm will be seen swimming from right to left. The photographs were taken at intervals of $\frac{1}{20}$ sec. over a grid of 1 in. squares. The positions of the four wave crests, numbered 1, 2, 3, 4, are marked in each photograph. It will be seen that they move forward from right to left.

* J. Gray, *J. exp Biol.* XVI (1939), 9.

The velocities relative to fixed co-ordinates of all the waves are close to 0.47×20 cm./sec., while the velocity of the head of the animal is 0.088×20 cm./sec. The value of V/U is therefore

$$\frac{V}{U} = \frac{0.088}{-0.47+0.088} = -0.23.$$

The value of B/λ is found to be approximately 0.18. The point $B/\lambda = 0.18$, $V/U = -0.23$ is shown in figure 10 as the point W. It will be seen that this point corresponds with a value $[C_D]_p/C_f$ rather less than 0.5, so that the longitudinal force

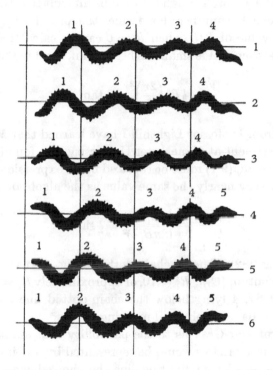

Fig. 11. *Nereis diversicolor*, 20 frames per second (J. Gray).

coefficient is more than twice as great as the transverse coefficient. It is doubtful whether the type of roughness which this animal possesses would confer on it the property that C_f is more than twice $[C_D]_p$ if the roughness consists of rigid projections. In fact the roughness consists of flexible projections which the animal moves in the same direction as that of the surface itself in places where that surface is moving fastest, i.e. at the outside of the bends in its body. This increases the longitudinal force coefficient. The details of the movement were explained by Gray. The way in which each projection swings from left to right while the crest of a wave passes under it can be seen in fig. 11.

9. LIMITATIONS TO THE APPLICATION OF THE ANALYSIS

In attempting to apply the analysis of the swimming of smooth animals it is necessary to bear in mind two kinds of limitation. In many animals the waves of displacement increase in amplitude as they pass from head to tail. It seems likely that such animals swim more efficiently than would be predicted by the analysis, but this question is not discussed here. The assumption that the force on each element of the flexible cylinder is the same as that on a similar portion of a long straight cylinder is likely to be inaccurate either when the diameter of the cylinder is comparable with the wave-length of the motion or when the Reynolds number associated with it is very low. Though it would be difficult to estimate theoretically the lower limit of Reynolds number above which the formulae might be expected to apply, the limiting value of V/U when $R_1 = 0$ can be calculated. So far only the analysis for the case when B/λ is small has been given,* and in that case

$$\frac{V}{U} = \tfrac{1}{2}B^2\left(\frac{2\pi}{\lambda}\right)^2 = \tfrac{1}{2}\tan^2\alpha. \tag{9.1}$$

In a private letter from Professor Lighthill I have learned that Mr. G. J. Hancock, a worker in his department at Manchester University, has found that when B/λ is not small but d/λ is small, (9.1) must be replaced by an expression involving elliptic functions which has very nearly the same value as the algebraic expression

$$\frac{V}{U} = \frac{\tfrac{1}{2}(2\pi B/\lambda)^2}{1+(2\pi B/\lambda)^2} = \tfrac{1}{2}\sin^2\alpha. \tag{9.2}$$

The line representing (9.2) is shown chain-dotted in fig. 4. It will be seen that this line lies close to the contour $[C_D]_p\, R_1^{\frac{1}{2}} = 10$, or approximately $R_1 = 100$ over its whole length. The part of fig. 4 lying below the chain-dotted line has been shaded to show that points in it have no physical meaning.

On consulting Professor Gray as to the possibility that animals may be found whose swimming characteristics would be represented by a point in the swimming diagram near the chain-dotted limiting line, he showed me photographs, here reproduced as fig. 12, which he had taken of a ceratopogonid larva. This animal is about 1 cm. long. The time interval between successive frames in fig. 12 is $\frac{1}{40}$ sec. On measuring the photographs the following mean values were found:

$$V = 2.17\,\text{cm./sec.}, \quad U = 7.37\,\text{cm./sec.}, \quad \lambda = 0.92\,\text{cm.}, \quad d = 0.038\,\text{cm.},$$

$$B = 0.09\,\text{cm.},$$

so that

$$R_1 = \frac{(7.37)\,(0.038)}{0.011} = 25.5.$$

* Paper 17 above.

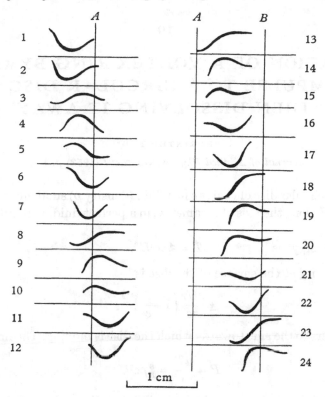

Fig. 12. Ceratopogonid larva swimming, 40 frames per second (J. Gray).

If $[C_D]_p$ is taken as $1 \cdot 0$, $[C_D]_p R_1^{\frac{1}{2}} = 5$. The contour $[C_D]_p R_1^{\frac{1}{2}} = 5$ lies within the shaded portion of fig. 4. The point representing the observations, namely,

$$\left(\frac{B}{\lambda} = \frac{0 \cdot 09}{0 \cdot 92} = 0 \cdot 1, \quad \frac{V}{U} = \frac{2 \cdot 17}{7 \cdot 37} = 0 \cdot 3 \right),$$

is shown as C in fig. 4. As would be expected it lies above the shaded portion.

My thanks are due to Professor James Gray for permitting the publication of figs. 5, 7 and 12, which have not previously been published, and of fig. 11, which is taken from a paper to which reference is made in the text. I am also grateful to him for suggestions made in the course of the work.

19

FORMATION OF A VORTEX RING BY GIVING AN IMPULSE TO A CIRCULAR DISC AND THEN DISSOLVING IT AWAY

REPRINTED FROM

Journal of Applied Physics, vol. XXIV (1953) p. 104

Consider a flat circular disc of radius c in a fluid (density ρ) suddenly moved forward with velocity U along the axis of symmetry, in a perfect fluid, without separation of flow.

The kinetic energy* is

$$T = \tfrac{4}{3}\rho c^3 U^2.$$

(1)

The radial velocity at the surface of the disc is

$$\pm \frac{2Ur}{\pi c}\left(1 - \frac{r^2}{c^2}\right)^{-\frac{1}{2}},$$

(2)

the + sign refers to the side towards which the disc is moving. The impulse is

$$P = \frac{dT}{dU} = \tfrac{8}{3}\rho c^3 U.$$

(3)

If the material of the disc were suddenly annihilated, the motion might be described as being due to a distribution of circular vortex lines over the plane of the disc. The total strength of the vortices in the circular plane annulus between r and $r + \delta r$ is, from (2),

$$\frac{4Ur}{\pi c}\left(1 - \frac{r^2}{c^2}\right)^{-\frac{1}{2}}\delta r.$$

(4)

The vortex elements at different radii will act on one another in such a way that the vortex lines where they are strongest in the part of the sheet near the edge, tend to roll up the weaker parts near the centre round them. Thus the vorticity will tend to concentrate inside a ring. If we assume that the vorticity in this ring becomes nearly constant inside a circle of radius a, i.e. that the vorticity is distributed in a core of mean radius R and radius of cross section a, we may express the energy T' in terms of R, a and K, the total circulation around the core. Lamb† gives the formulae for T' and V', the velocity of the ring, as

$$T' = \frac{K^2 R\rho}{2}\left\{\log\frac{8R}{a} - \frac{7}{4}\right\},$$

(5)

$$V' = \frac{K}{4\pi R}\left\{\log\frac{8R}{a} - \frac{1}{4}\right\}.$$

(6)

* Lamb, *Hydrodynamics* (Cambridge University Press, sixth edition), p. 139, Eq. (20).
† *Loc. cit.* p. 241.

The impulse is
$$P' = \pi\rho K R^2. \tag{7}$$

It seems that the redistribution of velocities involved in diffusing the vorticity within the core of the vortex ring is not likely to change the energy much or impulse at all, so that a useful approximation may be obtained by assuming that

$$T = T', \quad P = P'. \tag{8}$$

The rolling up of the vortex sheet does not change the circulation around the core so that

$$K = \int^c \frac{4Ur}{\pi c}\left(1 - \frac{r^2}{c^2}\right)^{-\frac{1}{2}} r\,dr = \frac{4Uc}{\pi}. \tag{9}$$

From (3), (7), (8),
$$\tfrac{8}{3}\rho c^3 U = \pi\rho K R^2. \tag{10}$$

From (9) and (10),
$$\frac{R^2}{c^2} = \frac{2}{3}, \quad \text{or} \quad R = 0.816c. \tag{11}$$

From (1), (5) and (8),
$$\log\frac{8R}{a} = \frac{7}{4} + \frac{2}{K^2 R\rho}\cdot\frac{4}{3}\rho c^3 U^2 = \frac{7}{4} + \frac{8}{3}\frac{c^3 U^2}{K^2 R}.$$

Using (9) and (11),
$$\log\frac{8R}{a} = \frac{7}{4} + \frac{1}{6}\frac{\pi^2 c}{R} = \frac{7}{4} + 2.01 = 3.76, \tag{12}$$

so that
$$\frac{R}{a} = 5.37 \quad \text{or} \quad \frac{a}{c} = \frac{0.816}{5.37} = 0.152. \tag{12a}$$

From (6), (9) and (12),
$$V' = \frac{Uc}{\pi^2 R}(3.76 - 0.25) = 0.436U. \tag{13}$$

The three parameters V', a, R, of the vortex ring have therefore been determined in terms of the two parameters U and c of the disc from which it has been formed.

20

AN EXPERIMENTAL STUDY
OF STANDING WAVES

REPRINTED FROM

Proceedings of the Royal Society, A, vol. CCXVIII (1953), pp. 44–59

The experiments here described were designed to test experimentally some conclusions about free standing waves recently reached analytically by Penney & Price. A close approximation to free oscillations was produced in a tank by wave makers operating with small amplitude and at frequencies where great amplification occurred, owing to resonance. The amplitude-frequency curve proved to consist of two non-intersecting branches, a result which can be explained theoretically.

A striking prediction made by Penney & Price was that when the height of the crests of standing waves reaches about 0·15 wave-length they will become pointed, in the form of a 90° ridge. Higher waves were expected to be unstable because the downward acceleration of the free surface near the crest would exceed that of gravity. The experimental conditions necessary for producing a crest in the form of an angled ridge were found and the wave photographed in this condition. Good agreement was found with the calculated form of the profile of the highest wave, which had an angle very near to 90°.

The predicted instability for two-dimensional waves was found to begin at the moment the crest became a sharp ridge. It rapidly assumed a three-dimensional character which was revealed by two photographic techniques. Even when the amplitude of oscillation of the wave makers was only 0·85°, violent types of instability developed which produced effects that are here recorded.

1. INTRODUCTION

In the mathematical analysis of water waves of small amplitude there is little difference between progressive waves, which travel without change of form in one direction, and standing waves which can be regarded as the result of superposing two sets of progressive waves of equal small amplitude travelling in opposite directions. It can be shown that free-standing waves of this type can exist in a rectangular tank with vertical ends. When the amplitude of waves is not small mathematical discussion becomes difficult, but it has been shown that finite progressive waves can be generated irrotationally and that the highest possible waves of this type would have sharp crests contained within an angle of 120°.

No analysis of standing waves of finite amplitude was avilable till Penney & Price* developed a method for analysing the free oscillations which can exist in water contained between two parallel vertical plane walls. They found an expansion in Fourier series for the velocity potential of the motion such that each term satisfied the boundary condition for standing waves and was multiplied by a factor which varied with time. This factor was also expressed as a Fourier series

* W. G. Penney & A. T. Price, *Phil. Trans. Roy. Soc.* A, CCXLIV (1952), 254.

containing a fundamental frequency n and all integral multiples of it. These frequencies were common to all terms in the velocity potential expansion, so that the motion was periodic with frequency n.

The analysis of Penney & Price contained one arbitrary number, A, which determined both the frequency, n, and the ratio amplitude/wave-length. The frequency of standing waves of small amplitude and wave-length λ is $(g/2\pi\lambda)^{\frac{1}{2}}$. The lowest frequency of oscillation between walls separated by a distance L is the same as that of waves of length $2L$, but these oscillations are anti-symmetrical. The lowest frequency for symmetrical oscillations of small amplitude is

$$n_0 = \left(\frac{g}{2\pi L}\right)^{\frac{1}{2}}.$$

If H is the height of the crest above the undisturbed surface when it reaches its maximum height, and D is the depth below this surface of the trough at its maximum, Penney & Price's calculations give H/L, D/L and n/n_0 in terms of the arbitrary parameter A. Taking their expansions to the fifth terms in the Fourier expansions they obtained the expressions

$$\frac{H}{L} = \frac{1}{2\pi}\{A + \tfrac{1}{2}A^2 + \tfrac{13}{32}A^3 + \tfrac{145}{672}A^4 + 0.116A^5\}, \tag{1}$$

$$(n/n_0)^2 = 1 - \tfrac{1}{4}A^2 - \tfrac{13}{128}A^4. \tag{2}$$

From (1) and (2) the values given in table 1 were calculated.

Table 1

A	n/n_0	H/L
0	1·000	0
0·1	0·9987	0·0168
0·2	0·9949	0·0356
0·3	0·9883	0·0570
0·4	0·9785	0·0816
0·5	0·9650	0·1103
0·592	0·9487	0·1411

The figures in the last line of table 1 for $A = 0.592$ correspond with the highest possible periodic wave which, according to Penney & Price, has the following properties:

(*a*) The crest rises to a sharp edge of angle 90°.

(*b*) The downward acceleration of the fluid at the crest when it is at its highest point is equal to g, the acceleration of gravity. When this occurs the surface may become unstable near the pointed crest.

(*c*) At the instant when the wave crest reaches its maximum height the fluid is everywhere at rest. This is also true for waves of all amplitudes.

(*d*) The wave profile is that which is reproduced in fig. 1.

In discussing the possibility that the crest of the highest wave will reach a sharp angle Penney & Price* give a general argument, which does not seem to depend on the manner in which the angle is formed, to show that if such an angle is formed it must be a right angle. They point out that if the distribution of pressure near the point can be expressed in ascending integral powers of the co-ordinates, free surfaces meeting at the point must meet at right angles, each being at 45° to the vertical. While this is undoubtedly true I have been unable to follow Penney & Price's arguments tending to show that the pressure distribution will necessarily have the mathematical form they have assumed.

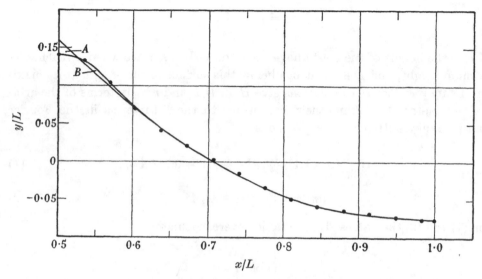

Fig. 1. Half of wave profile calculated by Penney & Price. The points represent the five-term approximation.

It is clearly possible to imagine that a wedge of water with any given vertical angle can be released from rest. If at some later instant the velocity at every point in the field is reversed, a field of flow will then exist which will subsequently produce the same angle as that which existed originally. For this reason it is difficult to understand how an argument which does not appear to depend on the mode of formation of the sharp-angle crest can be used to show that the angle must be a right angle when it is formed in the course of a free oscillation.

Penney & Price's profile of the highest standing wave

The five-term approximation used by Penney & Price gives the profile shown in fig. 1, p. 272, of their paper. Points taken from that figure are here reproduced in fig. 1. If a curve is drawn through these points it is found to have a slightly wavy appearance, but a smooth curve with only slowly changing curvature, such as

* *Loc. cit.* pp. 272–3.

that shown in fig. 1, can be drawn so that it passes very close to the five-term profile except near the crest, which cannot be pointed in this approximation. Penney & Price continued the smooth profile in the neighbourhood of the crest in the full line of fig. 1. They found that this continuation must slope at an angle which is very close to 45° when it reaches the wave crest. Though at first sight it might be thought that this process leaves open the possibility of considerable artistic licence, it should be remembered that the smooth curve must be drawn so that the total area between the wave profile and the undisturbed water surface is zero. Since the five-term approximation necessarily has this property, the smooth curve must be so drawn that the area shown in fig. 1 as A must be equal to the area shown as B. This consideration limits the possible maximum value of H/L to a small range of values close to $H/L = 0\cdot160$. Since the acceleration at the crest, calculated by the five-term approximation, with $A = 0\cdot592$, was equal to g at $H/L = 0\cdot141$, it seems that a particle which rises during a periodic oscillation above $H/L = 0\cdot141$ is likely to experience a downward acceleration greater than g. It seems therefore that to determine the pointed profile accurately the process described above for finding the height of the crest should be applied to a five-term profile corresponding to a value of A slightly less than $0\cdot592$; but it would not be possible to improve on the accuracy of Penney & Price's method without using more than five terms.

The slope of the smooth profile shown in fig. 1 was measured using a protractor and found to be 44° at the crest, corresponding to a crest angle of 92°. This is so near to a right angle that it seems highly probable that if an analysis of the motion near the point could be carried out Penney & Price's prediction would be found to be correct.

It has been pointed out that if the crest becomes pointed the acceleration there is equal to g, since the pressure gradient must be zero. If higher periodic waves can exist the downward acceleration at the crest must be greater than g during a short time when it is near its maximum value. There is nothing in Penney & Price's analysis to show that such waves cannot be calculated using their method and assuming a value for A greater than $0\cdot592$. Such waves would be unstable in the sense that disturbances would grow during that part of each period in which the downward acceleration near the crest was greater than g, but it is not obvious that any particular disturbance would be greater at the end of a complete period than it was at the beginning. When a liquid with a free surface is subjected to vertical vibrations corrugations appear. The theory of this kind of instability has recently been analysed by Dr F. Ursell, who pointed out to me that instability can occur under conditions when the greatest vertical acceleration is much less than g. If instability of the type associated with vertical oscillation of a plane horizontal surface can occur in a fluid which is oscillating as a standing wave, the waves calculated by Penney & Price might be unstable when the maximum value of H/L was less than the value for the pointed crest.

In view of these uncertainties, it seemed worth while to try to produce experimentally the wave contemplated analytically by Penney & Price. In real fluids the free oscillations imagined by mathematicians cannot be produced, owing to the damping effects of viscosity, but forced waves of constant amplitude can be produced; and if the frequency of the forcing agent is very close to that of a free oscillation, the mode will be very nearly the same as that of a free wave.

It is curious that though many experiments have been carried out using wave makers to produce progressive waves, no one seems to have succeeded in producing the highest wave for which a sharp 120°* crest is predicted. This is probably because a progressive wave of large amplitude could only be produced by a wave maker which would not only oscillate through a large amplitude, but its shape would have to be controlled to conform with the variable motions of particles at different depths and it would not oscillate in a simple harmonic motion. On the other hand, if a wave maker in a tank is made to oscillate with frequency close to that of a free mode, this mode should be excited even if the motion of the wave maker approximates only very roughly to that of the fluid particles during a free oscillation.

2. Experimental apparatus

A tank 4 m. long, 14·8 cm. wide and 23 cm. deep was available. Since most of Penney & Price's results were concerned with water of infinite depth it was necessary to shorten the tank till the ratio of depth to length was so great that the effect of the bottom could be neglected. This was effected by making movable vertical partitions, which could be laid on the bottom of the tank so as to form with the glass sides, a wave tank of variable length.

The apparatus is shown in fig. 2. The tank A is shown shorter than its actual length. The two partitions consist of two L-shaped pieces C, hinged at D to wave-making paddles E. The height of the vertical limb of the L-shaped pieces C was 5 cm. The width of the paddles was 14·3 cm., so that there was a gap of 2·5 mm. each side between the glass walls of the tank and the edge of the partitions. It was soon found that waves produced in the section between the paddles were affected by the oscillating flow of water through these gaps. This difficulty was overcome by slitting a rubber tube down one side and slipping it down between the partition and the glass side. In this way a water-lubricated seal was formed which completely stopped the fluctuating flow round the partition. These water seals are not shown in fig. 2, but a section of the edge of a partition and the seal is shown in a small inset below the main sketch.

The wave-making paddles E were caused to oscillate by means of wires F which joined at G after passing over pulleys H. These wires were pulled back and forth by a wire J, the other end of which was connected through a ball-bearing collar to a pin K. This pin could be fixed at any point of a slide L which formed the diameter of a wheel M. This wheel was driven by a motor N, the speed of which could be varied. The wires F, F and J were kept taut by springs O. A revolution counter,

* G. G. Stokes, *Collected Papers*, v (1880), 62.

not shown in fig. 2, was attached to the axle of the wheel M so that the frequency, n, could be measured by taking the time for 100 oscillations.

In the course of the work it was found that the damping was very small so that sharp resonance occurred near the speed of free oscillations. It was necessary to maintain the speed constant to about one part in 250 during a run of 100 oscillations. When the length of the oscillating basin was 32·9 cm. the frequency of the natural oscillations was 2·17 c/sec., so that the time for 100 oscillations was 46·0 sec.

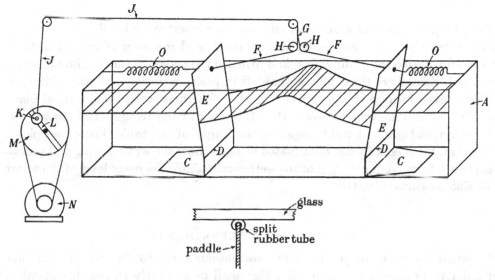

Fig. 2. Wave tank.

It was necessary therefore to ensure that the variation in timing of successive runs of 100 should not be greater than $\frac{1}{5}$ sec. Slow variations in the voltage of the electric supply to the motor N made it necessary to control the field current by hand during a run so that an instrument giving instantaneous readings of the angular velocity of M was needed. This was devised for me by Dr T. H. Ellison. A brass disc was fixed to the axis of M, and a powerful permanent magnet was fixed so that rotation of the disc gave rise to an e.m.f. between the centre and the circumference. The resulting current was tapped off by a sliding contact of brass which was identical with that of the disc and passed through a galvanometer. The readings of this instrument made it possible to control the frequency of the oscillations by means of a hand-operated rheostat to the desired degree of constancy.

When the wave maker was started in still water and its oscillations were maintained at constant frequency and constant small amplitude, the resulting forced waves at first grew with each succeeding stroke. Later they decreased till finally the beats died away and an oscillation of constant amplitude was produced. To measure the profiles of these waves three methods were used.

Method (*a*). Probes which just touched the wave at the highest point attained and under-water probes which just emerged from the surface at the lowest point were made and attached with appropriate vertical scales to a carriage which could slide on the glass walls of the tank. This method was not found to be very accurate.

Method (*b*). Photographs were taken by means of a camera 4 ft. from the side of the tank. The background was illuminated by means of a strong light and $\frac{1}{500}$ sec. exposures were taken at random. In some cases the instant of exposure was close to that at which the fluid was at rest and the crest at its maximum height.

Method (*c*). Glass plates ground on one side were inserted vertically before the in-stant when the maximum height was attained, and removed after that instant. The water wetted the ground glass and revealed the profile of the maximum height attained. Sometimes these plates were left in position during several oscillations. In each experiment two plates were inserted. The first was in the vertical plane parallel to, and midway between, the glass sides of the trough. The second was in the vertical plane at right angles to the length of the tank. These two plates were then hung on a frame, illuminated so as to show the wetted area, and photo-graphed. The maximum height of the wetted area above the mean level of the water was also measured directly.

3. RESONANCE EXPERIMENTS

The usual theory of simple harmonic oscillations leads to the expectation that in order to produce a forced mode which shall be as nearly as possible identical with a free mode the frequency of the wave maker should be as nearly as possible equal to that of the free oscillation. It is not possible to apply this principle directly in the case of standing waves, because the frequency of free oscillations depends on the wave amplitude and the amplitude which will be attained is not known beforehand. For this reason the eccentric pin (K, fig. 2) was set so that the ampli-tude of movement of the wave maker was constant and the maximum height of the wave at its crest was measured for a range of frequencies using the wetted plate method (*c*). The time taken for 100 oscillations was measured with a stop-watch.

In all the experiments here described the length L of the tank was 32·9 cm. The depth of the hinge of the wave makers was 10 cm. and the depth, d, of the water 15·5 cm. The amplitude was found by measuring the maximum extent of the move-ment of the top of the wave makers, which were 22·6 cm. long. The angular ampli-tude of their oscillations in degrees was therefore

$$\theta_0 = \frac{1}{2} \frac{\text{total travel of top of wave maker (cm.)}}{22\cdot6} \left(\frac{180}{\pi}\right). \qquad (3)$$

Reduction of results to non-dimensional form

The maximum height H of the crest above the level of the water before starting the wave maker can be measured. For comparison between results obtained in tanks of varying sizes as well as for comparison with theory, all length measurements were divided by the length L of the tank. Thus the maximum height H is expressed in the non-dimensional form H/L and horizontal distances, x, measured from one end of the tank are given as x/L. The crests occurred at $x/L = 0$, 0·5 and 1·0, and in some of the photographs (figs. 14, 16, 19, pl. 3) a scale for x/L is given. Vertical heights are expressed as y/H. These co-ordinates are used in fig. 1.

To express the measured frequency n in non-dimensional form it can be divided by n_0, the frequency for oscillations of small amplitude. The frequency n_0 can be calculated, using the formula for symmetrical oscillations of small amplitude

$$n_0^2 = \frac{g}{2\pi L}\tanh\frac{2\pi d}{L}. \tag{4}$$

In most of the experiments $d = 15\cdot5$ cm., $L = 32\cdot9$ cm., $g = 981$, so that

$$n_0 = 2\cdot173\,\text{c/sec.} \tag{5}$$

Oscillations of small amplitude could be excited by making the paddle perform two or three oscillations. After stopping the wave maker in its central position about 150 free oscillations could be counted. Several counts of 100 oscillations were timed, all of them at 46·1 or 46·0 sec. These correspond to $n_0 = 2\cdot169$ or $2\cdot174$. Thus, the observed frequency was within the limit of accuracy of the observation equal to that calculated in the classical manner, assuming that viscosity has no effect and that the water slips without resistance over solid boundaries. For comparison with Penny & Price's theoretical results the observed values of n were divided by $n_0 = 2\cdot173$.

Experimental results

Several series of measurements were made with a fixed setting of the amplitude pin (K, fig. 2). Crest heights H were measured over ranges of values of frequencies given by $0\cdot85 < n/n_0 < 1\cdot15$. The results for one such set in which the amplitude of the wave maker was $\theta_0 = 0\cdot716°$ are shown in fig. 3. Starting with $n/n_0 = 0\cdot885$ for which $H/L = 0\cdot006$, the frequency was very gradually increased. As would be expected H/L rose with increasing rapidity as the resonance frequency ($n/n_0 = 1$) was approached. It rose steadily till at $n/n_0 = 0\cdot968$, $H/L = 0\cdot025$. At this stage the slightest increase in frequency led to a very great increase in amplitude. This is represented on the resonance diagram of fig. 3 by the upward pointing arrow at $n/n_0 = 0\cdot968$. The increase at $n/n_0 = 0\cdot968$ from $H/L = 0\cdot025$ to $0\cdot12$ was observed to be accompanied by a reversal of phase of the forced wave in relation to that of the wave maker. If when the system was oscillating with $n/n_0 = 0\cdot968$ and $H/L = 0\cdot12$ the frequency was increased, it was found that H/L decreased till at the frequency of resonance for small oscillations, $n/n_0 = 1\cdot0$, H/L was only 0·058. Further increase in n/n_0 gave rise to further reduction in H/L till at $n/n_0 = 1\cdot10$, H/L was again reduced to $0\cdot006$. If, on the other hand, n/n_0 was very slowly reduced below the value at $0\cdot968$, it was found that H/L increased till at $n/n_0 = 0\cdot953$ it reached a value which was usually about $0\cdot15$. The smallest decrease in n/n_0 below $0\cdot53$ caused H/L to decrease from $0\cdot15$ to $0\cdot012$. This is indicated by the downward-pointing arrow at $n/n_0 = 0\cdot953$. It will be noticed in fig. 3 that the part of the

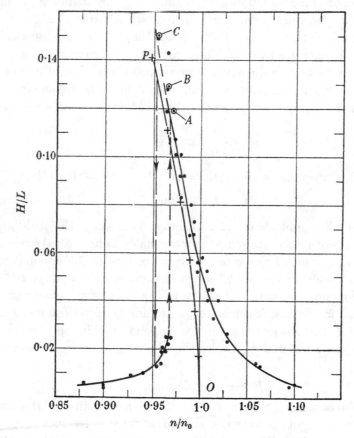

Fig. 3. Resonance curve when wave makers oscillate with amplitude $\theta_0 = 0\cdot716°$. Crosses are calculated points for free oscillations (table 1); points A, B and C are for motions shown in figs. 15, 16, 17 (pl. 3).

resonance curve which lies above $H/L = 0\cdot125$ is represented by a broken line. This is because the waves for which $H/L > 0\cdot125$ were observed to be three-dimensional in character, the crest at the centre of the tank being slightly higher than near the walls. This kind of instability is discussed in §4.

It will be seen in fig. 3 that the resonance curve consists of two branches which do not intersect. With the present experimental tank, and with wave-maker amplitude $0\cdot716°$, there is a range $0\cdot953 < n/n_0 < 0\cdot968$ in which either large waves corresponding with points on the upper branch or small waves corresponding with

PLATE 1

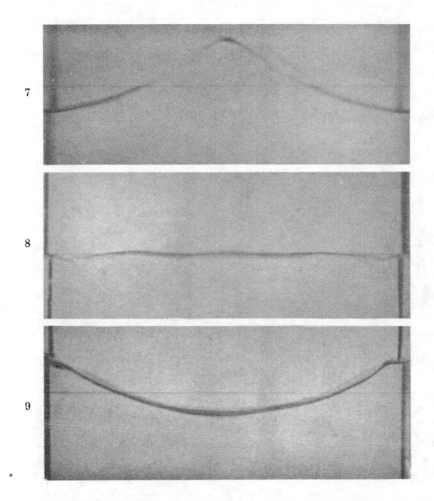

Figs. 7 to 9. Various phases of oscillation when $\theta_0 = 0.75°$, $n/n_0 = 0.965$.

PLATE 2

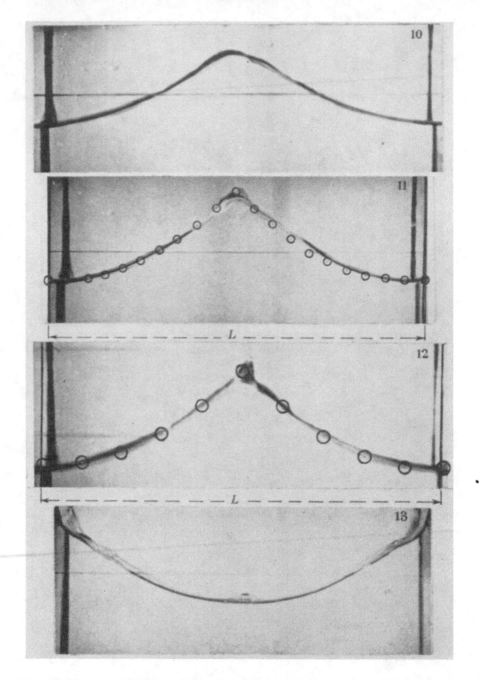

Figs. 10 to 13. Transient condition before three-dimensional motion starts, when $\theta_0 = 0\cdot75°$, $n/n_0 = 0\cdot955$.

Fig. 10. Crest approaching maximum height.

Figs. 11, 12. Circles show Penney & Price's curve transferred on the correct scale from fig. 1.

Fig. 13. Trough near greatest depth.

PLATE 3

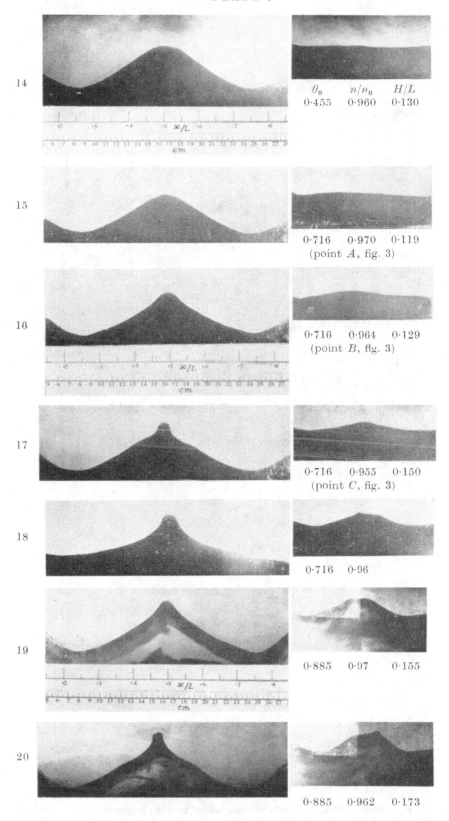

	θ_0	n/n_0	H/L
14	0·455	0·960	0·130
15	0·716	0·970	0·119
	(point A, fig. 3)		
16	0·716	0·964	0·129
	(point B, fig. 3)		
17	0·716	0·955	0·150
	(point C, fig. 3)		
18	0·716	0·96	
19	0·885	0·97	0·155
20	0·885	0·962	0·173

Figs. 14 to 20. Wetted areas on vertical ground-glass plates (method (c)). Left, plate parallel to sides of tank. Right, plate parallel to end of tank. Photographs set at correct relative levels. Fig. 18 shows the result of dipping the slide and quickly removing it.

PLATE 4

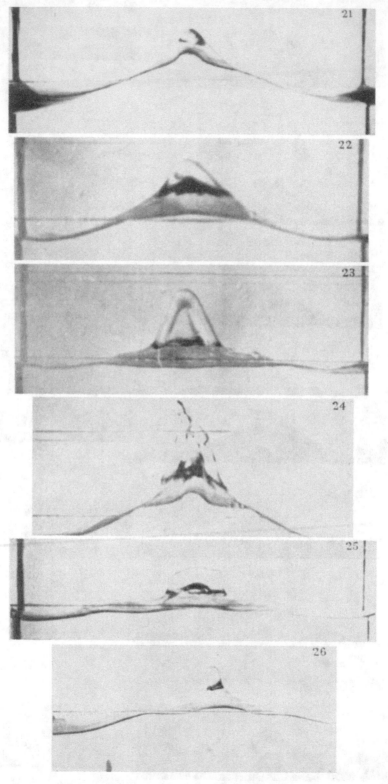

Fig. 21. $\theta_0 = 0\cdot885°$, $n/n_0 < 0\cdot962$, violent lateral motion beginning.
Figs. 22 to 26. Later phases of the violent motion with $\theta_0 = 0\cdot885°$, $n/n_0 < 0\cdot962$.

points on the lower branch can be generated. Which of these two alternative régimes is actually set up at any frequency in this range depends on whether the frequency has been attained by continuous decrease through $n = 0.968n_0$ or increase through $0.953n_0$.

Theoretical interpretation

These experimental results can be interpreted in the light of Penney & Price's calculations. Theoretical values of H/L for free oscillations and the corresponding values of n/n_0 found by using equations (1) and (2) are shown in fig. 3 by means of crosses. A curve has been drawn through these crosses which runs from the point O ($n/n_0 = 1.0$, $H/L = 0$) to the point P ($n/n_0 = 0.9487$, $H/L = 0.141$). The point P represents the highest crest for waves in which the downward acceleration is calculated to be everywhere less than g when the first five terms in the Fourier expansion are used.

Comparing the observed resonance curve with the theoretical frequency-amplitude curve for free undamped oscillations, it will be seen that they are related in much the same way that the well-known resonance curve for linear oscillating systems is related to the straight line at $n/n_0 = 1.0$, which represents the fact that in such systems all amplitudes are possible for free undamped oscillations, and all have the same frequency.

In fig. 4 two resonance curves are shown. Curve 1 has been calculated from the resonance equation for a linear oscillator with damping coefficient α which may be written

$$\frac{H}{L} = \gamma \left[\left(1 - \frac{n^2}{n_0^2} \right)^2 + \frac{\alpha^2}{4\pi^2} \frac{n^2}{n_0^2} \right]^{-\frac{1}{2}} . \tag{6}$$

The constants γ and α have been so chosen that the band width for any particular value of H/L is near that observed when the wave-maker amplitude was $0.716°$. In fact α was taken as 0.063 and γ as 0.0012.

Curve II, fig. 4, is derived from curve I by displacing the band width at each ordinate through a distance equal to the displacement of the corresponding point on the line OP, which is identical with the line OP in fig. 3. Thus, curve II on which lie the points $AKBCDJEFG$ is obtained from curve I by a strain, which involves only shear parallel to the abscissae. Though it is not suggested that the resonance curve for any particular non-linear oscillator can be obtained exactly in this way, it seems reasonable to suppose that the small departure from linearity in the restoring force which gives rise to the curvature of the line OP in figs. 3 and 4 would, when damping is small, distort the resonance curve in the way indicated in fig. 4.

Such a distortion will, if the damping is sufficiently small, ensure the existence of a small range of frequencies in which three wave amplitudes are possible for any given amplitude and frequency of the wave maker corresponding with the three points B, D, E (fig. 4) in which a line of constant frequency cuts the resonance curve. Of these it seems that B and E may represent stable states and D an unstable state. It is clear that if such a qualitative picture is correct, all the states represented on the line $AKBC$ (fig. 4) could be produced by slowly increasing the frequency of the wave maker, but as soon as the point C is reached where $\mathrm{d}H/\mathrm{d}n$ becomes infinite, the mode will change to that represented by F. If the frequency is further increased the successive states will be represented by points on the curve FG. If the frequencies are diminished the states will be represented by points on the portion FEJ of the

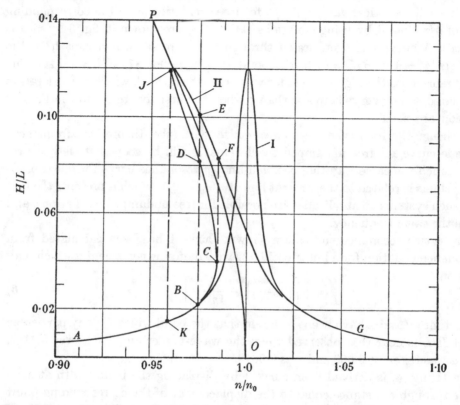

Fig. 4. Resonance curves calculated for idealised systems. Curve I for linear restoring force and damping proportional to velocity. Curve II for non-linear force but damping proportional to velocity.

curve. J is the second point where $\mathrm{d}H/\mathrm{d}n$ is infinite. If the frequency is reduced below that corresponding with J the amplitude will fall till it is represented by K and then the representative point will move back along the curve KA as the forcing frequency decreases. All the possible states of forced oscillations are represented by points on the two non-intersecting curves $AKBC$ and $JEFG$. The portion of the resonance curve between C and J does not represent possible steady states of resonance.

Referring to fig. 3, it will be seen that this qualitative picture of resonance for standing waves of finite amplitude agrees very closely with what is observed. It explains why all points corresponding with large amplification ($H/L > 0.026$) are on the high-frequency side of the undamped free oscillation line OP.

The consistency of this picture is emphasized by noting the relationship between the phase of the wave makers and that of the forced waves. It was observed that at the moment when the crest reaches its greatest height in the middle of the tank the wave makers were sloping towards one another when the representative front on the resonance curve was on the low-frequency branch and away from one another on the high-frequency branch. These states are represented diagrammatically in fig. 5.

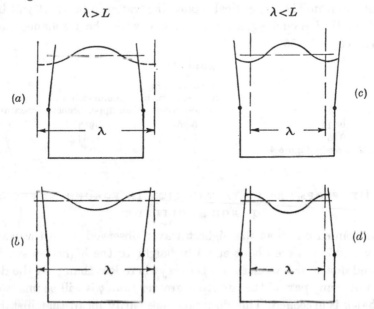

Fig. 5. Sketch showing relationship between phases of wave and wave maker.

Consider the motion of initially vertical lines of particles during an undamped free oscillation. If they lie inside the node they will slope inwards when the crest is in the middle of the tank. If they lie near, but outside, this nodal plane they will slope outwards. If the wave makers coincide with these lines of particles, the motion will be as in figs. 5(a) and (b) when the wave-length λ, corresponding with the free mode of frequency n, is greater than L, and as 5(c) and (d) if λ is less than L.

Thus the mode will be as in fig. 5(a) and (b) if the representative point is on the low-frequency branch of the resonance curve and as in (c) and (d) if on the high-frequency branch. For this reason high amplification occurs only when the mode is as in fig. 5(c) and (d).

Resonance experiments with reduced amplitude of wave makers

It will be noticed in fig. 3 that when $\theta_0 = 0.716°$ the crest rose, in some experiments, as high as $H = 0.15L$. This is higher than the value $H = 0.141L$ given by the five-term approximation for the height at which the greatest downward acceleration is equal to g. In order to find out whether there was any sign of instability when the acceleration is everywhere less than g the wave-maker amplitude was reduced till it was only $\theta_0 = 0.355°$.

Even with this small amplitude the wave height rose at its maximum to $H = 0.125L$. It was found that no instability developed and that the waves were, as far as the eye could tell, two-dimensional. The resonance curve was like that for $\theta_0 = 0.716°$, but the range of frequencies for which two values of H/L are possible was slightly different. These ranges are given in table 2 together with the range calculated from the theoretical model illustrated in fig. 4. It will be noticed that the theoretical model agrees very closely with the resonance phenomena observed when $\theta_0 = 0.355°$.

Table 2

θ_0	maximum n/n_0 on lower branch of resonance	minimum n/n_0 on upper branch	maximum H/L
$0.716°$	0.968	0.953	0.15
$0.335°$	0.980	0.956	0.125
theoretical model (figure 4)	0.978	0.96	0.12

4. Development of instability in forced waves of large amplitude

It has been pointed out that the highest waves observed when the wave-maker amplitude was $0.716°$ were about equal in height to the highest wave for which the calculated downward acceleration is everywhere less than g. If the downward acceleration over any part of the surface is greater than g it will be unstable during the time that g is exceeded. This does not necessarily mean that instability will be very marked immediately g is exceeded, for this excess may only be in existence for a very small proportion of each wave period. It would be difficult to investigate mathematically what happens when surface instability appears; accordingly, an experimental study, particularly of the early stage of instability, was undertaken.

The wave maker was set with amplitude $0.75°$ and frequency $n = 2.10$ c/sec. corresponding with $n/n_0 = 0.965$. The oscillation was in the mode represented by the upper branch of the resonance curve. In this condition a number of photographs (method (b), §2) were taken at random and those reproduced as figs. 7, 8, 9, pl. 1, have been selected as representing interesting phases of the motion. The motion appears to be very nearly two-dimensional. Fig. 7, which showed the highest crest, was measured, and it was found that $H/L = 0.12$. The value of H/L corresponding with $n/n_0 = 0.965$ in fig. 3 is also $H/L = 0.12$, so that this photograph must have been taken at or very close to the position of greatest displacement. Fig. 8 shows the form of the surface when it is nearly flat, and in fig. 9 the central trough has nearly reached its greatest depth.

The frequency was then reduced, keeping $\theta_0 = 0.75°$, till n/n_0 was between 0.95 and 0.96. The amplitude increased till the crest appeared to become nearly pointed, and if the wave-maker oscillation continued it ceased to be two-dimensional.

Comparison of profiles with calculation

To compare the highest two-dimensional waves with Penney & Price's calculations a number of runs were made. The waves were watched and photographs taken as soon as they appeared to reach their greatest amplitude as two-dimensional waves, i.e. just before they began to develop three-dimensional characteristics. Figs. 10 to 13, pl. 2, were taken under these conditions. It will be seen in figs. 11 and 12 that the crest is nearly sharp and is showing signs of instability in the sense that small protuberances are appearing at the crest. For this reason these profiles have been compared with Penney & Price's calculations. The centres of the circles in figs. 11 and 12 are points on the calculated profile which have been superposed on the photographs. These were obtained by altering the scale of fig. 1 so that it could be superposed on the photographs. Since the photographs were taken from a distance 121 cm. from the nearer glass wall and 135 cm. from the further wall, the length L would appear on different scales in the photograph according as it is measured on the nearer or farther wall. For comparison with Penney & Price's profile, the scale of fig. 1 was reduced so that the length $x = L$ on the photograph corresponded with a length 32·9 cm. placed in the mid-plane at a distance of $121 + 7·4 = 128·4$ cm. from the camera. In each photograph only the far edge of the wave maker is seen, the nearer edge being just outside the picture. The nearly vertical black lines seen on the edges of each photograph are the water-lubricated rubber seals. The breaks in them are, of course, due to the refraction of the water. The horizontal lines seen are black threads placed on the outside of the nearer glass wall to show approximately the position of the undisturbed water surface.

The scale length L on the photographs was found by multiplying the length between the farther edges of the wave makers by $135·8/128·4 = 1·057$. When Penney & Price's profile was reduced to this scale it was found that it was so close to the wave profile of photographs in figs. 11 and 12 that it was not possible to superpose it without confusion. For this reason the theoretical profiles were drawn on tracing paper and points pricked through on to the photographs. Circles were then drawn on figs. 11 and 12 round the pricked points.

It will be seen that the photographic profile in the case of fig. 12 is very close indeed to that calculated by Penney & Price, and that at the moment when instability just begins to appear while the wave is still two-dimensional the crest seems to be pointed with an angle which is very close to 90°.

5. DEVELOPMENT OF LATERAL INSTABILITY

It has already been stated that two-dimensional unstable motion can only be attained as a transitory phenomenon. If the wave makers are kept going at constant speed after the sharpest crest has been attained lateral motion sets in and the crest becomes wavy in the transverse direction perpendicular to the sides of the tank. The photographic technique (b) in which the profile was photographed from one side of the tank cannot be used in describing transverse instability. For this purpose recourse was had to the technique of method (c), §2, in which dry ground-glass plates were lowered into the water and the wetted area photographed. This method reveals the maximum height attained by the water surface during the time the plate was partially immersed. It could only give an instantaneous profile if the plate were immersed very rapidly and then immediately withdrawn. If the plate is lowered and kept in position for at least one period the true profiles of the crests which form at intervals of half a period at the middle and ends of the tank are revealed. Near the points $x = \frac{1}{4}L$ and $x = \frac{3}{4}L$ the water attains its maximum height at times which are intermediate between those at which the crests are at their highest.

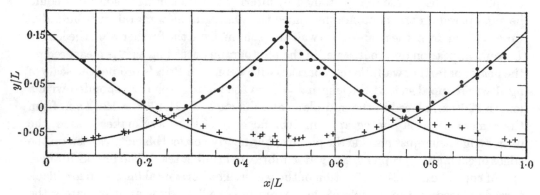

Fig. 6. Measurements with probes (method (a)). x, measured from one end of tank; y, measured from water surface before starting wave maker. Black circles, •, downward pointing probe; crosses, +, upward pointing probe.

This can be appreciated by referring to measurements made by method (a), in which the heights of probes which just touched the surface at its highest and lowest points were recorded. In fig. 6 measurements made with the downward-pointing probe are shown as circles and those made with the upward-pointing probe as crosses. The calculated profiles at the two instants when the water is at rest are also shown. Though the method is not very accurate on account of the difficulty of seeing when a fine probe just touches a surface which is only instantaneously at rest, it shows that the wetted plate, which traces the same curve as the downward-pointing probe, gives a correct instantaneous profile for the crests and for distances of about $0 \cdot 2L$ on each side of them. The troughs will not be revealed by the wetted plates, and the form of the wave in the two regions $0 \cdot 2L < x < 0 \cdot 3L$, $+0 \cdot 7L < x < +0 \cdot 8L$ is not revealed either by method (a) or by (c).

Results using method (c)

A frame was constructed which fitted over the tank and contained slots by means of which the ground-glass slides could be placed in a predetermined position. The ground-glass slides were usually kept in position for two or three oscillations and then withdrawn and placed in a frame set up at a fixed distance from a fixed camera. After withdrawing the longitudinal glass slide the transverse one was inserted, held in position for one or two oscillations, and then withdrawn and placed in the fixed frame above the wet longitudinal slide. The two were then photographed. Several examples of the results using this technique are shown in figs. 14 to 20, pl. 3. In each case the transverse profile at the centre of the tank, i.e. with the transverse plate along the crest, is shown to the right of the longitudinal wetted area. The two photographs are placed at the same level in each case. In some of them scales showing x (cm.) and x/L are given.

The first measurements were made under conditions which were expected to be stable. The amplitude of the wave maker was cut down to $0.455°$ and the frequency was maintained at $n/n_0 = 0.960$. A number of photographs were taken in which the measured values of H/L were 0.131, 0.130, 0.130, 0.129. In all of them the transverse section, taken at the mid-point of the tank, was practically flat so that the motion was two-dimensional. Fig. 14 shows one of these photographs.

The amplitude of the wave maker was then increased to $0.716°$. Photographs were taken at $n/n_0 = 0.970, 0.964$ and 0.955. These are shown in figs. 15, 16, 17. The values of H/L were 0.119, 0.129, 0.150. There is a slight sign of lateral motion in fig. 15 ($H/L = 0.110$) and this becomes more pronounced as H/L increases. Fig. 17 shows that instability in the form of a ridge rises from the crest. The transverse section in fig. 17 is very definitely highest in the middle, but a horizontal section of the ridge would be no longer in the transverse direction than in the longitudinal direction. Since $\theta_0 = 0.716°$ was the amplitude at which the resonance experiments recorded in fig. 3 were made, the points corresponding to figs. 15, 16 and 17 are marked as A, B and C on that diagram. Fig. 18 shows the result of putting a ground-glass slide in and then withdrawing it as rapidly as possible. The longitudinal photograph shows a figure which is probably intermediate in character between the photographs of method (*b*) and the wetted area figs. 14 to 17 produced by method (*c*).

The wave-maker amplitude was then increased till $\theta_0 = 0.885°$. A number of photographs were taken at frequency $n/n_0 = 0.962$. Instability similar to that noticed with $n = 0.716°$ was observed. Fig. 19 shows the instability at $H/L = 0.155$. It will be seen that it is of the same general nature as that shown in fig. 17, but the crest is higher and does not extend so far in a transverse direction. In fig. 20 where $H/K = 0.173$ the same type of instability is increasing.

With the amplitudes so far described the waves were essentially periodic with the frequency of the wave maker, but the slightest decrease in n/n_0 below 0.962 (retaining $\theta_0 = 0.885°$) brought a new phenomenon. The lateral motion began to increase very rapidly. The unstable crest which had consisted of a mass of fluid narrow in the direction of the length of the tank and broad in the transverse direction, became more nearly conical. The crests alternated, rising in the centre into a cone or on the two walls in the form of half-cones in successive periods of the wave maker. The frequency of the disturbance was therefore halved. The motion became very violent. Fig. 21, pl. 4, shows the beginning of this phase of the motion. This photograph (method (*b*)) shows the central conical crest rising in the middle of the tank. At the same time there are two transverse crests at the two walls at the two ends of the tank. These appear as a broadening of the black line representing the profile near the wave maker. When the same phase of the oscillation of the wave maker next occurs the highest crest will be at the two walls in the middle of the tank and transverse crests will appear at the middle of the ends of the tank. This motion rapidly developed a violent character, but it retained its property of being periodic with twice the period of the wave maker. Figs. 22 to 26, pl. 4, show various stages of this motion. The impression gained by observing this phenomenon was that when the height to which the unstable region at the crest was thrown became sufficiently great it moved as though it were a freely rising and falling body, and when a mass of fluid was thrown up in one oscillation it would meet, during its descent, the crest rising in the next oscillation, thus damping its motion. This explanation, however, is not complete, for the half-frequency motion with its lateral crests was so persistent that it seemed likely to be associated with some possible type of free oscillation. The amplitude of the wave makers was only 0·88 of a degree, and for so small a forcing movement to excite so violent a motion it is difficult to imagine any cause but synchronism between free and forced oscillations.

Though I have not been able to imagine any type of free oscillation with a mode similar to that excited at half the frequency of the wave maker, it was found necessary to avoid conditions suitable for setting up resonance in transverse modes. When, for instance, the length of the tank was made twice the width, the lowest transverse mode had a frequency equal to that of the lowest symmetrical longitudinal oscillation, and it was found impossible to excite two-dimensional waves of large amplitude.

These experiments were carried out in the Cavendish Laboratory, Cambridge.

21

DISPERSION OF SOLUBLE MATTER IN SOLVENT FLOWING SLOWLY THROUGH A TUBE

REPRINTED FROM

Proceedings of the Royal Society A, vol. CCXIX (1953), pp. 186–203

When a soluble substance is introduced into a fluid flowing slowly through a small-bore tube it spreads out under the combined action of molecular diffusion and the variation of velocity over the cross-section. It is shown analytically that the distribution of concentration produced in this way is centred on a point which moves with the mean speed of flow and is symmetrical about it in spite of the asymmetry of the flow. The dispersion along the tube is governed by a virtual coefficient of diffusivity which can be calculated from observed distributions of concentration. Since the analysis relates the longitudinal diffusivity to the coefficient of molecular diffusion, observations of concentration along a tube provide a new method for measuring diffusion coefficients. The coefficient so obtained was found, with potassium permanganate, to agree with that measured in other ways.

The results may be useful to physiologists who may wish to know how a soluble salt is dispersed in blood streams.

1. INTRODUCTION

If a conducting solution (e.g. brine) is injected into a tube through which water is flowing, the region in which it is concentrated moves downstream. At a fixed point the conductivity will rise as the solution reaches it, and if the conductivity is measured there the conductivity-time curve can be used as a means of measuring the stream velocity. If the injected material would remain concentrated in a small volume the method would be simple, but the stream velocity varies over the cross-section of a pipe. A part of the injected material which was initially near the centre of the tube would be carried to the measuring point faster than parts which were near the walls. To use the method as a means of measuring the mean speed of flow therefore it is necessary to know which point on the conductivity-time curve corresponds to this mean speed. This method has been used to measure the flow in large water mains where it is turbulent* and in small blood vessels where it may be non-turbulent.†

A similar method was used by Griffiths‡ in experiments designed to measure viscosity of water at very low speeds of flow. A drop of fluorescent solution was inserted as a marker or index in a stream of water flowing through a capillary

---

* C. M. Allen & E. A. Taylor, *Trans. Am. Soc. mech. Engrs*, XLV (1923).

† G. N. Stewart, *J. Physiol. Lond.* XV (1894), 1, and H. L. White, *Am. J. Physiol.* CLI (1947), 45.

‡ A. Griffiths, *Proc. phys. Soc.* XXIII (1911), 190.

tube. Griffiths found experimentally that the colouring matter spreads out in a symmetrical manner from a point which moves with the mean velocity of the water in the tube. Since the present paper contains an analysis of the situation observed by Griffiths it is worth quoting his theoretical reasoning on the subject. He wrote (p. 190):

'In this paper the full mathematical treatment is not attempted; but an elementary consideration, although not complete, will be of advantage in dealing with the experiments. It can easily be shown that if the intensity of colour were constant over a cross-section of the tube the colour would diffuse along the tube exactly as if the water travelled in a solid column. The intensity of the colour cannot be absolutely constant over a cross-section except in the theoretical case of a capillary tube of infinitely small bore but the experiments show that when the rate of flow is small the error involved in this assumption is not great. By stopping the supply of fluorescent solution and replacing it by water an approximately symmetrical column of colour of slowly increasing length can be obtained and when the slowly moving column is at a relatively long distance from the ends of the fine capillary it is obvious that the movement of the centre of the column must measure the mean speed of flow.'

The foregoing extract shows that Dr Griffiths had formed a good qualitative picture of the situation. The only parts of his statement that do not seem clear are those which follow the words 'It can easily be shown that...' and 'it is obvious that...'. The statement describes two experimental results which seem most remarkable though their discoverer does not comment on them in this sense. The first is that since water moves at twice the mean speed near the centre of the pipe and the patch of colour at the mean speed, the clear water in the middle must approach the colour path, absorb colour as it passes into it and then lose colour as it passes out, finally leaving the patch as perfectly clear water. The second remarkable feature is that the colour patch spreads out symmetrically from a point which moves with the mean speed of the fluid in spite of the fact that the distribution of velocity over the section, which gives rise to this dispersion, is highly unsymmetrical.

All the authors cited calibrated their apparatus, using other methods for measuring the flow, and thus determined empirically the point on the conductivity-time curve which corresponded with the mean speed of flow. In the present communication the way in which salts are dispersed along a tube through which fluid flows in steady motion will be discussed. In a later paper dispersion by turbulent flow in a pipe will be treated. The dispersion in steady flow is due to the combined action of convection parallel to the axis and molecular diffusion in the radial direction. It is of interest to consider, first, dispersion by convection alone, and then to introduce the effect of molecular diffusion. The result may be useful to physiologists who may wish to know how a soluble salt is dispersed in a blood vessel, but they may also be useful to physicists who wish to measure molecular diffusion coefficients.

2. DISPERSION BY CONVECTION ALONE

In a circular pipe of radius a the velocity u at distance r from the central line is

$$u = u_0(1 - r^2/a^2),$$ (1)

where u_0 is the maximum velocity at the axis. If the solute at time $t = 0$ is distributed symmetrically so that the concentration C is

$$C = f(x, r),$$

after time t the concentration will be

$$C = f(x - ut, r).$$ (2)

In the experiments to be described later the mean value, C_m, of the concentration over a cross-section of a tube was measured; C_m is defined by

$$C_m = \frac{2}{a^2} \int_0^a C r \, dr.$$ (3)

Some calculated distributions of C_m along a tube are given below. They will be needed later for comparison with experiments.

Case A 1. Initially the space between two planes $x = 0$ and $x = X$ (X/a being small) is filled with solute of concentration C_0. From (2) it will be seen that the amount which lies between r and $r + \delta r$ is constant during the flow and equal to $2\pi r C_0 X \delta r$. The solute will be distorted in time into the paraboloid

$$x = u_0 t(1 - r^2/a^2).$$ (4)

The total amount of the solute between x and $x + \delta x$ is therefore $2\pi r C_0 X (dr/dx) \delta x$, and from (4) $r \, dr/dx = -a^2/2u_0 t$, so that

$$C_m = \frac{1}{\pi a^2 \delta x} 2\pi C_0 X \delta x \frac{a^2}{2u_0 t} = \frac{C_0 X}{u_0 t}.$$ (5)

C_m therefore has the constant value $C_0 X/u_0 t$ in the interval $0 < x < u_0 t$ and is zero when $x < 0$ and when $x > u_0 t$.

Case A 2. Solute of constant concentration enters a tube which at time $t = 0$ contains only solvent.

Here
$$\left. \begin{aligned} C = C_0, \quad x < 0 \\ C = 0, \quad x > 0 \end{aligned} \right\} \quad \text{at time } t = 0.$$

This case can be solved by imagining that the constant initial concentration for $x < 0$ consists of a number of thin sections of the type imagined in case A 1. In this way it is found that

$$\left. \begin{aligned} C_m &= C_0, \quad x < 0, \\ C_m &= C_0\left(1 - \frac{x}{u_0 t}\right), \quad 0 < x < u_0 t, \\ C_m &= 0, \quad x > u_0 t. \end{aligned} \right\}$$ (6)

Case A 3. Solute confined initially to a length X so that

$$\left. \begin{array}{ll} C = 0, & x < 0 \\ C = C_0, & 0 < x < X \\ C = 0, & x > X \end{array} \right\} \text{ at time } t = 0.$$

This case can be obtained by superposing two examples of case A 2, namely

$$\left. \begin{array}{ll} C = C_0, & x < X \\ C = 0, & x > X \end{array} \right\} \text{ and } \left. \begin{array}{ll} C = -C_0, & x < 0, \\ C = 0, & x > 0. \end{array} \right\}$$

If $t < X/u_0$ the distribution is described by (7)

$$\left. \begin{array}{ll} C_m = 0, & x < 0, \\ C_m = C_0 x/u_0 t, & 0 < x < u_0 t, \\ C_m = C_0, & u_0 t < x < X, \\ C_m = C_0\left(1 - \dfrac{x-X}{u_0 t}\right), & X < x < X + u_0 t, \\ C_m = 0, & x > X + u_0 t. \end{array} \right\} \tag{7}$$

If $t > X/u_0$, it is given by (8)

$$\left. \begin{array}{ll} C_m = 0, & x < 0, \\ C_m = C_0(x/u_0 t), & 0 < x < X, \\ C_m = C_0(X/u_0 t), & X < x < u_0 t, \\ C_m = C_0\left(\dfrac{X + u_0 t - x}{u_0 t}\right), & u_0 t < x < u_0 t + X, \\ C = 0, & x > X + u_0 t. \end{array} \right\} \tag{8}$$

These three cases are illustrated in fig. 1. It will be noticed that A 1 is a limiting case of A 3 when X is small.

3. Effect of molecular diffusion on the dispersion

It will be assumed that the concentration is symmetrical about the central line of the pipe so that C is a function of r, x and t only. The equation for diffusion is

$$D\left(\frac{\partial^2 C}{\partial r^2} + \frac{1}{r}\frac{\partial C}{\partial r} + \frac{\partial^2 C}{\partial x^2}\right) = \frac{\partial C}{\partial t} + u_0\left(1 - \frac{r^2}{a^2}\right)\frac{\partial C}{\partial x}. \tag{9}$$

Here D, the coefficient of molecular diffusion, will be assumed independent of C. This is not a strictly accurate assumption when the formulae are to be applied to soluble salts, but the error introduced by it is not large, and if this assumption is not made the analysis of problems of diffusion becomes difficult. In all the cases which will be considered $\partial^2 C/\partial x^2$ is much less than $\partial^2 C/\partial r^2 + r^{-1}\partial C/\partial r$.

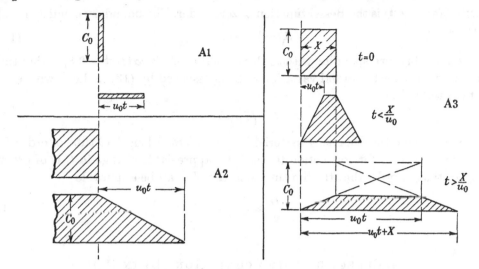

Fig. 1. Distribution of mean concentration in three cases in absence of molecular diffusion.

Writing
$$z = r/a, \tag{10}$$

(9) becomes
$$\frac{\partial^2 C}{\partial z^2} + \frac{1}{z} \frac{\partial C}{\partial z} = \frac{a^2}{D} \frac{\partial C}{\partial t} + \frac{a^2 u_0}{D} (1 - z^2) \frac{\partial C}{\partial x}. \tag{11}$$

The boundary condition which expresses the fact that the wall of the tube is impermeable is
$$\frac{\partial C}{\partial z} = 0, \quad \text{at} \quad z = 1. \tag{12}$$

It would be difficult to find a complete solution of (11) giving the value of C for all values of r, x and t when the distribution of C at time $t = 0$ is known; on the other hand, approximate solutions can be found which are valid in the following limiting conditions:

(A) The changes in C due to convective transport along the tube take place in a time which is so short that the effect of molecular diffusion may be neglected. The solutions already given in §2 are of this type.

(B) The time necessary for appreciable effects to appear, owing to convective transport, is long compared with the 'time of decay' during which radial variations of concentration are reduced to a fraction of their initial value through the action of molecular diffusion.

To find the conditions under which B may be expected to be valid it is necessary to calculate how rapidly a concentration which varies with r degenerates into a uniform concentration. The solutions of (11) for which $\partial C/\partial x = 0$ and the variables z and t are separated are of the form
$$C = e^{-\alpha t} J_0(az\alpha^{\frac{1}{2}}D^{-\frac{1}{2}}), \tag{13}$$

where $J_0(az\alpha^{\frac{1}{2}}D^{-\frac{1}{2}})$ is the Bessel function of zero order. The boundary condition (12) ensures that

$$J_1(a\alpha^{\frac{1}{2}}D^{-\frac{1}{2}}) = 0. \tag{14}$$

The root of (14) corresponding with the lowest value of α is $a\alpha^{\frac{1}{2}}D^{-\frac{1}{2}} = 3\cdot8$, so that the time necessary for the radial variation of C represented by (13) to die down to $1/e$ of its initial value is

$$t_1 = \frac{a^2}{(3\cdot8)^2 D}. \tag{15}$$

If at any time the dispersing material is spread over a length of tube of order L, the time necessary for convection to make an appreciable change in C is of order L/u_0, so that in order that the limiting condition B may be applicable

$$\frac{L}{u_0} \gg \frac{a^2}{3\cdot8^2 D}. \tag{16}$$

4. Effect of using condition (B) in (11)

Since molecular diffusion in the longitudinal direction has been neglected (the justification for this will be given later) the whole of the longitudinal transfer of C is due to convection. We shall consider the convection across a plane which moves at constant speed $\frac{1}{2}u_0$, i.e. with the mean speed of flow. Writing

$$x_1 = x - \tfrac{1}{2}u_0 t, \tag{17}$$

(11) becomes
$$\frac{\partial^2 C}{\partial z^2} + \frac{1}{z}\frac{\partial C}{\partial z} = \frac{a^2}{D}\frac{\partial C}{\partial t} + \frac{a^2 u_0}{D}(\tfrac{1}{2} - z^2)\frac{\partial C}{\partial x_1}. \tag{18}$$

Since the mean velocity across planes for which x_1 is constant is zero, the transfer of C across such planes depends only on the radial variation of C. If C were independent of x and condition (B) satisfied, it has been seen that any radial variation in C very rapidly disappears. The small radial variation in C can therefore be calculated from the equation

$$\frac{\partial^2 C}{\partial z^2} + \frac{1}{z}\frac{\partial C}{\partial z} = \frac{a^2 u_0}{D}(\tfrac{1}{2} - z^2)\frac{\partial C}{\partial x_1}, \tag{19}$$

and in that calculation $\partial C/\partial x_1$ may be taken as independent of z.

A solution of (19) which satisfies the condition $\partial C/\partial z = 0$ at $z = 1$ is

$$C = C_{x_1} + A(z^2 - \tfrac{1}{2}z^4), \tag{20}$$

where C_{x_1} is the value of C at $z = 0$ and A is a constant.

Substituting (20) in (19) it is found that

$$A = \frac{a^2 u_0}{8D}\frac{\partial C_{x_1}}{\partial x_1}. \tag{21}$$

The rate of transfer of C across the section at x_1 is

$$Q = -2\pi a^2 \int_0^1 u_0(\tfrac{1}{2} - z^2)\,Cz\,dz. \tag{22}$$

Inserting the value of C from (20) and (21), (22) becomes

$$Q = -\frac{\pi a^4 u_0^2}{192 D} \frac{\partial C_{x_1}}{\partial x_1}. \tag{23}$$

Since condition (B) is assumed to hold the radial variations in C are small compared with those in the longitudinal direction, and if C_m is the mean concentration over a section $\partial C_{x_1}/\partial x_1$ is indistinguishable from $\partial C_m/\partial x_1$, so that (23) may be written

$$Q = -\frac{\pi a^4 u_0^2}{192 D} \frac{\partial C_m}{\partial x_1}. \tag{24}$$

It will be seen therefore that C_m is dispersed *relative to a plane which moves with velocity* $\frac{1}{2} u_0$ exactly as though it were being diffused by a process which obeys the same law as molecular diffusion but with a diffusion coefficient k, where

$$k = \frac{a^2 u_0^2}{192 D}. \tag{25}$$

The fact that no material is lost in the process is expressed by the continuity equation for C_m, namely

$$\frac{\partial Q}{\partial x_1} = -\pi a^2 \frac{\partial C_m}{\partial t}, \tag{26}$$

where the symbol $\partial/\partial t$ here represents differentiation with respect to time at a point where x_1 is constant. Substituting for Q from (24) the equation governing longitudinal dispersion is

$$k \frac{\partial^2 C_m}{\partial x_1^2} = \frac{\partial C_m}{\partial t}. \tag{27}$$

5. Special cases

Two well-known solutions of (27) describe dispersion in cases which can conveniently be subjected to experimental verification. These are:

(B 1) Material of mass M concentrated at a point $x = 0$ at time $t = 0$.

(B 2) Dissolved material of uniform concentration C_0 allowed to enter the pipe at uniform rate at $x = 0$ starting at time $t = 0$. Initially the pipe is filled with solvent, only ($C = 0$).

Cases B 1 and B 2 correspond to cases A 1 and A 2 in §2, except that the molecular diffusion in the latter case was assumed to have only a negligible effect.

The solutions of (27) are:

Case (B 1):
$$C = \tfrac{1}{2} M a^{-2} \pi^{-\frac{3}{2}} k^{-\frac{1}{2}} t^{-\frac{1}{2}} e^{-x_1^2/4kt}. \tag{28}$$

Case (B 2):
$$\left. \begin{aligned} C/C_0 &= \tfrac{1}{2} + \tfrac{1}{2} \operatorname{erf}\left(\tfrac{1}{2} x_1 k^{-\frac{1}{2}} t^{-\frac{1}{2}}\right) \quad (x_1 < 0), \\ C/C_0 &= \tfrac{1}{2} - \tfrac{1}{2} \operatorname{erf}\left(\tfrac{1}{2} x_1 k^{-\frac{1}{2}} t^{-\frac{1}{2}}\right) \quad (x_1 > 0), \end{aligned} \right\} \tag{29}$$

where
$$\operatorname{erf} z = 2\pi^{-\frac{1}{2}} \int_0^z e^{-z^2} \, dz.$$

In case (B1) the length L_1 which contains 90% of the material is given by $\mathrm{erf}\,(\frac{1}{4}L_1 k^{-\frac{1}{2}}t^{-\frac{1}{2}}) = 0\cdot9$ and using tables this gives

$$L_1 = 4\cdot65k^{\frac{1}{2}}t^{\frac{1}{2}}. \tag{30}$$

If $C_{\max}$ is the maximum concentration at $x = \frac{1}{2}u_0 t$ the concentration at the ends of the length L is
$$0\cdot3126C_{\max}. \tag{31}$$

Similarly in case (B2), if L_2 is the length of the zone of transition in which C changes from $0\cdot9C_0$ to $0\cdot1C_0$, $\mathrm{erf}\,(\frac{1}{4}L_2 k^{-\frac{1}{2}}t^{-\frac{1}{2}}) = 0\cdot8$, and from tables

$$L_2 = 3\cdot62k^{\frac{1}{2}}t^{\frac{1}{2}}. \tag{32}$$

It will be noticed that as t increases both L_1 and L_2 increase proportionally to $t^{\frac{1}{2}}$, whereas the distances traversed by the particles of fluid are proportional to t. Eventually as t increases L_1 and L_2 will become small compared with $\frac{1}{2}u_0 t$. In case (B1) this means that in the central part of the pipe fluid which is free of the dissolved substance passes into the zone where the concentration is rising. The dissolved substance is then absorbed till C reaches its maximum value at $x = \frac{1}{2}u_0 t$. The fluid then passes through the region where C decreases with x and finally leaves this zone, having yielded up the whole of the dissolved substance which it had acquired. Analogous considerations apply to case B2.

This theoretical conclusion seemed so remarkable that I decided to set up apparatus to find out whether the predictions of the analysis could be verified experimentally. Since making these experiments Professor G. Temple, F.R.S., has called my attention to those of Griffiths,* in which it was shown, incidentally, that the phenomenon predicted does occur. Griffiths however was not concerned with describing it, but with the fact that it could be used for measuring slow flow.

6. Experimental technique

In order to satisfy the condition (B) of §3 it was necessary to use a tube of small bore. Two methods seemed suitable for determining C. If the dissolved substance is a good electrical conductor and the solvent (water) a bad one the conductivity can be determined as a function of time at the exit end of the pipe. This method had been used by several workers for determining the velocity of the blood stream in the arteries or veins of a living animal. It has also been used by engineers for determining the velocity in the large pipes which convey water to hydro-electric stations. It could not be used to determine C_m as a function of x at a fixed time. For this reason a colorometric method was devised which made it possible to do this without disturbing the fluid. The dissolved substance was potassium permanganate which is very strongly coloured. The pipe was a glass tube of approximately $0\cdot05$ cm. internal diameter and 152 cm. long. The initial concentration of the solution was 1% by weight of potassium permanganate and 99% of slightly acidulated water. This is so

* *Loc. cit.*

dark that it looks almost black when seen in a bottle, but in a pipe of 0·05 cm bore it is a transparent dark purple. A number of solutions of known concentrations were made by mixing the 1 % solution with various proportions of distilled water. A glass comparison tube of the same external and internal bore as the flow tube was prepared and filled successively with fluid of varying known concentrations. The comparison tube was placed in a light frame which could slide along the pipe and the position where the colours of the comparison tube and pipe were identical was

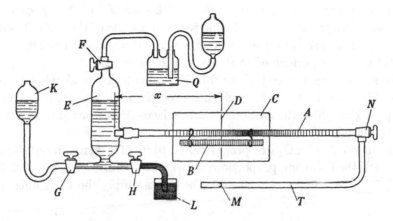

Fig. 2. Apparatus set up with experimental tube horizontal.

found. Thus x was determined as a function of C. This method had the advantage that comparisons were made only between colours whose spectrum and intensity were both identical at the determined position. No question arose as to how the intensity of colour varies with concentration.

The apparatus is shown diagrammatically in fig. 2. A is the pipe, B is the comparison tube. C is a ground-glass plate illuminated as uniformly as possible by means of a mirror which reflects either daylight or light from an electric bulb. D is a line ruled on C. To make a measurement B is filled with solution of known concentration and moved till the colour intensity is about the same as that in the pipe near its mid-point. The ground-glass plate C is then moved till the line D appears to cross the two tubes A and B at the point where their colours are identical. The distance x of the line D from the entry end of the pipe is then measured with a scale.

The flow through the pipe is controlled by a needle valve N or with a small bore capillary used as a leak at the exit end. The entry end projects into a glass chamber E out of which lead three tubes with glass taps, F, G, H. F is connected with a vessel Q by means of which pressure or suction can be applied. G is connected by a flexible tube to a funnel K filled with distilled water. H leads to a small tube which dips into a vessel, L, containing 1 % solution of $KMnO_4$.

The flow tube and chamber E are first washed with distilled water. Then leaving the pipe full and the valve N closed, the water is drained out through H and the chamber refilled with 1 % solution of $KMnO_4$. A pressure is then applied to the liquid in E. The valve N is then opened slightly and a stop-watch started. When the front of the coloured column is approaching the exit end of the pipe the needle valve is closed. It was found, as had been expected, that the molecular diffusion in the longitudinal direction was so small that no appreciable change in colour at a fixed spot occurs in several hours after the closing of N. The values of x corresponding with all the prepared comparison samples were then determined in the manner already described.

This procedure was used for comparison with case (B 2), §5. For case (B 1) the needle valve was opened for a short time to allow a little of the $KMnO_4$ solution to enter. It was then closed and the chamber E washed out and filled with water. The experiment was then performed in the same way as in case (B 2).

To measure the mean speed of the fluid and also to know at what moment the valve should be closed the water which flowed through the needle valve entered a second pipe T in which the motion of the meniscus M in front of the water column could be observed and measured.

A 1 % solution of $KMnO_4$ was prepared and parts of it were diluted so as to form solutions with the following proportions by weight of $KMnO_4$, $10^{-4} \times 1, 2, 3, 4, 6, 8,$ 10, 15, 20, 30, 40, 50, 60, 70, 80, 90, 100. The last is simply the 1 % solution.

7. Experimental results

The first experiments were designed to verify the conclusions of §2 when the effect of molecular diffusion is negligible. For this purpose it was necessary to carry them out so that the flow started and finished in a time small compared with $a^2/3 \cdot 8^2 D$. Since, for $KMnO_4$ in water, D is of order $0 \cdot 7 \times 10^{-5}$, $a^2/3 \cdot 8^2 D$ is $6 \cdot 2$ sec. for a tube $0 \cdot 5$ mm. bore and 25 sec. for a tube 1 mm. bore. For the first experiments a tube of approximately 1 mm. bore was chosen.

Case A 2. To verify experimentally the distribution of C_m predicted in equation (6) and illustrated in A 2, fig. 1, the reservoir E (fig. 2) was filled with the 1 % solution and the experimental tube with water. The valve N was opened for about $1\frac{1}{2}$ sec. and then closed. In this time the colour had travelled 65 cm. along the tube. The comparison tube was filled successively with the standard solutions and the corresponding distances x were measured. The results are given in table 1. These are plotted as A 2 in fig. 3. It will be seen that the experimental points lie very well on the line which was predicted in equation (6).

Cases A 1 *and A* 3. To verify the distributions calculated in equations (5) and (7), the flow tube was filled with water and the reservoir with 1 % $KMnO_4$. The valve was opened for an instant and a few centimetres of the tube thus filled with the solution. The reservoir was then washed out several times with water. The valve N was then again opened and closed after about $1\frac{1}{2}$ sec. The coloured column thus formed

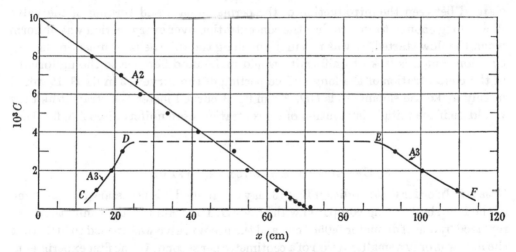

Fig. 3. Measurements of concentration when dispersion occurred in $1\frac{1}{2}$ sec.

appeared to be of uniform intensity over most of its length. The colorometric measurements given in table 1 were then made. They are shown as A 3 in fig. 3. It will be seen that the concentration rises between $x = 10$ to $x = 22$ cm. from $C = 0$ to $C = 0.0003$. From $x = 22$ to $x = 93$ cm. its colour was so nearly uniform that it was not possible to detect any variation in C. In this range C was between 0.0003 and 0.0004. This is represented in fig. 3 by the broken line at $C = 0.00035$. This experimental result is in good agreement with the prediction of equation (7) which is illustrated as A 3 in fig. 1.

Table 1

type of experiment	A 2 (fig. 3) $t = 1\frac{1}{2}$ sec. $a = 0.05$ cm.	A 3 (fig. 3) $t = 1\frac{1}{2}$ sec. $a = 0.05$ cm.
concentration, $10^4 C$ (g./ml.)	$C_0 = 0.01$ x (cm.)	x (cm.)
1	71.2	15.5 and 109.2
2	69.7	19.3 and 100.3
3	68.5	22.0 and 93.4
4	67.5	—
6	66.0	—
8	64.7	—
10	62.5	—
15	—	—
20	54.5	—
30	51.0	—
40	41.7	—
50	33.5	—
60	26.5	—
70	21.5	—
80	14.0	—
90	—	—

It will be noticed that the calculation for the case A 3 assumed that a section of length X is filled initially with solution of uniform concentration. The time which elapsed between the introduction of the permanganate and turning of the valve N was long enough to ensure that the concentration over every section was uniform when the flow started. It was not uniform along the tube as is assumed in the calculation, but this lack of uniformity would make no difference to the uniformity of the concentration of the long middle portion of the curve $A3$ in fig. 3. It would merely make the sloping ends CD, EF in fig. 3 curved instead of straight as they would be if the initial distribution of concentration were uniform (see $A3$, fig. 1).

8. EFFECT OF MOLECULAR DIFFUSION

Case B 2. In order that condition B of §3 may be satisfied it is necessary that the time of the flow shall be long compared with $a^2/3 \cdot 8^2 D$. To attain this the 1 mm. tube was replaced by one of diameter 0·0504 cm. and the needle valve was opened so little that the flow was only a small fraction of a centimetre per second. In the first experiments of this type the conditions were those of case $A2$, §2, except for the reduction in flow. In one case the flow was so slow that it took over 3 hr. to carry the colour 30 cm. The results of four experiments which ran for 4, 12, 240 and 11,220 sec. are given in table 2. To compare the observed distribution of C with that calculated theoretically the values of C/C_0 for the case when $t = 11,220$ sec. are plotted in fig. 4. Taking $\frac{1}{2}u_0 t = 31 \cdot 9$ cm., the value of x_1 in (29) is $x - 3 \cdot 19$. The value of $(4kt)^{-\frac{1}{2}}$ which gives best agreement with observations is $(4kt)^{-\frac{1}{2}} = 0 \cdot 552$. The curve

$$\frac{C}{C_0} = \tfrac{1}{2} \pm \operatorname{erf}\{0 \cdot 552(x - 31 \cdot 9)\} \tag{33}$$

is shown in fig. 4. It will be seen that the observed points fall very closely on the curve.

Table 2. *Experiments of type B* 2

time of flow (sec.)	about 4	12	240	11,220
$\frac{1}{2}u_0 t$ (cm.)	33	80	63·2	31·9
concentration $10^4 C$ (g./ml.)	x (cm.)			
1	63·5	134	80·4	34·75
2	62·4	129	79·5	34·4
4	58·5	120·4	77·0	34·0
6	56·4	117	76·0	33·85
8	55·0	114·5	75·0	33·70
10	53·1	111	74·0	33·65
20	47·8	101	71·1	33·0
30	44·4	93	68·0	32·55
40	40·0	89	66·7	32·4
50	33·8	82	63·8	31·7
60	29·5	73	60·5	31·4
70	24·0	63	58·0	31·3
80	17·4	56	53·7	30·8
90	11·6	41·0	45·0	30·0
100	—	—	—	—

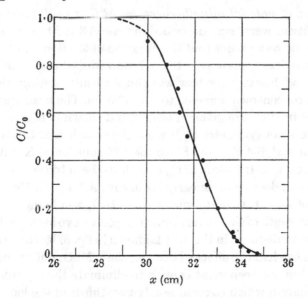

Fig. 4. Comparison between measured and theoretical distribution of concentration, $t = 11,200$ sec.

To demonstrate the effect of molecular diffusion on the distribution of concentration in the transition zone between clear water and the solution the observed values of C/C_0 are plotted against $x/\tfrac{1}{2}u_0 t$ in fig. 5. The strong tendency for molecular diffusion to prevent dispersion along the tube is shown very clearly. The limiting distribution as t tends to 0 is shown and the observed points for $t = 1\tfrac{1}{2}$ sec. have been transferred from fig. 3.

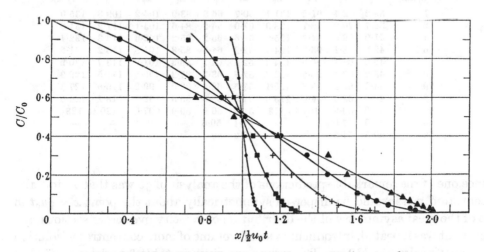

Fig. 5. Distributions of concentration about point $x = \tfrac{1}{2}u_0 t$.

symbol	●	■	+	●	▲
t (sec.)	11,220	240	12	4	$1\tfrac{1}{2}$

*Case B*1. *Dispersion of material initially concentrated in a small volume.* The initial experimental conditions were those described for case A 3 and the experiments were performed in the same way except that the duration of the flow was long compared with $a^2/3\cdot8^2D$. In the first experiment, 1 % solution was admitted and after filling the vessel E (fig. 2) with water the flow was run for 5 min. during which time the point of maximum concentration moved to $x = 52\cdot5$ cm. The measured positions of the standard concentrations are given in table 3 and shown in fig. 6. It will be seen that the (C, x) curve is not symmetrical. It seemed probable that this was due to the asymmetry of the initial distribution. Since the 1 % solution of $KMnO_4$ was introduced rapidly the initial distribution of C_m must have been triangular starting with $C_m = 0\cdot01$ at $x = 0$ and decreasing linearly (as in case A 2, fig. 1). By measuring the area of the curve of fig. 6 the total amount of $KMnO_4$ was found to be $\pi a^2 (0\cdot01)$ $\times (10\cdot5)$. The initial length of the column must therefore have been $2 \times 10\cdot5 = 21$ cm. The initial distribution deduced in this way is shown in fig. 6. It will be seen that the molecular diffusion has had the effect of clearing the $KMnO_4$ out of the first 39 cm., but the dispersion has not been great enough to eliminate the asymmetrical shape of the initial distribution which covered nearly two-thirds of the length of the dispersed distribution.

Table 3. *Experiments of type B*1

10⁴C (g./ml.)	fig. 6		fig. 7		fig. 8					
duration, t (sec.)	300		660		1740		2160		330	
$\frac{1}{2}u_0t$ (cm.)	52·5		110		58		94−58=36		122·5−94=28·5	
middle of colour (cm.)	52·5		110		58		94		122·5	
	x_1	x_2	x_1	x_2	x_1	x_2	x_1	x_2	x_1	x_2
1	39·2	72·4	91·2	131·1	50·7	66·8	83·9	104·2	109·8	135·0
2	39·7	70·9	92·1	128·3	51·0	66·2	84·6	103·1	111·2	134·0
4	41·0	69·5	94·7	125·9	51·6	65·2	85·2	102·6	112·3	132·4
6	42·0	68·1	95·3	124·7	52·2	64·7	85·8	101·8	113·2	131·6
8	42·5	68·1	96·0	124·3	52·4	64·1	86·3	101·2	113·7	130·9
10	43·2	67·1	96·5	122·7	52·7	63·7	87·2	100·8	114·5	129·9
20	44·2	65·2	99·8	120·1	53·5	62·6	88·0	99·2	116·6	127·3
30	46·0	63·0	104·4	115·6	54·6	62·0	85·5	98·7	118·2	124·0
40	48·0	59·8	107 to 113		54·9	60·9	89·9	97·0	120 to 123	
50	47·0	57·0	—	—	55·1	59·6	—	94·3	—	—
60	50·5	54·5	—	—	56·2	59·2	—	—	—	—

Since one of the remarkable predictions of the analysis of §5 was that an initially concentrated mass would be dispersed symmetrically about the point $x = \frac{1}{2}u_0t$ in spite of the great asymmetry of the distribution of velocity over cross-sections, the experiment was repeated, introducing a small volume of more concentrated solution and increasing $\frac{1}{2}u_0t$ to 110 cm. The results are given in table 3 and are plotted in fig. 7. It will be seen that the distribution of concentration about $x = 110$ cm. is very symmetrical.

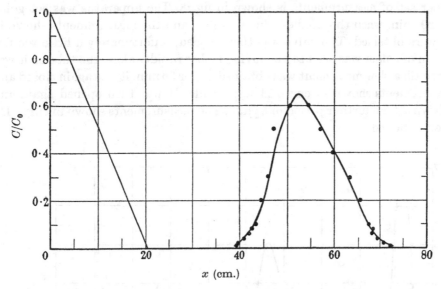

Fig. 6. Initially asymmetrical distribution of concentration which has not yet become symmetrical.

To compare these experimental results with the theoretical prediction of equation (28), the error curve

$$C = 0.0041\,e^{-(x-110)^2/121} \tag{34}$$

is shown in fig. 7. The constants in (34) have been chosen so that the error curve is as near to the experimental points as possible.

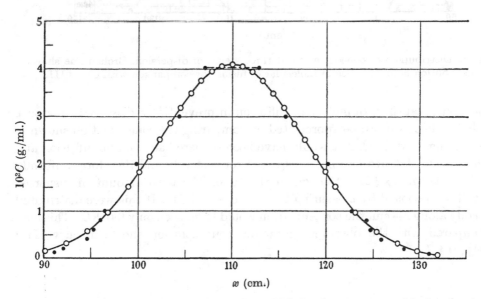

Fig. 7. Initially asymmetrical distribution which has become symmetrical owing to dispersion. o, $C = 0.0041 \exp\{-(110-x)^2/121\}$; ●, experiment; time = 11 min.

Another set of measurements is shown in fig. 8. The apparatus was set going first for 29 min. when the needle-value was shut and the measurements shown in curve I were obtained. The valve was then opened, unfortunately a little too far, the column of colour was seen to be moving rather rapidly along the tube so it was shut down till a slow movement was obtained. After 36 min. it was again closed and the measurements shown in curve II were made. It was then opened again and closed down after a further period of $5\frac{1}{2}$ min. The measurements shown in curve III were then obtained.

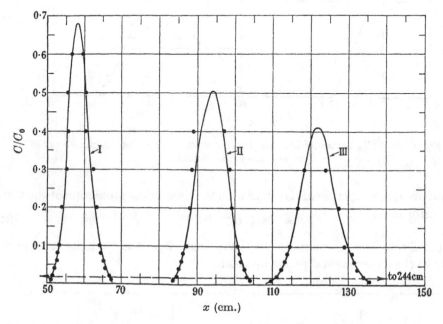

Fig. 8. Distributions of concentration at three stages of dispersion. Broken line shows distribution in the absence of molecular diffusion for comparison with curve III.

The very great effect of molecular diffusion in preventing a dissolved substance from being dispersed can be appreciated by comparing the concentration shown in curve III, fig. 8, with what it would have been if there had been no diffusion and the dissolved substance had been dispersed convectively and therefore uniformly (see (5)) through $2 \times 122 = 244$ cm. of the tube. The total amount of dispersed material can be found by measuring the area of curve III. If this were distributed uniformly along 244 cm. the concentration would have been only $0\cdot018\,C_0$. This may be compared with the observed maximum concentration shown in curve III, namely, $0\cdot4\,C_0$.

9. Calculation of molecular diffusion coefficient from measurements of longitudinal diffusivity

The work was undertaken to find out how a soluble substance is dispersed when injected into a stream of solvent flowing through a tube. The idea that it might form a basis for a simple method of measuring coefficients of molecular diffusion was not in my mind when the experiments were carried out, For that reason, the conditions necessary for that purpose were not in all cases satisfied. For instance, in the experiments illustrated in fig. 8, in the first two runs which carried the mean position of the colour to $x = 94$ cm., no precautions were taken to ensure the constancy of the flow. The last run which carried the colour to $x = 122 \cdot 5$ cm, in $5\frac{1}{8}$ min. was checked for constancy so that it can be used. The experiment illustrated in fig. 7 can be used because the spread is large compared with the length of column which would contain the whole of the dispersing salt at its initial concentration. Of the experiments given in table 2 only the one taking 240 sec. can be used. The first two must be rejected because the duration of flow (4 and 12 sec.) was too short for the theory to be applicable. The last experiment taking 11,220 sec. can hardly be expected to give an accurate result because when the tube is horizontal the effect of gravity acting on the slight difference in density between the 1 % solution and pure water is to increase the rate of longitudinal dispersion. This effect will be discussed in a subsequent paper on the combined effect of gravity and diffusion in which it will be shown that errors are to be expected when the time of diffusion is of order 10^4 sec. but are small when t is of order 10^3 sec. under the conditions of the experiments here described.

To use the experiments of type B2 to measure the diffusion coefficient D it is necessary to choose the parameter $(4kt)^{-\frac{1}{2}}$ in equation (29) so that the theoretical curve passes as nearly as possible through the observed points. This has been done for the case where $t = 11,220$ sec. (fig. 4). Applying the same method to the case given in column 4 of table 2 where $t = 240$ sec., the expression which gives the best fit is

$$\frac{C}{C_0} = \tfrac{1}{2} \pm \operatorname{erf}\{0 \cdot 0828(x-63)\}, \tag{35}$$

so that $4kt = 960\,k = (0 \cdot 0828)^{-2}$ and $k = 0 \cdot 152$. The mean velocity is

$$\tfrac{1}{2}u_0 = 63 \text{ cm.}/240 \text{ sec.},$$

so that $u_0 = 0 \cdot 524$ cm./sec. Using (25) and remembering that $a = 0 \cdot 0252$ cm.,

$$D = \frac{(0 \cdot 524)^2 (0 \cdot 0252)^2}{(192)(0 \cdot 152)} = 0 \cdot 60 \times 10^{-5} \text{ c.g.s. unit.} \tag{36}$$

To use experiments of type B1 the most accurate method would be to produce an error law distribution of concentration by running the apparatus for a time and then stopping it and making measurements. The flow would then be started again, run for a measured time and stopped again. If the curve fitted to the first set of observations is

$$C = A_1 e^{-\beta_1 (x - X_1)^2} \tag{37}$$

and the second is $$C = A_2 e^{-\beta_2 (x - X_2)^2},\qquad(38)$$

the fact that the same amount of the dispersed substance is present on both occasions leads to the condition that (37) and (38) have the same area, so that

$$\frac{\beta_2}{\beta_1} = \left(\frac{A_1}{A_2}\right)^2.\qquad(39)$$

If the flow had been running at the constant speed which it had between the two sets of measurements and the solution were initially very highly concentrated the first distribution would have been attained at time t_1 and the second at time t_2 where $t_2 - t_1 =$ actual time of running and

$$\frac{1}{4kt_1} = \beta_1 \quad \text{and} \quad \frac{1}{4kt_2} = \beta_2.$$

Hence, $$k(t_2 - t_1) = \frac{1}{4}\left(\frac{1}{\beta_2} - \frac{1}{\beta_1}\right).\qquad(40)$$

The error curves (28) which fit the observations of table 3 which are shown in fig. 8 have the following parameters:

$$\text{I:}\quad A_1 = 0{\cdot}65 C_0, \quad \beta_1 = 5{\cdot}86 \times 10^{-2},$$

$$\text{II:}\quad A_2 = 0{\cdot}51 C_0, \quad \beta_2 = 3{\cdot}72 \times 10^{-2},$$

$$\text{III:}\quad A_3 = 0{\cdot}41 C_0, \quad \beta_3 = 2{\cdot}30 \times 10^{-2}.$$

These do not exactly satisfy conditions (39) owing to experimental errors.

Applying (40) to II and III and inserting $t_2 - t_1 = 330$ sec. (table 3)

$$k = \frac{1}{4 \times 330}\left(\frac{100}{2{\cdot}30} - \frac{100}{3{\cdot}72}\right) = 1{\cdot}25 \times 10^{-2}.\qquad(41)$$

From table 3, $\frac{1}{2}u_0(t_2 - t_1) = 122{\cdot}5 - 94 = 28{\cdot}5$ cm., so that $u_0 = \frac{57}{330} = 0{\cdot}173$ cm./sec. Hence, from (25),

$$D = \frac{(0{\cdot}173)^2 (0{\cdot}0252)^2}{192(1{\cdot}25 \times 10^{-2})} = 0{\cdot}79 \times 10^{-5}\ \text{c.g.s. unit.}\qquad(42)$$

As another example the observation used in fig. 7 may be used. Here the concentration was not observed at any except the final position so that less accuracy may be expected. If the formulae (28) and (34) are used directly, assuming the start of the diffusion to be from a highly concentrated source, it is found from table 3 and fig. 7 that

$$k = \frac{121}{4(660)} = 0{\cdot}0459, \quad u_0 = \frac{2110}{660} = 0{\cdot}333\ \text{cm./sec.}$$

so that $$D = \frac{(0{\cdot}333)^2 (0{\cdot}0252)^2}{192(0{\cdot}0459)} = 0{\cdot}80 \times 10^{-5}\ \text{c.g.s. unit.}\qquad(43)$$

10. COMPARISON WITH PREVIOUS MEASUREMENTS

Measurements of the diffusion coefficient for $KMnO_4$ are quoted in Landolt & Börnstein's tables from Furth & Ullmann.* These figures are quoted in cm.^{2}/day. To reduce them to c.g.s. units they must be divided by 86,400, the number of seconds in 24 hr. The measurements were made at 18° C. which was approximately the temperature prevailing during the measurements described in the present paper. They covered a range $0 < C < 0.01$ g./ml. and in that range the diffusion coefficient ranged from 0.435×10^{-5} to 1.5×10^{-5}. They are shown in fig. 9 as circles and a smooth curve has been drawn to pass as nearly as possible through them.

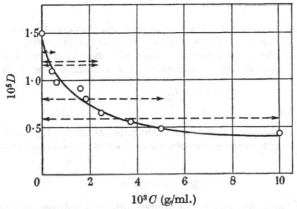

Fig. 9. Diffusion coefficient for $KMnO_4$ at 18° C. Comparison between present measurements (broken lines) and those of Furth & Ullmann (represented by circles).

In estimating the value of D which most nearly corresponds to the observed distribution of concentration, a constant D was assumed and the corresponding theoretical distributions of C were compared with those observed. It is possible to deduce from the present measurements shown in fig. 5 the variation of D with C but certain inaccuracies make the results of doubtful value. I hope later to describe improvements in the apparatus which will make it possible to give more accurate values of D. Some values of D obtained in the experiments here described are shown in fig. 9 and the range of values of C covered in the experiments are shown by means of broken lines. The experiment represented by the broken line at $D = 0.6 \times 10^{-5}$ is that of equation (36) and fig. 5. The experiment represented by the line at $D = 0.8 \times 10^{-5}$ is that of equation (43) and fig. 8. The lines at $D = 1.15$, 1.20 and 1.30×10^{-5} were obtained in experiments of type B 1 under conditions similar to those of fig. 8, but with smaller amounts of potassium permanganate. It will be seen that they are not inconsistent with the measurements of Ullmann, which probably give approximate values of C over the range $0 < C < 0.01$ g./ml., even though the validity of the theory used by this author in interpreting his experimental results may be questioned.

* R. Furth & E. Ullmann, *Kolloidzeitschrift*, XLI (1927), 307.

22

CONDITIONS UNDER WHICH DISPERSION OF A SOLUTE IN A STREAM OF SOLVENT CAN BE USED TO MEASURE MOLECULAR DIFFUSION

REPRINTED FROM

Proceedings of the Royal Society, A, vol. CCXXV (1954), pp. 473–7

It is shown that an assumption made in the author's previous discussion* can be presented as a result of analysis. The conditions under which an approximate solution of the equations for diffusion in a moving fluid can be used to interpret longitudinal dispersion of a solute in a stream of solvent flowing through a tube is given as

$$\frac{4L}{a} \gg \frac{Ua}{D} \gg 6 \cdot 9.$$

Here U is the mean velocity, a the radius of the tube, D the coefficient of molecular diffusion. L is the length of tube over which appreciable changes of concentration occur.

INTRODUCTION

In a recent paper,* I have discussed the dispersion of a soluble salt when injected into a stream of solvent flowing slowly through a tube. The distribution of concentration, C, of the soluble material depends on the balance between convection along the tube due to variation in velocity over the cross-section and radial molecular diffusion. Since the flow is laminar the velocity relative to the wall at a point distant r from the axis is $2U(1 - r^2/a^2)$, where U is the mean speed of flow and a the radius of the section. It is convenient in the present discussion to define concentration and velocity relative to axes which move with the mean flow. Relative to these axes, the velocity u is

$$u = U(1 - 2z^2), \tag{1}$$

where $z = r/a$, and the equation for diffusion is

$$\frac{\partial^2 C}{\partial z^2} + \frac{1}{z}\frac{\partial C}{\partial z} + a^2 \frac{\partial^2 C}{\partial x^2} = \frac{a^2}{D}\frac{\partial C}{\partial t} + \frac{a^2 U}{D}(1 - 2z^2)\frac{\partial C}{\partial x}, \tag{2}$$

where D is the coefficient of molecular diffusion. In general, the transfer of C along the tube by molecular diffusion is small compared with that produced by convection. It will be assumed therefore that $a^2(\partial^2 C/\partial x^2)$ is small compared with

$$\frac{\partial^2 C}{\partial z^2} + \frac{1}{z}\frac{\partial C}{\partial z}.$$

* Paper 21 above.

It will be seen later that the condition necessary for this to be true can be expressed in a simple manner. The transport equation will therefore be taken as

$$\frac{\partial^2 C}{\partial z^2} + \frac{1}{z}\frac{\partial C}{\partial z} = \frac{a^2}{D}\frac{\partial C}{\partial t} + \frac{a^2 U}{D}(1 - 2z^2)\frac{\partial C}{\partial x}. \tag{3}$$

Here $\partial/\partial t$ represents differentiation with respect to time at points fixed relative to axes moving with velocity U.

In the previous discussion the distribution of C in the case when $\partial C/\partial x = $ constant was found in the form

$$C = C_0 + \frac{a^2 U}{4D}\frac{\partial C_0}{\partial x}(z^2 - \tfrac{1}{2}z^4) \tag{4}$$

and $\partial C/\partial t = 0$, where C_0 is the concentration in the centre of the tube at $z = 0$.

In problems of transport along a tube the mean concentration C_m over any section is more significant than C_0.

C_m is defined by
$$C_m = 2\int_0^1 Cz\,dz, \tag{5}$$

and (4) may be modified to the form

$$C = C_m + \frac{a^2 U}{4D}\frac{\partial C_m}{\partial x}(-\tfrac{1}{3} + z^2 - \tfrac{1}{2}z^4) \tag{6}$$

by adding the constant which is necessary in order that (5) may be satisfied. As before, $\partial C/\partial t = 0$ in the case when $\partial C_m/\partial x$ is constant.

The expression (6) is a solution of (3) when $\partial C_m/\partial x$ is independent of x. The rate at which C is transported across a section is

$$Q = 2\pi a^2 \int_0^1 Cuz\,dz, \tag{7}$$

and inserting values of u and C from (1) and (6) it is found that

$$Q = -\pi a^2 \left(\frac{a^2 U^2}{48D}\right)\frac{\partial C_m}{\partial x}. \tag{8}$$

The rate of transfer of matter in a tube of radius a due to a diffusivity K is

$$-K\pi a^2 \frac{\partial C_m}{\partial x}. \tag{9}$$

Comparing (8) and (9), it will be seen that the combined effect of longitudinal convection and radial molecular diffusion is to give rise to a transfer across planes which move with the mean speed of flow which is equal to that which diffusivity

$$K = \frac{a^2 U^2}{48D} \tag{10}$$

would give to a stationary fluid.*

---

* The factor $\frac{1}{48}$ has been obtained by Westhaver in a problem concerning the migration of ions in an electric field against a slow flow in a capillary tube; see *J. Res. natn. Bur. Stand.* XXXVIII (1947), 169.

In the previous work it was assumed that the same relationship between Q and $\partial C_m/\partial x$ will exist, to a first approximation, even when $\partial C_m/\partial x$ is not constant, so that the dispersion relative to axes moving with speed U could be discussed by means of the equation

$$\frac{\partial C_m}{\partial t} = K\frac{\partial^2 C_m}{\partial x^2}. \tag{11}$$

It was realized that such an approximation would only be likely to be valid when the time necessary for a radial variation in C to die down owing to radial diffusion was much shorter than the time necessary for an appreciable change in C to occur through longitudinal convection, and this condition was expressed by the condition (paper 21 above, equation (16))

$$\frac{L}{U} \gg \frac{2a^2}{3\cdot 8^2 D}. \tag{12}$$

Here L is the longitudinal extent of the region in which $\partial C/\partial x$ is appreciable.

An alternative analysis leading to (11)

The intuitive method used in deriving (12) leaves much to be desired, and the nature of the assumption which justifies the critical step from (10) to (11) needs to be made clearer. For this purpose the effect of a small variation in $\partial C_m/\partial x$ on the distribution of C over a cross-section will be explored. By analogy with (6) it will be assumed that a solution may be taken in the form

$$C = C_m + \frac{a^2 U}{4D}\frac{\partial C_m}{\partial x}(-\tfrac{1}{3}+z^2-\tfrac{1}{2}z^2) + g(z)\frac{\partial^2 C_m}{\partial x^2}, \tag{13}$$

where $g(z)$ represents the perturbation in C due to the existence of a small finite value for $\partial^2 C_m/\partial x^2$. The problem is to determine $g(z)$ when $a^2 U/4DL$ is small.

Substituting (13) in (6)

$$\frac{\partial^2 C_m}{\partial x^2}\left[g''(z)+\frac{1}{z}g'(z)-\frac{a^4 U^2}{4D^2}(-\tfrac{1}{3}+\tfrac{5}{3}z^2-\tfrac{5}{2}z^4+z^6)\right]-\frac{a^2 U}{D}(1-2z^2)g(z)\frac{\partial^3 C_m}{\partial x^3}$$

$$=\frac{a^2}{D}\frac{\partial}{\partial t}\left\{C_m+\frac{a^2 U}{4D}(-\tfrac{1}{3}+z^2-\tfrac{1}{2}z^4)\frac{\partial C_m}{\partial x}+g(z)\frac{\partial^2 C_m}{\partial x^2}\right\}. \tag{14}$$

In order that (14) may be approximately of the same form as the diffusion equation it is necessary that

(i) the term $(a^2 U/D)(1-2z^2)g(z)\,\partial^3 C_m/\partial x^3$ may be small compared with the term in square brackets;

(ii) the terms containing $\partial C_m/\partial x$ and $\partial^2 C_m/\partial x^2$ in the bracket on the right-hand side of (14) shall be small compared with the term containing only C_m;

(iii) the expression inside the square bracket in (14) shall be independent of z.

(i) and (ii) are true when the length $a^2 U/D$ is small compared with the length of tube in which significant changes in C_m occur.

The equation for diffusion with coefficient of diffusion K' is

$$K'\frac{\partial^2 C_m}{\partial x^2} = \frac{\partial C_m}{\partial t}. \tag{15}$$

Comparing (14) and (15) and using only the significant terms, the equation for $g(z)$ is

$$g''(z) + \frac{1}{z}g'(z) - \alpha\left(-\tfrac{1}{3} + \tfrac{5}{3}z^2 - \tfrac{5}{2}z^4 + z^6\right) = \frac{K'a^2}{D}, \tag{16}$$

where

$$\alpha = \frac{a^4 U^2}{4D^2}. \tag{17}$$

The complementary function for (16) is

$$g(z) = A\ln z + B, \tag{18}$$

and $A = 0$ since C is finite at $z = 0$.

The expression

$$g(z) = Cz^8 + Dz^6 + Ez^4 + Fz^2 + B \tag{19}$$

is a particular integral of (16) if

$$C = \tfrac{1}{64}\alpha, \quad D = -\tfrac{5}{72}\alpha, \quad E = \tfrac{5}{48}\alpha, \quad F = \frac{K'a^2}{4D} - \tfrac{1}{12}\alpha. \tag{20}$$

The condition at the boundary $z = 1$ is $\partial C/\partial z = 0$. Hence from (6)

$$g'(z) = 0 \quad \text{at} \quad z = 1. \tag{21}$$

Inserting this condition in (19)

$$F = -4C - 3D - zE \tag{22}$$

and inserting the values of C, D and E from (20) this leads to the equation which determines K', namely,

$$K' = \frac{4D}{a^2}\left(\tfrac{1}{12} - \tfrac{1}{16}\right) \quad \text{or} \quad K' = \frac{a^2 U^2}{48D}. \tag{23}$$

Comparing (23) with (10) it will be seen that the same value is now obtained for the longitudinal diffusivity of the stream that was obtained making the intuitive assumptions.

To complete the calculation it is only necessary to determine the constant term B in (19).

In the definition of C_m given in (5) substitute the expression (13). This leads to

$$\int_0^1 zg(z)\,\mathrm{d}z = 0;$$

hence

$$\tfrac{1}{10}C + \tfrac{1}{8}D + \tfrac{1}{6}E + \tfrac{1}{4}F + \tfrac{1}{2}B = 0$$

or, using (20)

$$B = \frac{31}{64 \times 5 \times 9}\alpha,$$

and

$$g(z) = \frac{a^4 U^2}{16D^2}\left\{\tfrac{1}{16}z^8 - \tfrac{5}{18}z^6 + \tfrac{5}{12}z^4 - \tfrac{1}{4}z^2 + \frac{31}{16 \times 5 \times 9}\right\}. \tag{24}$$

The distribution represented by (24) will contribute something to the integral (7) for the transport over a section but it will be of a lower order of magnitude than that given in (8) if the length $a^2U/4D$ is small compared with the length of tube over which C has an appreciable value.

Conditions under which measurements of C_m may be applied to obtain values of D

In my previous paper it was suggested that measurements of K, made by measuring dispersion along a capillary tube, may be used for measuring the diffusion coefficient D using (10). In order that the expression $K = a^2U^2/48D$ may be used for a valid representation of dispersion in a tube, two conditions must be satisfied:

(i) In order that the longitudinal molecular diffusion may be negligible compared with the dispersive effect represented by K it is necessary that

$$D \ll \frac{a^2U^2}{48D},$$

or, approximately,

$$\frac{aU}{D} \gg \sqrt{48} \text{ or } 6.9. \tag{25}$$

(ii) $a^2U/4DL$ must be small. Here L is used to represent the distance in which the greater part of the change in concentration takes place. L^{-1} may be taken as a mean of $(1/C_m)(\partial C_m/\partial x)$ in that region.

Combining (i) and (ii) the condition to be satisfied in experiments is

$$\frac{4L}{a} \gg \frac{Ua}{D} \gg 6.9. \tag{26}$$

If ratios $10:1$ are permitted between the terms of the inequalities (26) the following are permissible:

$$\frac{Ua}{D} = 69 \quad \text{and} \quad \frac{4L}{a} = 690. \tag{27}$$

Thus for a tube of radius 0.025 cm., such as that used in my previous experiments, the least value of L for which (10) could be used accurately is

$$L = \frac{690}{4} \times 0.025 = 4.3 \text{ cm.}, \tag{28}$$

and to satisfy (27) when the diffusion coefficient was 1.0×10^{-5} the velocity U would have to be controlled at

$$U = \frac{69 \times 10^{-5}}{0.025} = 0.028 \text{ cm./sec.}$$

In my previous paper (p. 236 above), experiments were described in which potassium permanganate ($1.5 \times 10^{-5} > D > 0.5 \times 10^{-5}$) was introduced at various rates into a tube 0.054 cm. bore. The rates were $8.3, 6.7, 0.26, 0.0028$ cm./sec. Of these the first three satisfy the condition $Ua/D > 69$. The fourth does not satisfy this condition. At the other limit the first two do not satisfy the condition $L > 10(Ua^2/4D)$. The third nearly satisfies it if L is taken as the length of tube between the points where C is $\frac{1}{100}$ and $\frac{99}{100}$ of the full concentration. The fourth satisfies this condition. Thus only the third satisfies approximately both conditions, and in fact only the third was used in comparing the value of D, obtained from (10) with values given by previous observers.

23

THE TWO COEFFICIENTS OF VISCOSITY FOR A LIQUID CONTAINING AIR BUBBLES

REPRINTED FROM

Proceedings of the Royal Society, A, vol. CCXXVI (1954), pp. 34–9

Incompressible fluids possess only one coefficient of viscosity because, by definition, no changes in volume can occur. If such a fluid contains air bubbles it becomes compressible, and any changes in volume involves a contraction or expansion of the bubbles which is resisted by the ordinary viscosity of the surrounding fluid. The resulting second coefficient of viscosity is found to be $4\mu/3v$, where μ is the viscosity of the incompressible fluid and v the (small) proportion of the total volume which is occupied by the bubbles.

When the effect of compressibility in the fluid is taken into account, it is found that the volume viscosity of water containing air bubbles reaches a maximum value of 6700 times the viscosity of water when $v = 5 \times 10^{-5}$.

The viscosity of an incompressible fluid containing solid spheres was investigated by Einstein[†] who showed that the viscosity of the mixture is

$$\mu_1 = \mu(1 + 2 \cdot 5v), \tag{1}$$

where μ is the viscosity of the fluid and v the (small) proportion of the volume which is occupied by the spheres. Einstein's formula has been extended[‡] to cover the case where the solid spheres are replaced by drops of a liquid whose viscosity is μ_d. The modified expression is

$$\mu_1 = \mu \left\{ 1 + 2 \cdot 5v \left(\frac{\mu_d + \frac{2}{5}\mu}{\mu_d + \mu} \right) \right\}. \tag{2}$$

When the drops are air bubbles μ_d may be neglected compared with μ and (2) becomes

$$\mu_1 = \mu(1 + v). \tag{3}$$

When the drops as well as the outer fluid are incompressible no viscous effect due to volume changes can exist, but when the drops are replaced by air bubbles they confer on the fluid volume elasticity and also volume viscosity.

Consider a spherical bubble in a viscous fluid. When the pressure changes, the radius of the bubble changes and the equation of motion of the fluid outside it is

$$\frac{\partial p}{\partial r} = \mu \left(\frac{\partial^2 u}{\partial r^2} + \frac{2}{r}\frac{\partial u}{\partial r} - \frac{2u}{r^2} \right) - \rho \frac{\partial u}{\partial t} - \rho u \frac{\partial u}{\partial r}, \tag{4}$$

---

[*] *Editor's note.* Small changes have been made to the title and the summary to make them suit the paper in the present extended form which includes the note added subsequently.

[†] A. Einstein, *Annln Phys.* XIX (1906), 289.

[‡] Paper 10 above.

where u is the radial velocity. It will be assumed that the inertia terms $\rho\, \partial u/\partial t$ and $\rho u\, \partial u/\partial r$ are negligible compared with the term $\mu\nabla^2 u$. The equation of continuity is

$$u = u_0 a^2/r^2, \tag{5}$$

where a is the radius of the bubble and $u_0\ (= da/dt)$ is the velocity of expansion.

Using (5) it is found that

$$\frac{\partial^2 u}{\partial r^2} + \frac{2}{r}\frac{\partial u}{\partial r} - \frac{2u}{r^2} = 0,$$

so that

$$\frac{\partial p}{\partial r} = 0 \quad \text{and} \quad P = p_0, \tag{6}$$

where P is the pressure far from the bubble and p_0 the pressure just outside its surface. If unit mass of the fluid contains n bubbles of radius a, n is constant as the mean density varies. If v is the proportion of the volume occupied by bubbles

$$v = \tfrac{4}{3}\pi n a^3 \rho, \tag{7}$$

where ρ is the mean density of fluid and bubbles.

If σ is the density of the incompressible fluid, and the density of the air in the bubbles is neglected

$$\frac{1}{\rho} = \frac{1}{\sigma} + \tfrac{4}{3}\pi n a^3. \tag{8}$$

Hence

$$-\frac{1}{\rho^2}\frac{d\rho}{dt} = 4\pi n a^2 \frac{da}{dt} = 4\pi n a^2 u_0, \tag{9}$$

or

$$\frac{1}{\rho}\frac{d\rho}{dt} = -3v\frac{u_0}{a}. \tag{10}$$

The equation of continuity is

$$\frac{1}{\rho}\frac{d\rho}{dt} = -\Delta, \tag{11}$$

where

$$\Delta = \frac{\partial u}{\partial x} + \frac{\partial v}{\partial y} + \frac{\partial w}{\partial z}. \tag{12}$$

In many problems such as those concerned with the propagation of sound the small change, or increment, in density corresponding with a given change in pressure is needed. In such cases it is convenient to define the increments by means of primes, thus

$$P' = P - P_0, \quad \rho' = \rho - \rho_0, \tag{13}$$

and from (6)

$$P' = p_0'. \tag{14}$$

Inside the bubble the pressure is p_i and its initial value when the fluid was at rest p_{i0}, so that

$$p_i' = p_i - p_{i0}. \tag{15}$$

When at rest the difference between pressure inside and outside the bubble is balanced by a surface tension T. Hence

$$p_{i0} - P_0 = \frac{2T}{a}. \tag{16}$$

When the bubble is expanding, the normal force on the inner surface may be taken as p_i, but the normal tension on the outer surface is*

$$p_{rr} = -p_0 + 2\mu \left[\frac{\partial u}{\partial r}\right]_{r=a} = -\left\{P + 4\mu \frac{u_0}{a}\right\}. \tag{17}$$

Hence the equation for the balance of force at the surface is

$$p_i - P - 4\mu \left(\frac{u_0}{a}\right) = \frac{2T}{a + \delta a} = \frac{2T}{a} - \frac{2T}{a^2} \delta a, \tag{18}$$

where δa is the small increase in radius. Subtracting (16) from (18)

$$p_i' - P' = 4\mu \frac{u_0}{a} - 2T \frac{\delta a}{a^2}. \tag{19}$$

Since the bubbles are assumed small it is justifiable to assume that changes in volume are isothermal and the pressure changes with volume in accordance with Boyle's law, so that

$$p_i' = -3p_{i0} \frac{\delta a}{a} = -3\left(P_0 + \frac{2T}{a}\right) \frac{\delta a}{a}. \tag{20}$$

Eliminating p_i' from (20) and (19)

$$P' = -\left(P_0 + \frac{4}{3} \frac{T}{a}\right) \frac{3\delta a}{a} - 4\mu \frac{u_0}{a}. \tag{21}$$

From (10) and (11),

$$\frac{u_0}{a} = \frac{\Delta}{3v}, \tag{22}$$

and when $\delta a / a$ is small, from (9)

$$\frac{\rho'}{\rho} = -4\pi n a^2 \delta a = -\frac{3v \delta a}{a}, \tag{23}$$

and (21) becomes

$$P' = \left(P_0 + \frac{4}{3} \frac{T}{a}\right) \frac{\rho'}{v\rho_0} - \frac{4}{3}\mu \frac{\Delta}{v}. \tag{24}$$

It will be seen that the pressure P' is that which must be considered in any dynamical discussion of the motion of the complex medium, so that the equations of motion will be

$$p_{xx} = -P' + 2\mu \frac{\partial u}{\partial x}, \quad p_{yy} = -P' + 2\mu \frac{\partial v}{\partial y}, \quad p_{zz} = -P' + 2\mu \frac{\partial w}{\partial z}, \tag{25}$$

and substituting for P' for (24)

$$p_{xx} = -\left(P_0 + \frac{4}{3} \frac{T}{a}\right) \frac{\rho'}{v\rho_0} + \frac{4}{3}\mu \frac{\Delta}{v} + 2\mu \frac{\partial u}{\partial x}. \tag{26}$$

The equations proposed to take account of the second coefficient of viscosity μ_1' are

$$p_{xx} = -p + \mu_1' \Delta + 2\mu \frac{\partial u}{\partial x}. \tag{27}$$

* Lamb, *Hydrodynamics* (1932), p. 574.

Equation (26) is identical with (27) if (1) p is the pressure change corresponding with a slow change ρ' in density, i.e. p is the pressure which is determined from the density according to the physical equation of state, and (2)

$$\mu_1' = \frac{4}{3}\frac{\mu}{v}. \tag{28}$$

Thus the model consisting of an incompressible fluid of viscosity μ containing a small proportion v by volume of air bubbles, has a second coefficient of viscosity $\mu_1' = 4\mu/3v$.

It is interesting to compare the properties of this model with those actually observed in liquids and enumerated by Rosenhead* under seven headings.

(i) 'No correlation appears to exist between the first and second viscosities', μ and μ'. There is no correlation if μ and v vary. There would be an inverse correlation if only v varied.

(ii) 'The effect of variation of temperature in the case of water showed that the effects are the same for both viscosities', i.e. μ_1'/μ independent of temperature. If v is constant $\mu_1'/\mu = 4/3v$ = constant for the model.

(v) 'The viscosity ratio for most liquids so far investigated has a positive value.' $\mu_1'/\mu = 4/3v$ is necessarily positive in the model.

(vi) 'The value of the ratio differs widely.' Any value of $4/3v$ can be obtained by varying v.

(vii) 'For most liquids the value of the second viscosity appears to be much greater than that of the first.' This is true in the model when v is small, as is assumed in the analysis.

The missing numbers (iii) and (iv) are omitted because they do not define properties of μ'. The model might be expected to yield results in accordance with the statements in (iii) and (iv).

(Note added 15 May 1954)

During the Royal Society Discussion, I pointed out that my expression (28) for volume viscosity $\mu_1' = 4\mu/3v$ becomes infinite when $v \to 0$, i.e. when the volume of the bubbles becomes vanishingly small, and that this paradox is due to the assumption that the fluid surrounding the bubbles is incompressible. I said that I had made an approximate calculation of the effect of a small compressibility and had found that the part of the volume viscosity which is due to the bubbles vanishes instead of becoming infinite when $v \to 0$. I also pointed out that as v increases from zero μ_1' reaches a maximum and then decreases, and I illustrated this effect by means of a rough sketch on the blackboard. The diagram from which that sketch was taken is shown in fig. 1.

My calculations were cumbersome, and Dr R. O. Davies has independently made the more compact analysis which is given in his note† on my contribution. In that note he points out a fact which I failed to notice, namely, that the introduction

---

* *Proc. Roy. Soc.* A, ccxxvi (1954), 5.   † *Proc. Roy. Soc.* A, ccxxvi (1954), 39.

of compressibility in the fluid outside the bubbles entails the existence of a relaxation phenomenon. The volume viscosity of the mixture in Dr Davies' note is represented by ζ_{eff}, which would, in the notation of my contribution, be expressed as $\mu_1' + \frac{2}{3}\mu_1$. Introducing $y = \kappa_e/\kappa$, where κ_e and κ are the compressibilities of air and water respectively, his (4) may be written

$$\mu_1' + \tfrac{2}{3}\mu_1 = \left\{\mu' + \tfrac{2}{3}\mu + \frac{4}{3}\frac{\mu v y^2}{1+vy}\right\}\left(\frac{1}{1+vy}\right), \tag{A}$$

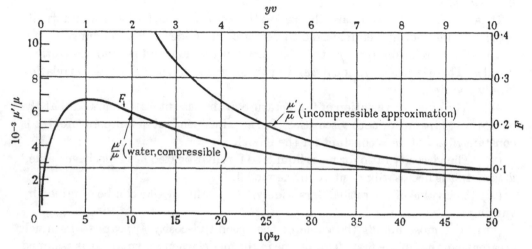

Fig. 1. Volume viscosity of water containing bubbles at atmospheric pressure.

where μ' is the second coefficient of viscosity of the fluid. For bubbles in water at atmospheric pressure κ_e may be taken as 10^{-6} if surface tension is neglected and the bubbles expand isothermally, or 0.7×10^{-6} if they expand adiabatically. For water $\kappa = 5 \times 10^{-11}$ so that $y = 2 \times 10^4$ approximately. Since μ' is of the same order of magnitude as μ, both may be neglected compared with $\frac{4}{3}\mu v y^2/(1+vy)$, so that (A) may be written

$$\mu_1' = \tfrac{4}{3}\mu y \frac{vy}{(1+vy)^2}. \tag{B}$$

Values of $vy/(1+vy)^2 = F$ are shown in fig. 1. It will be seen that F reaches its maximum value of 0.25 when $vy = 1$, so that the volume viscosity of water reaches its maximum value $\frac{1}{4}(\frac{4}{3}\mu)(2 \times 10^4) = 6.7 \times 10^3\mu$ when $v = 5 \times 10^{-5}$; this condition is reached when the bubbles are approximately 20 diameters apart. The values of μ_1'/μ are shown for water as a function of v by means of the scales below, and on the left of fig. 1.

24

THE ACTION OF A SURFACE CURRENT
USED AS A BREAKWATER

REPRINTED FROM

Proceedings of the Royal Society, A, vol. CCXXXI (1955), pp. 466–78

The conditions under which an outward-flowing surface current can prevent the passage of waves coming in from the sea are investigated mathematically. Two types of current are considered: (a) a current with uniform velocity extending to a depth h; (b) a current with velocity decreasing uniformly and vanishing at depth h. They have very similar effects. The mean velocity required to stop waves of given frequency is rather greater in case (a) than in case (b).

The water current produced by a curtain of air bubbles from a perforated tube on the sea bottom is investigated theoretically on the assumption that the bubbles are very small.

Evans* has measured the surface currents produced in a tank by a bubble curtain and finds them smaller than predicted. The discrepancy is partly due to the fact that the bubbles were not very small.

1. INTRODUCTION

The idea that a curtain of bubbles rising from a perforated pipe lying on the bottom of the sea might be effective as a shield against waves was put forward many years ago. In 1942, while discussing the mechanics of a bubble curtain with members of the Scientific Staff of the Admiralty, I pointed out that bubbles must produce a rising current of water which would spread into a horizontal current at the sea surface. It seemed to me more likely that the waves might be stopped by this current than by the direct action of the bubbles. The same idea has occurred to others. A history of the subject is given in the paper by Evans.*

To find out whether a surface current would be effective I calculated the relationship between wavelength and frequency when waves are propagated in a surface current of depth h and constant velocity U flowing over a sea of great depth which is at rest beneath the current. I found that when the current is directed against the oncoming waves it should stop all waves shorter than a certain critical length. These calculations, which were done in 1942, were not published, and it was only when Mr J. T. Evans made the experiments he now describes* that they were recovered from the Admiralty archives. They are given in §2.

Evans not only found that the waves are stopped by the adverse current produced by a curtain of bubbles, but he measured the distribution of velocity in the current. He found that the current is a maximum at the surface and decreases nearly linearly with depth so that it nearly disappears at a certain depth h. This particular distribution of velocity is one for which it is possible to calculate the frequency

* *Proc. Roy. Soc.* A, CCXXXI (1955), 457.

wavelength relationship. The calculation is given in §3. It is found that the effect of this kind of stream on waves is almost identical with that of a uniform current. When the depth is small compared with the wavelength the wave-stopping efficiency of a uniform stream is equivalent to that of a stream with a uniform gradient of velocity, provided the total flow of inertia of the two streams is the same in the two cases. A surface stream spreads downwards by entrainment of the water below it, and in this process the total momentum of the flow remains constant. In comparing the result of experiment with theory it is therefore not to be expected that the exact position in which the vertical distribution of horizontal velocity is measured will be important. It is in fact found that waves of small amplitude are stopped by an adverse current of approximately the strength calculated, and that the wave-stopping effect is the same whether the current is produced by bubbles or directly by a jet of water.

In my note to the Admiralty I estimated the form of the current which might be expected from the discharge of a known volume of very small bubbles at a given depth. This calculation is given in §4. In Evans' experiments the bubbles were not very small and had a vertical velocity through the water which was comparable with that of the rising water. The velocity of the horizontal current observed by Evans is smaller than the calculations predict when the assumption is made that the bubbles are very small.

2. EFFECT ON THE PROPAGATION OF SURFACE WAVES OF A UNIFORM CURRENT OF DEPTH h

Take as co-ordinates x, horizontal and positive in the direction of the current, and y vertical, the origin being in the undisturbed sea surface (fig. 1(a)). Waves of length $\lambda = 2\pi/k$ and frequency $\sigma/2\pi$ can be represented by velocity potentials ϕ_1 on the surface current and ϕ_2 in the sea below the current, where

$$\phi_1 = -Ux + (B_1 e^{ky} + B_2 e^{-ky})\, e^{i(kx-\sigma t)}, \tag{1}$$

$$\phi_2 = C e^{k(y+h)}\, e^{i(kx-\sigma t)}, \tag{2}$$

if the wave at the surface is represented by

$$y = a\, e^{i(kx-\sigma t)} \tag{3}$$

and that at the lower surface of the current by

$$y = -h + b\, e^{i(kx-\sigma t)}. \tag{4}$$

The choice of the form (2) for the disturbance below the surface layer necessarily implies that k is positive.

The continuity equations at the two surfaces are

$$ia(-\sigma + kU) = k(-B_1 + B_2), \tag{5}$$

$$ib(-\sigma + kU) = k(-B_1 e^{-kh} + B_2 e^{kh}), \tag{6}$$

$$i\sigma b = kC, \tag{7}$$

while the pressure conditions are

$$(B_1 + B_2)(-\sigma + kU) + iga = 0, \tag{8}$$

$$(B_1 e^{-kh} + B_2 e^{kh})(-\sigma + kU) + \sigma C = 0. \tag{9}$$

Eliminating a, b, B_1, B_2, C from (5) to (9) the resulting period equation is

$$\{(\sigma - kU)^4 - kg\sigma^2\}\sinh kh + (\sigma - kU)^2(\sigma^2 - kg)\cosh kh = 0. \tag{10}$$

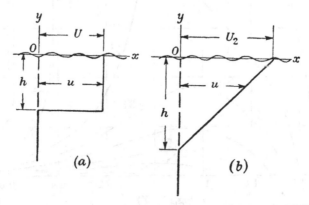

Fig. 1. Assumed distribution of velocity in surface currents.

It is convenient to express (10) in non-dimensional form. This may be accomplished in several ways. The one followed here is to take

$$Y = \frac{kU}{\sigma}, \quad \alpha = \frac{g}{U\sigma}, \quad Z = \frac{hg}{U^2}. \tag{11}$$

Equation (10) is then

$$\frac{(1 - Y)^2(1 - \alpha Y)}{(1 - Y)^4 - \alpha Y} = -\tanh\left(\frac{ZY}{\alpha}\right). \tag{12}$$

Since (12) is a transcendental equation it would be difficult to give a complete discussion of its roots, but since it involves only three quantities and since Z is uniquely determined for given values of α and Y, all possible combinations of real values of α, Y and Z can be represented on a plane diagram in which α and Y are co-ordinates and contours are drawn to represent constant values of Z. Since $\tanh(ZY/\alpha)$ lies between $+1$ and -1 the parts of this diagram for which the left-hand side of (12) is greater than $+1$ or less than -1 do not correspond with possible real solutions of (12).

Numerical solution of (12) can easily be obtained by taking a pair of values of α and Y and then calculating Z, using tables of inverse hyperbolic tangents. If, for instance, a constant value of α is chosen arbitrarily and the corresponding values of Z for a range of values of Y are calculated from (12) the results can be plotted as contours of constant α in a (Z, Y) diagram. Such a figure is shown covering a portion of the field in fig. 2. By cross-plotting contours of Z an (α, Y) diagram can be obtained; fig. 4 is an example.

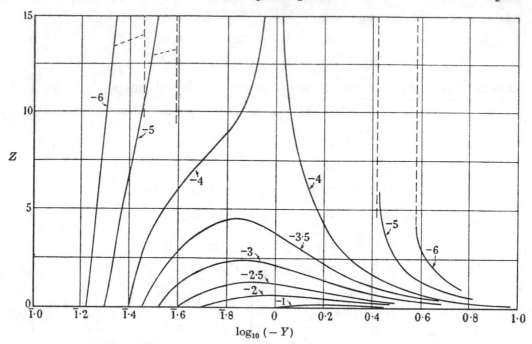

Fig. 2. Solutions of (12) for various values of α.

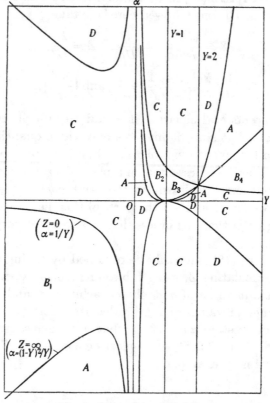

Fig. 3. Map of (α, Y) plane.

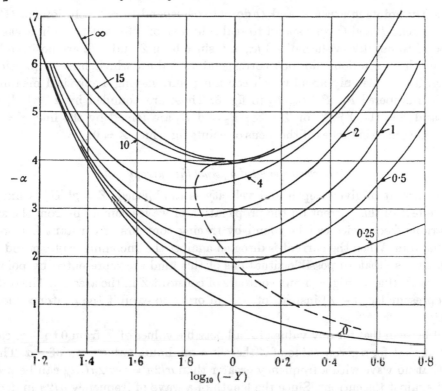

Fig. 4. Solutions of (12). The figures give the constant value of Z for each curve.

To construct these diagrams it is first necessary to map the areas in the (α, Y) field (fig. 3) which correspond with the real solutions of (12) that represent possible wave conditions. Since U and k are both positive, U because of the choice of axes, and k because of the choice of the form (2) to represent the disturbance below the surface current, both α and Y are positive if σ is positive. In this case the wave travels in the same direction as the surface current. In all cases for which the direction of propagation and the surface current are opposed both α and Y are negative. Writing

$$F = \frac{(1-Y)^2(1-\alpha Y)}{(1-Y)^4 - \alpha Y},$$

(12) becomes
$$-ZY\alpha^{-1} = \tanh^{-1} F. \tag{13}$$

The (α, Y) field may conveniently be divided into four kinds of area:

$$A, \quad \text{where} \quad -\infty < F < -1,$$
$$B, \quad \text{where} \quad -1 < F < 0,$$
$$C, \quad \text{where} \quad 0 < F < 1,$$
$$D, \quad \text{where} \quad 1 < F < \infty.$$

These are bounded by the curves $\alpha = Y^{-1}$, $\alpha = (1-Y)^2/Y$, $\alpha = (1-Y)^4/Y$ and they are shown in fig. 3.

Evidently points in areas A and D do not correspond to real solutions of (13), points in areas B and C correspond to real solutions of (13), but those in areas C correspond to negative values of ZY/α, and since both Z and Y/α are positive by definition these solutions do not correspond to values which have a physical meaning. Thus, the only areas which contain points having a physical meaning are those numbered B_1, B_2, B_3, B_4 in fig. 3. These are bounded by $\alpha = Y^{-1}$, or $Z = 0$, and $\alpha = (1 - Y)^2/Y$, or $Z = \infty$. B_2 and B_3 are joined by the line $Y = 1$, which is one of the branches of the locus of points for which $Z = 0$.

Waves propagated against the current

When a current of given depth and velocity is established, $Z = gh/U^2$ is fixed; and the effect of this current on the propagation of waves coming in from the sea with various frequencies can be found by tracing contours of constant Z in the (α, Y) diagram. When the current is directed against the oncoming waves α and Y are negative, so that all possible situations of this kind are represented by points in the area B_1 (fig. 3). Fig. 4 shows contours of constant Z in the area B_1. The ordinates represent $\log_{10}(-Y)$ instead of $-Y$ in order to exhibit the contours more distinctly.

It will be seen that for any value of Z all possible values of Y from 0 to $-\infty$ can occur, but that for a given value of Z there is a minimum value $-\alpha_m$ of $-\alpha$. This means that no wave with a frequency greater than $\sigma/2\pi = -g/2\pi U\alpha_m$ can be propagated against the current. Since the length of a wave of frequency $\sigma/2\pi$ in deep water is $2\pi g/\sigma^2$, no wave of length less than

$$\lambda_0 = 2\pi U^2 g^{-1} \alpha_m^2 \tag{14}$$

can be propagated when it reaches the opposite current. Each value of α_m corresponds with a definite value of Z and Y. These can be found graphically by measuring fig. 4, but they can also be found by minimizing Z using (12). In this way table 1 was constructed giving values of Y and Z for various values of α_m. The corresponding curve showing $-\alpha_m$ as a function of $-Y$ is shown by the broken line in fig. 4.

In using an artificially produced surface current as a breakwater the question might be asked is 'How fast must a current of given depth, h, flow in order that it may stop waves of length λ_0?' To answer this, one may notice from (14) that

$$\frac{\lambda_0}{2\pi h} = \frac{U^2}{gh}\alpha_m^2 = \frac{\alpha_m^2}{Z}. \tag{15}$$

Values of α_m^2/Z are given in column 4 of table 1 and their dependence on $-\alpha_m$ is shown in curve A of fig. 5. Suppose, for instance, one wishes to stop all waves of length less than λ_0 with a current of depth h, α_m^2/Z is then known from (15), and using fig. 5 the value of $-\alpha_m$ is read off. The speed of the necessary current is then

$$U = -\frac{1}{\alpha_m}\left(\frac{g\lambda_0}{2\pi}\right)^{\frac{1}{2}}. \tag{16}$$

For the particular case $\lambda_0 = 100$ ft. the depths of currents are given in column 6 and the corresponding velocity in column 5 of table 1.

It will be seen that if a current of 15 ft./sec. could be maintained over a sufficiently great distance it would only have to be 1·6 ft. deep to stop 100 ft. waves.

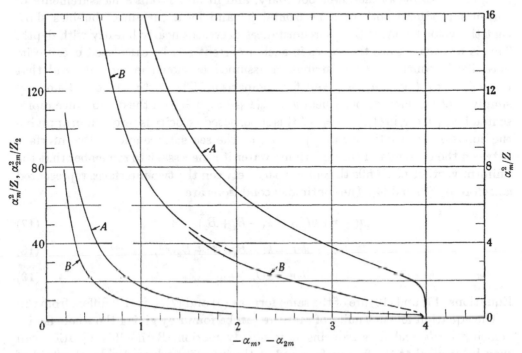

Fig. 5. Curves for finding velocity,

$$U = \frac{1}{\alpha_m} \left(\frac{g\lambda_0}{2\pi} \right)^{\frac{1}{2}},$$

of the current which will stop waves shorter than $\lambda_0 = 2\pi h(\alpha_m^2/Z)$. A, uniform current; B, current decreasing with depth.

Table 1. *Uniform current to depth h: minimum values of $-\alpha$ for constant values of Z*

				velocities for stopping 100 ft. waves	
$-\alpha_m$ (1)	$-Y$ (2)	Z (3)	α_m^2/Z (4)	U (ft./sec.) (5)	h (ft.) (6)
0·2	7·83	$1·75 \times 10^{-4}$	229	113	0·07
0·5	3·34	$5\ 35 \times 10^{-3}$	46·3	45	0·34
1·0	1·78	$5·82 \times 10^{-2}$	17·1	23	0·93
1·5	1·22	0·226	9·95	15	1·60
2·0	0·96	0·59	6·79	11	2·35
2·5	0·82	1·25	4·99	9·0	3·20
3·0	0·72	2·40	3·74	7·5	4·25
3·5	0·69	4·44	2·76	6·4	5·8
3·8	0·75	6·71	2·15	5·9	7·4
3·9	0·82	8·13	1·87	5·8	8·5
4·0	1·00	∞	—	5·6	∞

3. Waves in a Non-Uniform Surface Current

In §2 the waves which can be propagated over a surface current of uniform velocity and extending to a depth h are investigated. Such a distribution of velocity is highly unstable at the lower boundary, and in fact Evans's measurements of currents produced both by a horizontal jet and by a curtain of bubbles show that the velocity under these circumstances decreases nearly linearly with depth. This is fortunate since the waves in such a current can be calculated because its vorticity is constant. The distribution assumed is shown in fig. 1(b) and that treated in §2 (fig. 1(a)) is given for comparison. The disturbance is found by simply superposing a velocity field $u = U_2(1 + y/h)$, $v = 0$ on the disturbance represented by $\phi_1 + Ux$, in (1). At $y = -h$ this superposed velocity is zero, but in applying the equations of continuity and pressure at the sea surface and at the interface between the current and the underlying water it is necessary to remember that the uniform vorticity U_2/h fills the entire space between the two interfaces represented as before by (3) and (4). The continuity conditions are

$$ia(-\sigma + kU_2) = k(-B_1 + B_2),\tag{17}$$

$$-ib\sigma = k(-B_1 e^{-kh} + B_2 e^{kh}),\tag{18}$$

$$ib\sigma = kC.\tag{19}$$

Equations (17) and (19) are of the same form as (5) and (7), but (18) differs from (6).

The equation for continuity of pressure can be found by giving the whole field a velocity $-\sigma/k$ and thus reducing it to steady motion. Bernoulli's equation can then be applied at the free surface and at the interface which divides the fluid of uniform vorticity from the irrotationally moving water below. This process leads to

$$ia\left\{g - (\sigma - U_2 k)\frac{U_2}{kh}\right\} + (-\sigma + U_2 k)(B_1 + B_2) = 0\tag{20}$$

instead of (8), and

$$(B_1 e^{-kh} + B_2 e^{hh}) - C + i\frac{bU_2}{kh} = 0\tag{21}$$

instead of (9).

The constants B_1, B_2, C, a, b can now be eliminated. Two of the substitutions (11) namely $Y_2 = kU_2/\sigma$ and $\alpha_2 = g/\sigma U_2$ can now conveniently be made, and remembering that kh is non-dimensional the resulting period equation analogous to (12) is found to be

$$\alpha_2\left\{1 - \frac{2kh}{Y_2} - e^{-2kh}\right\} = (1 - Y_2)\left\{\frac{1 - e^{-2kh}}{kh} - \frac{1 + Y_2}{Y_2} - \frac{1 - Y_2}{Y_2^2}(2kh) + \frac{1 - Y_2}{Y_2}e^{-2kh}\right\}.\tag{22}$$

This equation is in a form which can be treated in the same way as (12) except that α_2 must be calculated for given values of Y_2 and kh. The value of $Z_2 = hg/U_2^2$ can then be calculated from

$$Z_2 = \frac{hg}{U_2^2} = hk\frac{g}{U\sigma}\frac{\sigma}{kU} = \frac{\alpha_2 hk}{Y_2}.\tag{23}$$

To obtain curves of constant Z_2 on the (α_2, Y_2) diagram it is necessary to cross-plot from curves which show how Z_2 varies with α_2 when Y_2 is kept constant and kh is varied. In this way the curves of fig. 6 were plotted for the same values of Z_2 as those of Z in fig. 4. Though fig. 6 is plotted using a natural scale for Y_2 and fig. 4 is plotted with a logarithmic scale for $-Y$ it will be seen that they both have the same characteristic; for a given value of Z, $-\alpha$ has a minimum value. The contours of constant Z_2 in the (α_2, Y_2) diagram are contained within the same limiting curves, $\alpha_2 = Y_2^{-1}$ for $Z_2 = 0$ and $\alpha_2 = (1 - Y_2)^2 / Y_2$ for Z_2 infinite, as in fig. 4. This is to be expected on physical grounds because both $Z = 0$ and $Z_2 = 0$ correspond with an infinitely thin surface layer which has no effect on the waves, while infinite values of Z and Z_2 both represent cases where the depth of the surface layer is infinite so that waves are merely convected by a stream moving at the velocity of the surface water.

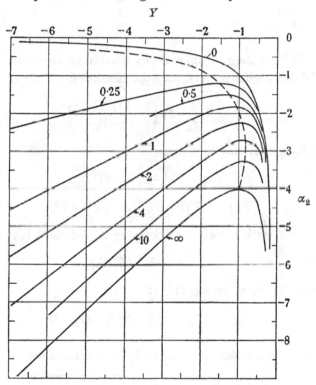

Fig. 6. Solutions of (22). Contours of constant Z for the same values as those of fig. 4.

To calculate the minimum values of α_2 is rather more difficult than in the case of a uniform current. For large values of kh they were taken graphically from the curves shown in fig. 6. The figures given in the lower sections of columns 1 and 3 of table 2 were obtained in this way. These minimum values, denoted by the symbol $-\alpha_{2m}$, could be obtained with fair accuracy for any given value of Z_2 but the corresponding graphical values of Y_2 given in column 2 of the lower section of table 2 are less accurate.

Table 2. *Minimum values of* $-\alpha_2$ *for constant* Z_2,
current decreasing with depth

$-\alpha_{2m}$	$-Y_2$	Z_2	α_2^2/Z_2
0·2	8	5×10^{-4}	80
0·33	5	$3·0 \times 10^{-3}$	36
0·42	4	$7·1 \times 10^{-3}$	25
0·58	3	$2·1 \times 10^{-2}$	16
0·91	2	0·093	8·9
1·25	1·5	0·257	6·1
1·95	1·0	1·01	3·7
1·2	—	0·25	5·75
1·52	—	0·50	4·63
1·84	—	1·0	3·39
2·25	—	2·0	2·52
2·70	—	4·0	1·82
3·30	—	10	1·09
4·0	—	∞	0

For values of kh less than 1, (22) was expanded in powers of kh. Neglecting terms containing k^2h^2 and higher powers, the approximation to (22) is

$$\alpha_2 = \frac{1}{Y_2} - kh \left(\frac{2 - \frac{4}{3}Y_2 + \frac{1}{3}Y^2}{1 - Y_2} \right). \tag{24}$$

Since $kh = Z_2 Y_2 / \alpha_2$, (24) may be written

$$\alpha_2 = Y_2^{-1} - Z_2 \alpha_2^{-1} F(Y_2), \tag{25}$$

where

$$F(Y) = Y(1 - Y)^{-1}(2 - \tfrac{4}{3}Y + \tfrac{1}{3}Y^2). \tag{26}$$

Differentiating (25) with respect to Y_2 keeping Z_2 constant it is found that when $[\partial \alpha_2 / \partial Y_2]_{Z_2 \text{ const}} = 0$,

$$\alpha_{2m} = - Z_2 Y_2^2 F'(Y_2) \tag{27}$$

and, substituting for Z_2 / α_2 from (27) in (25)

$$\alpha_{2m} = Y_2^{-1} + F(Y_2) Y_2^{-2} [F'(Y_2)]^{-1}. \tag{28}$$

Table 3. *Asymptotic formulae valid for large values of* $-Y$ *or* $-Y_2$

uniform current	current decreasing with depth
$\alpha_m = \tfrac{3}{2}Y^{-1}$	$\alpha_{2m} = \tfrac{3}{2}Y^{-1}$
$Z = \tfrac{3}{4}Y^{-4}$	$Z_2 = \tfrac{9}{4}Y_2^{-4}$
$\alpha_m^2 Z^{-1} = 3Y^2$	$\alpha_{2m}^2 Z_2^{-1} = Y_2^2$

In this way the values of $-\alpha_{2m}$ and the corresponding values of Z_2 and Y_2 given in the upper section of table 2 were found. The values of α_2^2/Z_2 are plotted in curves B, fig. 5, for comparison with the analogous curve A for the uniform surface current. The break in curve B separates the values obtained by the two methods which are given in the two sections of table 2.

To complete the calculation the asymptotic formulae for α_{2m} and the corresponding values of Z_2, Y_2 and α_{2m}^2/Z_2 valid for large values of Y_2 were found. These are given in table 3 and the analogous asymptotic formulae for the case of the uniform current are added for comparison.

4. COMPARISON OF THE EFFICIENCY AS BREAKWATERS OF THE TWO TYPES OF SURFACE CURRENT

When Mr J. T. Evans made his experiments* he had available my calculations of §2 but not those of §3, which I made after I had seen his measured distributions of velocity below the water surface. For this reason his comparison between his experiments and my theory was based on the assumption that the velocity which should be compared with my 'U' is the mean velocity taken over the depth to which the current penetrates. This is almost exactly half the maximum velocity U. To find out what error might be expected because the actual distribution is more like that contemplated in §3, I have compared the surface velocities U and U_2 required to stop a given wave of length λ_0. I have therefore taken

$$\alpha^2/Z = \alpha_2^2/Z_2 = \lambda_0/2\pi h.$$

To find the corresponding values of U and U_2, fig. 5 was used and the values of α_m and α_{2m} for a given value of $\lambda_0/2\pi$ were taken. These are given in columns 2 and 3 of table 4. The corresponding values of $U_2/U = a_m/\alpha_{2m}$ are given in column 4.

Table 4. *Comparison of velocity of currents of given depth capable of stopping waves of given length, U when current is uniform, U_2 when current decreases with depth*

$\lambda_0/2\pi h$	$-\alpha_m$	$-\alpha_{2m}$	$\dfrac{U_2}{U} = \dfrac{\alpha_m}{\alpha_{2m}}$
0	4·0	4·0	1·0
1·0	3·98	3·32	1·20
1·5	3·97	2·98	1·33
2·0	3·85	2·58	1·49
2·5	3·61	2·26	1·60
3·0	3·36	2·00	1·68
4·0	2·85	1·66	1·72
5·0	2·50	1·47	1·70
6·0	2·17	1·26	1·72
8·0	1·74	0·99	1·76
10·0	1·49	0·84	1·78
15·0	1·09	0·60	1·82
20·0	0·90	0·49	1·84
25	0·77	0·42	1·83
30	0·68	0·38	1·79
36	0·558	0·312	1·79
64	0·395	0·226	1·78
81	0·344	0·195	1·77
100	0·304	0·172	1·76
∞	—	—	1·73

* *Proc. Roy. Soc.* A, CCXXXI (1955), 457.

It will be seen that U_2 is always less than $2U$. In the limiting case of very deep currents for which α_2^2/Z is small $U = U_2$, as would be expected. At the other limit where α^2/Z is very large, i.e. the current is very shallow but has very high velocity, it is found from the asymptotic formulae that

$$U_2/U = \sqrt{3} = 1 \cdot 73.$$

This is the condition that when the total flow of momentum of very shallow currents is the same in the two cases of §§2 and 3 the stopping power of the currents is the same. It appears therefore that when Evans took the mean of a velocity distribution similar to that considered here and used it in formulae based on the solution for a uniform surface current, he slightly overestimated the 'theoretical' stopping velocity. His calculations were carried out for $\lambda_0/h = 10$ and 24. The corresponding values of α^2/Z are $10/2\pi$ and $24/2\pi$ or 1·6 and 3·8. In table 4 it will be seen that this corresponds with values of U_2/U in the range 1·4 to 1·7. Thus if the present calculations had been completed when Evans made his theoretical estimate he would have predicted velocities of a half of 1·4 to 1·7, or 0·7 to 0·85, of those which he used in his comparison with his experiments. I have no doubt that many causes not considered on the simple theory might more than neutralize this error.

5. Currents produced by a curtain of bubbles

If the bubbles are so small that they rise much more slowly through the surrounding water than the currents they produce, they may be expected to act simply by reducing the mean density of the mixture of water and bubbles. Each bubble, in fact, contributes a small amount of density deficiency and behaves in the same way as a small amount of heat in the rising air over a flame. If the weight of air in the bubbles is neglected compared with the weight of the water it displaces, a volume v of air bubbles gives to unit volume of water the same buoyancy, i.e. the same ratio of upward force to mass, that a rise in temperature ϑ, gives to air at absolute temperature T when

$$v = \vartheta/T. \tag{29}$$

Assuming that v is small, an analogy must exist between the vertical current produced by releasing bubbles at a constant rate from a point or line source and those produced by releasing heat. Bubbles released from a horizontal tube pierced by small holes will produce in water a distribution of vertical currents identical with those produced in air by release of H cal. cm.$^{-1}$ sec.$^{-1}$ from a horizontal line source provided

$$V = H/\rho s T. \tag{30}$$

Here V is the volume of bubbles emitted from unit length of the tube, ρ and s are the density and specific heat of air. This statement can be understood by considering the total upward flow of bubbles or heat over a horizontal plane. If w is the vertical velocity and x a horizontal co-ordinate

$$H = \int \rho s w \vartheta \, \mathrm{d}x \quad \text{and} \quad V = \int w v \, \mathrm{d}x. \tag{31}$$

If ρ and s are nearly constant, equations (31) are consistent in view of (29) and (30).

The currents above a horizontal line source of heat have been considered by Schmidt.* Schmidt showed that the heat spreads out linearly so that the width of the heated zone is proportional to the height z above the source. The vertical velocity w depends only on the angle which the line joining any point to the source makes with the vertical plane through it, so that

$$w = W_0 f(x/z). \tag{32}$$

It is convenient to take W_0 as the maximum vertical velocity on the middle of the rising column. Schmidt showed theoretically that W_0 must be proportional to $(Hg/\rho s T)^{\frac{1}{3}}$. This is found to be true experimentally, and measurements show that the constant of proportionality is about 1·9 so that $W_0 = 1·9 (Hg/\rho s T)^{\frac{1}{3}}$ and the angle of the wedge within which the rising column is contained is roughly $2 \tan^{-1} 0·28$. The analogy, (30), therefore shows that the maximum vertical current produced by the discharge of a volume V of bubbles per unit length per second should be

$$W_0 = 1·9(Vg)^{\frac{1}{3}}. \tag{33}$$

If no loss of energy occurs in the right-angled bend of the stream at the surface it might be expected that the horizontal velocity of the stream would be $1·9 (Vg)^{\frac{1}{3}}$ and the depth about $0·28D$, where D is the depth of the source of the bubbles. In Evans's experiments this is approximately 3 ft.

In the case shown in Evans's fig. 2 the rate of discharge of bubbles was 7·5 ft.³/ min. ft., and the observed surface speed was 1·75 ft./sec. The distribution of velocity is very nearly linear as in fig. 1 (b) and the length corresponding to h in fig. 1 (b) is about 10 in. Since 10 in. is roughly 0·28 of 3 ft. the depth of the horizontal current is about what would be expected from Schmidt's work. On the other hand, $V = 7·5/60 = 0·125$ ft.³/sec. ft. and taking g as 32 ft./sec.², $1·9 (Vg)^{\frac{1}{3}} = 3·0$ ft./sec. This is very much larger than the observed value 1·75 ft./sec. The difference may be accounted for partly by the fact that the bubbles were of such a size that they rose through the surrounding water at about 0·75 ft./sec., so that they spent a smaller time communicating upward momentum to the water than is implicitly assumed when the analogue equation (33) is used. It may also occur that there is some loss of energy at the point where the stream divides into two and changes direction through a right angle.

* *Z. angew. Math. Mech.* XXI (1941), 265.

25

FLUID FLOW BETWEEN POROUS ROLLERS*

REPRINTED FROM

Quarterly Journal of Mechanics and Applied Mathematics, vol. IX (1956), pp. 129–35

When a viscous fluid is entrained between two rollers which are separated by a small distance, a pressure is developed on the upstream side of the point of nearest approach and a suction on the downstream side. Theoretically these both become infinite when the rollers are in contact. If, however, the rollers are slightly porous these infinities are avoided. The analysis of this situation involves the solution of a differential equation which has some interest. The distribution of pressure is rather similar to that which occurs when impervious rollers are not in contact. The result may have industrial applications to paper-making machines and to the rollers used for painting walls.

1. INTRODUCTION

In certain industrial processes a viscous fluid is caught up between two rotating rollers or between a moving flat sheet and a roller which supports it. The pressure upstream of the point where the rollers most nearly touch first rises as this point is approached then falls to zero when the point is reached. Then after passing this point the pressure falls below that at great distances and subsequently rises.

The distribution† of pressure between cylindrical rollers which are not porous has been calculated† on the assumption that they are separated by a very narrow gap so that the equations of the theory of lubrication can be used. The pressure p can then be regarded as a function of x only, where the origin of coordinates is chosen at the point midway between the centres of two equal cylinders of radius R, and the axis of z is the line joining their centres (fig. 1). If U is the velocity of the surface of the rollers and $2h$ the small distance between them, the pressure p is

$$p = -\frac{4\mu U R^2 x}{(x^2 + 2Rh)^2},\tag{1}$$

where μ is the coefficient of viscosity. Since x is measured in the direction of motion the pressure is antisymmetrical, being a suction when x is positive, i.e. downstream of the point were the gap is narrowest. This suction is a maximum at the point $x = (\tfrac{2}{3}Rh)^{\frac{1}{2}}$ and is of magnitude

$$(-p)_{\max} = \tfrac{3}{16}\sqrt{6}\mu U R^{\frac{1}{2}}h^{-\frac{3}{2}}.$$

If h is decreased $(-p)_{\max}$ increases till, when the rollers come into contact, the formula predicts an infinite value. In fact‡ the suction increases till at a certain value depending on its physical properties, the fluid cavitates. There is however another

---

* With J. C. P. MILLER.

† H. M. Martin, *Engineering*, CII (1916), 119.

‡ W. H. Banks & C. C. Mill, *Proc. Roy. Soc.* A, CCXXIII (1954), 414.

possibility. The rollers themselves might be slightly porous and when the geometrical conditions are such that high suction would be produced by impermeable rollers, fluid might be sucked through their surfaces and so prevent the suction rising to such a level that cavitation would occur. It is of interest to discuss the flow in this case, partly to find out how porous the roller must be if cavitation is to be avoided but perhaps more so because the analysis presents some interesting mathematical features.

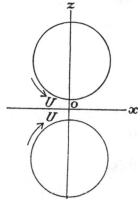
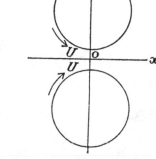
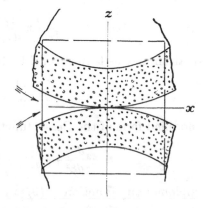

Fig. 1. Impervious rollers rotating at the same speed in opposite directions.

Fig. 2. Porous rollers in contact.

It will be assumed that the velocity W of the fluid through the porous surface of the rollers is related to the pressure p by the equation

$$W = kp/\mu; \tag{2}$$

k is then a porosity coefficient. It has the dimensions of a length. The geometry of the situation is indicated in fig. 2.

2. FLOW EQUATIONS

Since only the region close to the point of contact of the rollers will be considered it is sufficient to take $2t = x^2/R$ as the gap between them at the point x.

According to the theory of lubrication the velocity of flow at the point (x, z) is

$$u = A + \frac{1}{2\mu} z^2 \frac{dp}{dx}.$$

If U is the velocity of the surface of the rollers

$$A = U - \frac{x^4}{8\mu R^2} \frac{dp}{dx} \tag{3}$$

and the total flow Q is

$$Q = 2 \int^t u \, dz = \frac{x^2 U}{R} - \frac{1}{12} \frac{x^6}{R^3 \mu} \frac{dp}{dx}. \tag{4}$$

The equation of continuity is $\qquad \dfrac{dQ}{dx} = -2W$

so that, using (2), Q and W can be eliminated to give

$$\frac{1}{12\mu R^3}\frac{d}{dx}\left(x^6\frac{dp}{dx}\right) - \frac{2kp}{\mu} - \frac{2xU}{R} = 0, \tag{5}$$

and writing $\qquad x_1 = x(24kR^3)^{-\frac{1}{4}}, \quad y = -p\left(\dfrac{kR}{\mu U}\right)(24kR^3)^{-\frac{1}{4}} \tag{6}$

(5) becomes $\qquad \dfrac{d}{dx_1}\left(x_1^6\dfrac{dy}{dx_1}\right) - y + x_1 = 0. \tag{7}$

This equation may be transformed by the substitution

$$\xi = \frac{1}{2x_1^2}, \quad \eta = 5^{\frac{7}{4}}x_1^{\frac{5}{2}}y \tag{8}$$

into a modified form of the Bessel function equation, viz.

$$\xi^2\frac{d^2\eta}{d\xi^2} + \xi\frac{d\eta}{d\xi} - \eta\{(\tfrac{5}{4})^2 + \xi^2\} = -(\tfrac{5}{2})^{\frac{7}{4}}\xi^{\frac{1}{4}}. \tag{9}$$

The complementary function of (9) is

$$A'I_{\frac{5}{4}}(\xi) + B'I_{-\frac{5}{4}}(\xi). \tag{10}$$

To complete the solution it is necessary to find a particular integral of (9). For this purpose consider Struve's function*

$$L_\nu(\xi) = \sum_{m=0}^{\infty}\frac{(\tfrac{1}{2}\xi)^{\nu+2m+1}}{\Gamma(m+\tfrac{3}{2})\,\Gamma(m+\nu+\tfrac{3}{2})}, \tag{11}$$

which satisfies $\qquad \nabla_\nu L_\nu(\xi) = \dfrac{4(\tfrac{1}{2}\xi)^{\nu+1}}{\Gamma(\tfrac{1}{2})\,\Gamma(\nu+\tfrac{1}{2})}, \tag{12}$

where ∇_ν is written for the operator

$$\xi^2\frac{d^2}{d\xi^2} + \xi\frac{d}{d\xi} - \xi^2 - \nu^2. \tag{13}$$

Then when $\nu = \tfrac{5}{4}$ $\qquad \nabla_{\frac{5}{4}}[L_{\frac{5}{4}}(\xi)] = \dfrac{4(\tfrac{1}{2})^{\frac{9}{4}}\xi^{\frac{9}{4}}}{\Gamma(\tfrac{1}{2})\,\Gamma(\tfrac{7}{4})}. \tag{14}$

Also $\qquad \nabla_{\frac{5}{4}}(\xi^{\frac{1}{4}}) = (\tfrac{1}{16} - \tfrac{25}{16})\xi^{\frac{1}{4}} - \xi^{\frac{1}{4}+2} = -\tfrac{3}{2}\xi^{\frac{1}{4}} - \xi^{\frac{9}{4}}; \tag{15}$

the function $\qquad \eta = CL_{\frac{5}{4}}(\xi) + D\xi^{\frac{1}{4}} \tag{16}$

therefore satisfies (9), provided

$$D = \tfrac{2}{3}(\tfrac{5}{2})^{\frac{7}{4}}, \quad C = 2^{-\frac{5}{4}}5^{\frac{7}{4}}\Gamma(\tfrac{1}{2})\,\Gamma(\tfrac{3}{4})$$

so that a particular integral of (9) is

$$5^{\frac{7}{4}}\{2^{-\frac{5}{4}}\Gamma(\tfrac{1}{2})\,\Gamma(\tfrac{3}{4})L_{\frac{5}{4}}(\xi) + 2^{-\frac{3}{2}}3^{-1}\xi^{\frac{1}{4}}\} \tag{17}$$

* G. N. Watson, *Theory of Bessel Functions*, Cambridge University Press, 1922, p. 329.

and the complete solution of (7) is

$$y = (2\xi)^{\frac{1}{4}}[2^{-\frac{3}{2}}\Gamma(\tfrac{1}{2})\,\Gamma(\tfrac{3}{4})\,L_{\frac{1}{4}}(\xi) + 2^{-\frac{2}{3}}3^{-1}\xi^{\frac{3}{4}} + AI_{\frac{1}{4}}(\xi) + BI_{-\frac{1}{4}}(\xi)]. \qquad (18)$$

When the functions $L_{\frac{1}{4}}(\xi)$, $I_{\frac{1}{4}}(\xi)$, $I_{-\frac{1}{4}}(\xi)$ are expanded in powers of ξ the only term which does not vanish when $\xi \to 0$ is that which contains B. Since the physical conditions require that $p \to 0$ as $x \to \infty$, or $y = 0$ when $\xi = 0$, therefore $B = 0$. We next consider whether any value can be chosen for A which will make $y \to 0$ as $\xi \to \infty$. Both $I_{\frac{1}{4}}(\xi)$ and $L_{\frac{1}{4}}(\xi)$ tend to infinity as $\xi \to \infty$ but the difference between them is of a lower order of magnitude than either separately. This can be seen by writing down the expressions for $L_\nu(\xi)$ and $I_\nu(\xi)$ as definite integrals. These are*

$$L_\nu(\xi) = \frac{2(\tfrac{1}{2}\xi)^\nu}{\Gamma(\nu + \tfrac{1}{2})\,\Gamma(\tfrac{1}{2})} \int_0^{\frac{1}{2}\pi} \sinh\,(\xi\cos\theta)\sin^{2\nu}\theta\,d\theta \qquad (19)$$

and

$$I_\nu(\xi) = \frac{2(\tfrac{1}{2}\xi)^\nu}{\Gamma(\nu + \tfrac{1}{2})\,\Gamma(\tfrac{1}{2})} \int_0^{\frac{1}{2}\pi} \cosh\,(\xi\cos\theta)\sin^{2\nu}\theta\,d\theta; \qquad (20)$$

hence

$$L_\nu(\xi) - I_\nu(\xi) = \frac{-2(\tfrac{1}{2}\xi)^\nu}{\Gamma(\nu + \tfrac{1}{2})\,\Gamma(\tfrac{1}{2})} \int_0^{\frac{1}{2}\pi} e^{-\xi\cos\theta}\sin^{2\nu}\theta\,d\theta \qquad (21)$$

and the reason for the special character of this particular combination of $L_\nu(\xi)$ and $I_\nu(\xi)$ is obvious.

When ξ is large only the part of the range of θ which lies close to $\tfrac{1}{2}\pi$ contributes appreciably to the integral in (21). Writing $\phi = \tfrac{1}{2}\pi - \theta$ this integral is

$$\mathscr{I} = \int_0^{\frac{1}{2}\pi} e^{-\xi\sin\phi}\cos^{2\nu}\phi\,d\phi. \qquad (22)$$

To obtain the first two terms in the expansion of (22) in descending powers of ξ write $\sin\phi = \phi - \tfrac{1}{6}\phi^3 = \psi$ so that to the same order $\phi = \psi + \tfrac{1}{6}\psi^3$. Then

$$\mathscr{I} \sim \int_0^1 e^{-\xi\psi}\{1 + (\tfrac{1}{2} - \nu)\,\psi^2\}\,d\psi \sim \frac{1}{\xi} + \frac{(1 - 2\nu)}{\xi^3}. \qquad (23)$$

Hence, from (21)

$$L_{\frac{1}{4}}(\xi) - I_{\frac{1}{4}}(\xi) \sim \frac{-2^{-\frac{1}{4}}}{\Gamma(\tfrac{7}{4})\,\Gamma(\tfrac{1}{2})}[\xi^{\frac{3}{4}} - \tfrac{3}{2}\xi^{-\frac{7}{4}}]. \qquad (24)$$

If A is taken in (18) as

$$A = -2^{-\frac{3}{2}}\Gamma(\tfrac{1}{2})\Gamma(\tfrac{3}{4})$$

$L_{\frac{1}{4}}(\xi)$ and $I_{\frac{1}{4}}(\xi)$ occur in the combination of (21) and the complete solution is

$$y = (2\xi)^{\frac{1}{4}}[2^{-\frac{3}{2}}\Gamma(\tfrac{1}{2})\,\Gamma(\tfrac{3}{4})\,\{L_{\frac{1}{4}}(\xi) - I_{\frac{1}{4}}(\xi)\} + 2^{-\frac{2}{3}}3^{-1}\xi^{\frac{3}{4}}]. \qquad (25)$$

The limiting form of (25) as $\xi \to \infty$ is found by substituting from (24) in (25). It is

$$y \sim (2\xi)^{\frac{1}{4}}[-2^{-\frac{2}{3}}3^{-1}(\xi^{\frac{3}{4}} - \tfrac{3}{2}\xi^{-\frac{7}{4}}) + 2^{-\frac{2}{3}}3^{-1}\xi^{\frac{3}{4}}]. \qquad (26)$$

The term in $\xi^{\frac{3}{4}}$ vanishes leaving

$$y \sim (2\xi)^{\frac{1}{4}}(2\xi)^{-\frac{7}{4}} = (2\xi)^{-\frac{1}{2}} = x_1. \qquad (27)$$

* *Theory of Bessel Functions*, pp. 329 and 79.

Expressed in terms of x_1 (25) can be written

$$y = \frac{\Gamma(\frac{1}{2})\,\Gamma(\frac{3}{4})}{(2x_1)^{\frac{1}{2}}}\left\{L_{\frac{1}{4}}\left(\frac{1}{2x_1^2}\right) - I_{\frac{1}{4}}\left(\frac{1}{2x_1^2}\right)\right\} + \frac{1}{6x_1^3}. \tag{28}$$

This approximates to $y = x_1$ near $x_1 = 0$ and to $y = 1/6x_1^3$ as $x_1 \to \infty$.

It is of interest to notice that the pressure at $x = \infty$ is given by $yx_1^3 = \frac{1}{6}$. Using (6) this reduces to

$$p = -4\mu U R^2 x^{-3} \tag{29}$$

which is the same as the expression (2) for the pressure between impermeable rollers when $h = 0$.

3. DISTRIBUTION OF PRESSURE

Since the function $L_{\frac{1}{4}}(\xi)$ has not been tabulated y cannot be evaluated from the expression (25). For that reason the expansion of $L_{\frac{1}{4}}(\xi) - I_{\frac{1}{4}}(\xi)$ in power series was used to evaluate y. Substituting from (21) in (25)

$$y = -\frac{\sqrt{2}}{3}\,\xi^{\frac{5}{2}}\int_0^{\frac{1}{2}\pi} e^{-\xi\cos\theta}\sin^{\frac{5}{2}}\theta\,d\theta + \frac{1}{6}x_1^{-3}. \tag{30}$$

Expanding $e^{-\xi\cos\theta}$ in a power series and integrating

$$F(\xi) = \int_0^{\frac{1}{2}\pi} e^{-\xi\cos\theta}\sin^{\frac{5}{2}}\theta\,d\theta = \sum_{m=0}^{\infty}(-1)^m\xi^m\frac{\Gamma(\frac{7}{4})\,\Gamma(\frac{1}{2}m+\frac{1}{2})}{2m!\,\Gamma(\frac{1}{2}m+\frac{9}{4})}. \tag{31}$$

The terms in the series were computed to 6 decimals as far as needed for $\xi = 0\cdot2$, $0\cdot5$, $1\cdot0$, $2\cdot0$, $3\cdot0$, $4\cdot0$, and $6\cdot0$. For $\xi = 6$, the term in ξ^{22} was the last one needed. An independent computation agrees within a unit in the last figure given.

The computed values of y and x_1 are given in table 1.

Table 1

ξ	6·0	4·0	3·0	2·0	1·0	0·5	0·2
y	0·299	0·356	0·376	0·358	0·236	0·117	0·0366
x_1	0·289	0·354	0·408	0·500	0·707	1·000	1·581

Fig. 3 shows the distribution of pressure. It will be seen that at $x_1 = 0\cdot29$ and $x_1 = 0\cdot35$, y is very slightly greater than x_1, so that the asymptotic form (27) valid for x_1 small is only about 3% wrong at $x_1 = 0\cdot29$. The first attempt to solve (7) was made by assuming a power series for y and determining the coefficients by substituting it in (7). In this way the following series can be derived:

$$y = x_1 + 6x_1^5 + 300x_1^9 + 37800x_1^{13} + \dots. \tag{32}$$

It will be seen that though the first term is the same as that derived from the asymptotic expansion in terms of ξ the series is highly divergent. Finding this very rapid divergence and being unable to see any physical reason why the solution should not be regular in the neighbourhood of $x_1 = 0$ an attempt was made to solve equation (7) numerically, starting with the values of x_1 and dy/dx_1 given by the first two terms of (32) at $x_1 = 0.1$. Here again the attempt failed, giving values of y which appeared to increase continually with increasing x_1. When a descending power series was assumed for y the leading term was $\frac{1}{6}x_1^{-3}$ and the coefficients appeared to be convergent, but except for values of x_1 greater than 1, it did not yield good numerical agreement with the correct values given in table 1.

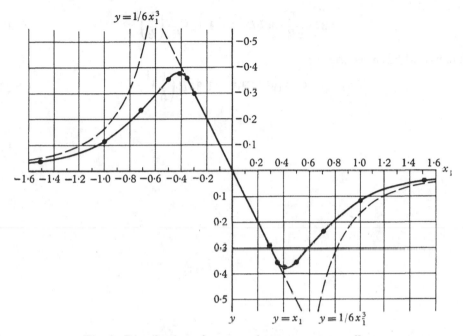

Fig. 3. Distribution of pressure between porous rollers.

4. Maximum suction

In fig. 3 it will be seen that the maximum suction occurs at $x_1 = 0.43$ approximately and that y is there 0.378 approximately. The suction is then

$$-p = (0.378 \times 24^{\frac{1}{4}}) k^{-\frac{1}{4}} R^{-\frac{1}{4}} \mu U = 0.84 \frac{\mu U}{R} \left(\frac{R}{k}\right)^{\frac{3}{4}} \tag{33}$$

and it occurs at distance

$$x = 0.43 (24kR^3)^{\frac{1}{4}} = 0.95R \left(\frac{k}{R}\right)^{\frac{1}{4}} \tag{34}$$

from the point of contact.

5. Straight porous band and impervious roller

If instead of two porous rollers the problem had been concerned with a straight porous band running on an impervious roller (a configuration which occurs with the rollers used for painting walls and in paper-making machinery) the analysis would have yielded equation (7) after the transformation

$$x_1 = x(96kR^3)^{-\frac{1}{4}}, \quad y = -\frac{pkR}{\mu U}(96kR^3)^{-\frac{1}{4}}$$

instead of (6). The maximum suction would have been

$$0\cdot378\frac{\mu U}{kR}(96kR^3)^{\frac{1}{4}} = 1\cdot18\frac{\mu U}{R}\left(\frac{R}{k}\right)^{\frac{3}{4}} \tag{35}$$

and it would have occured at

$$x = 0\cdot43(96kR^3)^{\frac{1}{4}} = 1\cdot35R\left(\frac{k}{R}\right)^{\frac{1}{4}}. \tag{36}$$

26

FLUID FLOW IN REGIONS BOUNDED BY POROUS SURFACES

REPRINTED FROM

Proceedings of the Royal Society, A, vol. ccxxxiv (1956), pp. 456–75

The aerodynamic effects of sucking away boundary layers or blowing air into them through a porous surface have been studied on the assumption that the rate of discharge through the pores is under the control of the experimenter. When all the fluid reaches the field of flow through the pores the pressure at any point in the field depends on the distribution over the porous surface of the flow through it, and the through-flow at any point of it depends on the pressure there. To describe flow of this kind mathematically is difficult and no case seems to have been discussed before. In § 1 a particular case, that of flow through a wedge, cylinder or cone made of a material the resistance of which is proportional to the square of the velocity through it, is treated by means of an integral equation. This equation is solved and the results reduced to a form which lends itself to experimental verification.

In § 2 experiments are described in which the physical conditions assumed in the analysis were very nearly attained and the theoretical conclusions then verified. The most striking result was the agreement, to within 1%, between the calculated and the measured discharge of water from a tank through an internal porous tube whose base was an orifice in the bottom. The measured distribution of velocity in the plane of the orifice also agreed with the calculations. Experiments with porous cones also yielded results agreeing with theory when the correct experimental conditions could be satisfied.

1. THEORETICAL DISCUSSION

Introduction

In recent years the effect of distributed suction or pressure on aerodynamic flow past a porous surface has been studied. In all cases it has been assumed that the pressure distribution in the field is known and does not depend on the effect of the air which passes through the porous surface. The amount of air passing through is assumed, therefore, to be known, and calculations are concerned with the effect of this suction or pressure on a thin boundary layer. In most cases, the still grosser assumption is made that the inflow or outflow depends only on the amount of suction or pressure applied to the back of the surface. In fact, this latter condition can be attained approximately by applying to the back of a porous surface of high resistance a suction much greater than the variations in pressure in the field of flow.

The present paper deals with some problems of flow past porous surfaces in which the whole aerodynamic field depends on the flow through them, and at the same time the through-flow depends only on the pressure in that field.

My attention was called to problems of this kind by Mr J. Mardon of the North-eastern Paper Products Co., of Quebec, who described to me the operation of a paper mill and enquired about the hydrodynamical situation in one part of it. Paper is made by running a watery suspension of fibres over a porous sheet through which the fluid drains. The mat of fibres which remains becomes paper when fully drained and dry. Originally, sheets of paper were made separately in hand-operated draining devices, but in a paper mill the operation is continuous. The suspension flows through a long horizontal orifice on to a moving band made of wire gauze. The liquid drains through this band as it passes over a long series of rollers and finally over a vacuum vessel which removes most of the water that is left. In the space between the rollers, liquid can drain away under the action of the very small pressure due to the action of gravity on a layer of the suspension whose depth ranges from 1 or 2 cm. to fractions of a millimetre. At each roller, hydrodynamic conditions are such that a large suction is created in a small region under the band and downstream of its point of contact with the roller. A large part of the drainage occurs there. The question is: how great is this suction and how is it distributed under the moving band?

Before discussing the paper-mill problem, it is useful to consider some simpler ones involving the same physical principles and similar mathematical treatment. A simple problem of this type is that in which a fluid is forced into the wedge-shaped space between two porous sheets or into a narrow-angled cone through porous walls. This problem is discussed here not because it has any practical interest, but because it seems to be of a different type from problems of flow with vorticity which have been solved before, and because it is possible to test experimentally the results of analytical predictions based on simple physical assumptions relating to the properties of porous sheets.

Properties of porous sheets

It is necessary first to define the properties of porous sheets. Wire gauze has the property that it refracts a stream of fluid passing through it, and different arrangements of wire may product different refractive properties. When the resistance is great, however, all gauzes and porous sheets absorb the transverse component of velocity on the inflow side and deliver it from the outflow surface at right angles to the sheet. The porous sheets here considered will be assumed to possess this property.

In the first instance, the resistance law will be assumed in the general form

$$W = F(s), \tag{1}$$

where W is the velocity perpendicular to the porous sheet and s the difference in pressure between the two sides. The particular cases of this law which are most frequently met with are (a) the square law of resistance

$$s = c(\tfrac{1}{2}\rho W^2) \quad \text{or} \quad W = (2s/\rho c)^{\frac{1}{2}}, \tag{2a}$$

where c is a coefficient of resistance; and (b) resistance proportional to velocity

$$s = \mu W/\kappa \quad \text{or} \quad W = \kappa s/\mu, \tag{2b}$$

where μ is the viscosity and κ a coefficient which has the dimension of a length. Wire gauze and perforated sheets have approximately the property (a) and material with very fine pores has properties more like (b).

The geometry of the situation investigated is shown in fig. 1b, where O is the vertex of a cone of small semi-vertical angle α or the trace of the line of intersection of two planes OA and OB. Axes Ox, Oy are taken along and perpendicular to the axis or plane of symmetry. It will be assumed that the angle α is small, so that the velocity parallel to the axis x is much greater than that at right angles to it.

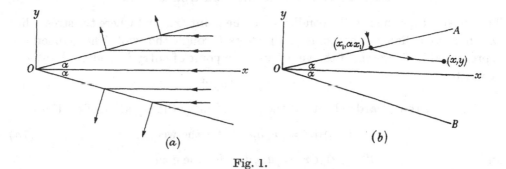

Fig. 1.

The simplest case is the trivial one indicated in fig. 1a, in which the pressure inside the cone is greater than that outside it. The flow can then be parallel to the axis and uniform through the wedge or cone. If u represents velocity parallel to Ox, then u is negative and equal, say, to $-U$. The condition of continuity is then

$$U \sin \alpha = W, \tag{3}$$

and for the two porosity laws (2a) and 2b) this leads to

$$U = \left(\frac{2s}{\rho}\right)^{\frac{1}{2}} \left(\frac{1}{\sqrt{c}\sin\alpha}\right) \quad \text{in case (2a),} \tag{4a}$$

$$U = \kappa s/(\mu \sin \alpha) \quad \text{in case (2b).} \tag{4b}$$

Pressure lower inside wedge or cone

When the uniform flow into the wedge or cone (fig. 1a) is maintained over one section (in the case of the cone, the base), each stream line will be parallel to the axis and the constant velocity will be maintained till the porous surface is reached, when the component parallel to the generators of the cone will be absorbed and the normal component will persist through the material and flow into the fluid surrounding its outer surface. This must escape and cause some variations in pressure outside the cone, but the condition that α is small makes it legitimate to neglect variations in pressure outside. It is for that reason that the 'trivial' case of figure 1a, which depends on the pressure difference between the inside and the outside being constant, is a legitimate approximation when considering conditions inside the wedge or cone. When the pressure is higher outside than inside, the fluid enters through the walls with a velocity which is much smaller than the longitudinal velocity, which it must acquire to escape through the open end. There must therefore be a pressure gradient along the axis. It will be assumed that the pressure gradient perpendicular to the axis is small compared with the longitudinal one. Since the pressure inside is less than that outside, the pressure difference will be represented by a suction or negative pressure $s(x)$. If $_xu_y$ is the velocity at the point (x, y) Bernoulli's equation is

$$s(x) - \tfrac{1}{2}\rho(_xu_y)^2 = \text{constant along a stream line.} \tag{5}$$

The constant depends on the conditions at the point $(x_1, \alpha x_1)$ where the stream line which passes through the point (x, y) enters through the wall. The geometry is represented in figure 1 b. If the velocity at the point of entry is neglected, (5) may be written

$$s(x) - s(x_1) = \tfrac{1}{2}\rho \,_xu_y^2. \tag{6}$$

If $W(x_1)$ is the inward velocity at the point of entry the equation of continuity is

$$W(x_1)\,dx_1 = {}_xu_y\,dy \qquad \text{for the wedge,} \tag{7a}$$

and

$$W(x_1)\,\alpha x_1\,dx_1 = {}_xu_y\,y\,dy \quad \text{for the cone.} \tag{7b}$$

There is, however, another continuity condition to be satisfied. The fluid must fill the wedge or cone for all values of x. This may be expressed by integrating (7a) or (7b). Substituting for $_xu_y$ from (6) and W from (1)

$$\alpha x = \int_0^x dy = (\tfrac{1}{2}\rho)^{\frac{1}{2}} \int_0^x \frac{F(s_1)\,dy_1}{\{s(x) - s(x_1)\}^{\frac{1}{2}}} \qquad \text{for the wedge,} \tag{8a}$$

$$\tfrac{1}{2}\alpha^2 x^2 = \int_0^x y\,dy = \alpha(\tfrac{1}{2}\rho)^{\frac{1}{2}} \int_0^x \frac{F(s_1)\,x_1\,dx_1}{\{s(x) - s(x_1)\}^{\frac{1}{2}}} \quad \text{for the cone.} \tag{8b}$$

(8a) or (8b) are the integral equations which contain the solution of the problem. In the case where $F(s_1)$ has the form (2a) it is possible to solve them. Substituting this form, (8a) and (8b) assume the forms

$$\alpha x = c^{-\frac{1}{2}} \int_0^x \left[\frac{s(x_1)}{s(x) - s(x_1)}\right]^{\frac{1}{2}} dx_1 \tag{9a}$$

or

$$\tfrac{1}{2}\alpha x^2 = c^{-\frac{1}{2}} \int_0^x \left[\frac{s(x_1)}{s(x) - s(x_1)}\right]^{\frac{1}{2}} x_1\,dx_1. \tag{9b}$$

Writing

$$\frac{s(x_1)}{s(x)} = \zeta, \quad \frac{x_1}{x} = \eta, \tag{10}$$

these become

$$\alpha c^{\frac{1}{2}} = \int_0^1 \left(\frac{\zeta}{1-\zeta}\right)^{\frac{1}{2}} d\eta \tag{11a}$$

and

$$\alpha c^{\frac{1}{2}} = 2 \int_0^1 \left(\frac{\zeta}{1-\zeta}\right)^{\frac{1}{2}} \eta\,d\eta. \tag{11b}$$

These are both satisfied by taking $\zeta = \eta^{1/m}$, provided that for the wedge

$$\alpha c^{\frac{1}{2}} = m \int_0^1 \zeta^{m-\frac{1}{2}}(1-\zeta)^{-\frac{1}{2}}\,d\zeta = mB(m+\tfrac{1}{2}, \tfrac{1}{2}), \tag{12a}$$

and for the cone

$$\alpha c^{\frac{1}{2}} = 2m \int_0^1 \zeta^{2m-\frac{1}{2}}(1-\zeta)^{-\frac{1}{2}}\,d\zeta = 2mB(2m+\tfrac{1}{2}, \tfrac{1}{2}),\tag{12b}$$

where $B(p,q)$ is the beta function $\Gamma(p)\,\Gamma(q)/\Gamma(p+q)$. The fact that the final equations (12a) and (12b) do not contain x or $s(x)$ shows that ζ is a function of η. The incomplete beta function $B_\zeta(p,q) = \int_0^\zeta \zeta^{p-1}(1-\zeta)^{q-1}\,d\zeta$, which will be required later, has been tabulated by Pearson* in the form $I_\zeta(p,q) = B_\zeta(p,q)/B(p,q)$, which ranges from 0 to 1 as ζ varies from 0 to 1. The value of $B(p,q)$ is given at the head of each column in Pearson's tables. Some corresponding values of $\alpha\sqrt{c}$ and M for the cone are given at the top of table 1.

Distribution of velocity

Bernoulli's equation may be written

$$_x u_y = \left(\frac{2s(x)}{\rho}\right)^{\frac{1}{2}} (1-\zeta)^{\frac{1}{2}}.\tag{13}$$

To find the value of y at which the particular stream line which enters at x_1 passes through the plane x, the integrals (8a) and (8b) must be taken out only to the point where the ζ stream line crosses the plane x; thus, instead of (12a),

$$\frac{yc^{\frac{1}{2}}}{x} = \int_0^\eta \left(\frac{\zeta}{1-\zeta}\right)^{\frac{1}{2}}\,d\eta = mB_\zeta(m+\tfrac{1}{2}, \tfrac{1}{2}),\tag{14}$$

and dividing the two sides of (14) by the corresponding sides of (12a),

$$\frac{y}{\alpha x} = I_\zeta(m+\tfrac{1}{2}, \tfrac{1}{2}),\tag{15a}$$

and similarly for the cone, $$\frac{y^2}{\alpha^2 x^2} = I_\zeta(2m+\tfrac{1}{2}, \tfrac{1}{2}).\tag{15b}$$

Using (13) and Pearson's tables with (15a) or (15b) it is therefore possible to calculate the distribution of velocity for the wedge or the cone. The cone is the more interesting case because it is more convenient for experimental work. It is convenient to use the symbol S for the suction applied at the base of the wedge or cone; u then varies from $(2S/\rho)^{\frac{1}{2}}$ at the centre to 0 at the porous wall.

Limiting case when $m = 0$

In this case $\alpha = 0$, so that the wedge degenerates into two parallel planes and the cone into a cylinder. Since the origin of x recedes to an infinite distance from the base of the cone, it is convenient to define a co-ordinate ξ as the distance from the base.

* *Table of the Incomplete Beta Function*. Biometrika Office, University College, London, 1934.

If $2b$ is the distance apart of the planes or the diameter of the cylinder, then

$$\eta = \lim_{\alpha \to 0}\left(\frac{b\alpha^{-1} - \xi}{b\alpha^{-1}}\right) = \lim_{\alpha \to 0}\left(1 - \frac{\xi\alpha}{b}\right).$$

If a suction S is applied at $\xi = 0$ the suction at ξ is

$$s(\xi) = S\eta^{1/m} = S\lim_{\alpha \to 0}\left(1 - \frac{\xi\alpha}{b}\right)^{1/m}.$$

For parallel planes, $(12a)$ shows that

$$\lim_{\alpha \to 0}\left(\frac{\alpha c^{\frac{1}{2}}}{m}\right) = B(\tfrac{1}{2}, \tfrac{1}{2}) = \pi, \quad \text{so that} \quad \frac{1}{m} \to \frac{\pi}{\alpha c^{\frac{1}{2}}}, \tag{16}$$

and for the cylinder $1/m \to 2\pi/\alpha c^{\frac{1}{2}}$. Hence for the parallel planes

$$s(\xi) = S\lim_{\alpha \to 0}\left(1 - \frac{\xi\alpha}{b}\right)^{\pi/\alpha c^{\frac{1}{2}}} = Se^{-\pi\xi/bc^{\frac{1}{2}}} \tag{17a}$$

and for the cylinder

$$s(\xi) = Se^{-2\pi\xi/bc^{\frac{1}{2}}}. \tag{17b}$$

The distribution of velocity when $m = 0$

The value of ζ at distance y from the axis on the plane $\xi = 0$ is given by

$$\frac{y}{b} = I_\zeta(\tfrac{1}{2}, \tfrac{1}{2}) \quad \text{for the parallel planes,} \tag{18a}$$

$$\frac{y^2}{b^2} = I_\zeta(\tfrac{1}{2}, \tfrac{1}{2}) \quad \text{for the cone.} \tag{18b}$$

Since

$$B_\zeta(\tfrac{1}{2}, \tfrac{1}{2}) = 2\sin^{-1}\zeta^{\frac{1}{2}}, \quad I_\zeta(\tfrac{1}{2}, \tfrac{1}{2}) = \frac{2}{\pi}\sin^{-1}\zeta^{\frac{1}{2}},$$

hence using (13)

$$u\left(\frac{\rho}{2S}\right)^{\frac{1}{2}} = \cos\frac{\pi y}{2b} \quad \text{for the parallel planes,} \tag{19a}$$

$$u\left(\frac{\rho}{2S}\right)^{\frac{1}{2}} = \cos\left(\frac{\pi y^2}{2b^2}\right) \quad \text{for the cylinder.} \tag{19b}$$

This distribution is shown in fig. 2 in the curve for $\alpha\sqrt{c} = 0$ and in fig. 6.

Distribution of velocity in porous cones for some values of $\alpha\sqrt{c}$

The distribution of velocity at the base of the cone depends on $\alpha\sqrt{c}$. It is of interest to calculate it for various values of this parameter. In table 1 values of ζ and $u(\rho/2S)^{\frac{1}{2}} = (1 - \zeta)^{\frac{1}{2}}$ are given in the first and second columns. At the head of each of the other columns are given the corresponding values of $\alpha\sqrt{c}$ and m determined from $(12b)$. Values of $y/\alpha x$ are found using $(15b)$ and Pearson's tables of $I_\zeta(2m + \tfrac{1}{2}, \tfrac{1}{2})$. These are given in the third and further columns of table 1 and are displayed in fig. 2.

It will be seen that when $\alpha = 0$, i.e. the case of the porous cylinder, the velocity in the central area is very nearly uniform and equal to $(2S/\rho)^{\frac{1}{2}}$. In the outer part the velocity falls to zero at $y = b$ and its distribution is rather like that in a boundary layer. As $\alpha\sqrt{c}$ increases, the distribution in the central region changes; u rapidly decreases as y increases from zero. Though theoretically there is always a velocity

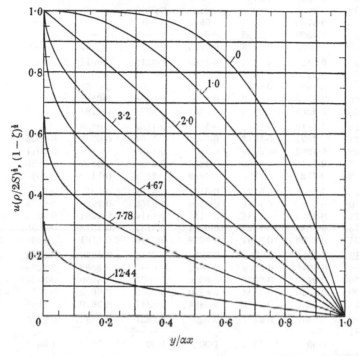

Fig. 2. Radial distribution of velocity through the base of a cone. The figures on the curves indicate the values of $\alpha\sqrt{c}$ for which calculations were made.

$(2S/\rho)^{\frac{1}{2}}$ on the axis, the region where the flow is greater than say $\frac{1}{2}(2S/\rho)^{\frac{1}{2}}$ gets exceedingly small. Thus at $\alpha\sqrt{c} = 7\cdot7$ it only extends to a radius $0\cdot027b$, i.e. over an area $1/1400$ of the area of the base of the cone. The reason for this is that the suction at any point is $S\eta^{1/m}$, so that when m is large it is nearly equal to S except for a small region near the vertex of the cone where $\eta^{1/m}$ differs appreciably from $1\cdot0$.

Discharge coefficient

When water flows through a circular hole of area A in, say, the flat base of a tank, the stream, which fills the hole as it leaves the tank, afterwards contracts to an area CA, where C is called the contraction coefficient, and at that stage it has velocity $(2S/\rho)^{\frac{1}{2}}$, where S is the pressure drop between the base of the tank and outside atmosphere. Thus the rate of discharge is $Q = CA(2S/\rho)^{\frac{1}{2}}$. C is therefore the ratio of the actual rate of discharge to what it would be if the full velocity head were attained in the plane of the orifice. For this reason it has also been called a discharge

Table 1. *Velocity of efflux, u, at radius y, expressed in non-dimensional units, $(1-\zeta)^{\frac{1}{2}} = u(\rho/2S)^{\frac{1}{2}}$. Columns 3 to 9 give values of $y/\alpha x$*

ζ	$(1-\zeta)^{\frac{1}{2}}$	$\alpha\sqrt{c} = 0$ $m = 0$	1·00 0·25	2·0 0·75	3·20 1·75	4·67 3·5	7·78 9·75	12·44 24·75
0·01	0·9950	0·2525	0·0767	0·0061	—	—	—	—
0·02	0·9899	0·3005	0·1002	0·0123	0·0000	—	—	—
0·05	0·9747	0·3790	0·1591	0·0309	0·0013	—	—	—
0·10	0·9487	0·4526	0·2265	0·0623	0·0053		—	—
0·15	0·9220	0·5032	0·2793	0·0943	0·0122	0·0003	—	—
0·20	0·8944	0·5433	0·3250	0·1269	0·0219	0·0011	—	—
0·25	0·8660	0·5774	0·3661	0·1603	0·0345	0·0026	—	—
0·30	0·8367	0·6075	0·4039	0·1945	0·0504	0·0053	—	—
0·35	0·8062	0·6348	0·4403	0·2295	0·0696	0·0096	—	—
0·40	0·7746	0·6603	0·4743	0·2655	0·0922	0·0161	—	—
0·45	0·7416	0·6842	0·5082	0·3025	0·1186	0·0256	—	—
0·50	0·7071	0·7071	0·5413	0·3407	0·1491	0·0387	0·0009	—
0·55	0·6708	0·7293	0·5738	0·3804	0·1836	0·0567	0·0011	—
0·60	0·6325	0·7510	0·6062	0·4217	0·2230	0·0802	0·0026	—
0·65	0·5916	0·7727	0·6389	0·4649	0·2676	0·1109	0·0061	—
0·70	0·5477	0·7944	0·6725	0·5104	0·3181	0·1511	0·0131	—
0·75	0·5000	0·8165	0·7071	0·5590	0·3756	0·2024	0·0273	—
0·80	0·4472	0·8395	0·7434	0·6115	0·4416	0·2681	0·0546	$2·5 \times 10^{-6}$
0·85	0·3873	0·8642	0·7827	0·6694	0·5187	0·3528	0·1058	$5·8 \times 10^{-5}$
0·88	0·3464	0·8803	0·8084	0·7079	0·5716	0·4161	0·1568	$3·6 \times 10^{-4}$
0·90	0·3162	0·8916	0·8269	0·7358	0·6110	0·4650	0·2032	$1·2 \times 10^{-3}$
0·92	0·2828	0·9041	0·8468	0·7662	0·6545	0·5208	0·2636	$4·0 \times 10^{-3}$
0·94	0·2450	0·9179	0·8692	0·8000	0·7037	0·5860	0·3434	$1·31 \times 10^{-2}$
0·96	0·2000	0·9338	0·8944	0·8390	0·7613	0·6645	0·4518	$4·39 \times 10^{-2}$
0·98	0·1414	0·9536	0·9266	0·8884	0·8347	0·7670	0·6096	0·156
0·99	0·1000	0·9676	0·9487	0·9222	0·8851	0·8382	0·7270	0·317
0·995	0·0707	—	—	—	—	—	—	—
1·000	0	1·00	1·00	1·00	1·00	1·00	1·00	1·00

coefficient. Such coefficients can be calculated for the flow through cones and wedges of porous material. In the degenerate cases $\alpha = 0$, the calculation is very simple. The rate of discharge through a cylindrical porous sheet is

$$Q = 2\pi \int_0^b uy\,dy = \left(\frac{2S}{\rho}\right)^{\frac{1}{2}} \pi \left(\frac{2b^2}{\pi}\right). \qquad (20)$$

The discharge coefficient is

$$C = Q/(2S/\rho)^{\frac{1}{2}}\pi b^2 = 2/\pi = 0·6366. \qquad (21)$$

Attention may be drawn to two points in connexion with this result. First, the discharge does not depend on the porosity of the cylinder walls. If c is increased the length of the cylinder through which water is drawn increases, but the rate of discharge remains constant. The second point is that the value of C for a circular hole in a flat plate is 0·60, which is nearly the same as 0·6366. This must be regarded as an accidental coincidence.

Discharge coefficients when $m \neq 0$

The total flow is
$$Q = 2\pi \int_0^b uy \, dy.$$

Using (7b), this is

$$Q = 2\pi\alpha \int_0^{b/a} W(x_1) x_1 \, dx_1 = 2\pi\alpha \left(\frac{b^2}{a_2}\right)\left(\frac{2S}{\rho}\right)^{\frac{1}{2}} \int_0^1 \zeta^{\frac{1}{2}}\eta \, d\eta \tag{22}$$

and

$$\int_0^1 \zeta^{\frac{1}{2}}\eta \, d\eta = \int_0^1 \eta^{1+1/2m} \, d\eta = \frac{2m}{4m+1}. \tag{23}$$

If the discharge coefficient is C,

$$C = \frac{Q}{\pi b^2}\left(\frac{\rho}{2S}\right)^{\frac{1}{2}} = \frac{2}{\alpha \sqrt{c}}\left(\frac{2m}{4m+1}\right). \tag{24}$$

For $\alpha \sqrt{c} = 0$, $\lim\limits_{m\to 0}\dfrac{m}{\alpha\sqrt{c}} = \dfrac{1}{2\pi}$ so that $C = \dfrac{2}{\pi}$ which agrees with (21).

For other values of $\alpha\sqrt{c}$ it is necessary to use (12b). Table 2 gives some corresponding values of m, $\alpha\sqrt{c}$, $4m/(4m+1)$ and C.

Table 2. *Discharge coefficient, C, for efflux through porous cones*

m	$2m+\frac{1}{2}$	$\alpha\sqrt{c}$	$\dfrac{4m}{4m+1}$	C
0	0·5	0	0	0·637
0·1	0·7	0·50	0·286	0·570
0·25	1·0	1·00	0·500	0·500
0·5	1·5	1·57	0·667	0·424
0·75	2·0	2·00	0·750	0·375
1·75	4·0	3·20	0·875	0·273
3·5	7·5	4·61	0·933	0·203
9·75	20	7·78	0·974	0·125
24·75	50	12·44	0·990	0·080

When m is great the discharge coefficient is nearly equal to $1/\alpha\sqrt{c}$. This is the coefficient which would be calculated if it were assumed that the flow within the cone is parallel to the axis, and the whole drop in pressure occurs at the porous surface, as it is in the 'trivial' case where the flow is in the reverse direction. Though the value of c is very close to what would be found on the assumption of uniform flow, it will be seen in fig. 2 that there is no tendency for the actual flow to be uniform.

Flow into a porous cylinder when resistance is not proportional to W^2

When the resistance is not proportional to W^2, I have found no solutions of the integral equations (8a) or (8b) except in the special case

$$\alpha = 0, \quad W = F(s) = As^t, \tag{25}$$

which represents the flow into a cylindrical porous tube when the resistance is proportional to $W^{1/t}$, t being a positive number greater than $\frac{1}{2}$.

Using the co-ordinate ξ measured from a fixed, but so far unspecified point on the axis of the cylinder, as is done in (17b), (8b) becomes

$$\tfrac{1}{2}b = (\tfrac{1}{2}\rho)^{\frac{1}{2}} \int_{\xi}^{\infty} \frac{A[s(\xi_1)]^t}{[s(\xi) - s(\xi_1)]^{\frac{1}{2}}}\, d\xi_1. \tag{26}$$

A solution of (26) can be found by assuming

$$s(\xi) = B\xi^n. \tag{27}$$

The value of the integral in (26) is independent of ξ provided

$$nt + 1 = \tfrac{1}{2}n \quad \text{or} \quad n = 2/(1 - 2t). \tag{28}$$

Writing

$$\chi = (\xi_1/\xi)^{\frac{1}{2}n}, \tag{29}$$

(26) becomes

$$\tfrac{1}{2}b = -(\tfrac{1}{2}\rho)^{\frac{1}{2}} AB^{-1/n}\left(\frac{2}{n}\right) \int_0^1 \frac{d\chi}{(1 - \chi^2)^{\frac{1}{2}}}, \tag{30}$$

so that

$$AB^{-1/n} = -\frac{nb}{2\pi}(\tfrac{1}{2}\rho)^{-\frac{1}{2}}. \tag{31}$$

Equation (31) determines B and (28) determines n in terms of the physical data A and t, but to complete the solution the position of the origin of co-ordinates $\xi = 0$, in relation to a point in the cylinder where s has a known value S is required. The origin is at distance ξ_0 from this point where

$$\xi_0 = (S/B)^{1/n}. \tag{32}$$

To find the distribution of velocity over the section the same procedure as that used to derive (15b) may be used. The velocity u at $\xi = \xi_0$, where $s = S$ and at radius y is given by

$$u = \left(\frac{2}{\rho}\right)^{\frac{1}{2}} [S - s(\xi_2)]^{\frac{1}{2}}, \tag{33}$$

where ξ_2 is the value of ξ at which the stream line which passes through y entered the tube.

y is related to ξ_2 by the equation

$$2\pi yu\, dy = 2\pi b W\, d\xi_2, \tag{34}$$

or

$$\tfrac{1}{2}y^2 = b(\tfrac{1}{2}\rho)^{\frac{1}{2}} \int_{\xi_2}^{\infty} \frac{F(s_1)\, d\xi_1}{[S - s(\xi_1)]^{\frac{1}{2}}}$$

$$= -b(\tfrac{1}{2}\rho)^{\frac{1}{2}} (AB^{-1/n})\frac{2}{n} \int_0^{\chi_2} \frac{d\chi}{(1 - \chi^2)^{\frac{1}{2}}}, \tag{35}$$

where

$$\chi_2 = (\xi_2/\xi_0)^{\frac{1}{2}n} = [s(\xi_2)/S]^{\frac{1}{2}}. \tag{36}$$

Substituting for $AB^{-1/n}$ from (31) and integrating it is found that (35) reduces to

$$\frac{\pi}{2}\left(\frac{y}{b}\right)^2 = \sin^{-1}\chi_2 = \cos^{-1}\left[1 - \frac{s(\xi_2)}{S}\right]^{\frac{1}{2}}. \tag{37}$$

Hence, from (33),
$$u \left(\frac{\rho}{2S} \right)^{\frac{1}{2}} = \cos \frac{\pi y^0}{2b^2};$$
(38)

thus (38) is identical with (19b). The distribution of velocity at a section where a given suction S is applied is therefore the same for all values of A and t, i.e. for all porous materials in which the resistance is related to the velocity through it by a power law. The distribution of velocity along the cylinder depends very much on A and t.

As a corollary it will be noticed that the discharge coefficient has the same value, $2/\pi$, for all porous tubes for which the very general resistance law (25) applies.

Conditions when $\alpha \neq 0$ and porous resistance not proportional to W^2

I have been unable to find solutions of the integral equations (8a) or (8b) when α is not zero and W is not proportional to $\sqrt{s}$. Thinking that perhaps one could get an indication of the nature of the flow by simple momentum methods, I tried assuming that the fluid mixes so completely after coming from the porous surface that the velocity u is constant over each section. In the case of a narrow cone this leads to the equations:

$$\text{continuity,} \quad 2xW = \alpha \frac{\mathrm{d}}{\mathrm{d}x}(ux^2),$$
(39)

$$\text{momentum,} \quad \frac{\mathrm{d}}{\mathrm{d}x}(x^2 u^2) = \frac{1}{\rho} x^2 \frac{\mathrm{d}s}{\mathrm{d}x}.$$
(40)

If the porosity equation (2a) is used and $(2s/\rho c)^{\frac{1}{2}}$ is substituted for W in (39), then (39) and (40) may be solved by making the substitutions

$$s = A x^{1/m}, \quad u = B x^n;$$
(41)

(39) yields
$$2^{\frac{3}{2}} \rho^{-\frac{1}{2}} c^{-\frac{1}{2}} A^{\frac{1}{2}} x^{1/2m+1} = B\alpha(n+2) x^{n+1},$$
(42)

(40) yields
$$(2n+2) B^2 x^{2n+1} = \rho^{-1} A x^{1/m+1},$$
(43)

(42) and (43) are consistent if
$$mn = \tfrac{1}{2},$$
(44)

and also
$$\frac{B^2}{A} = \frac{1}{\rho m(2n+2)} = \frac{8}{\rho c(n+2)^2 \alpha^2}$$
(45)

or
$$\alpha \sqrt{c} = \frac{4\sqrt{2m}}{4m+1}(1+2m)^{\frac{1}{2}}.$$
(46)

This is an equation for determining m when α and c are known. It is analogous to (12b) for the case when there is no mixing.

The discharge coefficient is

$$C = u \sqrt{\frac{g}{2S}} = B x^n \sqrt{\frac{\rho}{2A x^{1/m}}} = B \sqrt{\frac{\rho}{2A}},$$
(47)

or, using (43),
$$C = (2+4m)^{-\frac{1}{2}}.$$
(48)

The variation of C with $\alpha\sqrt{c}$ in the two cases of no mixture and complete mixture are shown in fig. 3. It will be seen that the mixture slightly increases the discharge coefficient. Thus the solution is of the same type as before, though the equation for determining m as a function of $\alpha\sqrt{c}$ is not identical with (12b). When the porosity equation (2b) is used and $s\kappa/\mu$ is substituted for W in (39) the equation resulting from the elimination of s between (39) and (40) is difficult, and I have been unable to make any headway in trying to solve it.

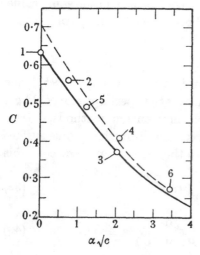

Fig. 3. Discharge coefficients, C. ○, experimental; —, no mixing after entering cone, theoretical; - - -, complete mixture over transverse sections of cone, theoretical.

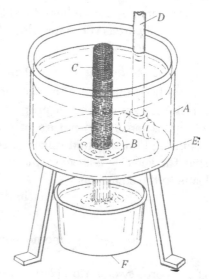

Fig. 4. Arrangement of experimental tank.

2. HYDRAULIC EXPERIMENTS

Experiments with a porous cylinder

In considering how far the theoretical results obtained by solving the integral equation may be expected to represent physically obtainable flow conditions it was noticed that in the cylindrical case the discharge coefficient and also the distribution of velocity over cross-sections (but not the distribution along the cylinder) depend only on the assumption that the stream after entering through one area of porous wall does not mix appreciably with streams which come from other points. They do not even depend on the law of resistance to flow through the porous surface, at any rate if it is a power law. If complete mixture occurs the distribution of velocity over a cross-section would be uniform and the discharge coefficient would depend on the law of resistance in the walls. If there is little mixing the discharge coefficient should be $2/\pi$ and the distribution of velocity over the cross-section should be in accordance with (19b) and (38). This distribution is shown in fig. 2 as the curve for $\alpha\sqrt{c} = 0$.

To construct a cylindrical tube whose walls have uniform resistance a simple method would be to wrap a perforated sheet, such as the zinc sheets of which meat safes are constructed, on a cylindrical rod. The flow into such a tube, however, might be expected to mix, since jets from individual holes might well penetrate far into the flow unless the diameters of these holes were very minute compared with that of the tube. For this reason a uniformly porous material with fine interstices was sought. Copper wire gauze with a square mesh consisting of wires 0·023 cm. in diameter spaced 0·067 cm. apart was chosen. A tube 3·33 cm. in internal diameter and 45 cm. long was first made from this material by winding a sheet of it once round a 3·33 cm. rod. To avoid asymmetry of flow there was no overlap. The gauze was found to retain its cylindrical form when fine copper wire rings were spaced every few centimetres along its length. The bottom of the gauze cylinder passed through a hole 3·50 cm. diameter in a thin flat plate. The ends were slit, bent back over the plate and soldered to it. In this way a cylinder of uniform porosity was made. The plate could be fixed so that it closed a hole in the bottom of a cylindrical tank 36 in. in diameter × 18 in. deep. This is shown in fig. 4, where A is the tank, B the plate to which the gauze cylinder C is soldered. The plate B is fixed to the bottom of the tank by nuts and bolts which pass through both and through a watertight rubber washer.

The tank was set up in such a position in the Engineering Laboratory at Cambridge that a bucket could be placed below it to catch the outflow. It was supplied with water through a pipe D which was connected to a large supply tank. The water in the tank A was maintained at constant level by means of a spillway, which is not shown in fig. 4.

The supply entered through the vertical 2 in. pipe D (fig. 4) placed near the edge of the tank, and the considerable swirl and turbulence which resulted could not be cured by perforated plates or radial vanes. The trouble, however, was cured by first fixing a T-piece at the bottom of the pipe D. A long fabric tube or stocking was made by cutting 18 ft. from a roll of a fabric known as 'mutton cloth'. This fabric which is sold in garages as a cleaning material, is woven in the form of long tube. Half of the 18 ft. length was then drawn through the other half, forming a double-walled 9 ft. tube and the ends fastened over the two arms of the T-piece. The bight was laid round the bottom of the tank (E, fig. 4). This so reduced the energy of the stream flowing into the tank that no swirl or turbulence could be detected. It also had the advantage of removing any small pieces of grit or rust entering the system.

To measure the total discharge through the wire gauze cylinder a large bucket was placed under it (F, fig. 4). The jet was deflected to a drain while the bucket was placed in position. The deflector was then rapidly removed and then rapidly replaced when the bucket was nearly full. It was found that the time during which the jet was filling the bucket could be measured correctly to $\frac{1}{4}$ sec. The bucket was then weighed. At the greatest rates of discharge the time to fill the bucket nearly full was about 8 sec. so that an accuracy of between 1 and 2 % could be attained.

The calculated discharge coefficient is $2/\pi = 0.6366$. In the first experiments it was found that the measured discharge coefficients ranged from 0.66 to 0.67. When the velocity distribution over the exit plane was measured by the method to be described later and plotted on the diagram of fig. 2, the experimental points were found to lie outside the theoretical curve for $\alpha\sqrt{c} = 0$. It seemed that the reason for these discrepancies must be that one of the simplifying assumptions made in the analysis must be wrong. These were (a) that the resistance of the gauze is sufficient to ensure that the velocity of the water is negligible when it enters through the pores and (b) that the streams entering at different positions along the gauze tube do not mix and therefore that the Bernoulli constant for each stream line remains constant.

Table 3. *Measurements of discharge coefficients for efflux through a porous cone*

head, h (cm.)	weight of water (lb.)	time (sec.)	discharge coefficient C
23.9	21.5	8.2	0.632
24.05	25.0	9.4	0.636
24.05	23.2	8.8	0.634
42.9	28.1	8.0	0.632
42.9	27.6	7.8	0.638

It seemed probable that the value of c for the gauze used (afterwards measured and found to be about 4) was too small to justify the use of assumption (a). If assumption (b) is valid, then better agreement between theory and experiment would be expected if a more highly resisting porous cylinder could be constructed. This end was achieved by wrapping the gauze three times round the 3.33 cm. cylinder. The resulting resistance coefficient was measured by the method described later and found to be approximately 12 in the range of Reynolds numbers used. This is large enough to justify the use of assumption (a). The wire gauze cylinder was then set up in the cylindrical tank and the measurements given in table 3 were made.

It will be seen that the agreement with the theoretical value 0.6366 is very good.

Velocity distribution

To measure the distribution of velocity over the base of the porous tube an upward-pointing Pitot tube was traversed across the base, or exit plane. The arrangement is shown in fig. 5, where D is the Pitot and E is a traversing tube which slides in a box B which is bolted to the bottom A of the tank. The porous cylinder or cone was bolted to the top of the box B. Fig. 5 shows a porous cone C. F is a side tube for measuring the pressure in the box B and G is an orifice which could be chosen so as to control the pressure in B at any desired value. Three manometer tubes, I, II, III were fixed to the side of the tank. Tube I was connected to the side tube F (fig. 5), II to the sliding Pitot E and III through a syphon to the main body of water in the tank.

If h_1, h_2, h_3 are the levels in the three manometers the velocity u at the top of the

Pitot tube is taken as $u = \lfloor 2g(h_2 - h_1) \rfloor^{\frac{1}{2}}$. The quantity represented by S in the analysis is $g\rho(h_3 - h_1)$, so that

$$u \left(\frac{\rho}{2S}\right)^{\frac{1}{2}} = \left(\frac{h_2 - h_1}{h_3 - h_1}\right)^{\frac{1}{2}}. \tag{49}$$

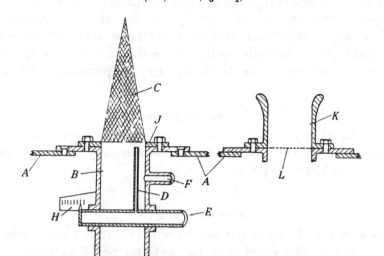

Fig. 5. Apparatus for measuring distribution of velocity (left) and resistance coefficient, c (right).

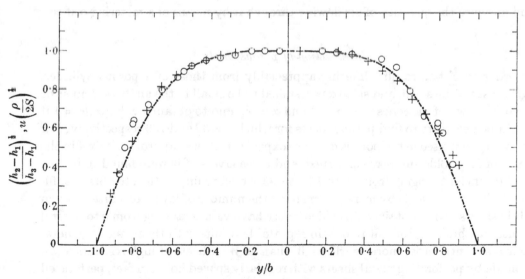

Fig. 6. —·—· calculated distribution of velocity (equation (19b)), $u(\rho/2S)^{\frac{1}{2}}$; $\bigcirc$ and $+$, measured velocities $\{h_2 - h_1/(h_3 - h_1)\}^{\frac{1}{2}}$.

Several Pitot traverses were made, reading the position of the Pitot on a scale, shown as H in fig. 5, outside the box B. The reading of this scale when the Pitot was central was subtracted in each case from the reading on the scale H and the result taken as y. The internal diameter of the tube (or the base of the cone in later experiments) was measured and taken as $2b$. The values of y/b found in this way and the corresponding values of $\{h_2 - h_1/(h_3 - h_1)\}^{\frac{1}{2}}$ are displayed in fig. 6.

To compare these observations with the results obtained theoretically the calculated velocity distribution of (19b) and (38) is also shown. It will be seen that the agreement is good. This indicates that under the conditions of this experiment there is little mixing between the streams entering the porous cylinder at different heights.

Experiments with porous cones

For cylinders the theory indicates that it is not critically important to know the value of c, but for experiments with cones it is necessary to know $\alpha\sqrt{c}$ so that measurements of c had to be made.

Measurement of c

Two methods were used. In one the cone shown as C in fig. 5 was replaced by the faired mouthpiece K. This could be bolted on to the box B so that a sheet of the material (L, fig. 5) to be tested was stretched over the orifice. The operation of the apparatus for measuring c was identical with that for measuring C for the cylinder or cones. The other method was to set the piece of sheet being tested across a tube or small wind tunnel, measure the pressure difference between the two sides of the sheet and measure the quantity of air flowing through it by means of a metering orifice.

Construction of porous sheets

In order to obtain results differing appreciably from those of a porous cylinder, $\alpha\sqrt{c}$ must not be small, and since α is assumed to be small in the analysis, c must be large. For two of the cones used $\alpha = \frac{1}{6}$ and $\alpha = \frac{1}{13}$, and to obtain a value $\alpha\sqrt{c} = 1$ it would be necessary to find porous sheets for which $c = 36$ and 169 respectively. For the wire gauze used in the porous cylinder experiments c was approximately 4 in the range of Reynolds numbers concerned, and three layers of it were found to have a total coefficient ranging from 12 to 15. It was therefore unpractical to obtain sufficiently large resistance by merely increasing the number of layers of gauze. Woven fabric sheets can be obtained for which c may have values ranging from 0 to several thousands, but it is difficult to obtain repeatable results with them, and in general their resistance varies more nearly as W than as W^2. Sheets of high resistance can be made by perforating metal sheets with regularly spaced holes, in fact, perforated sheets of sheet iron and zinc with holes in hexagonal array can be bought. Their resistance depends on β, the area of holes per unit area of sheet. For sheets with holes of diameter d spaced out in hexagonal array so that nearest neighbours are at distance l apart, $\beta = 0\cdot907d^2/l^2$. The resistance coefficients of such sheets depend little on the Reynolds number within the range which is significant in the experiments to be described. It was found* that over a large range of values of β the experimental results were represented by

$$c + 1 = A/\beta^2, \tag{50}$$

where A is a number which varies slowly with β over a range roughly from 0·9 when β is large to 1·5 when β is small. Perforated zinc was obtained for which $d = 0\cdot174$ cm., $l = 0\cdot265$ cm., so that $\beta = 0\cdot384$. For this value of β the 'A' in (50) was found* to be about 1·0, so that c would be expected to be about 5·8. In fact, the value found experimentally with this particular perforated sheet was about $c = 6$. To obtain a material of greater resistance this zinc sheet was laid over an unperforated copper sheet, and every alternate hole was bored with a drill 0·174 cm. diameter. This treatment produced a sheet for which $\beta = 0\cdot096$. Using a value $A = 1\cdot5$ which has been found for sheet with approximately this value of β, the expected value of c was 160.

Measurements of c were made by replacing the porous cone or cylinder (shown at C in figs. 4 and 5) by the faired mouthpiece K (fig. 5). The piece of material being tested was gripped between this mouthpiece and the box B at the bottom of the tank in which the experiments with the porous cylinder had been made.

The rate of discharge under the head $h_3 - h_1$ was measured. If this is Q and the area of the exposed part of the porous sheet is πb^2,

$$c + 1 = 2h(h_3 - h_1)\left(\frac{\pi b^2}{Q}\right)^2. \tag{51}$$

The perforated sheet for which $\beta = 0\cdot096$ was tested and values of c were found to vary very little as the mean velocity W varied. For the Reynolds number $R = Wd/\nu = 350$, c varied from 156 to 158; for $R = 240$, c varied from 156 to 158. These values are closer to that predicted by the formula $c = 1\cdot5\beta^{-2} - 1$ (namely, 160) than the accuracy of the experiments would justify one in expecting.

Though the resistance of the perforated copper sheet, for which $c = 158$, was great enough for the purpose intended, jets projected inwards through holes 0·174 cm. in diameter might be expected to cause much mixing of streams inside a cone made from it. For this reason it was thought that a compound sheet consisting of a layer or layers of wire gauze laid over or under the copper sheet might have a high resistance without producing jets which would penetrate as far as those issuing from unobstructed holes. Compound sheets of this kind were tested. It was found that their resistance is not as a rule the sum of the resistances of the components. The resistance of a compound sheet consisting of two similar gauze sheets, for instance, depended to a certain extent on whether they were laid with their wires parallel or at 45° to one another.

The most striking effect of this kind occurred when three sheets of gauze for each of which $c = 4$ were laid above the perforated copper sheet for which $c = 158$. For

---

* G. I. Taylor & R. M. Davies, *Rep. Memo. aeronaut. Res. Comm., Lond.*, no. 2237 (1944) (paper 45 in volume III).

a downward-moving stream it was found that if the copper sheet was above the gauze c was of order 350, whereas if the gauze was above the sheet c was of order 140. This great difference made me suspect that cavitation was occurring in one case and perhaps not in the other, for no such big differences were found in experiments with these compound sheets using air as the working fluid. Great precautions were taken to prevent the accidental inclusion of air bubbles between the layers of the porous surfaces, but cavitation might occur in spite of them.

Since the measurements of resistance were made at approximately the same pressures as the experiments with conical porous sheets which are to be described, the values of c determined in experiments with water were used in comparison with theoretical predictions rather than experiments with air when there was any difference between them.

Cones constructed

Cone I was made of the perforated copper sheet ($\beta = 0.096$). The semi-vertical angle was $\sin^{-1}\frac{1}{6}$. This particular angle was chosen because after cutting a 60° sector from the flat perforated sheet it could be folded into a cone containing only circular holes; the holes which had been bisected in the cutting out of the sheet formed complete holes when the cone was folded into its final form.

Cone II was similar to cone I except that it was cut from the perforated zinc sheet for which $\beta = 0.384$.

Cone III was made by folding a narrow sector of the zinc sheet ($\beta = 0.384$) to form a cone of semi-vertical angle 4·4°, or $\alpha = 0.077$. On the generator where the cut edges joined there were some holes smaller, some larger, and some the same size as those in the uncut parts of the sheet.

The external diameters of the bases of all these cones were 3·5 cm. in diameter, so that they would all fit into the same hole (J, fig. 5) in the bottom of the tank.

Besides these rigid metal cones, conical sheaths of wire gauze and of fabric were made which could be fitted inside or outside them. No measurements were made with cone I when covered with gauze outside because of the uncertainty about the resistance of this arrangement.

Discharge coefficients

To compare experimental with theoretical results it is necessary to measure the rate of discharge, the pressure drop and the radius of the base of the cone. When the cones were used without internal gauze sheaths the physical quantity taken as equal to the radius '$2b$' of the theoretical analysis was the actual diameter of the inside of the base of the cone, and this was 3·4 cm. in all cases. When the cone was lined with wire gauze sheaths the diameter of a cylindrical disc which would just enter the base of the inner sheath was taken as $2b$. When three gauze liners were used $2b$ was found to be 3·1 cm. Since c varies slightly with the water velocity, W, an estimated mean value had to be used to correspond with the mean velocity through the gauze. The mean velocity through the base is $Q/\pi b^2$, where Q is the measured volume of discharge in cm.^{3}/sec. The mean velocity through the gauze was therefore taken to be

$$W_m = \alpha Q/\pi b^2 \text{ cm./sec.} \tag{52}$$

In comparing observation with theory the measured rate of discharge was used to obtain W_m and the corresponding value of c was taken from a curve representing the results of experiments on the resistance of flat sheets at varying velocities. The value of $\alpha\sqrt{c}$ so obtained was used as the abscissa for an experimental point on the diagram of fig. 3. The corresponding ordinates were

$$C = \frac{Q}{\pi b^2 \{2g(h_1 - h_2)\}^{\frac{1}{2}}}. \tag{53}$$

The results of various experiments are shown in fig. 3, together with the theoretical curves assuming no mixing and complete mixing over planes perpendicular to the axis of the cones.

Referring to the numbered experimental points:

Point 1 represents the experiments with a cylinder made of three thicknesses of gauze. It fits the theoretical curve corresponding to no mixture with great accuracy (cf. table 3).

Point 2 represents the results with cone II covered outside with three closely fitting cones of wire gauze. The mean velocity of discharge $Q/\pi b^2 = 109\cdot8$ cm./sec., so that $W = 109\cdot8 \times \frac{1}{6} = 18\cdot3$ cm./sec. At this speed the value of c measured in the apparatus shown in fig. 5 with three thicknesses of gauze over perforated zinc sheet was $20\cdot5$ so that $\alpha\sqrt{c} = \frac{1}{6}\sqrt{20\cdot5} = 0\cdot75$. The measured value of C at this rate of discharge was $0\cdot564$. It will be seen that the experimental point lies about half-way between the two theoretical curves.

Point 3 represents results with cone I with three gauze liners inside it. Here the radius b could not be determined very accurately because the base opening was not quite circular. Measurements of $2b$ varying from $3\cdot0$ to $3\cdot1$ cm. were obtained and discharges ranging from 733 to 739 cm.^{3}/sec. were obtained under a head of $37\cdot35$ cm. of water. Using $2b = 3\cdot05$ cm. this corresponds with $u = 100\cdot8$, $W = 16\cdot8$ cm./sec. At this speed values of c ranging from 132 to 165 were obtained, varying with the relative directions of the wires in the three layers. With these data C is found to be $0\cdot37$ and $\alpha\sqrt{c}$ varies from $1\cdot93$ to $2\cdot12$. Here again the point lies on the theoretical curve based on the assumption that there is no mixture.

Point 4 represents results with cone I without gauze liners. Here three measurements of C give $0\cdot410$, $0\cdot404$, $0\cdot407$ and $c = 158$, so that $\alpha\sqrt{c} = 2\cdot09$. It will be seen that the point is well above the line corresponding to no mixing and in fact is very close to the curve corresponding to instantaneous mixing over cross-sections of the cone.

Point 5 is for cone III covered with two conical sleeves of the open fabric 'butter muslin'. With this it was found that at a head of 42 cm. the mean velocity over the base was 141 cm./sec. so that $C = 0\cdot49$. The mean velocity through the surface was

$0.077\,(141) = 10.9$ cm./sec. At that speed it was found that $c = 247$, so that the corresponding point in fig. 3 is about half-way between the curves for no mixing and complete mixing.

Point 6 is for cone III covered with a conical sock make from a fine linen handkerchief. In this case experiments showed that the resistance is nearly proportional to the velocity. At a head of 43.1 cm. it was found that the mean speed of flow through the base was 80 cm./sec., so that $C = 80/\sqrt{(2g \times 43.1)} = 0.275$. The mean velocity of flow through the surface is therefore $0.077 \times 80 = 6.2$ cm./sec., and at this speed measurements gave $C = 2000$ so that $\alpha\sqrt{c} = 3.44$.

Distribution of velocity

The distribution of velocity over the base of the cone was measured in the same way as for the cylinder. Two such measurements are shown in figs. 7 and 8. Fig. 7 corresponds to point 2 in fig. 3 when $\alpha\sqrt{c} = 0.75$. The theoretical distributions for $\alpha\sqrt{c} = 0$ and $\alpha\sqrt{c} = 1.0$ are shown. The observed points lie between the curves except near $y = b$ where they fall below them. In this case the gauze was above the cone II. It might be expected that the flow would be disturbed in this case because the jets coming from the holes in the cone are unobstructed.

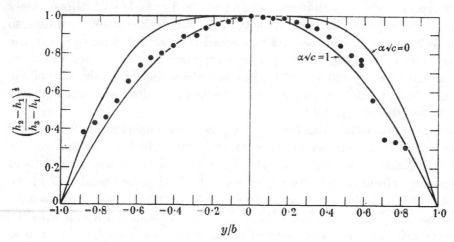

Fig. 7. Distribution of velocity of efflux over base of cone ($\alpha = \frac{1}{6}$, $c = 20.5$, $\alpha\sqrt{c} = 0.75$). Calculated curves for $\alpha\sqrt{c} = 0$ and 1.

Fig. 8 shows the distribution of velocity over the base in the case where the gauze liners lie inside the perforated conical sheet so that jets from the perforations might be broken up by the gauze. This case is that for which the discharge coefficient is represented by the point 3, fig. 3. The value of $\alpha\sqrt{c}$ lies in the range 1.9 to 2.1 and the distribution of velocity has been calculated (see fig. 2) for the case $\alpha\sqrt{c} = 2$. This theoretical distribution is shown in fig. 8. It will be seen that the observed points lie close to it except for those close to the centre of the base.

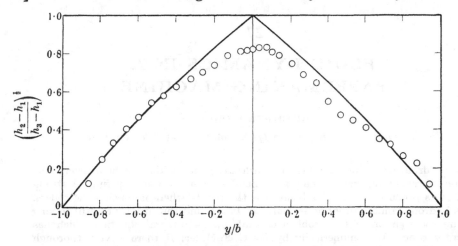

Fig. 8. Distribution of velocity of efflux over base of cone ($\alpha = \frac{1}{8}$, $c = 132$ to 165, $\alpha\sqrt{c} = 1\cdot93$ to $2\cdot12$). Full line is the theoretical distribution for $\alpha\sqrt{c} = 2\cdot0$.

The curious feature of the theoretical curve for $\alpha\sqrt{c} = 2$ that it is pointed in the middle is due to the assumption that there is no mixing, and the closeness of the experimental points to the theoretical ones in this region indicates that this ideal condition has nearly, but not quite, been attained.

27

FLUID DYNAMICS IN A PAPERMAKING MACHINE

REPRINTED FROM

Proceedings of the Royal Society, A, vol. CCXLII (1957), pp. 1–15

Water is drained from paper pulp in a Fourdrinier papermaking machine through the action of rollers over which the porous band supporting the pulp runs. A strong suction is produced under this band near the points where it touches the rollers. This hydrodynamical situation is analysed. It is found that if no turbulent mixing occurs the solution of the problem depends on an integral equation which has recently been solved numerically by Mr G. F. Miller. If there is very thorough mixing, qualitatively similar distributions of suction are obtained, but the maximum suction is increased.

Comparison with a previous partial analysis by P. E. Wrist confirms the correctness of his physical ideas but reveals an inconsistency in their application. Experiments are described in which predictions based on Wrist's ideas, but under conditions where the inconsistency does not arise, are verified. These experiments were extended to include conditions similar to those in a Fourdrinier machine. The distribution of suction was measured and compared with that calculated using the integral equation.

In a recent paper* it was stated that when a steady motion is set up in a space bounded by porous walls the flow can sometimes be described by means of an integral equation. The special problem treated, that of flow into a porous cylinder or cone was chosen because this equation can be solved in that particular case in terms of functions which have been tabulated. The investigation started, however, in an attempt to solve another special problem, namely, that presented by the drainage of water from paper pulp in a Fourdrinier papermaking machine. This was mentioned in the introduction to the previous paper, but no details of the analysis were given and the relevant integral equation had not then been solved.

In the Fourdrinier machine paper pulp, a mixture of fibres with some 500 times their volume of water, is forced through a long narrow orifice to form a horizontal jet which falls on to a moving horizontal porous band. This band is driven at a speed which is nearly the same as that of the jet. The water drains through the porous band, or 'wire' as it is called, and leaves a mat of fibres which becomes paper when it is dry. If the wire were stationary the pulp would drain away slowly under the action of gravity, but in modern machines the speed of the wire is such that the time available is very short compared with that which would be needed if gravity alone were acting. In fact, almost the whole of the drainage is due to suction which is produced hydrodynamically under the wire where it passes over supporting rollers. Fig. 1 is a very much simplified sketch of a paper-machine showing the 'head-

---

* Paper 26 above.

box' A from which the pulp jet B is projected; C is the wire supported on rollers $D_1 D_2 D_3$. E is a suction box designed to remove any water which remains when the hydrodynamical action at the rollers is no longer effective and F is the paper being stripped off the wire. The geometry of the situation to be analysed is shown in fig. 2 which is an enlarged view of the region inside the rectangle of broken lines shown

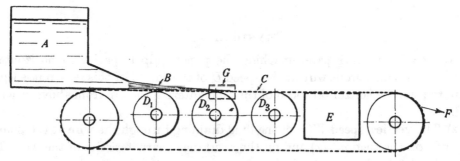

Fig. 1. Sketch of the wet end of a Fourdrinier papermaking machine.

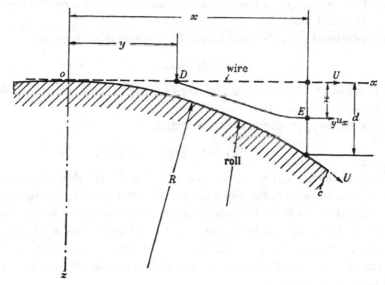

Fig. 2. Neighbourhood of point of contact of wire and roll.

at G in fig. 1. The porous band or wire ox (fig. 2) moves horizontally with velocity U over a cylindrical roller oc, touching it at the point o which will be taken as the origin of co-ordinates ox (horizontal) and oz (vertical). Since suction is only appreciable in the narrow region—or nip—between wire and roller which is close to o, it is sufficiently accurate for the purpose of the present discussion to take the pressure at any point in the nip as a function of x only. The small variations in the vertical direction due to gravity, and those associated with vertical components of acceleration will be ignored and the pressure taken as $P - s(x)$, where P is the atmospheric pressure and $s(x)$ is the suction produced at x by motion of the band and roller.

Since the nip is narrow its depth may be taken as

$$d = x^2/2R, \tag{1}$$

where R is the radius of the cylinder roller.

PHYSICAL DATA

(1) In the Fourdrinier paper-machine the paper pulp is projected horizontally on to the moving porous wire at the speed U of the wire. For this reason it is legitimate to take the horizontal speed of the fluid entering the nip from above as equal to U.

(2) The vertical speed $W(x)$ of the fluid draining through the wire at the point x depends on the suction $s(x)$ and on the resistance characteristics of the wire. The general resistance law

$$W = \phi(s) \tag{2}$$

will be used in developing the theoretical discussion, but two particular cases will be discussed in detail, namely

(i) W proportional to s. This law is usually expressed in the form

$$W = \phi(s) = \kappa s/\mu, \tag{3}$$

where κ is a constant having the dimensions of length, and μ is the coefficient of viscosity of the fluid in the nip.

(ii) The resistance is proportional to W^2. Here

$$s = c(\tfrac{1}{2}\rho W^2) \quad \text{or} \quad \phi(s) = (2s/\rho c)^{\frac{1}{2}}, \tag{4}$$

where c is a non-dimensional resistance coefficient and ρ the density of the fluid.

(3) The fluid which enters the nip will mix to a certain extent with fluid which has already entered at points upstream. Since there is no way in which the amount of this turbulent mixing can be estimated, two extreme cases will be treated; (a) there is no mixture, and (b) the mixing is so rapid through each narrow vertical section of the nip that the fluid velocity u may be considered constant though its depth and to vary with x only.

(4) The direct effect of viscosity on the flow in the nip in the non-turbulent case must be small at the speeds at which the paper machines are operated. It will be disregarded.

(5) The friction of the fluid on the cylindrical roller will be neglected. The roller is moving with velocity U and it will be found that the fluid near it must move with velocity greater than U if friction is neglected. The effect of friction might be to form a thin retarded layer near the surface of the roller, and the neglect of friction amounts to assuming that this retarded layer does not mix with the fluid in the nip sufficiently to alter the character of the flow.

ANALYSIS

The methods of analysis in the two cases (*a*) and (*b*) are so different that they will be considered independently and only compared when numerical results have been attained.

Case (*a*): *no turbulent mixing*

Here the velocity varies but the pressure may be taken as constant through the depth of the nip. Along a stream line such as that shown as *DE* in fig. 2 the pressure and velocity are related by the Bernoulli equation

$$s(x) - s(y) = \tfrac{1}{2}\rho(_y u_x^2 - U^2). \tag{5}$$

Here y is the co-ordinate of the point of entry, namely D fig. 2, and $_y u_x$ is the velocity at x of the fluid which entered at y. The vertical velocity is everywhere small compared with the horizontal component and is neglected in (5).

Consider two neighbouring stream lines which start from the underside of the wire at y and $y + \delta y$, respectively. At distance x from O their depths are, say, z and $z - \delta z$, and since all the fluid which enters with downward velocity $W(y)$ between y and $y + \delta y$ flows horizontally with velocity $_y u_x$ through the strip of height δz, the equation of continuity which represents the fact that no fluid disappears *en route* is

$$W(y)\,\delta y = {}_y u_x \delta z, \tag{6}$$

where $W(y)$ is the vertical velocity at the point D of fig. 2.

It remains to express the connexion between $W(y)$ and $s(y)$, that is the law of hydraulic resistance of the wire and overlying paper pulp. For the moment this may be left in the general form (2), namely $W = \phi(s)$ where $\phi(s)$ represents the result of experiments in which the drainage rate W is measured as a function of s, the applied suction. Eliminating $W(y)$ between (6) and (2) and substituting for $_y u_x$ from (5),

$$\delta z = \frac{\phi\{s(y)\}\,\delta y}{[U^2 + (2/\rho)\{s(x) - s(y)\}]^{\frac{1}{2}}}. \tag{7}$$

This equation expresses, in differential form and in terms of the still-unknown distribution of suction under the wire, the depth below it at station x of the stream line which entered at y.

There is still one more condition which must be satisfied. The fluid must fill the space between the wire and the roll. Using (1), this is expressed by the equation

$$d = \frac{x^2}{2R} = \int_0^d dz \tag{8}$$

and, using (7), this becomes

$$\frac{x^2}{2R} = \int_0^x \frac{\phi\{s(y)\}\,dy}{[U^2 + (2/\rho)\{s(x) - s(y)\}]^{\frac{1}{2}}}. \tag{9}$$

Solutions of (9)

When the resistance of the porous band is proportional to W and $\kappa s(y)/\mu$ is substituted from (3) in (9) the resulting equation may be rendered non-dimensional by the substitutions

$$x = \kappa\rho R U x'/\mu, \quad s(x) = \tfrac{1}{2}\rho U^2 f(x'),$$
$$y = \kappa\rho R U y'/\mu, \quad s(y) = \tfrac{1}{2}\rho U^2 f(y'); \qquad (10)$$

(9) then becomes

$$x'^2 = \int_0^{x'} \frac{f(y)'\,\mathrm{d}y'}{[1+f(x')-f(y')]^{\frac{1}{2}}}. \qquad (11)$$

When the resistance is proportional to W^2 and (4) is substituted in (9) the resulting equation may be made non-dimensional by the substitutions

$$x = 2RC^{-\frac{1}{2}}x', \quad s(x) = \tfrac{1}{2}\rho U^2 f(x'),$$
$$y = 2RC^{-\frac{1}{2}}y', \quad s(y) = \tfrac{1}{2}\rho U^2 f(y'); \qquad (12)$$

(9) then becomes

$$x'^2 = \int_0^{x'} \left[\frac{f(y')}{1+f(x')-f(y')}\right]^{\frac{1}{2}} \mathrm{d}y'. \qquad (13)$$

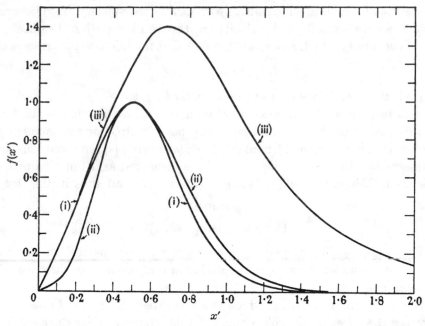

Fig. 3. Distributions of suction: curve (i) resistance proportional to W; (ii) resistance proportional to W^2; (iii) complete mixture in vertical section.

The solutions of (11) and (13) which satisfy the boundary conditions suitable for the present problems, namely, $f(x') = 0$ at $x' = 0$ and as x' tends to infinity, have been found and tabulated by Mr G. F. Miller.* They are shown by means of curves in fig. 3 where curve (i) is the solution of (11) and (ii) of (13).

* *Proc. Roy. Soc.* A, ccxxxvi (1956), 529.

Previous analysis by P. E. Wrist

It will be seen in fig. 3 that in both the cases represented by (11) and (13) respectively, $f(x')$ reaches its maximum value of $1 \cdot 0$ at the point $x' = \frac{1}{2}$. Substituting from (10) or (12), it will be seen that the maximum suction is $\frac{1}{2}\rho U^2$ and that it occurs at the point $x = \frac{1}{2}\kappa\rho RU/\mu$ when the resistance law is linear, and at the point $x = R/\sqrt{c}$ when the resistance is proportional to W^2.

This result had already been predicted by P. E. Wrist* whose argument is of interest, even though it necessarily involves two mutually incompatible assumptions, because it involves only simple physical reasoning. To explain the production of suction, Wrist points out that as the wire moves away from the point of contact with the roller, the two points which had been in contact remain almost in a vertical line and separate with velocity xU/R. If the particles of fluid after passing through the wire remain in this vertical line, the downward velocity W of the fluid entering through the wire is equal to xU/R. Wrist's first assumption is that this is in fact the case, so that the suction can be calculated as that which would be produced in a cylinder by the downward motion of a piston if the top of it were covered by a porous sheet. When the resistance is proportional to W (3) leads to the prediction that the suction is $s = \mu Ux/R\kappa$ or, in the non-dimensional units of (10),

$$f(x') = 2x'. \tag{14}$$

Wrist's physical assumption that W is equal to xU/R necessarily involves the assumption that there is no lateral pressure gradient, i.e. that $ds/dx = 0$. This is inconsistent with (14), which leads to $ds/dx = \mu U/R\kappa$. Nevertheless, when x is small (14) is in agreement to a first approximation with Miller's solution.

Wrist's first assumption would lead to the prediction that $s(x)$ should increase indefinitely with x till the geometrical approximation (1) on which (11) is founded becomes invalid. This would involve a greater suction than is in fact observed. To explain why the suction stops increasing Wrist makes a second assumption, namely, that the suction cannot increase beyond the point at which the inertia of a stream moving at velocity U could just reach atmospheric pressure against an adverse pressure gradient. This second assumption is incompatible with Wrist's first assumption that there is no pressure gradient, yet it predicts the existence of a maximum suction of $\frac{1}{2}\rho U^2$ just as the present analysis does, and it predicts that this maximum suction will occur at the same point, $x' = \frac{1}{2}$.

The reason why Wrist's calculation is in agreement with the present work at this point is that when $s(x)$ is a maximum, $ds/dx = 0$; so that Wrist's assumption that at the wire $W = xU/R$, which is true when there is no horizontal pressure gradient, is in fact true at the point where s is a maximum.

* *Pulp Pap. Mag. Can.* LV (1954), 115.

Distribution of velocity through the depth of the nip

The distribution of velocity in a vertical section can be calculated because at a fixed value of x both $_y u_x$, and the depth z of a stream line starting at distance y from 0 are functions of y. It is convenient to use the non-dimensional variables

$$z' = z/d, \quad u' = {_y u_x}/U, \tag{15}$$

where d is the depth of the point on the roll which is vertically below x. At the wire $u' = 1$ and $z' = 0$. At the roll $z' = 1$.

From (5)

$$u' = \{1 + f(x') - f(y')\}^{\frac{1}{2}} \tag{16}$$

and by integrating (7)

$$z' = \int_{y'}^{x'} \frac{f(y') \, \mathrm{d}y'}{u'} \bigg/ \int_0^{x'} \frac{f(y') \, \mathrm{d}y'}{u'}. \tag{17}$$

The distributions of velocity were calculated for two stations, station A at $x' = 1\cdot3$, and station B at $x' = 0\cdot5$. It will be seen in curve (i), fig. 3 that station B is where the maximum suction $\frac{1}{2}\rho U^2$ occurs, while at station A the suction has fallen to $0\cdot017$ ($\frac{1}{2}\rho U^2$). The values of z' and u' were calculated from (16) and (17) using the figures given by Miller in table 1 of his paper and integrating numerically. The results are shown in fig. 4. In order to display the two distributions of velocity so that they can be compared, they have been drawn with the same vertical scale. It will be noticed that at $x' = 1\cdot3$, where the suction has fallen to $0\cdot017$ of its maximum value, the distribution of velocity is very far from uniform. Midway between the wire and the roll the velocity is only about $\frac{1}{7}U$. Thus as the pressure rises to the atmospheric value the fluid tends to cling to the wire and the roll leaving the centre region with very little velocity.

Total amount of drainage

The total rate of drainage, q, is the volume discharged per second per unit length of the roller. This is

$$q = \int_0^\infty W \, \mathrm{d}x = \frac{\kappa}{\mu} \int_0^\infty s(x) \, \mathrm{d}x$$

or, using (10),

$$q = \frac{\kappa}{\mu} (\tfrac{1}{2}\rho U^2) \left(\frac{\kappa \rho R U^2}{\mu} \right) \int_0^\infty f(x') \, \mathrm{d}x'.$$

Integrating numerically the figures given in Miller's table

$$\int_0^\infty f(x') \, \mathrm{d}x' = 0\cdot59, \tag{18}$$

so that

$$q = 0\cdot295 \kappa^2 R \rho^2 U^3 / \mu^2. \tag{19}$$

This may be compared with Wrist's value

$$q = \tfrac{1}{8} D^2 R \rho^2 U^3. \tag{20}$$

Since the 'D' of (20) is identical with κ/μ in (19) the formulae are, as would be expected, of the same form. They differ in that (19) predicts rather more than twice as much drainage as (20). This difference is mainly due to the fact that Wrist's simple

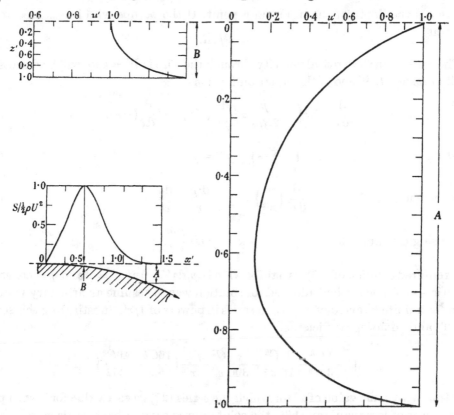

Fig. 4. Distribution of velocity through the depth of the nip at two
sections: A, where $x' = 1·3$; B, where $x' = 0·5$.

theory does not deal with the part of the fluid which drains through the wire in the
portion of the suction region where the suction is decreasing from its maximum value
to zero. In the analysis here given rather more than half the drainage occurs there.

Case (b): Complete turbulent mixing without surface friction

When there is no turbulent mixing it is possible to use Bernoulli's equation to
connect the pressure and velocity along a stream line. The equation cannot be used
when there is mixing so that the arguments which led to the prediction that the
maximum suction should be $\frac{1}{2}\rho U^2$ are no longer valid. If the mixing over any
vertical section of the nip is complete the velocity is constant through its depth so
that u is a function of x only. The momentum equation for the flow in the nip is then

$$\frac{\mathrm{d}}{\mathrm{d}x}(t\rho u)^2 = t\frac{\mathrm{d}s}{\mathrm{d}x} + \rho W U \tag{21}$$

and the equation of continuity is

$$\frac{\mathrm{d}}{\mathrm{d}x}(tu) = W, \tag{22}$$

where t is the depth of the nip at any point. If the porous wire is assumed to be straight

$$t = x^2/2R. \tag{23}$$

The case when the vertical velocity through the wire is $W = \kappa s/\mu$ will be discussed. Eliminating t, W and s, the equation for u is

$$\frac{d}{dx}(x^2 u^2) = \frac{\mu}{2R\rho\kappa} x^2 \frac{d^2}{dx^2}(x^2 u) + U \frac{d}{dx}(x^2 u). \tag{24}$$

Writing

$$x = x'\left(\frac{\kappa R\rho U}{\mu}\right), \quad u' = \frac{u}{U'}, \quad y = u'x'^2 \tag{25}$$

(24) becomes

$$\frac{d}{dx'}\left(\frac{y^2}{x'^2}\right) = \tfrac{1}{2}x'^2 \frac{d^2 y}{dx'^2} + \frac{dy}{dx'} \tag{26}$$

and, using (22) and (23),

$$s = Wu/\kappa = \tfrac{1}{2}\rho U^2 \frac{dy}{dx'}. \tag{27}$$

The required solution of (26) is that for which $dy/dx' = 0$ at $x' = 0$ and $y = $ constant when $x' \to \infty$. The method adopted for solution was to assume an arbitrary value C for y for infinite x', and develop y as a series in powers of $1/x'$. Substituting this series in (26) and equating coefficients,

$$y = C - \frac{1}{3}\frac{C^2}{x'^2} - \frac{1}{10}\frac{C^2}{x'^4} - \frac{8}{300}\frac{C^2}{x'^5} - \frac{1}{x'^6}\left(\frac{16C^2}{2520} - \frac{40C^3}{252}\right) \cdots \tag{28}$$

Starting with some value of x' for which the series (28) gave a value for y with probable error of less than, say, 1 %, the solution was carried backwards step by step using numerical methods to calculate y for decreasing values of x'. For $C = 4$, and $C = 2 \cdot 0$ it was found that values of y corresponding to $x = 0$ were positive. For $C = 1 \cdot 2$ it was found that $y = 0$ at a positive value of x'. For some value of C in the range $1 \cdot 2 < C < 2 \cdot 0$ it was expected, therefore, that the solution would pass through $x' = y = 0$. I am indebted to Dr J.C.P. Miller and the Cambridge Mathematical Laboratory for determining this value, which is $C = 1 \cdot 5072$. The values of y and also of $dy/dx' = S/\tfrac{1}{2}\rho U^2$ are given in table 1 and are shown in curve (iii) of fig. 3 for comparison with those calculated on the assumption that there is no mixing or friction (curve (i), fig. 3).

It will be seen that the effect of mixture is to increase greatly the maximum suction from $\tfrac{1}{2}\rho U^2$ to $1 \cdot 4(\tfrac{1}{2}\rho U^2)$. In experiments with full-scale and with model Fourdinier papermaking machines it has been found that the maximum suction is in fact nearly equal to $\tfrac{1}{2}\rho U^2$. This could be taken to indicate that turbulent mixing within the nip is unimportant, but if turbulent mixing does in fact take place the occurrence of $\tfrac{1}{2}\rho U^2$ instead of a maximum suction $1 \cdot 4(\tfrac{1}{2}\rho U^2)$ would indicate that friction of the fluid in the part of the nip, where it is moving faster than the wire reduces the maximum suction. The fact that the measured maximum suction is close to $\tfrac{1}{2}\rho U^2$ would then be an accidental circumstance due to the friction happening to reduce the suction to this value.

Table 1. *Solution of equation* $\dfrac{d}{dx'}\left(\dfrac{y^2}{x'^2}\right) = \tfrac{1}{2}x'^2\dfrac{d^2y}{dx'^2}+\dfrac{dy}{dx'}$

x'	y	$S/\tfrac{1}{2}\rho U^2$	x'	y	$S/\tfrac{1}{2}\rho U^2$
0	0	0	0·70	0·5620	1·4019
0·04	0·0016	0·0816	0·75	0·6317	1·3838
0·08	0·0046	0·1666	0·80	0·7000	1·3435
0·12	0·0150	0·2555	0·85	0·7658	1·2856
0·16	0·0271	0·3481	0·94	0·8283	1·2148
0·20	0·0429	0·4450	0·98	0·8871	1·1356
0·24	0·0627	0·5463	1·00	0·9418	1·0522
0·28	0·0867	0·6514	1·10	1·0386	0·8847
0·32	0·1149	0·7592	1·20	1·1192	0·7299
0·36	0·1474	0·8676	1·30	1·1853	0·5959
0·40	0·1842	0·9730	1·40	1·2391	0·4838
0·44	0·2252	1·0743	1·50	1·2827	0·3926
0·48	0·2701	1·1661	1·60	1·3182	0·3192
0·50	0·2938	1·2076	1·70	1·3471	0·2605
0·55	0·3565	1·2960	1·80	1·3707	0·2137
0·60	0·4230	1·3590	1·90	1·3901	0·1762
0·65	0·4920	1·3943	2·00	1·4062	0·1462
			∞	1·5072	0

SOME LABORATORY EXPERIMENTS

Wriot's conjecture that the suction induced where a porous surface moving with velocity U separates from a solid surface will not produce a suction greater than $\tfrac{1}{2}\rho U^2$ can be illustrated in experiments designed to avoid the inconsistency which has been shown to mar the direct application of his principle to the hydrodynamic conditions in an idealized papermaking machine. The principle adopted was to separate the part of the field of flow where the fluid is sucked through the porous band from the part where the suction is reduced by expansion under conditions represented by Bernoulli's equation. To attain this end the narrow space into which the water flowed through the band must be V-shaped, for in such a channel a uniform inflow through the band is consistent with a uniform outflow. Apparatus was constructed on this principle in the Cavendish Laboratory. Fig. 5 indicates the geometry of the situation and fig. 6 is a sketch showing how the apparatus worked. Two similar endless bands of porous material were mounted on four rollers. They were immersed in a tank with a glass bottom and driven at the same speed U in the directions indicated. Two of the rollers were spaced so that the two bands are in contact at one point and the angle 2α between them could be varied. The bands were mounted between two sheets of Perspex which covered the triangular space between them so that the motion could be two-dimensional and these sheets were placed well above the bottom of the tank and below the surface of the water contained in it so as to permit easy access of water to the space between the two straight portions of each endless band. Experience in hydrodynamic experiments led me to expect that the water discharged at the open end of the V-shaped channel with velocity V might experience a rise in pressure as high as $\tfrac{1}{2}\rho V^2$, because the surfaces bounding

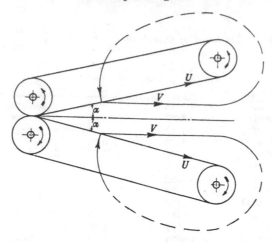

Fig. 5. Experiment with two straight diverging porous bands.

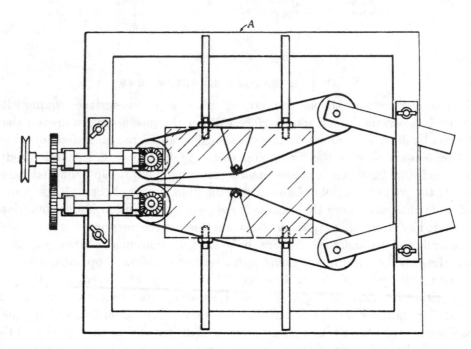

Fig. 6. Plan of apparatus. *A* marks side from which photograph of fig. 17 was taken.

the channel itself would be moving faster than the water during its expansion, so that reverse flow at the boundary would not occur. If this recovery of pressure were sufficient to force enough water through the porous bands to fill the channel with fluid moving with velocity *V*, the circulation indicated by the arrows in fig. 5 would be set up. If the pressure recovery were not great enough water would not flow into the V-shaped space through the bands. The bands, however, would be expected to produce boundary layers of outward-moving fluid and to supply this an inward-moving current would be necessary in the middle of the V-shaped space. There should therefore be a critical angle $2\alpha_{crit}$ between the bands such that if the actual angle $2\alpha < 2\alpha_{crit}$ the circulation of fig. 5 will be set up and the flow between the bands will be outwards. If $2\alpha > 2\alpha_{crit}$ the flow should be inwards.

Calculation of critical angle

Two cases will be considered

(i) The resistance is proportional to W so that the suction $s = \mu W/\kappa$. Since $W = V \sin \alpha$ the critical value occurs when

$$\mu W/\kappa = \mu V \sin \alpha/\kappa = \tfrac{1}{2}\rho V^2$$

or
$$\sin \alpha_{crit} = \tfrac{1}{2}V\kappa\rho/\mu. \tag{29}$$

(ii) The resistance is proportional to W^2 so that a resistance coefficient c equal to $2s/\rho W^2$ can be defined. In that case the critical condition is

$$\tfrac{1}{2}\rho V^2 = \tfrac{1}{2}\rho W^2/c$$

so that
$$\alpha_{crit} = 1/\sqrt{c}. \tag{30}$$

The apparatus was small, the bands being only 2 in. wide. The glass-bottom tank with the spur wheels which drove the bands in opposite directions at the same speed and the Perspex sheets which were expected to limit the flow to two dimensions are shown in fig. 6. The bands were made of the artificial rubber-like plastic Neoprene 0·1 cm. thick, and to make them porous they were pierced by holes 0·26 cm. in diameter in a regular hexagonal pattern. The distance between neighbours was 0·548 cm. The ratio β, of the area of holes to area of band, was 0·20. The resistance of perforated sheets of this form has been studied and found to be nearly proportional to W^2.* The resistance coefficient

$$c = \frac{(\text{pressure difference between two sides of sheet})}{\tfrac{1}{2}\rho W^2}$$

has been found to be of the form $c = (\beta^{-1}C^{-1} - 1)^2$, where C is a contraction coefficient which varies slowly with β and is approximately 0·7 for $\beta = 0·2$. Thus previous experiments led me to expect a value for c of about 37. Measurements made with the actual band used gave $c = 33$. If the resistance coefficient depends only on the normal component of velocity the theoretical prediction based on Wrist's principle is therefore that when the angle $2\alpha_{crit}$ between the bands is less than

$$2/\sqrt{33} = 0·348\,\text{rad} = 20° \tag{31}$$

the circulation indicated in fig. 5 should be possible.

In fact the resistance coefficient is considerably increased when the relative angle W/U between the incident stream and the band is small so that the critical angle

* B. Eckert & F. Pfluger, *Tech. Memo. natn. advis. Comm. Aeronaut., Wash.,* no. 1003 (1942); G. I. Taylor & R. M. Davies, *Rep. Memo. aeronaut. Res. Comm., Lond.,* no. 2237 (1944) (paper 45 in volume III).

would be expected to be considerably less than 20°. To find out whether the circulation had been established, coloured fluid was injected from a capillary tube at a
point within the space between the bands. The stream leaving the capillary was
photographed by means of a camera pointing downwards, the field being illuminated
by a photoflash bulb placed under a translucent screen on which the tank rested.

Fig. 7, pl. 1, is a photograph taken when $2\alpha = 12°$. The capillary with its curved
end is seen in the middle of the field and the stream of coloured fluid flowing outwards away from the apex of the wedge between the bands. Thus, in this case the
predicted circulation had been established.

In fig. 8, pl. 1, $2\alpha = 23°$. The flow is inwards towards the apex of wedge. The
direction of flow has been reversed. This is because skin friction between the
band and the water in the wedge gives rise to an outward-flowing boundary layer
at the surface and the fluid needed to replace the water carried off in this layer must
flow inwards towards the apex.

In fig. 9, pl. 1, $2\alpha = 22°$, but the coloured liquid is delivered just upstream
of the point where the bands nearly touch. The colour is carried along in the boundary layers near each band of a roll.

It has been seen in figs. 7 and 8 that at $2\alpha = 12°$ the circulation was established,
while at $2\alpha = 22°$ it was not. By changing the angle 2α slowly between these angles
it was found that the critical angle was about $2\alpha = 15°$. This is not quite in agreement with the theoretical critical angle $2\alpha = 20 \cdot 0°$ given in (31), though the
prediction that a critical angle exists is fully verified. The discrepancy is possibly
due to the fact that the relative angle of incidence of the stream passing through
the band is very oblique and at oblique angles the resistance coefficient is greater
than at normal incidence.

Further experiments

Having established experimentally that Wrist's principle leads to a prediction
which is verified experimentally, it seemed natural to modify the apparatus to
make it represent more nearly what happens in the nip between wire and roll in
a paper-machine.

In the nip the angle between wire and roll increases with distance from their line
of contact. The first modification was to insert guides shown in fig. 6 so that the
bands first expanded at an angle below the critical and later at a greater angle.
Figs. 10 and 11, pl. 1, show experiments with the apparatus in this condition.
In fig. 10 the colour was injected in the central part of the wider wedge. The photograph shows that there was hardly any motion there. Since there was no indication
of what had happened to the comparatively high speed flow from the narrow
wedge, colour was injected just up-stream of the point of contact of the two bands.
The result is shown in fig. 11. It was found that the colour moved out from the
narrow wedge into the wider space at practically the same speed it had acquired
while travelling up the narrow wedge. This can be seen in fig. 11, where the colour
appears as a continuous sheet stretching between the Neoprene bands. Such a
sheet could only be formed if the fluid which emerged from the narrow-angled

PLATE 1

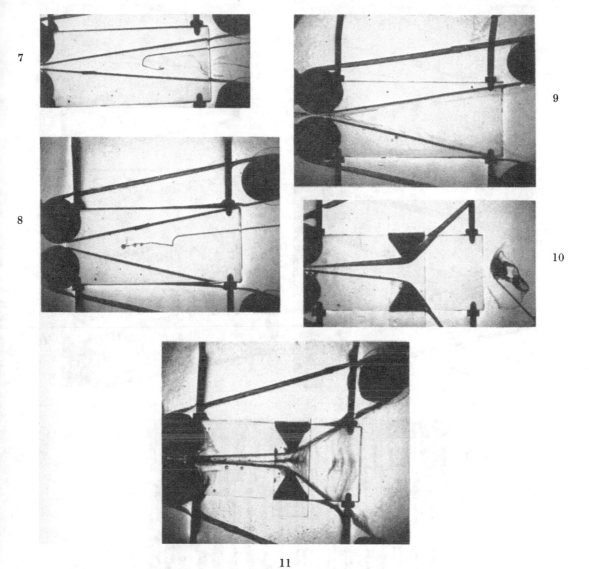

Fig. 7. Outflow when divergence of porous band is less than the critical angle.
Fig. 8. Inflow when divergence is greater than critical angle.
Fig. 9. Boundary layer when divergence is greater than critical angle.
Fig. 10. Water nearly motionless in most of the wide angle channel.
Fig. 11. Outward velocity acquired in narrow channel flows with undiminished velocity in narrow sheet into wide channel.

(*Facing p.* 308)

PLATE 2

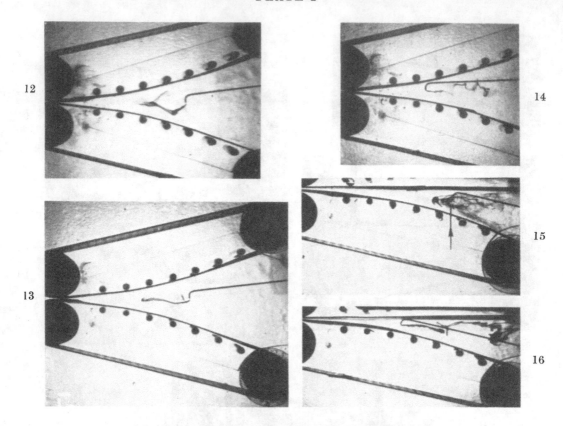

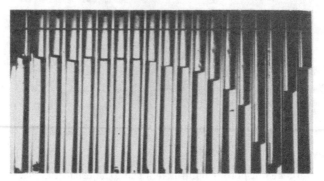

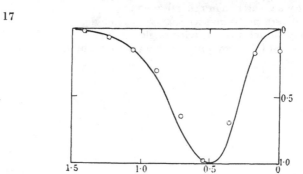

wedge as a vertical sheet contracted vertically and expanded laterally into a horizontal sheet. (If a jet were forced out from a narrow-angled channel into a much wider channel with fixed impervious walls it would either leave both walls or leave one and adhere to the other.)

To make the apparatus more nearly representative of the nip in a paper-machine a number of guides were made so that the bands could travel in a nearly circular path. These are seen in figs. 12 to 16, pl. 2, as black circles. They were in fact short lengths of glass tube rolling on brass rods inserted in the Perspex top plate. In figs. 12 and 13 the colour was inserted between the fourth and fifth guide rod. It moved slowly inwards till it reached a point just inside the third guide and then turned and flowed in a turbulent stream in the reverse direction. In fig. 14 the colour was inserted to the left of the point where it turned back in the two previous cases. It moves outwards in a fairly steady stream until it reaches the third guide where it seems to become unsteady.

In figs. 15 and 16 a straight unperforated Neoprene band has been substituted for one of the curved perforated ones. In fig. 15 the colour was injected beyond the fifth guide rod at the point indicated by an arrow, it flows first towards the point of contact of the bands and then turns at a point between the fourth and fifth guides leaving clear water ahead. In fig. 16 the colour is injected just beyond the third guide rod. There is an outward current which begins to flow nearly parallel to the flat surface and then becomes variable in direction about halfway between the fourth and fifth guide rod.

Distribution of pressure

To measure the distribution of pressure in the nip the upper Perspex sheet used in the apparatus with which figs. 10 to 16 were taken was replaced by a brass plate down the centre of which pressure holes 0·05 cm. bore were spaced at intervals 0·5 cm. apart. Above this plate were mounted a number of small glass tubes each 0·3 cm. bore. Each of these was connected to a pressure hole at the bottom and they led through an airtight seal into a common air reservoir at the top. The pressure in this reservoir could be controlled so as to bring the level of the water in the tubes to a convenient height. The apparatus was set going, the speed of the bands timed by a stop-watch in order to measure U, and the row of tubes photographed using a flash-bulb. A record is shown in the upper half of fig. 17, pl. 2. The direction of motion of the bands was of course the same as when figs. 10 to 16 were taken,

---

CAPTIONS FOR PLATE 2

Figs. 12, 13, 14. Outward flow in narrowest part of the nip and inward flow farther out.

Fig. 15. Marking fluid released at position marked by arrow.

Fig. 16. Marking fluid moves outwards but becomes unsteady near 5th roller.

Fig. 17. Above: distribution of suction revealed by multitube manometer. Below: calculated distribution with experimental points. (Scales chosen arbitrarily so that experimental curve has at least one point, that of maximum suction, on the theoretical curve.)

but the camera was placed on the side of the tank (indicated by A in fig. 6) from which they appeared to move from right to left. The apparent discrepancy between fig. 17 and figs. 10 to 16 is due simply to the fact that the former is seen in elevation and the latter as plans. A thread laid horizontally over the tubes is seen near the top of the photograph.

Quantitative comparison between the theoretical and observed distributions of pressure were attempted, but it was found that though the maximum suction is in fact comparable with, though less than, $\frac{1}{2}\rho U^2$ the distance from the point of contact of the bands to the point where the suction was a maximum was considerably less than the amount calculated, assuming that the resistance of the band is the same as that measured when the flow is normal. It is true that the coefficient of resistance of a perforated sheet increases when the relative angle of incidence of the water approaching the band becomes oblique, and this might account for the discrepancy in position of maximum suction, but other causes might also contribute so that quantitative comparisons with results obtained using the present apparatus are of little value. A qualitative comparison can be made by choosing values for V/U and c so that maximum suction agrees with theory both in magnitude and position; this amounts to choosing the horizontal and vertical scales to get the best fit. Since the resistance of a perforated sheet is nearly proportional to W^2 the appropriate curve for comparison is (ii) of fig. 3. This is shown (inverted for comparison with the manometer readings) in the lower section of fig. 17. Manometer readings measured on a photograph similar to the one shown in the upper half of fig. 17 and with the horizontal and vertical scales adjusted in the manner just described are also marked. It will be seen that the distribution of suction is similar to what was expected on theoretical grounds though it was not possible to make valid quantitative comparisons.

28

THE PENETRATION OF A FLUID INTO A POROUS MEDIUM OR HELE-SHAW CELL CONTAINING A MORE VISCOUS LIQUID†

REPRINTED FROM

Proceedings of the Royal Society, A, vol. CCXLV (1958), pp. 312–29

When a viscous fluid filling the voids in a porous medium is driven forwards by the pressure of another driving fluid, the interface between them is liable to be unstable if the driving fluid is the less viscous of the two. This condition occurs in oil fields. To describe the normal modes of small disturbances from a plane interface and their rate of growth, it is necessary to know, or to assume one knows, the conditions which must be satisfied at the interface. The simplest assumption, that the fluids remain completely separated along a definite interface, leads to formulae which are analogous to known expressions developed by scientists working in the oil industry, and analogous to expressions representing the instability of accelerated interfaces between fluids of different densities. In the latter case the instability develops into round-ended fingers of less dense fluid penetrating into the more dense one. Experiments in which a viscous fluid confined between closely spaced parallel sheets of glass, a Hele-Shaw cell, is driven out by a less viscous one reveal a similar state. The motion in a Hele-Shaw cell is mathematically analogous to two-dimensional flow in a porous medium.

Analysis which assumes continuity of pressure through the interface shows that a flow is possible in which equally spaced fingers advance steadily. The ratio $\lambda =$ (width of finger)/(spacing of fingers) appears as the parameter in a singly infinite set of such motions, all of which appear equally possible. Experiments in which various fluids were forced into a narrow Hele-Shaw cell showed that single fingers can be produced, and that unless the flow is very slow $\lambda =$ (width of finger)/(width of channel) is close to $\frac{1}{2}$, so that behind the tips of the advancing fingers the widths of the two columns of fluid are equal. When $\lambda = \frac{1}{2}$ the calculated form of the fingers is very close to that which is registered photographically in the Hele-Shaw cell, but at very slow speeds where the measured value of λ increased from $\frac{1}{2}$ to the limit $1 \cdot 0$ as the speed decreased to zero, there were considerable differences. Assuming that these might be due to surface tension, experiments were made in which a fluid of small viscosity, air or water, displaced a much more viscous oil. It is to be expected in that case that λ would be a function of $\mu U/T$ only, where μ is the viscosity, U the speed of advance and T the interfacial tension. This was verified using air as the less viscous fluid penetrating two oils of viscosities $0 \cdot 30$ and $4 \cdot 5$ poises.

1. THE STABILITY OF THE INTERFACE BETWEEN TWO FLUIDS IN A POROUS MEDIUM

It has been pointed out‡ and verified experimentally§ that when two superposed fluids of different densities and negligible viscosities are accelerated in a direction

---

† With P. G. SAFFMAN.
‡ G. I. Taylor, *Proc. Roy. Soc.* A, CCI (1950), 192 (paper 56 in volume III).
§ D. J. Lewis, *Proc. Roy. Soc.* A, CCII (1950), 81.

perpendicular to their interface, this surface is stable or unstable for small deviations according as the acceleration is directed from the more dense to the less dense fluid or vice versa.

An analogous instability can occur when two superposed viscous fluids are forced by gravity and an imposed pressure gradient through a porous medium. If the steady state is one of uniform motion with velocity V vertically upwards and the interface between the two fluids is horizontal, then it can be shown that the interface is stable for small deviations from the steady state if

$$\left(\frac{\mu_2}{k_2} - \frac{\mu_1}{k_1}\right) V + (\rho_2 - \rho_1)g > 0, \tag{1}$$

and unstable if
$$\left(\frac{\mu_2}{k_2} - \frac{\mu_1}{k_1}\right) V + (\rho_2 - \rho_1)g < 0, \tag{2}$$

where the suffix 1 refers to the upper fluid and the suffix 2 to the lower. The motion of the fluids through the medium is here supposed to be governed by Darcy's law which asserts that the velocity of the fluid is given by

$$\mathbf{u} = -\frac{k}{\mu}\,\mathrm{grad}\,(p + \rho gx) = \mathrm{grad}\,\phi, \quad \text{say}, \tag{3}$$

where $\mathbf{u}$ denotes the velocity, μ the viscosity and ρ the density, k is the permeability of the medium to the fluid, g the acceleration due to gravity, x the vertical height above some horizontal plane, and ϕ is called the velocity potential.

To describe a disturbance of the surface of separation, take rectangular co-ordinates (x, y, z), the instantaneous position of the undisturbed interface coinciding with the plane $x = 0$. Suppose the interface is deformed slightly into a wave-like corrugation of wavelength $2\pi/n$ described by

$$x = a\,\mathrm{e}^{iny + \sigma t}. \tag{4}$$

Assuming that the fluids are incompressible and that the medium is of uniform porosity, the equation of continuity satisfied by the velocity field is $\mathrm{div}\,\mathbf{u} = 0$ and the velocity potential therefore satisfies Laplace's equation $\nabla^2\phi = 0$.

It is now necessary to make some assumption about the nature of the motion in the vicinity of the interface because, in fact, a sharp interface between the two fluids does not exist but there is, rather, an ill-defined transition region in which the two fluids intermingle. This region is often very thick and we shall assume that the fluids do not interpenetrate to any marked extent and that the width of the transition zone is small compared with the length scale of the motion. It is then reasonable to assume for the purpose of mathematical analysis that the two fluids are separated by a sharp interface, across which the normal component of velocity and the pressure are continuous (surface tension or any other similar effect is neglected).†

† The assumption that one fluid completely expels the other may sometimes be relaxed and the analysis can be modified to treat cases in which a proportion of one fluid is left behind to be surrounded by the oncoming fluid. The continuity of normal velocity across the interface no longer holds, but provided that the mixture of fluids can be regarded as homogeneous and the

It follows from the continuity of normal velocity that the velocity potentials in the upper and lower fluid satisfy on $x = 0$, to the first order in the deviation

$$\frac{\partial \phi_1}{\partial x} = \frac{\partial \phi_2}{\partial x} = V + a\sigma e^{iny+\sigma t}. \tag{5}$$

Hence, $\phi_1 = Vx - (a\sigma/n) e^{iny-nx+\sigma t}$

and $\phi_2 = Vx + (a\sigma/n) e^{iny+nx+\sigma t}$,

these being the appropriate solutions of $\nabla^2\phi = 0$ which satisfy (5) and for which the disturbance vanishes at infinity.

The pressure, p_1, in the upper fluid is $-(\mu_1/k_1)\phi_1 - \rho_1 gx$ and that, p_2, in the lower fluid is $-(\mu_2/k_2)\phi_2 - \rho_2 gx$. Equating the values of p_1 and p_2 on the interface (4), we find that, to the first order in the deviation, σ must satisfy

$$\frac{\sigma}{n}\left(\frac{\mu_1}{k_1} + \frac{\mu_2}{k_2}\right) = (\rho_1 - \rho_2)g + \left(\frac{\mu_1}{k_1} - \frac{\mu_2}{k_2}\right)V. \tag{6}$$

If the right-hand side of (6) is positive, then σ is positive and the amplitude of the deviation increases at an exponential rate and the interface is then unstable to small disturbances. If the right-hand side is negative, the deviation is damped at an exponential rate and the motion is stable to small disturbances. Thus, the results (1) and (2) are true for all wavelengths and consequently for all types of small disturbance. They can also be put as follows. When two superposed fluids of different viscosities are forced through a porous medium in a direction perpendicular to their interface, this surface is stable or unstable to small deviations according as the direction of motion is directed from the more viscous to the less viscous fluid or vice versa, whatever the relative densities of the fluids, provided that the velocity is sufficiently large.

It appears that this result is not essentially new and that mining engineers and geologists have long been aware of it. In certain types of oil wells the oil taken out of the ground is replaced by encroaching water which comes from the expansion of a large water accumulation or seepage from the surface. Since oil is lighter than water but somewhat more viscous, it follows from (2) that when the velocity of extraction is too large the interface will become unstable. It is indeed observed in practice that when the velocity of extraction is too large, long tongues or cones of water penetrate the oil and it comes out of the well mixed with water. This phenomenon is known in the literature as 'water tonguing or coning'. However, earlier writers† do not appear to have considered explicitly the stability of the interface, but rather the conditions necessary in certain cases for a steady interface to exist.

proportion left behind the interface is constant, then it is found that the interface moves as if it separated two fluids whose viscosities and densities differ from those of the original two fluids but which completely expel one another. The values of these viscosities and densities depend upon the physical properties of the mixture and the proportion left behind. In this connection, see also the Appendix.

† See, for example, D. N. Dietz, *Proc. Acad. Sci. Amst.* B, LVI (1953), 83, and R. E. Kidder, *J. appl. Phys.* XXVII (1956), 867.

2. An analogue for two-dimensional flow
in a porous medium

The motion of fluid in a porous medium according to Darcy's law can be derived from a potential $\phi = -(k/\mu)(p + \rho gx)$. Motion in two dimensions can therefore be studied experimentally by means of an analogue devised by Hele-Shaw.[†] This makes use of the result that the motion of a viscous fluid, between two fixed parallel plates which are sufficiently close together, is such that the components of the mean velocity across the stratum are

$$u = -\frac{b^2}{12\mu}\left(\frac{\partial p}{\partial x} + \rho g\right), \quad v = -\frac{b^2}{12\mu}\frac{\partial p}{\partial y}, \tag{7}$$

where b denotes the distance between the plates.[‡] The plates are here taken as vertical, the x-axis is vertically upwards, the y-axis parallel to the plates, and the z-axis perpendicular to the plates; u and v are the components of mean velocity in the x- and y-directions, respectively.

These are the equations satisfied by the velocity in a porous media of permeability $b^2/12$ and there is thus a direct analogy between two-dimensional flow in a porous medium and the flow between parallel plates, the velocity in the former case corresponding to the mean velocity in the latter. In this way, for example, the streamlines for the flow around bodies of arbitrary shape can be determined experimentally.

The analogue can also be used to reproduce experimentally the (two-dimensional) motion of the interface between two fluids in a porous medium. Consider the motion between parallel plates of two immiscible fluids of viscosities μ_1, μ_2 and densities ρ_1, ρ_2, respectively, and suppose the direction of motion is away from fluid 2 towards fluid 1. Now the fluid 1 is not necessarily completely expelled or replaced by fluid 2, a film of fluid 1 may wet the plates and adhere to them, while a tongue of fluid 2 advances along the middle of the gap between the plates. The thickness of the tongue may be taken as a fraction t, say, of the gap between the plates and the analysis which follows is applicable and the analogue valid provided that t is constant.

In experiments to be described later, one of the fluids (fluid 2) was air and in order to find out what proportion of fluid 1 was left behind after the interface (or the tip of the meniscus) has passed, subsidiary experiments were made in which a measured volume of air was blown centrally into the narrow space (0·09 cm.) between the flat base of a metal vessel containing oil or glycerine and a circular flat disc. The rather irregular outline of the bubble so formed was photographed and the area contained within it determined. In several such trials it was found that the volume of the bubble divided by the area gave a thickness which was always less than 0·09 cm. but was never more than $1\frac{1}{2}$ % less. The rate at which the air was blown into the apparatus had to be kept low for otherwise the instability which gives rise to the irregular outlines of the interface made it difficult to measure the

† *Trans. Instn nav. Archit.* XL (1898), 21.
‡ H. Lamb, *Hydrodynamics*, Cambridge University Press, 1932.

area. The velocity of the interface was, however, of the same order as those occurring in the experiments to be described later. In discussing those experiments, it is therefore legitimate to assume that $t = 1$ and the outlines which were observed and photographed represent interfaces completely separating the two fluids.

The thickness of the film of liquid left behind when a bubble moves in a capillary tube has been investigated[†] and shown to depend on the non-dimensional parameter $\mu U/T$, where μ is the viscosity, U is the velocity of the bubble and T the surface tension. The value of the parameter in our experiments was such that only a small fraction of the fluid would be expected to remain behind.

The mean velocity across the stratum of fluid 1 ahead of the interface is given by (assuming b to be sufficiently small)

$$\mathbf{u}_1 = -(b^2/12\mu_1)\,\mathrm{grad}\,(p + \rho_1 gx) = \mathrm{grad}\,\phi_1, \quad \text{say}, \tag{8}$$

taking the plates vertical and the x-axis vertically upwards. Taking $t = 1$ and fluid 1 to be completely expelled, the mean velocity across the stratum of fluid 2 is given by

$$\mathbf{u}_2 = -(b^2/12\mu_2)\,\mathrm{grad}\,(p + \rho_2 gx) = \mathrm{grad}\,\phi_2, \quad \text{say}. \tag{9}$$

Now when b is small, the width of the projection of the meniscus on to the plates is small, and expressions (8) and (9) can be supposed to hold (for the purposes of the analysis) right up to the interface which may be regarded as a sharp line. It follows from continuity that the components of $\mathbf{u}_1$ and $\mathbf{u}_2$ normal to the interface are continuous across the interface; if surface tension effects are negligible the pressure is constant across the interface (see, further, §5 below), and the motion of the two fluids will then reproduce the two-dimensional motion of the interface between two fluids of viscosities μ_1 and μ_2 in a porous medium of permeability $b^2/12$.

The analogue is still valid when $t \neq 1$, provided that t is constant. For this case, the interface can be identified with the tip of the meniscus between the two fluids. The modifications that are required are given in the appendix.

The considerations of §1 apply to the motion in the Hele-Shaw cell and it will be noticed that when the cell is vertical there is a critical velocity for the interface, which separates unstable from stable condition. This is given by (1) and (2) with k_1 and k_2 replaced by $b^2/12$. When the Hele-Shaw apparatus is set horizontal the analysis applies with g put equal to zero, and it follows that the interface is always unstable when the less viscous fluid is driving the more viscous.

3. Effect of surface tension on stability in the Hele-Shaw cell

The effect of surface tension on the stability of the interface may depend on a variety of physical conditions. The simplest assumption is to take the pressure drop through the interface as $T(2/b + 1/R)$, where R is the radius of curvature of the projection on the planes bounding the cell of the tip of the meniscus. In discussing

[†] F. Fairbrother & A. E. Stubbs, *J. chem. Soc.* I (1935), 527.

the stability of a plane interface in the Hele-Shaw apparatus, this may be taken as $T(2/b + \mathrm{d}^2x/\mathrm{d}y^2)$, where x is given by (4), and it is easily seen that (6) is thereby altered to

$$\frac{12}{b^2}\sigma(\mu_1 + \mu_2) = \frac{2\pi}{l}\left\{\frac{12V}{b^2}(\mu_1 - \mu_2) + g(\rho_1 - \rho_2)\right\} - \frac{8\pi^3 T}{l^3}, \qquad (10)$$

where σ is the amplification factor of disturbances of wavelength $l = 2\pi/n$.

It will be seen that this effect of surface tension is to limit the range of disturbances which are unstable to those of wavelength greater than

$$l_{\mathrm{crit}} = 2\pi T^{\frac{1}{2}}b\{12V(\mu_1 - \mu_2) + b^2 g(\rho_1 - \rho_2)\}^{-\frac{1}{2}}. \qquad (11)$$

That surface tension should have this effect was pointed out to us by Dr Chuoke in a similar connection. The amplification factor is a maximum for disturbances of wavelength $\sqrt{3}l_{\mathrm{crit}}$.

4. Experiments using the Hele-Shaw cell

It was shown experimentally by Lewis† that the unstable accelerating interface between two fluids of different density develops in such a way that long fingers of the less dense fluid penetrate into the more dense one, and beyond the level to which these fingers have penetrated into the more dense fluid the acceleration has the same value it would have if the interface had remained plane. Analogous results are obtained when the Hele-Shaw apparatus contains two immiscible fluids.

In the first apparatus, two pieces of commercially flat plate glass were separated by strips of rubber 0·09 cm. thick laid along their long edges. The channel 0·09 × 12 × 38 cm. thus formed was connected at its two ends with vessels containing the two fluids. The pressure gradient along the channel was produced by applying air pressure or suction to the airspace above the fluid in one of the end vessels and the pressure at the other was maintained at that of the atmosphere. The apparatus which is shown in the sketch (fig. 1) could be used either vertically or horizontally and the meniscus of the interface could be photographed as a sharp line. The three-way cock shown in fig. 1 on the left side of the top of the front view made it possible to change the pressure in the air chamber rapidly from pressure to suction. The first experiments were made with the apparatus vertical and with glycerine as the more viscous and air as the less viscous fluid. In this case the critical velocity is downwards and the unstable disturbances are to be expected when the air is above if the velocity is downwards and greater than this. If the glycerine lies above the air the flow becomes stable when the downwards velocity is greater than the critical.

In the experiment shown in fig. 2, pl. 1, the fluid was sucked up to near the top of the Hele-Shaw channel and allowed to fall with a velocity less than the critical and then maintained at rest for a short time. The fluid left behind on the glass then gathered itself together to form vertical streaks which produced on the interface a small variation in level. The air pressure was then turned on; this is indicated by

† *Proc. Roy. Soc.* A, ccii (1950), 81.

the mercury manometer seen to the left of the channel. The photograph was taken by a flash bulb after the glycerine had been forced downwards through a few centimetres. It will be seen that the instability has already manifested itself. In this experiment V was 0·1 cm./sec. and the critical wavelength given by (11) was 1·2 cm.; the average wavelength of the disturbance shown in fig. 2 was 2·2 cm., so that instability was to be expected.

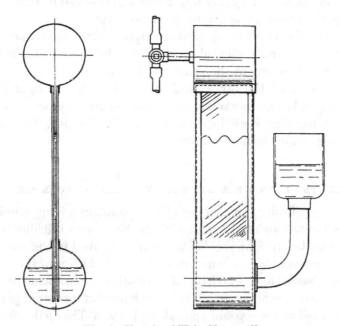

Fig. 1. Sketch of Hele-Shaw cell.

Fig. 3, pl. 1, shows a later stage of the instability (not in the same experiment as that of fig. 2). In this case the pressure was turned on when the nearly straight interface was at the level of the top of the mark seen projecting from the left-hand side of the channel. A characteristic feature of the later stages of the growth of 'instability' into 'fingers', is shown in fig. 3, namely the tendency of the fingers to space themselves so that the width of the air fingers and the columns of fluid between them are of approximately the same breadth. The development of these fingers is very similar to those recorded by Lewis for the later states of instability of an accelerated interface, but it differs from them in that the air fingers in Lewis's experiment were separated by narrow columns of fluid.

The reason for the narrow columns in Lewis's experiment was that the water left behind after the passage of the front of 'fingers' is in a field of uniform pressure, and therefore moves uniformly at the speed at which it was passed by the front, while the fluid ahead of this front is accelerating away from it. There is thus a longitudinal rate of strain in the columns so that they must continually decrease in thickness as the front leaves them.

Fig. 4, pl. 1, shows another characteristic feature, the inhibiting effect on the growth of its neighbours that happens when the end of one of the fingers gets ahead of them. On the right-hand side of this photograph can be seen three fingers which started to grow at the same time. The middle finger, however, was slightly larger than its neighbours and at the stage shown in fig. 4 has almost completely inhibited their growth and as it passed them it spread laterally. The inhibiting effect on the growth of neighbours by any finger which gets ahead of them also occurred in Lewis's experiments and was due to the same cause.

In attempting to form a mathematical description of the mechanics of the formation of 'fingers' we were naturally led to consider an infinite set of equal and equally spaced fingers all advancing at the same speed. Since each finger is then identical mathematically with all the others and the fluid on the straight lines halfway between neighbours has no transverse component of velocity, we considered only a single finger propagating itself in a channel of fixed width. The details of the analysis will be given in the next section.

5. Penetration of a single 'finger' into a channel

In this section we consider the analysis of the motion of a long bubble or 'finger' of fluid moving through an infinite channel in the Hele-Shaw cell filled with a viscous fluid and bounded by straight parallel walls. As described in the previous section, the motion of these bubbles is connected with the mechanics of the formation and propagation of 'fingers' and, by virtue of the analogue, also bears on the question of how long it would take for a vertical tube or channel of saturated porous material to drain when closed at the top and open at the bottom. The analogous problem for inviscid liquids of the motion of bubbles through liquids in tubes and channels has been studied experimentally and theoretically in some detail.† The present problem is of particular interest, since it is possible to obtain exact solutions of the equations of motion in closed form and compare them with experiment.

A bubble of fluid of viscosity μ_2 and density ρ_2 is supposed to be moving steadily through a vertical channel in the Hele-Shaw cell filled with fluid of viscosity μ_1 and density ρ_1. In fig. 5, BC and FE are the walls of the channel, AOG is the surface of the bubble or interface between the two fluids, the x-axis is taken vertically

† See, for example, R. M. Davies & G. I. Taylor, *Proc. Roy. Soc.* A, cc (1950), 375 (paper 52 in volume III), and P. R. Garabedian, *Proc. Roy. Soc.* A, ccxli (1957), 423.

CAPTIONS FOR PLATE 1

Fig. 2. Interface between air and glycerine at an early stage of the instability.

Fig. 3. Development of instability.

Fig. 4. Inhibiting effect of a finger which gets ahead of its neighbour.

Fig. 8. An air finger advancing into glycerine.

Fig. 9. Enlarged forward portion of the finger shown in fig. 8. ○, points calculated from (17) using $\lambda = \frac{1}{2}$.

PLATE 1

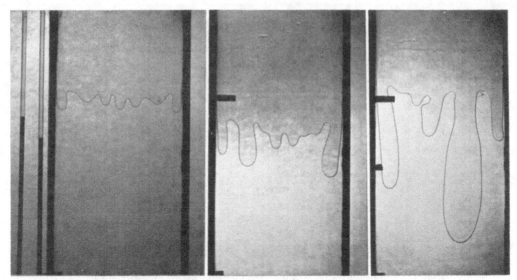

Fig. 2.   Fig. 3.   Fig. 4.

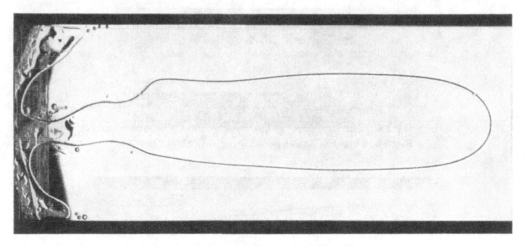

Fig. 8.

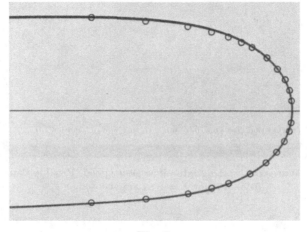

Fig. 9.

PLATE 2

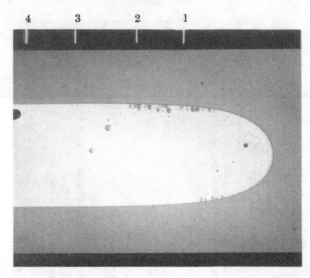

Fig. 11. Finger of oil ($\mu = 4 \cdot 5$ P) penetrating glycerine ($\mu = 9$ P).

Fig. 12. Finger of water penetrating oil (Shell Diala).

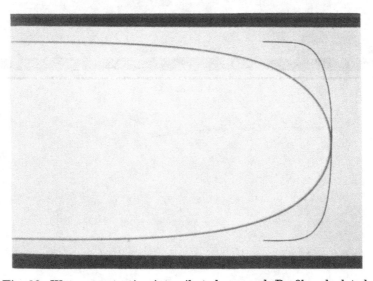

Fig. 13. Water penetrating into oil at slow speed. Profile calculated
from (17) for $\lambda = 0 \cdot 87$ superposed.

upwards along the centre of the channel, the y-axis horizontal and perpendicular to the walls, and the origin is at the nose of the bubble which is supposed to be of infinite extent and symmetrical about the centre of the channel. The velocity of the bubble is denoted by U and the velocity of the fluid at infinity in front of the bubble by V. The walls of the channel are taken as $y = \pm 1$, and the width of the bubble at infinity as 2λ, where λ is a parameter which is for the present unspecified except that it lies between 0 and 1. Suffixes 1 and 2 refer to quantities outside and inside the bubble, respectively.

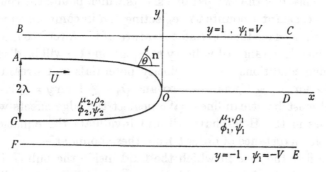

Fig. 5. Finger moving into a channel.

The mean velocity across the stratum is given by equations (8) and (9) and the equation of continuity takes the form

$$\frac{\partial u}{\partial x} + \frac{\partial v}{\partial y} = 0,$$

where u and v are the components of mean velocity parallel to the axes. A stream function ψ can be defined by

$$u = \frac{\partial \phi}{\partial x} = \frac{\partial \psi}{\partial y}, \quad v = \frac{\partial \phi}{\partial y} = -\frac{\partial \psi}{\partial x},$$

and it follows from the Cauchy–Riemann equations that $w = \phi + i\psi$ is an analytic function of $z = x + iy$.

It will be assumed that the experiment described in §2 is valid, i.e. one fluid completely expels the other, and that the meniscus separating the two fluids is a sharp line, so that the components of mean velocity normal to the interface are continuous, i.e.

$$\frac{\partial \phi_1}{\partial n} = \frac{\partial \phi_2}{\partial n} = U \cos \theta, \tag{12}$$

where θ is the angle between the x-axis and the outward normal $\mathbf{n}$ to the interface. Now $\partial \phi / \partial n = \partial \psi / \partial s$ where $\partial / \partial s$ denotes differentiation along the interface, and $\cos \theta = \partial y / \partial s$, from which it follows that

$$\psi_1 = \psi_2 = Uy \tag{13}$$

on the surface of the bubble.

If the pressure change due to surface tension at the interface is ignored, the equation for continuity of pressure is

$$(12/b^2)(\mu_1\phi_1 - \mu_2\phi_2) = g(\rho_2 - \rho_1)x. \tag{14}$$

The same equation is valid if the change in pressure on passing through the interface is constant, since an arbitrary constant may be added to the velocity potential. If the fluid wets the plane surfaces it might be expected that under static conditions the pressure drop would be $T(2/b + 1/R)$, where R is the radius of curvature of the meniscus on the plane of the two parallel sheets which bound the cell. The assumption that this is constant amounts to neglecting $1/R$ in comparison with $2/b$ and to assuming that the surface tension and curvature of the interface are identical with their static values. The range of validity of equation (14) will be discussed later.

The remaining conditions on the velocity potentials and stream functions are $\phi_1 \sim Vx$ as $x \to +\infty$, $\phi_2 \sim Ux$ as $x \to -\infty$, and $\psi_1 = \pm V$ on $y = \pm 1$, since the walls of the channel must be streamlines. (We neglect the edge effects which occur at solid boundaries in the Hele-Shaw cell and invalidate the equations of motion (8) and (9) within a distance of order b from these boundaries.)

We examine first the case in which the fluid inside the bubble is of negligible viscosity and density, and suppose also that gravity forces are negligible, i.e. that the imposed pressure gradient which moves the fluid is large compared with that due to gravity. This case corresponds to the experimental arrangement described in §7 below when a bubble of air is blown through glycerine.

Equation (14) now reduces to $\phi_1 = 0$ on the bubble surface. Further, the flow becomes uniform a long way behind the nose of the bubble, i.e. as $x \to -\infty$, and since ϕ_1 is zero on the interface, $\phi_1 \to -\infty$ and the fluid is at rest at $x = -\infty$. Hence, the stream function has the same value at A as at B (and also at G as at F) and it follows from (13) that

$$V = \lambda U. \tag{15}$$

Equation (15) gives the velocity of the bubble in terms of its width and the velocity at infinity; the latter velocity will be determined by the external means by which the motion is generated.

The shape of the interface is as yet unknown and to solve this free boundary problem, we transform into the ϕ, ψ plane and consider $x + iy$ as an analytic function of $\phi + i\psi$ (for brevity we drop the suffix 1).† In fig. 6, corresponding points in the physical and potential planes are marked with the same letter; the exterior of the bubble transforms into the semi-infinite strip

$$\phi > 0, \quad -V < \psi < V,$$

the surface of the bubble to the ϕ-axis between $\psi = \pm V$, and the walls of the channel to $\psi = \pm V$.

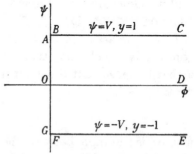

Fig. 6. The potential plane for motion in a channel.

† We are grateful to Mr F. Ursell for first suggesting this transformation to us.

Now y is a harmonic function of ϕ and ψ which has the values -1 on FE, 0 on DO, $+1$ on BC, $y = \psi/U$ on GA (the surface of the bubble), and $y \to \psi/V$ as $\phi \to +\infty$. Take, therefore,

$$y = \frac{\psi}{V} + \sum_1^\infty A_n \sin \frac{n\pi\psi}{V} \exp\left(-\frac{n\pi\phi}{V}\right).$$

This is a harmonic function satisfying all the boundary conditions, provided

$$\frac{\psi}{U} = \frac{\psi}{V} + \sum_1^\infty A_n \sin \frac{n\pi\psi}{V}, \quad \text{for} \quad -V < \psi < V.$$

Calculating the coefficients by the usual method of Fourier series, we find that

$$A_n = -\frac{2}{\pi}\left(1 - \frac{V}{U}\right) = -\frac{2}{\pi}(1 - \lambda),$$

and therefore

$$z = \frac{w}{V} + \frac{2}{\pi}(1 - \lambda)\ln\frac{1 + \exp(-\pi w/V)}{2}. \tag{16}$$

This equation determines the complex potential implicitly as a function of z. The pressure then follows from the relation

$$p = -\frac{b^2}{12\mu}\phi + \text{const.}$$

The parametric equation of the interface is obtained by putting $\phi = 0$ in equation (16) and it follows that the bubble surface is

$$x = \frac{1 - \lambda}{\pi}\ln\frac{1 + \cos(\pi y/\lambda)}{2}. \tag{17}$$

This completes the solution for the case of a bubble of fluid of negligible viscosity when effects due to gravity forces and departures from equation (14) are unimportant.

Before discussing this further, we give briefly the solution for the more general case in which the viscosity of the fluid in the bubble is not neglected and pressure gradients due to gravity are taken into account. The equations of motion for the fluid inside the bubble can be satisfied by taking

$$\phi_2 + i\psi_2 = U(x + iy). \tag{18}$$

The boundary condition for ϕ_1 then becomes $\phi_1 = -U^*x$ on the bubble, where

$$\frac{12\mu_1 U^*}{b^2} = g(\rho_1 - \rho_2) - \frac{12\mu_2 U}{b^2}. \tag{19}$$

Define now

$$W = \Phi + i\Psi = w_1 + U^*z. \tag{20}$$

On the interface, $\Phi = 0$, $\Psi = (U + U^*)y$; further $\Psi = \pm(V + U^*)$ on $y = \pm 1$ and $W \to (V + U^*)x$ as $x \to +\infty$. The problem is thus reduced to the simpler case considered above and the complex potential for the motion outside the bubble is given by (15) and (16) with V replaced by $V + U^*$, U by $U + U^*$, and w by W. The bubble surface corresponds to $\Phi = 0$ and is therefore given by (17) also. It is worth noting that if the asymptotic width at infinity is fixed, then the shape of the bubble is independent of the physical properties of the fluids. The width 2λ of the bubble is given in terms of U (the velocity of the bubble), V (the velocity at infinity), and the physical properties of the fluids by

$$\lambda = \frac{V + U^*}{U + U^*}.$$

The maximum velocity of propagation again corresponds to $\lambda = 0$ and is

$$U_{max} = \frac{\mu_1}{\mu_2} V + \frac{b^2 g}{12\mu_2} (\rho_1 - \rho_2).$$

6. NON-UNIQUENESS OF THE SOLUTION

There is nothing in the preceding mathematical analysis to determine the width of the bubble and the value of λ, i.e. the fraction of the channel occupied by the bubble after the nose has passed. In other words, if only the velocity at infinity ahead of the bubble is specified (this is equivalent to specifying the pressures driving the less viscous finger into the more viscous fluid), then the free-boundary problem does not have a unique solution and there are an infinite number of possible steady shapes, each with a different velocity of propagation. These shapes are the members of the family of curves given by (17) for values of λ between 0 and 1. The velocity of propagation of the bubble corresponding to each of these shapes is related to the velocity at infinity by (15). The two extreme members are given by $\lambda = 1$ when the interface extends in a straight line across the whole width of the channel, and $\lambda = 0$ when the bubble has zero width and propagates with infinite velocity. The calculated shapes for $\lambda = 0.2$, 0.5 and 0.8 are shown in fig. 7.

It was pointed out recently by Garabedian† that the analogous free-boundary problem for the propagation of an air bubble through a vertical tube or channel containing an inviscid liquid also does not possess a unique solution, and that there are many possible 'equilibrium' shapes which a bubble can have, each again corresponding to a different velocity of rise. (Garabedian did not demonstrate explicitly the multiplicity of solutions and it is of some interest that this can be done for the problem we are considering.) He gave arguments based on the hypothesis of a maximum rate of loss of potential energy to show that in practice only one of these possible shapes would occur, the one occurring being that with the maximum velocity of propagation.

These arguments do not apply to motion in a porous medium or a Hele-Shaw cell, since the motion is then dissipative and in any case the maximum velocity of propagation according to the analysis is infinite, but we should still expect on the grounds of physical experience that only one of all the possible shapes would occur

† *Proc. Roy. Soc.* A, ccxli (1957), 423.

in practice. It should be noted that the hypothesis of a maximum or minimum rate of dissipation of energy by viscosity does not determine a unique value for λ, since this rate for a channel of finite, but very large length is proportional to the product of the velocity at infinity and the pressure difference between the ends of the channel, and it follows from (16) that the value of this product becomes independent of λ as the length of the channel tends to infinity. (The terms involving λ, which tend to zero as the length increases, are monotonic in λ and a hypothesis of maximum or minimum dissipation would in any case give $\lambda = 1$ or $\lambda = 0$.)

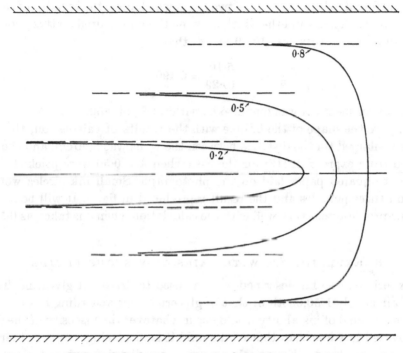

Fig. 7. Calculated profiles for $\lambda = 0\cdot2$, $0\cdot5$ and $0\cdot8$.

Our experiments with the Hele-Shaw cell indicate that, as the speed of flow for any given fluid increases, λ rapidly decreases to $\lambda = \frac{1}{2}$ and remains close to this value over a large range of speeds, till at high speeds of flow the tongue or finger of the advancing fluid itself breaks down and divides into smaller fingers.

7. EXPERIMENTS IN CHANNELS

The apparatus with which the photographs of figs. 2 to 4 were taken was not suitable for producing single steadily moving bubbles of the type considered in the analysis. It was not long enough to permit one of the fingers to grow at the expense of the others and thus be propagated as a single steadily moving column. The apparatus was therefore modified by inserting liners between the plates which limited the width of the viscous fluid channel to 5·6 cm., leaving the length (38 cm.) and thickness (0·09 cm.) unchanged, and it was used horizontally. With this width it was possible to obtain bubbles which for several inches behind the advancing meniscus were parallel sided. Fig. 8, pl. 1 is a photograph of one. It will be seen that the 'finger' starting from an artificial initial disturbance, produced by blowing a small central air bubble close to the initially straight meniscus, swells out to a definite breadth and is then propagated without further change down the channel. The outer edge of the outline of the bubble represents the outer edge of the meniscus and, measuring the breadth of this at its widest section on the enlarged photograph of fig. 8, it was found to be 5·10 cm. The two liners which limit the breadth of the channel appear black in fig. 8 and the width between them measured on this photograph with a reading microscope was 10·29 cm., so that

$$\lambda = \frac{5 \cdot 10}{10 \cdot 29} = 0 \cdot 496.$$

In several experiments, λ was found to be within 2 % of 0·50.

To compare the shape of the bubble with the results of calculation, the photograph was enlarged till the distance between the walls was 20·0 cm. and the points calculated from expression (17) for the case when $\lambda = 0 \cdot 50$ were pricked through transparent squared paper laid on the photograph. Small ink circles were then centred on these pinholes and the result reproduced in fig. 9. It will be seen that the bubble profile agrees very well with the calculation when λ is taken as 0·50.

8. EXPERIMENTS WITH PAIRS OF VISCOUS FLUIDS

In the experiments so far described, air was used to drive out glycerine. The viscosity of air is only 1/50,000th of that of glycerine, but according to our calculations the same kind of instability would occur whatever the viscosity of the driving fluid, provided it is less than that of the fluid it is displacing. On the other hand, since we have no clue on theoretical grounds as to what determines the value of λ which will occur, it seemed possible that it might depend on the ratio of the viscosities, μ_2/μ_1. For this reason experiments were made in which Shell oil mixture of viscosity 2·75 P was used to drive coloured glycerine of viscosity 8 P through our channel. It was found that when the two viscosities are comparable the length of run required before the driving fluid penetrates into the more viscous fluid in the form of a steadily moving 'finger' was much greater than when one is air. For this reason, a new channel was built of Perspex $2 \cdot 54 \times 91 \times 0 \cdot 08$ cm. This is shown in fig. 10. The camera A points vertically downwards at the channel B which is illuminated by a flash-bulb D placed under a tracing-paper screen C. The driving pressure was produced by raising a water vessel K so as to increase the air pressure in the bottle J. This raised the pressure above the fluid contained in the upstream reservoir L. The more viscous fluid was supplied from a vessel F, so as to fill the upstream reservoir E. If the pressure in E were maintained constant, as it would be if the pipe connecting E and F were left open, the velocity of the finger would increase as the length of the column of more viscous fluid decreased. To maintain a more constant velocity the cock G between E and F was closed during an experiment and the fluid in E escaped to the atmosphere through a needle valve H, the resistance of which was greater than that of the channel itself. By measuring the velocity of the head of the advancing finger before and after passing under the camera A, it was possible to obtain a good estimate of the velocity at the moment the photograph was taken.

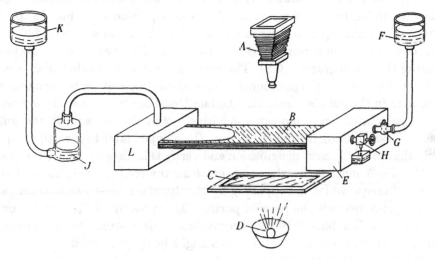

Fig. 10. Arrangement of long Hele-Shaw cell for photographing fingers.

Fig. 11, pl. 2, shows a 'finger' of the oil penetrating at 1 mm./sec. into the tinted glycerine. It will be seen that the outline is very similar to that obtained with air and glycerine. Measurements of λ = (width of finger)/(width of channel) were made at four sections which are marked on the photograph. These gave the following results:

position	1	2	3	4
λ	0·485	0·502	0·508	0·514

It will be seen that after running a distance of 70 cm., i.e. 28 times the width of the channel, the finger has nearly, but not quite, settled down to a constant width of a little more than half the width of the channel.

Fig. 12, pl. 2, covers a larger part of the channel and shows water penetrating into oil in the form of a very uniform finger.

9. Effect on the Shape of the Bubble of Surface Stress at the Interface

The good agreement between the photograph of the bubble shape (fig. 9) and that calculated for the case when $\lambda = \frac{1}{2}$ is an indication that the boundary condition (14), which expresses the assumption that the pressure difference between the two sides of the interface is constant over its length, is justifiable at the flow speed used. When experiments were made at much lower speeds, however, it was found that the bubble was wider but that the wider forms do not conform to the corresponding profiles calculated for the same value of λ from equation (17). Fig. 13, pl. 2, for instance, shows water penetrating into an oil (Shell Diala) at a speed 0·226 cm./sec. in the channel of dimensions $0·08 \times 2·54 \times 91$ cm. The value of λ obtained by measuring this photograph is 0·87. The contour for $\lambda = 0·87$ calculated using (17) is shown superposed on the photograph. The calculated contour is one along which the pressure in the fluid is constant. The fact that the observed contour is less flat than the calculated curve indicates intuitively that the pressure in the fluid increases on passing along it from the vertex. (It is hoped later to investigate quantitatively the actual pressure distribution and verify this directly.) Since the pressure inside the finger must be nearly constant along the contour, this means that the pressure difference which is supported by the interface must decrease on passing from the vertex towards the parallel portion. This pressure difference can only be attributed to surface tension or some equivalent surface stress. Under that general heading, however, various physical effects might be distinguished.

If, for instance, they are due to a constant surface tension of a fluid which wets the walls of the channel and leaves only a negligible amount of the penetrated fluid behind after the interface has passed, one would naturally look to the curvature of the meniscus in the plane midway between the parallel sheets. If, on the other hand, the interface has a finite angle of contact it might be necessary to study how that angle of contact varies when the interface moves over the plane surface. Possible differences between surface tension measured statically and that which acts over newly formed surfaces should also be studied. These things involve further experimental work, some of which is now being carried out at the Cavendish Laboratory and it may be some time before the results are available. In the present note we propose to discuss only those which afford justification for the use of the boundary condition (14) in the theoretical discussion.

Confining attention to the case when a viscous fluid of interfacial tension T is driven slowly by a fluid of much smaller viscosity such as water or air, it is to be expected on dimensional grounds that the shape of the meniscus in a channel of given shape should be a function of $\mu U/T$ only, when the shape of the channel is fixed and the fluid wets the flat slides of the channel. It was found that neither water nor glycerine wet the Perspex which was used to construct the channel; accordingly two oils of very different viscosities, which have in static experiments nearly the same interfacial tension, were used to test whether λ is a function of $\mu U/T$ only.

The two oils used were Shell Diala ($\mu = 0.30\,\text{P}$, $\rho = 0.875\,\text{g./cm.}^3$) and Shell Talpa ($\mu = 4.5\,\text{P}$, $\rho = 0.90\,\text{g./cm.}^3$), and measurements were made using both water and air as the less viscous fluid. The viscosities were determined by measuring the flow through a Veridia accurate bore tube of $1.0\,\text{mm.}$ bore at $20°\,\text{C}$. To measure the interfacial tension between these oils and air a tube of $0.5\,\text{mm.}$ bore was dipped into the oil and the height to which the meniscus rose was observed. The surface tension of Diala was found to be between 27 and $30\,\text{dyn/cm.}$ at $20\,°\text{C}$, and that of Talpa slightly higher. The interfacial tension between the oils and water was measured by two methods. In the first a vertical short length of $0.5\,\text{mm.}$ tube was lowered through a thick layer of oil lying above water, till the upper end was submerged in the oil and the lower end penetrated into the water. The oil wetted the tube and a meniscus convex upwards was formed in the tube. The depth of this below the level of the oil/water interface was measured. The high viscosities of the oils and the comparatively small difference in density of the oil and water made the meniscus move slowly, but consistent results were obtained even with the more viscous oil when an hour or so was allowed to elapse between inserting the tube and measuring the height of the interface.

In this way interfacial tensions between 13.5 and $15\,\text{dyn/cm.}$ were measured for the Diala/water interface. The other method was to insert a longer piece of $\frac{1}{2}\,\text{mm.}$ tube through the oil into the water. There were then two interfaces in the tube: an oil/air interface concave upwards and an oil/water interface downwards. The positions of these in relation to the levels of the flat oil and water surfaces outside the tube were measured. Knowing the densities of oil and water, and assuming that the oil wets the glass as it appeared to do, we obtain with this method the difference between the surface tensions at the two interfaces. Independent measurements gave the surface tension at the oil/air interface. This second method gave values for the interfacial tensions of Diala/water and Talpa/water of between 14 and 16.5.

The results of the experiments with water are shown in fig. 14 in which the measured values of $\lambda = (\text{width of finger})/(\text{width of channel})$ are plotted as ordinates and $\mu U/T$ as abscissae. It will be seen that in spite of a $15:1$ ratio in the viscosities the points obtained with the two oils appear to fall nearly on the same curve. Some of the points were obtained by direct measurement after the finger had been formed for some time while others were obtained by measuring the photographs. It will be noticed that the direct measurement points, particularly for the larger values of $\mu U/T$, correspond with rather smaller values of λ than those derived photographically. This seems to be because at the higher values of $\mu U/T$ a small amount of the fluid is left behind after the passage of the air finger, and this fluid then flows slowly outward towards the sides of the channel thus slightly reducing the width of the air finger.

The most interesting feature of the results exhibited in fig. 14, and of similar sets of observations using air instead of water as the finger, or using oil penetrating into glycerine, is that in all cases the value of λ rapidly decreases as $\mu U/T$ increases till it reaches a value which is very close to $\frac{1}{2}$. We have never measured values of λ appreciably less than $\frac{1}{2}$.

As $\mu U/T$ increases, the effect of surface tension in determining the shape of the interface decreases relatively to that of viscous stress. The fact that λ tends rapidly to $\frac{1}{2}$ as $\mu U/T$ increases, is an indication that when the physical conditions are such that the boundary condition (14) can legitimately be used, the only one of the set of shapes described by (17) which can actually occur is that for which $\lambda = \frac{1}{2}$. The probability that this conclusion is correct is strengthened by the very close agreement between the shape of the contour shown in the photograph of fig. 9 with that calculated using $\lambda = \frac{1}{2}$ in (17).

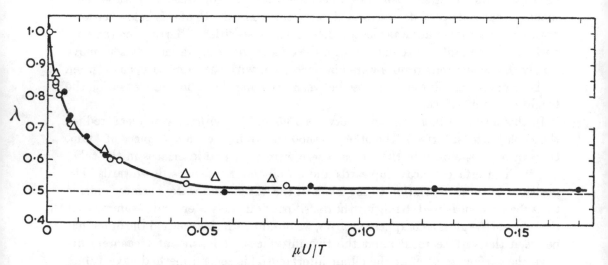

Fig. 14. Measured values of λ for water penetrating into two oils: $\triangle$, Diala (photographic measurements); $\bigcirc$, Diala (direct measurements); $\bullet$, Talpa (direct measurements).†

We have found no theoretical reason for this deduction from observation, but we have noticed a few purely analytical features which distinguish the particular shape corresponding with $\lambda = \frac{1}{2}$ from those corresponding with other values of λ. These have been omitted from the present paper owing to lack of any obvious physical meaning.

It is perhaps worth mentioning, in conclusion, that we have investigated theoretically the stability for infinitesimal disturbance of the shapes given by (17) for the case in which the fluid inside the bubble is of negligible viscosity. The not entirely unexpected result was found that if surface tension effects are neglected, i.e. if the velocity potential is taken as constant along the perturbed surface, then all the shapes are unstable, whatever the value of λ. The analysis gives no indication about why the shape with $\lambda = \frac{1}{2}$ is observed in practice.

† *Editor's note.* The numbers on the abscissa scale in the original publication of this figure were incorrect.

Appendix

In this appendix we give for completeness the modifications of the analysis of §2 which are required when the penetrating fluid (fluid 2) does not completely expel the other (fluid 1), but a tongue of fluid 2, occupying a constant fraction t of the stratum between the plates bounding the Hele-Shaw cell, advances along the gap. It is assumed that the gap is small so that the mean velocity may be calculated by assuming that the motion is everywhere parallel to the plates, and that derivatives of the velocity in directions other than normal to the plates can be neglected in comparison with those along the normal. Using the co-ordinate system introduced in §2, the mean velocity in fluid 1 ahead of the tongue is

$$\mathbf{u}_1 = -\frac{b^2}{12\mu_1}\operatorname{grad}(p+\rho_1 gx). \tag{A 1}$$

After the passage of the tongue, the mean velocity of fluid 2 is

$$\mathbf{u}_2 = -\frac{b^2}{12\mu_2}\left(t^2+\frac{3\mu_2}{2\mu_1}(1-t^2)\right)\operatorname{grad}(p+\rho_2' gx), \tag{A 2}$$

where

$$\rho_2' = \rho_2\left[1+\frac{3\mu_2(1-t)^2(\rho_1-\rho_2)}{\mu_1\{2\mu_1 t^2+3\mu_2(1-t^2)\}}\right]. \tag{A 3}$$

The mean velocity of the film of fluid 1 left adhering to the plates is

$$\mathbf{u}_{12} = -\frac{b^2}{12\mu_1}(1-t)(1+\tfrac{1}{2}t)\operatorname{grad}(p+\rho_1 gx) - \frac{b^2 t(1-t)}{8\mu_1}\operatorname{grad}(\rho_2-\rho_1)gx. \tag{A 4}$$

The interface is taken as the tip of the meniscus, which advances with velocity $U = \mathbf{n}.\mathbf{u}_2$ normal to itself, where $\mathbf{n}$ is the outwards normal to the projection of the tip of the meniscus on the plates. The equation of continuity across the interface is

$$\mathbf{n}.\mathbf{u}_1 = t\mathbf{n}.\mathbf{u}_2 + (1-t)\mathbf{n}.\mathbf{u}_{12}. \tag{A 5}$$

If we define $\mathbf{u}_1'$ by

$$\mathbf{u}_1' = -\frac{b^2}{12\mu_1 A}\operatorname{grad}(p+\rho_1(1+B)gx), \tag{A 6}$$

where

$$A = t+\frac{\mu_2(1-t)^2(1+\tfrac{1}{2}t)}{\mu_1 t+\tfrac{3}{2}\mu_2(1-t^2)}, \quad B = (1-t)^2\left\{(1+\tfrac{1}{2}t)\left(\frac{\rho_2'}{\rho_1}-1\right)-\tfrac{3}{2}t\left(\frac{\rho_2}{\rho_1}-1\right)\right\},$$

then it can be shown that $\qquad U = \mathbf{n}.\mathbf{u}_1',$ at the interface.

Hence the motion of the interface in the case in which a constant fraction t of fluid 2 penetrates fluid 1 corresponds, according to the analysis, with the motion of the interface between a fluid viscosity μ_2 and density ρ_2', which completely expels a fluid of viscosity $A\mu_1$ and density $\rho_1(1+B)$. Thus, the kinematics of the motion are unaltered by a constant fraction of the penetrated fluid being left adhering to the plates bounding the cell.

29

A NOTE ON THE MOTION OF BUBBLES IN A HELE-SHAW CELL AND POROUS MEDIUM†

REPRINTED FROM

Quarterly Journal of Mechanics and Applied Mathematics, vol. XII (1959), pp. 265–79

Exact solutions are presented for the steady motion of a symmetrical bubble through a parallel-sided channel in a Hele-Shaw cell containing a viscous liquid. The solutions also describe two-dimensional motion in a porous medium, since the two cases are mathematically analogous. It is shown that the solutions are not mathematically unique and a purely analytical criterion for specifying a particular solution is given. When the bubble is large, these particular solutions are such that the maximum width of the bubble is almost half that of the channel. Solutions are also presented for the motion of large asymmetric bubbles through the channel. The three-dimensional motion of bubbles in a porous medium of infinite extent is also considered briefly.

1. Introduction

The motion of the interface between two immiscible viscous fluids in a Hele-Shaw cell, that is, in the region between two parallel, closely spaced flat plates, was recently described in a paper by Saffman and Taylor.‡ In particular, they studied the propagation of a long finger of fluid through a parallel-sided channel in a Hele-Shaw apparatus containing a more viscous liquid. The boundary conditions at the interface depend on the interfacial tension as well as on the viscosities of the two fluids but the exact nature of this dependence is unknown. To calculate the flow in the Hele-Shaw apparatus, however, only the change in pressure on passing through the interface and the thickness of the two layers of the more viscous fluid which are left behind adhering to the flat plates after the interface has passed are required. It was shown that the simplifying assumption that both of these are constant leads to an exact solution in closed form for the propagation of a semi-infinite finger.

Actually, it was found that there exists a whole family of solutions corresponding to different values of the velocity of the finger and its asymptotic width. An experimental investigation showed that the width of fingers when the flow was not very slow was close to one-half that of the channel, and excellent agreement was found between the observed shape and the calculated one which was half the channel width. (At very low flow rates, the observed shapes had a width greater than one-half that of the channel and did not conform with the calculated shapes having the same width. In this case it appears that the assumption that the pressure change across the interface is constant is not valid.)

---

† With P. G. Saffman.
‡ *Proc. Roy. Soc.* A, ccxlv (1958), 312 (paper 28 above).

The primary purpose of the present note is to give the mathematical solution of the problem in which a bubble of finite size, i.e. a bubble bounded by a closed contour as opposed to the infinite ones considered by Saffman and Taylor, moves steadily through a parallel-sided channel in a Hele-Shaw cell containing another viscous fluid, under the same simplifying assumptions about the boundary conditions at the interface. Such bubbles sometimes occurred in the experiments previously described, although it was usually desirable to remove them, and the solution may not be without practical interest. However, it appears that the range of circumstances for which the solution is a reasonably close description of the actual motion is somewhat more restricted than for the long fingers. This is because, for reasons not yet fully understood, the physical conditions at the rear of the bubble, where the interface is retreating, may be different from those at the front where the interface is advancing, except in the case of very slow motion where very little fluid is left behind after the passage of the interface; but then the assumption of a constant pressure drop across the interface becomes less accurate.

The analysis, therefore, is probably mainly of academic interest, but nevertheless may give a fair approximation to the shape of the front of the bubble where the assumptions are likely to be reasonable; this approximation will be better the larger the bubble, and a long finger can indeed be regarded as the front of a very large bubble. Also, for the case of very slow motion and a bubble whose dimensions are small compared with the width of the channel, the pressure change across the interface due to surface tension and the interfacial curvature in the plane of the flat sides of the cell can be taken into account and the analysis should not be without physical significance.

It is found that for bubbles of given area and for given conditions at infinity the solution is not unique, and that there exists a whole family of possible solutions. It is worth noting, however, that for the case in which the viscosity of the fluid inside the bubble and hydrostatic pressure gradients are negligible, a particular solution can be singled out by imposing arbitrarily the condition that the product of the velocity and maximum width of the bubble is a minimum. For very large bubbles, whose front portion resembles a long finger, the width given in this way is half that of the channel, in agreement with the experiments, but a satisfactory physical explanation of this result has not been found. For very small bubbles, the solution selected in this way is that bounded by a circular contour.

The analysis of bubbles of finite size is restricted, for the sake of simplicity, to bubbles which are symmetrical about the centre line of the channel. Solutions will be presented, however, for the steady motion of very large asymmetric bubbles, that is, of long fingers similar to those considered previously. The general asymmetric case seems to be capable of solution, but it did not appear to be of sufficient interest to justify the labour involved.

The motion in a Hele-Shaw cell is mathematically analogous to two-dimensional flow in a porous medium in which the motion is governed by Darcy's law. Simple closed solutions exist for the three-dimensional motion of bubbles in a porous medium of infinite extent, according to which the bubbles are ellipsoids of revolution. It is noted in section 7 that the particular solution for which the bubble is a sphere is singled out by a minimum condition similar to that just mentioned above.

2. Motion of a bubble in a Hele-Shaw cell

In this section the analysis of the steady motion of a bubble in a parallel sided channel in a Hele-Shaw cell is presented for the case in which the bubble is symmetrical about the centre line of the channel. We suppose first that the viscosity of the fluid inside the bubble is negligible and that pressure gradients due to gravity may be ignored (this will be so if the cell is horizontal). The mean velocity across the stratum of the viscous fluid through which the bubble is moving is given by†

$$u = -\frac{b^2}{12\mu}\frac{\partial p}{\partial x} = \frac{\partial \phi}{\partial x} = \frac{\partial \psi}{\partial y}, \quad v = -\frac{b^2}{12\mu}\frac{\partial p}{\partial y} = \frac{\partial \phi}{\partial y} = -\frac{\partial \psi}{\partial x}, \tag{1}$$

where x and y are coordinates in the plane of the plates bounding the cell, u and v are the components of mean velocity parallel to these axes, p is the pressure, μ the viscosity, $\phi = -(b^2/12\mu)p$ is a velocity potential, ψ is the stream function (which exists by virtue of the equation of continuity), and the gap between the plates b is assumed sufficiently small for (1) to hold. (Equations (1) are satisfied by the two-dimensional flow in a porous medium of permeability $b^2/12$.)

Taking the walls of the channel as $y = \pm 1$ and supposing the viscous fluid has unit velocity at infinity in front of and behind the bubble, ϕ and ψ satisfy the conditions

$$\psi = \pm 1 \quad \text{on} \quad y = \pm 1, \quad \phi \to x \quad \text{as} \quad x \to \pm \infty \tag{2}$$

(edge effects which invalidate (1) within a distance of order b from the channel walls are neglected).

To obtain boundary conditions satisfied by ϕ and ψ on the interface, we first assume that the pressure drop across the interface is constant. When the fluid inside is of negligible viscosity, the pressure inside is constant and hence

$$\phi = \text{const} = 0, \quad \text{say}, \tag{3}$$

on the interface. The other assumption is that there is no film of viscous fluid left behind between the bubble and the plates bounding the cell, i.e. the advancing interface expels all the viscous fluid in front of it, from which it follows that the velocity of the interface normal to itself is equal to the component of the mean velocity in the same direction. Expressed analytically, this is

$$\frac{\partial \phi}{\partial n} = \frac{\partial \psi}{\partial s} = U\frac{\partial y}{\partial s}, \tag{4}$$

where n, s denote differentiation in the normal and tangential directions, respectively, and U is the velocity of the bubble. Integrating (4) and noting that the arbitrary

† See, for example, *Hydrodynamics*, by H. Lamb, Cambridge University Press (1932), § 330.

constant of integration is zero because the bubble is supposed to be symmetrical about $y = 0$ gives, finally,

$$\psi = Uy \quad \text{on} \quad \phi = 0. \tag{5}$$

A discussion of these assumptions is given in paper 28 above. It is also shown there how, by a simple transformation, the solution for the case in which there is a film of *constant* thickness between the bubble and the plates may be obtained from that in which the films are of zero thickness.

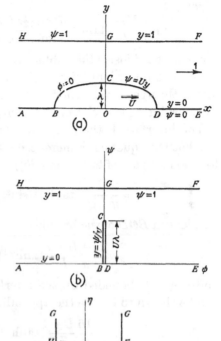

Now it follows from (1) that $w = \phi + i\psi$ is an analytic function of $z = x + iy$, and the solution of the problem is therefore effected by constructing an analytic function of z which satisfies the boundary conditions expressed in (2) and (5). In fig. 1(a) is shown the upper half of the physical plane; the maximum half-width of the bubble is denoted by λ. The potential plane, the boundaries of which are given by (2) and (5), is shown in fig. 1(b), corresponding points being marked with the same letter. Note that we must have $U\lambda < 1$ in order that the potential and physical planes should be simple images of one another. The transformation

$$i \cos \zeta = \tanh\left(\tfrac{1}{2}\pi w\right) \cot\left(\tfrac{1}{2}\pi U\lambda\right),$$

where $\zeta = \xi + i\eta$, transforms the potential plane into the semi-infinite strip $\eta > 0$, $-\tfrac{1}{2}\pi < \xi < \tfrac{1}{2}\pi$, shown in fig. 1(c), corresponding points again being marked with the same letter. At the bubble interface, where $\phi = 0$ and $\eta = 0$, this transformation may be written as

$$\psi = \frac{2}{\pi} \tan^{-1}\left(\cos\xi \tan \tfrac{1}{2}\pi U\lambda\right).$$

Fig. 1. The motion of a bubble through a channel: (a) physical plane, (b) potential plane, (c) transformed potential plane.

Writing $z = w + z_1$, $z_1(\zeta) = x_1(\xi, \eta) + iy_1(\xi, \eta)$, we find that the boundary conditions (2) and (5) become in the ζ-plane,

$$y_1 = 0 \quad \text{on} \quad \xi = \pm\tfrac{1}{2}\pi, \tag{6}$$

and

$$y_1 = \psi\left(\frac{1}{U} - 1\right) = -\frac{4}{\pi}\frac{U-1}{U}\,\text{im} \tanh^{-1}\left(i\,e^{i\xi} \tan \tfrac{1}{4}\pi U\lambda\right). \tag{7}$$

Here we have used the general formula

$$\tanh^{-1}(a + ib) = \tfrac{1}{2}\tanh^{-1}\left(\frac{2a}{1 + a^2 + b^2}\right) + \tfrac{1}{2}i\tan^{-1}\left(\frac{2b}{1 - a^2 - b^2}\right).$$

It may now be verified directly that

$$z_1 = -\frac{4}{\pi}\frac{U-1}{U}\tanh^{-1}(i\,e^{i\zeta}\tan\tfrac{1}{4}\pi U\lambda) \tag{8}$$

is regular in the strip and satisfies (6) and (7). The velocity potential and stream function are therefore given implicitly in terms of x and y by (expressing ζ in terms of w)

$$z = w + \frac{4}{\pi}\frac{U-1}{U}\tanh^{-1}\left[\frac{\tan(\tfrac{1}{4}\pi U\lambda)}{\tan(\tfrac{1}{2}\pi U\lambda)}\{(\tanh^2\tfrac{1}{2}\pi w+\tan^2\tfrac{1}{4}\pi U\lambda)^{\frac{1}{2}}-\tanh\tfrac{1}{2}\pi w\}\right]. \tag{9}$$

Since the fluid inside the bubble is of negligible viscosity, the bubble will move faster than the fluid at infinity, i.e. $U > 1$. Mathematically, this result arises as the condition that the transformation (9) be one-one (as may be seen without difficulty by considering the relation between x and ϕ on the axis $\psi = 0$).

The interface is the image in the physical plane of $\phi = 0$, $-U\lambda < \psi < U\lambda$, and has the equation (remembering that $\psi = Uy$ and making use of the expression for the real part of $\tanh^{-1}(a+ib)$)

$$x = \frac{2}{\pi}\frac{U-1}{U}\tanh^{-1}\{\sin^2(\tfrac{1}{2}\pi U\lambda)-\cos^2(\tfrac{1}{2}\pi U\lambda)\tan^2(\tfrac{1}{2}\pi Uy)\}^{\frac{1}{2}}. \tag{10}$$

The length BOD of the bubble is

$$\frac{4}{\pi}\frac{U-1}{U}\tanh^{-1}\{\sinh(\tfrac{1}{2}\pi U\lambda)\} = L, \quad \text{say,}$$

and the area bounded by the interface, i.e. by the closed contour given by (10), can be shown to be (see the appendix)

$$\frac{16}{\pi}\frac{U-1}{U^2}\tanh^{-1}\{\tan^2(\tfrac{1}{4}\pi U\lambda)\} = S, \quad \text{say.} \tag{11}$$

The analytical solution (9) contains two parameters, U (the velocity of the bubble) and λ (its maximum half-width), and it is clear that specifying the area S of the bubble provides only one relation between them. That is, if the area of the bubble and the velocity at infinity are specified, then there exists a whole family of solutions and possible bubble shapes corresponding to the various combinations of U and λ which satisfy (11). The extreme shapes are $\lambda = 0$, for which $U = \infty$, $L \to \infty$, and the bubble has zero thickness; and $\lambda = 1$, for which $U = 1$ (since $U\lambda \not> 1$) and the velocity everywhere is equal to the velocity at infinity. In the latter case, the bubble is bounded by the channel walls and the lines $x = \pm\tfrac{1}{2}L = \pm\tfrac{1}{4}S$. The shapes intermediate between these are ovals of length L and width 2λ, moving with velocities between $U = 1$ and $U = \infty$.

If $U\lambda \to 1$, keeping U fixed, the area of the bubble becomes very large and it is easily seen, changing the origin to the vertex of the bubble, that we recover the solution obtained in paper 28 above for the motion of a long finger whose asymptotic width is λ times that of the channel, namely,

$$z = w + \frac{2}{\pi}(1-\lambda)\log\tfrac{1}{2}\{1+\exp(-\pi w)\}. \tag{12}$$

3. The hypothesis of minimum $U\lambda$

Although the motion of a bubble in a channel does not possess, when formulated as above, a mathematically unique solution, we should expect it to be physically unique and that there is some mechanism, not accounted for in the analysis, to single out one particular solution. The search for some such effect has so far been unsuccessful, but several features of the analysis which distinguish a particular solution have been noticed. One of these is as follows.

Writing (11) in the form

$$S = \frac{16}{\pi} \frac{\lambda(U\lambda - \lambda)}{U^2\lambda^2} \tanh^{-1}\{\tan^2(\tfrac{1}{4}\pi U\lambda)\},$$

it follows that S is a maximum for given $U\lambda$ and varying λ when $\lambda = \tfrac{1}{2}U\lambda$, i.e. when $U = 2$. Alternatively, $U\lambda$ is a minimum for given S and varying U when $U = 2$. Thus, if we make the hypothesis that the motion is such that $U\lambda$ is a minimum, then the bubble whose velocity is twice that at infinity is selected, and the corresponding value of λ is given by (11) with $U = 2$.

The product $U\lambda$ of the velocity of the bubble and half its maximum width does not have a clear physical significance, although it may be identified intuitively with the rate at which fluid is pushed aside by the bubble. We have in fact been unable to place the hypothesis on a sound physical basis (attempts to relate it to a minimum rate of energy dissipation fail), but it has the interesting feature that when the bubble is large and $U\lambda$ is close to 1, it gives a value for λ close to $\tfrac{1}{2}$, and this is the value observed experimentally with long fingers.

It is perhaps worth mentioning, as a curiosity, that the shape for $U = 2$ arises as the unique solution of a problem in the steady two-dimensional flow of an electric current in a uniform conducting medium. Suppose [see fig. 1 (*a*)] that HF and AOE are two parallel electrodes at unit potential difference, separated by a medium of uniform conductivity, and that a perfect conductor of given area is laid on AE, the boundary of this conductor being $BCDOB$ in the figure. If we now determine the shape of this conductor so that the amount of current entering it, i.e. flowing across the curvilinear boundary BCD, is a minimum, then it turns out (the details of the analysis are omitted for brevity) that the conductor has the same shape as the bubble of the same area for which $U = 2$.

Another unique feature of the solution for a semi-infinite finger which is half the channel width may perhaps also be mentioned here. As has been pointed out by Zhuralev,† the velocity potential due to the superposition of a uniform stream of inviscid fluid in a channel bounded by parallel walls $y = \pm 1$ and a two-dimensional source of strength m placed midway between the walls at the origin of coordinates is

$$w = qz + \frac{m}{2\pi} \log (2 \sinh \tfrac{1}{2}\pi z), \tag{13}$$

† *Zap. Leningr. gorn. Inst.* **XXXIII** (1956), 54.

where q is the velocity of the stream (a different coordinate system is used by Zhuralev and the expression given there is accordingly different).

It will be noticed that (12) and (13) would be closely similar if z and w were interchanged in one of them. Now (12) may be written

$$e^{-\pi w} - 2e^{-\pi w/2(1-\lambda)} e^{\pi z/2(1-\lambda)} + 1 = 0$$

which can be solved explicitly for w when $\lambda = \frac{1}{2}$, giving

$$w = \tfrac{1}{2}z' + \frac{1}{\pi} \log \left(2 \sinh \tfrac{1}{2}\pi z' \right),$$

where $z' = z - (\log 2)/\pi$. This last expression is identical with (13) when $m = 4q$, except for a change of origin and a multiplicative constant. That is, the potential for a finger which is half the channel width is identical with that arising from the super-position of a uniform stream and source of appropriate strengths; and only the finger with $\lambda = \frac{1}{2}$ can be obtained in this simple way.† (It can be shown that the potential for a finger for which $\lambda \neq \frac{1}{2}$ arises from a rather complicated system of sources, sinks, and dipoles superposed on a uniform stream.)

Further, the dividing streamline of the flow due to a source in a uniform stream can be shown to have the equation

$$x = \frac{1}{\pi} \log \left(\frac{\sin \pi y/\lambda}{\sin (1-\lambda) \pi y/\lambda} \right),$$

where $2\lambda = 4m/(4q+m)$ is its asymptotic width; equation (13) can be interpreted as giving the flow past a fixed obstacle of this shape in the channel. Similarly, by taking axes fixed relative to the finger, we can deduce from (12) the flow past a fixed obstacle having the same shape as the finger, this shape having the equation‡

$$x = \frac{1-\lambda}{\pi} \log \frac{1}{2} \left(1 + \cos \frac{\pi y}{\lambda} \right) = \frac{2(1-\lambda)}{\pi} \log \left(\cos \frac{\pi y}{2\lambda} \right).$$

When $\lambda = \frac{1}{2}$, the shapes of these obstacles are identical.

4. The motion of small bubbles

When λ is small compared with 1 and U is finite, the dimensions of the bubble are small compared with the width of the channel, and (10) reduces to

$$\frac{x^2}{(U-1)^2} + y^2 = \lambda^2. \tag{14}$$

† This result for the potential of the motion outside the finger remains true if the fluid inside the finger is not of negligible viscosity, the effect of gravity is taken into account and it is supposed that layers of fluid of constant thickness are left behind after the passage of the interface. However, the values of q and m will depend upon the particular circumstances.

‡ See paper 28 above.

That is, small bubbles are ellipses with axis ratio $U - 1$. The solution for a small bubble is obtained, in effect, by letting the width of the channel tend to infinity, so that (14) gives the shape of a bubble in an unbounded Hele-Shaw cell. The corresponding complex potential is

$$z = \frac{U-1}{U}(w^2 + U^2\lambda^2)^{\frac{1}{2}} + \frac{w}{U}.$$

The bubble of given size with a minimum value of $U\lambda$ is a circle of radius λ moving with twice the velocity at infinity. Now the analysis has been based on the assumption that the pressure drop across the interface is constant. This pressure drop consists of a term T/R, where T is the interfacial tension and R is the radius of curvature of the projection of the interface onto the plates bounding the cell, plus a term due to the curvature of the meniscus in a plane perpendicular to the bounding plates. A necessary condition for this assumption to be valid is that variations in T/R should be small compared with changes in the pressure of the fluid surrounding the bubble, that is, small compared with $12\mu UR/b^2$ (this condition may not be sufficient). When the bubble is small, this condition may not be satisfied, but it is interesting that the circular bubble remains an exact solution when the T/R term is taken into account, since R is then constant.

This effect of surface tension tends to make the perimeter of the bubble as short as possible, and therefore to make the bubble circular. It is noteworthy that in this case this effect of surface tension and the condition of minimum $U\lambda$ give the same result.

Visual observations of small bubbles which occurred during the experiments reported by Saffman and Taylor indicated that they were circular when sufficiently small. When larger, they were often pear-shaped with a rounded front and pointed back; some on the other hand seemed to be ovoid with the sharper end pointed in the direction of motion. These phenomena are probably due to the physical conditions at the retreating interface over the back of the bubble being different from those at the front; there is, however, no clear evidence on this matter.

5. BUBBLES OF FLUID WITH NON-ZERO VISCOSITY

The previous analysis has been concerned entirely with the case in which the viscosity of the fluid inside the bubble is negligible and gravity forces are ignored. We shall now suppose that the fluid inside the bubble has a non-zero viscosity, that the x-axis is vertically upwards parallel to the channel walls, and that gravity forces are acting. The mean velocity across the stratum between the plates is now derived from a velocity potential

$$\phi = -\frac{b^2}{12\mu}(p + \rho gx),$$

where ρ is the density of the fluid and g is the acceleration of gravity.

We shall denote quantities inside the bubble by the suffix 2 and quantities outside by the suffix 1. The boundary conditions (2), (3), and (5) now become (still making the same simplifying assumptions about the physical conditions at the interface)

$$\psi_1 = \pm V \quad \text{on} \quad y = \pm 1, \quad \phi_1 \to Vx \quad \text{as} \quad x \to \pm \infty, \tag{2)'}$$

where the velocity at infinity is now denoted by V; also

$$\mu_1 \phi_1 + \frac{b^2 \rho_1 gx}{12\mu} = \mu_2 \phi_2 + \frac{b^2 \rho_2 gx}{12\mu}, \tag{3)'}$$

$$\psi_1 = \psi_2 = Uy, \tag{5)'}$$

on the interface.

It follows from (5)' that the motion inside the bubble is given by

$$\phi_2 = Ux, \quad \psi_2 = Uy. \tag{15}$$

After some algebra we find that Φ and Ψ defined by

$$W = \Phi + i\Psi = \frac{\phi_1 + i\psi_1 - U^*z}{V - U^*}, \tag{16}$$

where

$$U^* = \frac{\mu_2 U}{\mu_1} - \frac{b^2 g}{12\mu_1}(\rho_1 - \rho_2), \tag{17}$$

satisfy (2), (3), and (5) with U replaced by

$$\frac{U - U^*}{V - U^*} = U', \quad \text{say.}$$

The expressions (9) and (10) with U replaced by U' therefore give the relation between z and W and also the equation of the interface. In particular, it is to be noted that the family of possible shapes is independent of the physical properties of the fluid.

The particular solution with $U' = 2$ (and λ close to $\frac{1}{2}$ for large bubbles) is now obtained by minimizing $U'\lambda$ for given bubble size. U^* is the velocity with which the fluid outside the bubble would move under the action of the actual pressure gradient inside the bubble, and the hypothesis of minimum $U\lambda$ has to be replaced by an hypothesis in which velocities are measured relative to U^*. The physical significance of this result is not clear.

6. Penetration of an asymmetrical finger into a channel

In this section we shall give solutions of the equations of motion which describe the steady motion of bubbles not symmetrical about the centre line of the channel for the case in which the bubble is very large and may be regarded as a semi-infinite finger. The motion of symmetrical fingers was considered by Saffman and Taylor,† and it was stated that there exists a singly infinite family of solutions for the problem of the penetration of a finger into a channel filled with a more viscous fluid. This

† Paper 28 above.

statement is in fact erroneous, and was made because it was not realized that further asymmetrical solutions existed. Actually, there exists a doubly infinite family of steady solutions, of which the symmetrical ones form a singly infinite sub-set, and it was thought worthwhile describing these here.

It has already been noted that the question as to why long fingers formed experimentally are half the channel width, when their velocity is not too slow, remains unanswered, although several criteria of a purely mathematical nature have been discovered; the further question now arises as to why these fingers are apparently symmetrical, and a satisfactory answer has yet to be found.

It will be supposed in the analysis that the viscosity of the fluid inside the finger is negligible and that gravity may be neglected. The solutions for the more general case may be obtained immediately by methods identical with those employed in section 5 above. (It is also assumed that no fluid is left behind adhering to the flat plates bounding the Hele-Shaw apparatus after the interface has passed by. Solutions for the case in which films of constant thickness are left behind may be derived in the manner described by Saffman and Taylor.) We employ the same notation here as in section 2, except that the origin of coordinates will no longer be the centre of the bubble.

The velocity potential satisfies

$$\phi = 0 \tag{3}$$

on the interface, by virtue of the assumption that the pressure drop across it is constant. For the stream function, the boundary condition (4) now integrates to

$$\psi = U(y - y_0) \tag{18}$$

where U is the velocity of propagation of the finger and y_0 is a constant whose value is as yet unspecified. (y_0 may be taken positive without loss of generality, solutions with y_0 negative being mirror images in $y = 0$ of those with y_0 positive.) Taking unit velocity at infinity ahead of the finger, we have the further conditions $\phi \sim x$, $\psi \sim y$ as $x \to +\infty$.

The symmetrical solutions are obtained by putting $y_0 = 0$. We have shown†
that the complex potential for the steady motion of a symmetric semi-infinite finger is given by equation (12), where λ is the ratio of the asymptotic width of the finger to that of the channel and the velocity of propagation $U = 1/\lambda$. The equation of the interface (i.e. the image of $\phi = 0$) may be written in the form

$$x = \frac{2(1 - U^{-1})}{\pi} \log (\cos \tfrac{1}{2}\pi U y), \tag{19}$$

the origin of co-ordinates being taken at the nose of the finger. The potential plane (see fig. 6 of Saffman and Taylor's paper) is the semi-infinite strip $\phi > 0$, $-1 < \psi < 1$; the line $\phi = 0$, $-1 < \psi < 1$ is the image of the interface, and the lines $\phi > 0$, $\psi = \pm 1$ of the channel walls.

† See paper 28 above.

The method by which the symmetrical solutions are obtained may be extended with little difficulty to deal with the asymmetrical case in which the boundary condition (18) applies. We have that x and y are conjugate harmonic functions of ϕ and ψ such that $y = \pm 1$ on $\psi = \pm 1$, $\phi > 0$; $y \sim \psi$ as $\phi \to +\infty$; and $y = U^{-1}\psi + y_0$ on $\phi = 0$, $-1 < \psi < 1$. Take therefore

$$y = \psi + \sum_{n=1}^{\infty} A_n e^{-n\pi\phi} \sin n\pi\psi + \sum_{n=1}^{\infty} B_n e^{-\frac{1}{2}n\pi(\psi+1)} \sin \tfrac{1}{2}n\pi(\psi+1),$$

thereby satisfying the boundary conditions on $\psi = \pm 1$ and at $\phi = \infty$. The boundary condition on $\phi = 0$ is satisfied if

$$U^{-1}\psi + y_0 = \psi + \sum_1^{\infty} A_n \sin n\pi\psi + \sum_1^{\infty} B_n \sin \tfrac{1}{2}n\pi(\psi+1) \quad (-1 < \psi < 1).$$

Calculating A_n and B_n by the usual method of Fourier series, we find that

$$A_n = -\frac{2}{n\pi}(1 - U^{-1}), \quad B_n = \frac{2y_0}{n\pi}\{1 - (-1)^n\}.$$

The series for $z = x + iy$ may be summed, giving finally

$$z = w + \frac{2}{\pi}(1 - U^{-1})\log \tfrac{1}{2}(1 + e^{-\pi w}) + \frac{2y_0}{\pi}\log \left(\frac{1 + i\,e^{-\frac{1}{2}\pi w}}{1 - i\,e^{-\frac{1}{2}\pi w}}\right). \tag{20}$$

The interface is the image of $\phi = 0$ and has the equation (remembering that $\psi = U(y - y_0)$)

$$x = \frac{2}{\pi}(1 - U^{-1})\log \cos \tfrac{1}{2}\pi U(y - y_0) + \frac{2y_0}{\pi}\log \tan \{\tfrac{1}{4}\pi + \tfrac{1}{4}\pi U(y - y_0)\}. \tag{21}$$

Now it may be verified that the transformation (20) is one-one and free from singularities if $U > 1$ and $y_0 + U^{-1} < 1$. The former of these conditions is clear physically because the fluid in the finger is less viscous than that in the channel. The latter condition arises from the fact that all the streamlines must originate on the interface (simply by continuity, because the velocity of the fluid at $x = -\infty$ is zero since conditions must be uniform along the straight sides of the finger; hence $\phi = 0$ for all y at $x = -\infty$). Hence, the value of y at the point where the streamline $\psi = 1$ meets the interface must be less than one, and by (18) this value is $y_0 + U^{-1}$. y_0 is in fact the value of y at the point where the streamline $\psi = 0$ meets the interface.

As regards the shape of the interface given by (21), we note that

as $\qquad y \to y_0 + U^{-1}, \; x \sim (1 - U^{-1} - y_0)\log(y_0 + U^{-1} - y) \to -\infty;$

and as $\qquad y \to y_0 - U^{-1}, \; x \sim (1 - U^{-1} + y_0)\log(y - U^{-1} + y_0) \to -\infty.$

The interface therefore has the two asymptotes $y = y_0 + U^{-1}$, $y = y_0 - U^{-1}$, and lies between them, being nearer the wall $y = 1$ than the wall $y = -1$. The asymptotic width of the finger is $2U^{-1} = 2\lambda$, say. The interface for the values $U = 2$, $y_0 = \tfrac{1}{4}$ is sketched in fig. 2.

If $y_0 + U^{-1} = 1$, the interface intersects the wall $y = 1$ at right angles. In this case, the interface is half of a symmetrical finger between the walls $y = -1$ and $y = 3$, as is clear from the fact that (21) may then be written in the form

$$\tfrac{1}{2}x = \frac{2}{\pi}(1 - U^{-1})\log \cos \tfrac{1}{4}\pi U(y-1).$$

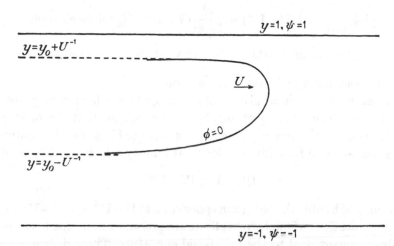

Fig. 2. An asymmetric finger penetrating a channel. The interface is calculated from (21) with $U = 2$, $y_0 = \tfrac{1}{4}$.

There is no feature of the analysis which specifies the values of λ (or U) and y_0, and therefore there is a doubly infinite family of mathematical solutions for the penetration of a finger into a channel containing a more viscous fluid. It remains to explain why the fingers obtained in the experiments previously described appear to conform with the shape calculated from $\lambda = \tfrac{1}{2}$, $y_0 = 0$.

7. BUBBLES IN THREE DIMENSIONS

Provided that the same simplifying assumptions are made about the physical conditions at the interface, the motion in three dimensions of a bubble in a uniform unbounded porous medium, filled with viscous fluid moving uniformly at infinity, and in which the motion is governed by Darcy's law, possesses simple solutions in which the interface is an ellipsoid of revolution. These solutions are briefly referred to in an article by Polubarinova-Kochina and Falkovich,† but since they are apparently not well-known, it was thought worthwhile to present the results in a more general form than hitherto, and to point out the lack of uniqueness.

If $\mathbf{V}$ denotes the velocity at infinity, $\mathbf{U}$ the velocity of the bubble, $\mathbf{g}$ the acceleration due to gravity, and

$$\mathbf{U}^* = \frac{k}{\mu_1}\Big\{(\rho_2 - \rho_1)\,\mathbf{g} - \frac{\mu_2}{k}\mathbf{U}\Big\},$$

† *Adv. appl. Mech.* II (1951), 153.

where k is the permeability of the medium and the suffixes 1 and 2 refer to the fluid outside and inside the bubble, respectively, then the boundary conditions that the pressure and normal velocity are continuous across the interface can be satisfied by taking the bubble as any ellipsoid of revolution with axis parallel to $\mathbf{U} - \mathbf{U}^*$. The relation between $\mathbf{U}$ and $\mathbf{V}$ is

$$\left(\tfrac{1}{2}\log\frac{1+e}{1-e} - e\right)(\mathbf{U}-\mathbf{U}^*) = \frac{e^3}{1-e^2}(\mathbf{V}-\mathbf{U}^*) \quad \text{(prolate ellipsoid)},$$

$$(e - (1-e^2)^{\frac{1}{2}}\sin^{-1}e)(\mathbf{U}-\mathbf{U}^*) = e^3(\mathbf{V}-\mathbf{U}^*) \quad \text{(oblate ellipsoid)},$$

where e is the eccentricity of a meridian section.

The solution for a bubble of given volume is again mathematically not unique, but a particular solution is singled out by the hypothesis that the product of the equatorial area A and the velocity measured relative to $\mathbf{U}^*$ should be minimum for a bubble of given size and for a given velocity at infinity relative to $\mathbf{U}^*$, i.e. that

$$A|\mathbf{U}-\mathbf{U}^*|/|\mathbf{V}-\mathbf{U}^*|$$

is a minimum. The bubble shape is then spherical and $\mathbf{U}-\mathbf{U}^* = 3(\mathbf{V}-\mathbf{U}^*)$; this solution remains exact if interfacial stress effects are present whose effect is to produce a pressure drop proportional to the interfacial curvature. There does not appear to be at present any experimental evidence on the motion of bubbles in porous media.

APPENDIX

To calculate the area bounded by the contour given by equation (10), we note that on the interface (where $\phi = 0$ and $\psi = Uy$),

$$x_1 = x - \phi = x \quad \text{and} \quad y_1 = y - \psi = -y(U-1).$$

Hence, we have from (8)

$$x - (U-1)y = -\frac{4}{\pi}\frac{U-1}{U}\tanh^{-1}\{i\,e^{i\xi}\tan(\tfrac{1}{4}\pi U\lambda)\}$$

$$= -\frac{4}{\pi}\frac{U-1}{U}\sum_{n=0}^{\infty}\frac{i(-1)^n e^{i(2n+1)\xi}\tan^{2n+1}(\tfrac{1}{4}\pi U\lambda)}{2n+1}$$

on the interface (remembering that $\eta = 0$ there). The equation of the interface may therefore be expressed parametrically by

$$x = \frac{4}{\pi}\frac{U-1}{U}\sum_{0}^{\infty}(-1)^n\frac{\sin(2n+1)\xi\,\tan^{2n+1}(\tfrac{1}{4}\pi U\lambda)}{2n+1},$$

$$y = \frac{4}{\pi U}\sum_{0}^{\infty}(-1)^n\frac{\cos(2n+1)\xi\,\tan^{2n+1}(\tfrac{1}{4}\pi U\lambda)}{2n+1},$$

and the complete contour corresponds with $-\pi < \xi < \pi$. These series are uniformly convergent if $\tan\left(\frac{1}{4}n U\lambda\right) < 1$, which is the case if the bubble dimensions are finite. We have now

$$S = \int_{-\pi}^{\pi} y \frac{dx}{d\xi}\,d\xi = \frac{16}{\pi}\frac{U-1}{U^2}\sum_0^{\infty}\frac{\tan^{4n+2}(\frac{1}{4}\pi U\lambda)}{2n+1} = \frac{16}{\pi}\frac{U-1}{U^2}\tanh^{-1}(\tan\tfrac{1}{4}\pi U\lambda)^2,$$

by virtue of the relations

$$\int_{-\pi}^{\pi} \cos m\xi \cos n\xi\, d\xi = 0 \quad \text{if} \quad m \neq n$$
$$= \pi \quad \text{if} \quad m = n.$$

30

THE DYNAMICS OF THIN SHEETS OF FLUID
I. WATER BELLS

REPRINTED FROM

Proceedings of the Royal Society, A, vol. CCLIII (1959), pp. 289–95

A simple solution of the equations representing the shape of an axially symmetric sheet of fluid is given. The shapes so calculated are compared with photographs of a 'water bell' produced by placing a plane or conical obstruction in the centre of a jet of water. The effect of air friction is evident in one of the photographs, and calculation, using a formula due to Howarth, shows that it should have been expected, though previous discussions of water bells have assumed it negligible.

The dynamics of thin sheets of fluid was studied as long ago as 1833 by Savart* who showed that it is possible to produce a bell-like sheet by placing a disc-shaped obstruction in the path of a vertical cylindrical jet of water or by making two such jets collide. The water spreads out horizontally and falls under the action of gravity. If the bell-shaped sheet so formed falls on a horizontal surface of water or a solid sheet it imprisons a volume of air and by supplying or taking away air from this volume a large number of shapes can be attained. The experimental work of Savart does not seem to have been followed up till Boussinesq† gave the differential equations for such sheets and solved them in a few cases by numerical methods.

More recently the subject has been taken up again. Bond‡ measured the dimensions of a sheet projected horizontally by the impact of two vertical jets. This method was also used by Savart. Bond deduced from these measurements the surface tension of water under dynamic conditions.

Boussinesq and Bond considered only cases where the pressure on the two sides of the film were equal. More recently Hopwood§ has described some remarkable shapes attained by water bells, when there is a difference in pressure between the inside and outside of the bell. This work was followed up by Lance & Perry‖ who set up the equations for describing such water bells and obtained a few particular solutions to them by numerical computation. As they point out, it is unlikely that analytical solutions will be found. On the other hand, a solution applicable when conditions are such that the effect of gravity is small can be found by neglecting the terms involving g in the full equations. Since this does not seem to be referred to in any of the papers to which I have had access, it is given below and experiments are

---

* *Annls. chim. Phys.* LIX (1833), 55 and 113.

† *C. R. Acad. Sci.*, Paris, LXIX (1869), 45 and 128.

‡ *Proc. phys. Soc.* XLVII (1935), 549.

§ *Proc. phys. Soc.* B, LXV (1952), 2.

‖ *Proc. phys. Soc.* B, LXVI (1953), 1067.

described in which the shapes of water bells made under the appropriate conditions were photographed and compared with these calculated shapes.

It will be assumed that the surface tension, T, is constant and viscosity is negligible, and that the sheet is projected in the form of a thin-walled cone of semi-vertical angle ϕ_0 having a vertical axis. The relevant physical variables are p, the pressure difference between the inside and outside of the bell; Q, the volume of fluid projected per second; u_0, the velocity of projection and g, gravity.

Since the surface tension is constant the equation for the fluid velocity u, neglecting air friction, is

$$u^2 = u_0^2 + 2gx, \tag{1}$$

where u_0 is the initial velocity and x the depth below the point when $u = u_0$. If we take y as the distance from the axis, the equation for the balance of inertia at right angles to the stream is

$$\frac{2T}{r_c} + \frac{2T\cos\phi}{y} - p + g\rho t\sin\phi - \frac{u^2\rho t}{r_c} = 0, \tag{2}$$

where ϕ is the slope of the surface to the vertical, t the thickness of the sheet at any point and r_c the radius of curvature of a meridian section. The equation of continuity is

$$2\pi y u t = Q. \tag{3}$$

Equations (1) and (2) can be expressed in non-dimensional form by setting

$$\frac{R_c}{r_c} = \frac{X}{x} = \frac{Y}{y} = \frac{S}{s} = \frac{4\pi T}{\rho Q u_0} = \frac{1}{R} \quad \text{and} \quad U = \frac{u}{u_0}, \tag{4}$$

where s is the length of arc of a meridian section, starting from the axis. Equation (1) becomes

$$U^2 = 1 + 2\beta X, \tag{5}$$

where

$$\beta = \frac{g\rho Q}{4\pi u_0 T}; \tag{6}$$

and since

$$\frac{1}{r_c} = -\frac{d\phi}{ds}, $$

(2) becomes

$$-\frac{d\phi}{dS}\left(1 - \frac{U}{y}\right) + \frac{\cos\phi}{Y} - \alpha + \beta\frac{\sin\phi}{UY} = 0, \tag{7}$$

where

$$\alpha = \frac{\rho u_0 Q p}{8\pi T^2}. \tag{8}$$

Equations (5) and (7) together with the geometrical equation

$$\sin\phi = dY/dS \tag{9}$$

are sufficient to determine the shape of a water bell. They are identical with those used by Lance & Perry except that theirs were not expressed in non-dimensional form, so that the effects of the terms depending on gravity and pressure, namely, those containing β and α, were not so easily distinguished.

It will be noticed that only two physical parameters, α and β, are involved in the equations. The only other parameter of the whole body of solutions is ϕ_0, the value of ϕ at the axis.

Solution of (7) when $\alpha = 0$ and β is negligible

In this case (5) shows that $U = 1$. Hence with the use of (9), (7) is reduced to

$$\left(1 - \frac{1}{Y}\right) \sin \phi \frac{d\phi}{dY} = \frac{\cos \phi}{Y} \tag{10}$$

which, on integration and with the condition $\phi = \phi_0$ when $Y = 0$, yields

$$(1 - Y) \cos \phi = \cos \phi_0. \tag{11}$$

For the purpose of tracing the meridian sections it is convenient to express (11) in terms of X and Y, by integrating $\tan \phi = dY/dX$ using (11)

$$X = \cos \phi_0 [\cosh^{-1}(\sec \phi_0) - \cosh^{-1}\{(1 - Y) \sec \phi_0\}]. \tag{12}$$

Expression (12) applies in the range $0 < x < \cos \phi_0 \cosh^{-1}(\sec \phi_0)$. At the upper limit $1 - Y = \cos \phi_0$ and $dY/dX = 0$ but d^2Y/dX^2 is finite. The solution can be continued into the region

$$\cos \phi_0 \cosh^{-1}(\sec \phi_0) < X < 2 \cos \phi_0 \cosh^{-1}(\sec \phi_0)$$

and the meridian section is symmetrical about the plane

$$X = \cos \phi_0 \cosh^{-1}(\sec \phi_0)$$

till at the upper limit of this section $Y = 0$ again.

The meridian sections for $\cos \phi_0 = 0$, $0 \cdot 1$, $0 \cdot 5$, $0 \cdot 8$ are shown in fig. 1, which represents in fact the sequence of shapes which would arise if a jet could be deflected by, say, a series of conical obstacles with the same energy loss at each. The distance from the vertex at which the sheet converges again onto the axis increases to a maximum value corresponding with $\phi_0 = 60°$. The broken curve in fig. 1 is the locus of the maximum radii of the sheets.

Experimental production of conical sheets

Sheets can be projected in an initially conical form by placing a conical impacter in the path of a cylindrical stream. To obtain a sheet comparable with those described by (12) it is necessary that β shall be small. If no energy were lost at impact $Q = \pi a^2 u_0$, where a is the radius of the jet before impact, so that $\beta = g\rho a^2/4T$. The experiments to be described were made with an orifice $0 \cdot 298$ cm. in diameter so that if T is taken as 73 dyn/cm. and the jet is of the same diameter as the orifice, $\beta = 0 \cdot 074$. In fact β will be rather larger than this by the factor $\sqrt{(2gH)/u}$, where H is the head of water used in forming the jet. This factor was in the experiments to be described of order $1 \cdot 25$ so that β was $0 \cdot 092$. Though β would not be negligible in

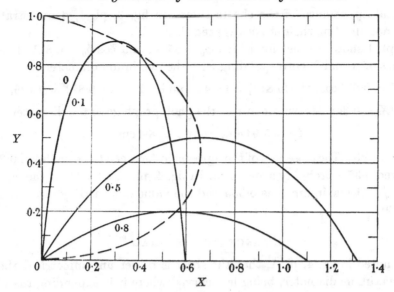

Fig. 1. Calculated meridian sections of water bells. The numbers give
the values of $\cos\phi_0$.

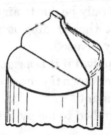

Fig. 2. Tool for making orifice.

any deductions from accurate measurements made for the purpose of determining T, it is sufficiently small to justify the use of (12) in interpreting the experiments.

The orifice of the jet was made in brass using a hardened steel tool in the form shown in the sketch fig. 2. The tube through which the water approached the orifice was 1·91 cm. in diameter and contained a number of discs made of copper-wire gauze to reduce turbulence. The water was supplied from a movable tank connected to the water supply and furnished with an overflow sill to keep the head constant. The impacters were either small flat or conical discs from 0·5 to 1·2 cm. in diameter and in order to ensure that the pressure inside the sheet was equal to the atmospheric pressure the sheet was allowed to fall on the outside of a co-axial tube which also carried the impacter. In order to afford visible proof that the effect of gravity on the shape of the sheet was not large the experiments were carried out with a horizontal jet. The effect of gravity would tend to make the lower half of the sheet sag away from the axis while the upper half would tend towards it. The symmetrical appearance of the sheets shown in fig. 3, pl. 1, demonstrates that gravity is not affecting their shapes appreciably.

Fig. 3, pl. 1 shows a sheet for which $\phi_0 = 35 \cdot 8°$, and $\cos \phi_0 = 0 \cdot 811$. The calculated values of X and Y corresponding with the maximum diameter are

$$X = 0 \cdot 811 \cosh^{-1}(1/0 \cdot 811) = 0 \cdot 544 \quad \text{and} \quad Y = 1 - 0 \cdot 811 = 0 \cdot 189,$$

so that $Y/X = 0 \cdot 348$. Measurements of the photograph gave the diameter

$$2y = 5 \cdot 94 \,\text{cm.} \quad \text{at} \quad x = 8 \cdot 4 \,\text{cm.},$$

so that $y/x = 0 \cdot 35$. The closeness of the agreement between the calculated $0 \cdot 348$ and the observed $0 \cdot 35$ is probably accidental. In fig. 4, pl. 1, $\phi_0 = 52 \cdot 8°$ the calculated value of Y/X at maximum Y is $0 \cdot 655$ and the value obtained by measurement of fig. 4 is $0 \cdot 60$.

Effect of air drag

It will be noted in fig. 4, pl. 1, that the sheet is rather unsymmetrical about the plane of maximum diameter, being less curved where it is expanding than it is at the same distance from the axis in the contracting half of the sheet. This could be accounted for by a decrease in u or an increase in T with distance from the vertex. An increase in surface tension with time seems to be thermodynamically impossible so that u must decrease. This can only be due to air drag. In previous work on water bells[*] it has been assumed that the air drag is negligible. To calculate the air drag, boundary-layer theory may be used.

The calculation is greatly simplified if it is assumed that the tangential reaction of the air is not great enough to slow down the sheet appreciably so that the boundary condition at the sheet is $u = \text{constant}$.

The necessary analysis was supplied to me by Professor L. Howarth, F.R.S., and is contained in an appendix.

The change in velocity of the sheet due to the air friction F acting over the area contained between sections of the bell is given by

$$\rho Q \,du = -2\pi y F \,ds. \tag{13}$$

Assuming that the change in velocity is small compared with u_0 the initial velocity at $y = 0$, u in Howarth's expression for F (appendix, (A 8)) can be replaced by u_0 and (13) becomes

$$-\frac{\delta u}{u_0} = \frac{u_0 - u}{u_0} = \frac{2\pi}{\rho Q u_0} \left(\frac{0 \cdot 6275}{2^{\frac{1}{2}}} \right) \rho_a u_0^{\frac{3}{2}} \nu_a^{\frac{1}{2}} \int_0^s \frac{y^2}{\left[\int_0^s y^2 \,ds \right]^{\frac{1}{2}}} \,ds. \tag{14}$$

Using (14), this may be expressed non-dimensionally in the form

$$-\frac{\delta u}{u_0} = 2 \cdot 788 \frac{\rho_a}{\rho} \left(\frac{\nu_a}{R u_0} \right)^{\frac{1}{2}} \left(\frac{R^2 u_0}{Q} \right) \int_0^S \frac{Y^2}{\left[\int_0^S Y^2 \,dS \right]^{\frac{1}{2}}} \,dS. \tag{15}$$

[*] W. N. Bond, *Proc. phys. Soc.* XLVII (1935), 549.

PLATE 1

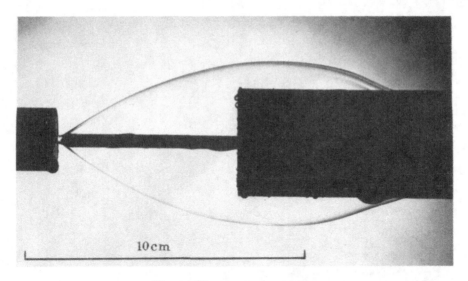

Fig. 3. Water bell, $\phi_0 = 35\cdot8°$.

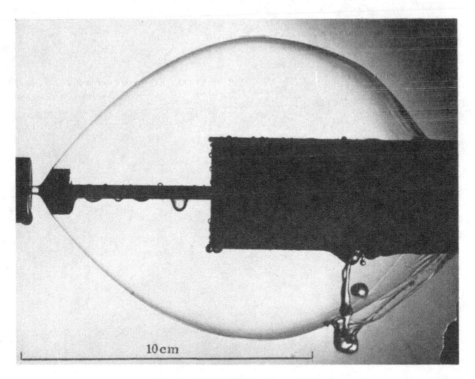

Fig. 4. Water bell, $\phi_0 = 52\cdot8°$.

The factor 2·788 is written for $2\pi(0.6275)\,2^{-\frac{1}{2}}$. This expression gives the decrease in velocity due to the friction on one side only of the sheet. The integral on the right is non-dimensional and could be calculated for any given value of ϕ_0 by means of (11). It can be seen without further calculation that the value of the integral will rapidly decrease as ϕ_0 decreases and this explains why the effect of air friction seen in fig. 4 for $\phi_0 = 52.8°$ is greater than it is for the water bell of fig. 3 for which $\phi_0 = 35.8°$.

To find the order of magnitude of the effect of air friction the simplest case, $\phi_0 = \frac{1}{2}\pi$, will be discussed numerically. This happens to be the case where the effect of air friction is greatest. The water bell degenerates to a flat disc of radius R and the integral in (15) is $2\sqrt{(r/3R)}$.

In experiments to be described in part III a flat sheet approximately 11·7 cm. radius was produced by impact of a stream at head 112·8 cm. onto a disc 1·2 cm. in diameter. In this case the value of R was shown to be 12·0 cm. Q was 33·2 cm.³/sec. The maximum radius when $\phi_0 = \frac{1}{2}\pi$ is $R = \rho u Q / 4\pi T$ and using the figure given above together with $T = 73$ c.g.s., u was found to be 331 cm./sec. With these values,

$$(\nu_a/Ru)^{\frac{1}{2}} = 5.9 \times 10^{-3}, \quad R^2 u/Q = 1.43 \times 10^3 \quad \text{and} \quad \rho_a/\rho = \tfrac{1}{800}$$

so that at $r = R$, (15) becomes

$$-\frac{\delta u}{u_0} = 2.788 \left(\tfrac{1}{800}\right) (5.9 \times 10^{-3}) (1.43 \times 10^3) \left(\frac{2}{\sqrt{3}}\right) = 0.034.$$

It appears therefore that in this case 3·4 % of the velocity has been lost by air friction on one side of the sheet. Since the sheet has two faces the loss of velocity owing to air friction is of order 6·8 %.

APPENDIX BY L. HOWARTH, F.R.S.

The boundary-layer problem posed by the foregoing experiments is effectively to determine the boundary layer in the air adjacent to a surface of revolution along which liquid flows with constant velocity u_0 in the direction of the meridians.

The pressure gradient is absent in the boundary layer and the equations take on the form*

$$u\frac{\partial u}{\partial s} + w\frac{\partial u}{\partial \zeta} = \nu_a \frac{\partial^2 u}{\partial \zeta^2}, \tag{A1}$$

$$yu = \frac{\partial \psi}{\partial \zeta}, \quad yw = -\frac{\partial \psi}{\partial s}, \tag{A2}$$

where ν_a is the viscosity of air, s is measured along a meridian of the surface of the body of revolution, ζ normal to it and $y = y(s)$ is the radius of the boundary. The boundary conditions are $u = u_0$, $w = 0$ when $\zeta = 0$, $u \to 0$ as $\zeta \to \infty$.

* See *Modern Developments in Fluid Dynamics*, edited by S. Goldstein, Oxford University Press (1938), p. 130.

It follows at once from Mangler's transformation* that if we write

$$x = \int_0^s [y(s)]^2 \, ds, \quad z = y\zeta, \quad \overline{\psi}(x,z) = \psi(s,\zeta), \tag{A 3}$$

the equations are reduced to the two-dimensional form

$$\overline{u}\frac{\partial \overline{u}}{\partial x} + \overline{w}\frac{\partial \overline{u}}{\partial z} = \nu_a \frac{\partial^2 \overline{u}}{\partial z^2}, \tag{A 4}$$

$$\overline{u} = \frac{\partial \overline{\psi}}{\partial z}, \quad \overline{w} = -\frac{\partial \overline{\psi}}{\partial x}, \tag{A 5}$$

with $\overline{u} = u_0$, $\overline{w} = 0$ at $z = 0$ and $\overline{u} = 0$ at $z = \infty$.

The solution is obtained by putting

$$\frac{\overline{u}}{u_0} = f'(\theta), \quad \text{where} \quad \theta = \left(\frac{u_0}{2\nu_a x}\right)^{\frac{1}{2}} z, \tag{A 6}$$

from which it follows that $\qquad f''' + ff'' = 0 \tag{A 7}$

with $f' = 1, f = 0$ at $\theta = 0, f' \to 0$ as $\theta \to \infty$.

The skin friction for the original problem with the boundary of revolution is therefore

$$F = \mu \left(\frac{\partial u}{\partial \zeta}\right)_0 = \frac{\rho_a u_0^{\frac{3}{2}} \nu_a^{\frac{1}{2}} f''(0)}{2^{\frac{1}{2}}} \frac{y}{\left[\int_0^s y^2 \, ds\right]^{\frac{1}{2}}}. \tag{A 8}$$

A first approximation to $f''(0)$ can be obtained, following Hartree, by setting the term f in (A 7) equal to a constant a. Solution of the resulting linear equation gives $f = (1 - e^{-a\theta})/a$ and so for consistency for large θ we must take $a = 1$.

A second approximation is then found by replacing the term f in (A 7) by $(1 - e^{-\theta})$. The resulting linear equation has a solution

$$\frac{u}{u_0} = f'(\theta) = \frac{1 - \exp(-e^{-\theta})}{1 - e^{-1}}$$

and this gives $f''(0) = -1/(e-1) = -0.58$.

A very rough numerical integration by Adams' method suggests that

$$f''(0) = -0.63,$$

while Pohlausen's method applied to (A 4) leads to a value $-(\frac{23}{63})^{\frac{1}{2}} = -0.60$, for a quartic profile.

[After the above rough estimates had been made Dr M. H. Rogers kindly offered to compute the function f to seven-figure accuracy on the Southampton Pegasus machine. He finds $f''(0) = -0.6275538$.]

* *Z. angew. Math. Mech.* xxviii (1948), 97.

31

THE DYNAMICS OF THIN SHEETS OF FLUID
II. WAVES ON FLUID SHEETS

REPRINTED FROM

Proceedings of the Royal Society, A, vol. CCLIII (1959), pp. 296–312

It is shown that capillary waves are of two kinds, symmetrical waves in which the displacements of opposite surfaces are in opposite directions, and antisymmetrical waves in which the displacements are in the same direction. Any disturbance can be regarded as composed of these two types of wave. The antisymmetrical waves are non-dispersive. In a sheet of uniform thickness a moving point disturbance produces two narrow line-like waves. In a radially expanding sheet a fixed disturbance point produces two narrow disturbances in the form of cardioids.

It is shown theoretically that a finite change in direction of flow can occur at a cardioid which therefore assumes the form of a sharp edge. A method was found for producing and photographing a sheet with a sharp edge in the form of a cardioid.

The symmetric waves are very different, they are highly dispersive and are propagated much more slowly than the antisymmetrical waves. Experimentally a point disturbance produces both kinds of wave simultaneously. Reflection photographs show the antisymmetrical waves, while the schlieren method is needed to reveal the symmetrical waves. The symmetrical waves produced in a moving sheet by a point disturbance are parabolas when the sheet is uniform in thickness, and of a more complicated form when the sheet is expanding. The predicted wave patterns agree with those revealed by the schlieren photographs.

INTRODUCTION

Capillary waves on flat and cylindrical fluid surfaces have been studied extensively. Less attention has been paid to sheets of fluid bounded by two parallel planes. Squire* has studied the instability which arises in a moving sheet owing to the reaction of the surrounding air. This instability is of importance in technical applications, particularly in studying how a sheet breaks up into drops when projected at high speed into air. The existence of steady water bells such as those illustrated in part I shows that there are cases where this instability is unimportant and in the following pages the properties of wave systems due to capillarity alone will be studied, effects due to air reaction and gravity being neglected.

Small disturbances of the surface of deep water can be analysed into their Fourier components so that it is necessary only to study those in which the velocity potential and the surface displacement are simple harmonic functions involving only one wavelength. Four numbers are required to describe the most general disturbance of this type at any instant. These might, for instance, be an amplitude and a phase both for the velocity potential and for the surface displacement. The well-known analyses of this case show that the motion can be described by regarding the initial

* *Br. J. appl. Phys.* IV (1953), 163.

distribution of velocity and displacement as being due to the superposition of two waves moving at equal speeds in opposite directions. Such pairs of waves require four numbers for their complete specification and the pressure condition is satisfied at the surface if the wave speed has a certain value.

When the same kind of analysis is applied to a sheet of uniform thickness it is found that eight numbers are required for a complete specification of the most general simple harmonic disturbance in two dimensions. These can be resolved into four waves, two in which the displacement on the two faces of the sheet are equal in magnitude and phase, and two in which they are equal in magnitude but opposite in phase. Each of these four component waves is specified by two numbers, an amplitude and a phase, so that eight numbers are required.

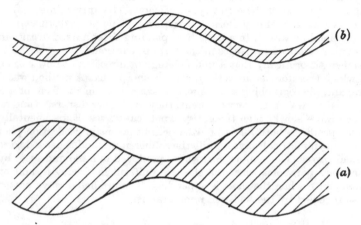

Fig. 1. Sketch of (*a*) symmetrical, and (*b*) antisymmetrical waves.

As in the case of a single surface the relationship between the velocity potential and surface displacement can be chosen for each component so that the pressure condition is satisfied if the wave propagates itself at a certain speed. It is found, however, that two wave speeds are involved, one, (*b*), appropriate to waves in which the displacement of opposite surfaces of the sheet are in phase and the other, (*a*), in which they are in opposite phase. These will be described as antisymmetrical and symmetrical waves. The two types are shown in the sketch (fig. 1).

SYMMETRICAL WAVES

If $\lambda = 2\pi/\kappa$ is the wavelength, τ time, t the thickness of the sheet and w_s the velocity of symmetrical waves, the velocity potential

$$\phi = a w_s \cosh \kappa z \cos \kappa(x - w_s \tau) \tag{1}$$

corresponds to displacements of the two surfaces

$$\eta = \pm a \sinh \tfrac{1}{2}\kappa t \sin \kappa(x - w_s \tau). \tag{2}$$

The pressure condition to be satisfied is

$$\rho\phi = -T\partial^2\eta/\partial x^2 \quad \text{at} \quad z = \tfrac{1}{2}t + a\sinh\tfrac{1}{2}\kappa t \sin\kappa(x-w_s\tau), \tag{3}$$

where T is the surface tension.

This leads to the equation for wave velocity

$$w_s^2 = (2T/\rho t)\{\tfrac{1}{2}\kappa t \tanh(\tfrac{1}{2}\kappa t)\} \tag{4}$$

and the components of velocity in the fluid

are

$$-\partial\phi/\partial x = aw_s\kappa\cosh\kappa z \sin\kappa(x-w_s\tau) \tag{5}$$

and

$$-\partial\phi/\partial z = aw_s\kappa\sinh\kappa z \cos\kappa(x-w_s\tau). \tag{6}$$

ANTISYMMETRICAL WAVES

If w_a is the wave velocity in this case the velocity potential is

$$\phi = aw_a\sinh\kappa z \cos\kappa(x-w_a\tau) \tag{7}$$

and the displacement of both surfaces

$$\eta = a\cosh\tfrac{1}{2}\kappa t \sin\kappa(x-w_a\tau). \tag{8}$$

The pressure condition leads to

$$w_a^2 = (2T/\rho t)\{\tfrac{1}{2}\kappa t \coth(\tfrac{1}{2}\kappa t)\} \tag{9}$$

and the velocity components of the fluid are

$$-\partial\phi/\partial x = aw_a\kappa\sinh\kappa z \sin\kappa(x-w_a\tau), \tag{10}$$

$$-\partial\phi/\partial z = aw_a\kappa\cosh\kappa z \cos\kappa(x-w_a\tau). \tag{11}$$

κt SMALL

When κt is large, $w_a = w_s$ and each is the velocity of capillary waves on water of great depth. In most of the sheets to be discussed t is of order 5 to 100μ so that for any visible wave κt is very small. This leads to great simplicity in the discussion of both types of wave.

In symmetrical waves (5) and (6) show that the component of velocity parallel with the surface of the sheet is large compared with that at right angles to it so that a treatment which neglects the latter velocity is a legitimate approximation. This will be the basis of further treatment of these waves.

In the antisymmetrical waves, (10) and (11) show that the component of velocity along the sheet is small compared with the velocity at right angles to it. This leads to the simplified treatment in which the waves are considered as exactly analogous to waves in a string stretched with tension $2T$ and mass per unit length ρt. In fact when κt is small $\tfrac{1}{2}\kappa t \coth\tfrac{1}{2}\kappa t = 1$ and (9) gives

$$w_a^2 = 2T/\rho t. \tag{12}$$

The velocity of antisymmetrical waves is therefore independent of wavelength.

STATIONARY ANTISYMMETRICAL WAVES IN A MOVING SHEET

The fact that antisymmetrical waves are propagated at speed $(2T/\rho t)^{\frac{1}{2}}$ whatever their wavelength suggests an analogy with sound waves. In a sheet moving with velocity u they will be at rest in space if lines of constant phase are at angle ψ to the direction of flow provided

$$\sin \psi = (2T/\rho t u^2)^{\frac{1}{2}}. \tag{13}$$

The number $2T/\rho t u^2$ has been defined as the Weber number, W, and ψ is analogous to the 'Mach angle' in supersonic flow.

In a sheet of uniform thickness moving with velocity u therefore, a small fixed disturbing obstacle would be expected to produce disturbances limited to the close neighbourhood of the two Mach lines at angles $\pm \sin^{-1}(W^{\frac{1}{2}})$ to the direction of flow and in experiments to be described later it is shown that this is a correct prediction.

When the sheet is expanding radially so that t is inversely proportional to the distance from the centre, straight waves of the type considered in the last section are no longer possible. It would be expected, however, that waves of lengths small compared with that in which appreciable variation in t occurs would be propagated at the speed of waves in a sheet of uniform thickness equal to the local value of t. With the assumption that this is true, the direction of lines of constant phase in waves on an expanding sheet will be at rest at angle ψ to the radii if

$$\sin^2 \psi = \frac{2T}{\rho u^2 t} = \frac{4\pi T r}{\rho u Q} = \frac{r}{R}, \tag{14}$$

where

$$R = \rho u Q/4\pi T \tag{15}$$

and Q is the volume flowing outwards per second.

It will be seen that no waves can remain at rest outside $r = R$ and that at that radius lines of constant phase are at right angles to the radii. Comparing this with (4) in part I,* it will be seen that R is also the limiting radius of the series of water bells described by (12) when the bell becomes two flat sheets joined together at the radius R.

In polar co-ordinates (r, θ) the wavelets presented by (14) are

$$2r/R = 1 - \cos(\theta - \theta_0). \tag{16}$$

θ_0 is a constant introduced by integration and it represents physically the angular position of an arbitrary point from which the wave might be regarded as starting. The curves (16) are cardioids which are the loci of points on a circle of radius $\frac{1}{4}R$ when it rolls on a fixed circle of the same radius. They can be drawn mechanically and a set of eight of them corresponding to eight equally spaced values of θ_0 is shown in fig. 2.

* Paper 30 above.

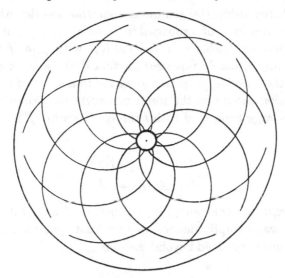

Fig. 2. Eight cardioids.

CALCULATION OF ANTISYMMETRICAL WAVES IN AN EXPANDING SHEET

The equation of motion for a steady disturbance which involves a displacement η from the plane of the sheet is

$$2T\left[\frac{\partial^2\eta}{\partial r^2}+\frac{1}{r}\frac{\partial\eta}{\partial r}+\frac{1}{r^2}\frac{\partial^2\eta}{\partial\theta^2}\right]-\rho t u^2\frac{\partial^2\eta}{\partial r^2}=0. \tag{17}$$

Writing $x = r/R$ this becomes

$$\left(1-\frac{1}{x}\right)\frac{\partial^2\eta}{\partial x^2}+\frac{1}{x}\frac{\partial\eta}{\partial x}+\frac{1}{x^2}\frac{\partial^2\eta}{\partial\theta^2}=0. \tag{18}$$

If all such disturbances could be regarded as being due to superposition of small cardioid disturbances, a solution of (18) might be expected in the form

$$\eta = \Phi(x)f\{\theta - \cos^{-1}(2x-1)\} \tag{19}$$

because the function $\theta - \cos^{-1}(2x-1)$ is equal to θ_0 which is constant along a cardioid. To test whether this form is a solution, (19) was substituted in (18), and the terms rearranged, giving

$$\left[\frac{\Phi}{x^2}-\frac{\Phi}{x^2}\right]f''+[2\Phi'x^{-\frac{3}{2}}(1-x)^{\frac{1}{2}}+\tfrac{1}{2}\Phi x^{-\frac{5}{2}}(1-x)^{-\frac{1}{2}}-\Phi x^{-\frac{3}{2}}(1-x)^{-\frac{1}{2}}]f'$$

$$+\left[\frac{1-x}{x}\Phi''-\frac{1}{x}\Phi'\right]f = 0. \tag{20}$$

In order that (20) may be satisfied each of the expressions multiplying f, f' and f'' must vanish. These conditions are not consistent, so that the form (19) is inadmissible as a general solution of (18). On the other hand, the existence of the cardioid

23-2

wavelets was predicted using the condition that the wavelength or scale of the disturbance in the direction of propagation is small compared with R. If δ is this length f' must be comparable with $(R/\delta)f$ and f'' with $(R/\delta)^2 f$. The vanishing of the coefficient of f'' in (20) merely expresses the fact that waves of short wavelength are propagated at the speed appropriate to waves in a sheet of uniform thickness. It gives no information about how the amplitude of the disturbance varies along the cardioid. Since f' is large compared with f the coefficient of f' in (20) must vanish. This reduces to

$$\frac{\Phi'}{\Phi} = \frac{1}{4x(1-x)}, \tag{21}$$

so that

$$\Phi = Ax^{\frac{1}{4}}/(1-x)^{\frac{1}{4}}, \tag{22}$$

where A is the integration constant. Fig. 3 shows Φ/A as a function of x. It will be noticed that the wave amplitude increases with x throughout its range, but the increase only becomes very rapid beyond $x = 0.9$.

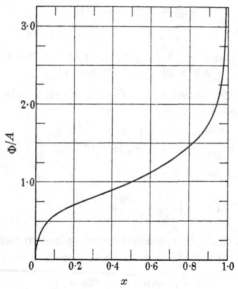

Fig. 3. Amplitude of cardioid disturbance Φ as a function of $x (= r/R)$. $\Phi = Ax^{\frac{1}{4}}/(1-x)^{\frac{1}{4}}$.

EXPERIMENTS ON ANTISYMMETRICAL WAVES IN AN EXPANDING SHEET

The apparatus already described in part I was set so that the jet was directed vertically upwards. A flat impact disc 1·2 cm. in diameter was mounted on the chuck of a lathe and eight equally spaced radial nicks were cut in its face by a cutting tool in a slide rest set slightly obliquely. This is shown in fig. 4.

The object of preparing such a disc was to produce an easily recognizable wave pattern for comparison with the cardioid diagram of fig. 2. When the sheet produced by this impactor is looked at very obliquely a wave system could be seen, but since the object of the experiment was to compare the wave system produced with the predicted pattern it was necessary to arrange the lighting so that the system could be seen in a camera placed on or near the axis of symmetry. It was found to be difficult to arrange any system of illumination which revealed more than a short arc of the wave by reflected light when the camera was set accurately on the axis, but by placing it a short distance off centre, larger arcs could be seen.

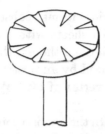

Fig. 4. Impact disc with eight grooves.

The method finally adopted for photographing the waves was to place the camera with its lens 30 in. above the level of the impact disc and offset from the vertical through its centre. The camera was set with the plates horizontal so that the picture would not be distorted even though the image was not centred on the vertical through the lens. To reveal the whole of the wave system it was necessary to move the source of light over a limited field at a level halfway between the camera and the sheet and take a time exposure. Since the continual movement of the ragged edge of the sheet might be expected to make the whole wave pattern rather unsteady, a metal ring 19 cm. in diameter was placed so that it intercepted the fluid before reaching the free edge. The photograph (fig. 5, pl. 1) shows the wave system enclosed in the limiting ring. The exposure in this case lasted for 45 sec. Fig. 6, pl. 1, shows the part of the pattern which could be seen by means of two photoflashes which operated simultaneously.

To compare the waves with those predicted in (16) the negative of fig. 2 was projected on to fig. 5 and the magnification adjusted so as to make one of the cardioids lie as nearly as possible on one of the waves. It was found that the two could be made to coincide over their whole length till they met the limiting ring. It will be seen from (14) that this is a proof that T/u is invariable along the radii at any rate to the accuracy with which a true cardioid could be compared with the photograph.

EXISTENCE OF STEADY FLOW ROUND A SHARP CORNER

The wavelets which can lie at rest on a moving sheet have some properties analogous to those of the wavelets in an expanding supersonic flow. It is interesting to compare them. In a supersonic stream the change in the direction of flow which occurs at a wavelet lying at the 'Mach angle' to the stream is necessarily accompanied by a change in velocity. This involves a change in Mach angle. This change as well as the change in direction makes it impossible for neighbouring wavelets or 'characteristics' to lie parallel to one another. Thus a finite sudden change in velocity cannot occur on a Mach line. Antisymmetrical wavelets in a sheet which involve a small change in the direction of motion of particles do not involve a change in particle velocity. For this reason the cardioids in the expanding sheet could lie very close together, and in fact a finite change in direction of motion should be possible at a cardioid.

To test this experimentally, various forms of obstruction to an expanding jet were tried, and some of them produced sheets with a sharp edge along a cardioid. It was found difficult to photograph them though they were easy to see. Figs. 7 and 8, pl. 1, show a sharp edge produced by an obstruction in the form of a sheet cut into the form of a triangle with a vertex of 140°. When in position each arm sloped at 20° to the horizontal.

In fig. 7 the camera was level with the unobstructed sheet which therefore appears as a horizontal upper edge. The cardioid being in this plane is not distinguishable, but the edge where the stream turns through a sharp angle at the cardoid is very clear. The source of light in this picture was a flash bulb below the level of the sheet and obscured by the obstruction. The rather mottled appearance is due to drops leaving the far edge of the sheet and seen through it. Fig. 8 was taken by a camera facing downwards at an angle of 26·5° to the horizontal. The source of light was a flash in a diffusing reflector which was placed so as to reflect light into the camera lens from as large a part of the unobstructed sheet as possible. The cardioid forms the nearer edge of the bright area and the outline of the smooth sheet formed by the water after it has turned through a sharp angle at the edge is also visible. The spots are drops which have become detached from the ragged far edge of the unobstructed sheet and are seen through it rather out of focus in the background. They mar the beautiful smooth appearance of the curved sheet as seen by the eye, but they do not obscure the sharpness of the edge between it and the unobstructed part of the sheet. The black vertical line seen across the middle of the unobstructed sheet in fig. 8 is the rod which holds the impact disc. The disc itself is obscured by the obstruction. Figs. 9 and 10, pl. 1, show a top view of a similar sharp edge at which the sheet turned sharply through about 45°. The piece of metal which anchored the edge of the sheet sloped downwards at about 30°. Fig. 9 is lighted from below and fig. 10 from above. The discontinuous appearance of the sharp edge in fig. 10 is due to the fact that though two extended sources of light and a mirror were used to illuminate the edge, parts of it did not reflect any light into the camera lens.

SYMMETRICAL WAVES ON A THIN SHEET OF UNIFORM THICKNESS

It has been pointed out in (5) and (6) that when κt is small the component of velocity parallel to the sheet in symmetrical waves is large compared with the component perpendicular to it and is nearly uniform through its thickness. The disturbance equations will be formulated in a simple form which assumes constant velocity v through the sheet, and neglects components at right angles to it. Taking $\pm \eta$ as the displacements of the two surfaces of the sheet, y the co-ordinate in the direction of propagation and τ time, the relevant equations, linearised so as to apply to waves of small amplitude, are

$$p = -T\frac{\partial^2 \eta}{\partial y^2} \quad \text{(surface tension condition)},$$

$$-\frac{1}{\rho}\frac{\partial p}{\partial y} = \frac{\partial v}{\partial \tau} \quad \text{(equation of motion)},$$

$$-t\frac{\partial v}{\partial y} = 2\frac{\partial \eta}{\partial \tau} \quad \text{(continuity)}.$$

From these the wave equation is found to be

$$\frac{Tt}{2\rho}\frac{\partial^4 \eta}{\partial y^4} + \frac{\partial^2 \eta}{\partial \tau^2} = 0. \tag{23}$$

Solutions of (23) corresponding with infinite trains of waves are of the form $\eta \propto \cos \kappa(y - w_s\tau)$ provided

$$Tt\kappa^2/2\rho = w_s^2, \tag{24}$$

which is identical with the limiting form of (4) as $\kappa t \to 0$.

Symmetrical waves are dispersive so that it is more difficult to discuss the wave system produced by a moving point source than it is for antisymmetrical waves. An approximate treatment can be given when the point disturbance moves through the sheet with velocity comparable with $(2T/\rho t)^{\frac{1}{2}}$ and therefore large compared with the velocity of symmetrical waves. In such a case the waves lie at a small angle to the direction of motion of the point source of disturbance and it is legitimate to use an approximation introduced by Rayleigh while discussing boundary-layer problems. In this approximation the waves at distance x behind a source moving with velocity u are considered as identical with those produced by a momentary impact at all points along the line of motion at time x/u after the impact. In a dispersive wave pattern the magnitude of the disturbance and the phase at any point will depend on details of the flow very close to its origin, but a short distance away the waves are likely to assume a form which does not depend on these details. For this reason a similarity solution of (23) was sought in which the scale increases with time. The form of the solution for an infinite train of simple waves suggested that the similarity variable might be y^2/τ and the form

$$\eta = A\tau^{-\frac{1}{2}}f(y^2\tau^{-1}) \tag{25}$$

was tried. It gives $\qquad \partial^2 \eta/\partial \tau^2 = \frac{3}{4}\tau^{-\frac{5}{2}}f + 3y^2\tau^{-\frac{7}{2}}f' + y^4\tau^{-\frac{9}{2}}f''$

and $\qquad \partial^4 \eta/\partial y^4 = 12\tau^{-\frac{5}{2}}f'' + 48y^2\tau^{-\frac{7}{2}}f''' + 16y^4\tau^{-\frac{9}{2}}f^{\mathrm{iv}}.$

Substituting these in (23), it is found that (25) is valid if

$$16f'' + \frac{2\rho}{Ti}f = 0,\tag{26}$$

so that

$$\eta = A\tau^{-\frac{1}{2}}\cos\left\{\left(\frac{\rho}{8Tt}\right)^{\frac{1}{2}}y^2\tau^{-1}+\epsilon\right\},\tag{27}$$

where A and ϵ are constants of integration.

Substituting x/u for τ in (27), a possible steady wave pattern is found, namely

$$\eta = A\left(\frac{u}{x}\right)^{\frac{1}{2}}\cos\left\{\frac{y^2u}{x}\left(\frac{\rho}{8Tt}\right)^{\frac{1}{2}}+\epsilon\right\}.\tag{28}$$

The lines of constant phase are parabolas with their common axes lying downstream from the origin. Equation (27) is not the complete solution of the problem presented by a sudden disturbance applied along a line to a sheet which was previously at rest. It involves an infinite amount of energy, but this does not invalidate the solution as giving a correct representation of the geometry of the lines of constant phase. A similar situation arises in the case of gravity waves. The infinite energy associated with the simple wave system (27) is due to the fact that this system extends to infinity. In fact, an impulsive generation of waves is not limited to an infinitesimally small area or to an instant of time and the pattern produced could be regarded as a superposition of the wave patterns of the simple type represented by (27). Close to the origin the waves reinforce one another but far from the origin they interfere.

The wave pattern due to the motion of a pressure point on a sheet of uniform thickness can also be calculated by a method due to Rayleigh.* It is assumed that the lines of constant phase are geometrically similar to one another, the centre of symmetry being the pressure point. Since the angle ψ is then constant along any radius and for a given sheet ψ depends only on λ, the phase must increase linearly along any radius from the pressure point. Also, since tangents to lines of constant phase at points along this radius are parallel to one another the perpendicular p_ϵ from the origin to the tangent to a line whose phase is ϵ is

$$p_\epsilon = \lambda(\epsilon-\epsilon_0)/2\pi,\tag{29}$$

where ϵ_0 is the phase on the streamline through the disturbance point. To find the equation to this phase line therefore $2\pi p_\epsilon/(\epsilon-\epsilon_0)$ must be substituted for λ in the equation representing the connection between λ and ψ. Thus with the use of (4), the equation to the phase line of phase ϵ is

$$\sin^2\psi = W\left\{\frac{t(\epsilon-\epsilon_0)}{2p_\epsilon}\tanh\frac{t(\epsilon-\epsilon_0)}{2p_\epsilon}\right\}.\tag{30}$$

Equation (30) is not limited to the case when ψ is small but it is difficult to transform an equation between ψ and p_ϵ into a polar or Cartesian form. For any assigned

* *Proc. Lond. math. Soc.* xv (1893), 69.

PLATE 1

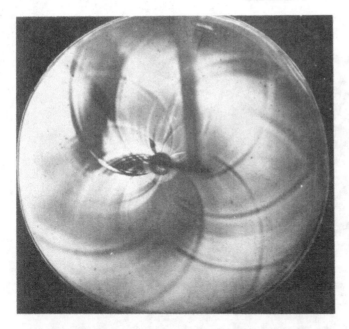

Fig. 6. Eight cardioids of
flash exposure.

Fig. 5. Eight cardioid waves, exposure 45 sec.

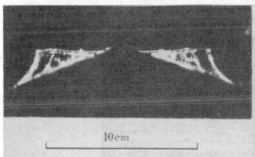

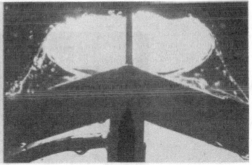

Fig. 7. Sheet with sharp edge
seen from side.

Fig. 8. Sheet with sharp edge
seen obliquely.

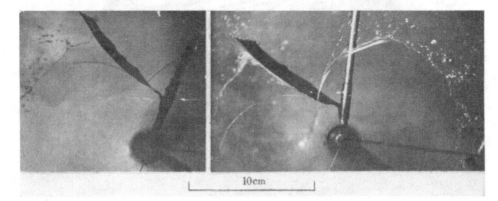

Fig. 9. Sheet with sharp edge
lighted from below.

Fig. 10. Sheet with sharp edge lighted
from above.

PLATE 2

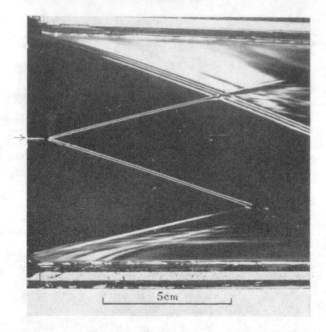

Fig. 12. Antisymmetrical waves in a sheet of uniform thickness. The arrow marks the position of the glass capillary which directs the air jet.

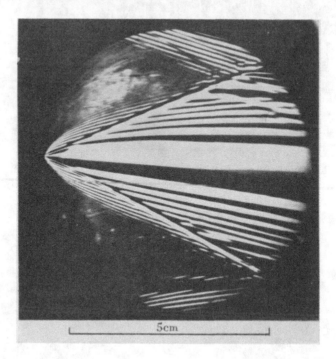

Fig. 13. Symmetrical waves in a vertical sheet flowing horizontally.

PLATE 3

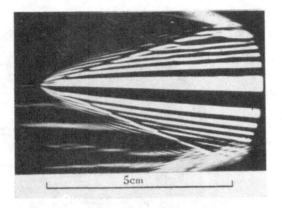

Fig. 14. Symmetrical waves in a horizontal sheet.

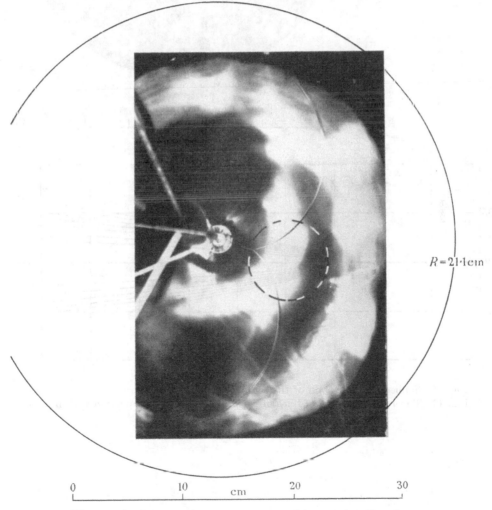

Fig. 15. Antisymmetrical waves produced by a point disturbance.

PLATE 4

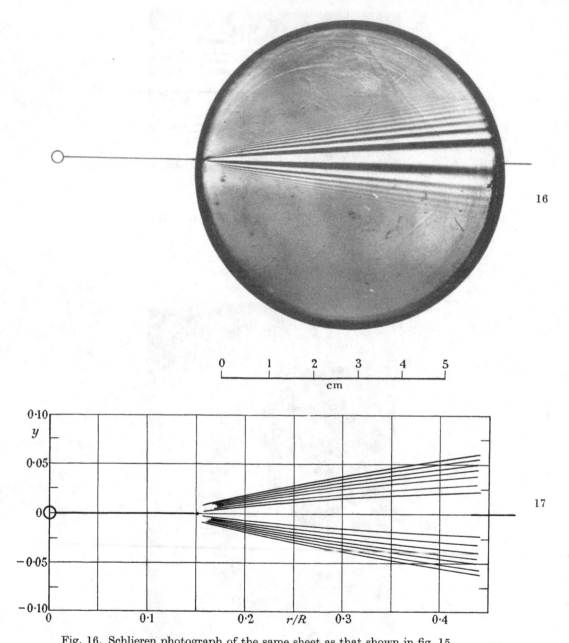

Fig. 16. Schlieren photograph of the same sheet as that shown in fig. 15.
Fig. 17. Theoretical pattern of lines of constant phase for comparison with fig. 16.

value of p_ϵ, ψ can be calculated and the curve of constant phase can be constructed graphically as the envelope of the tangents is calculated. In the limiting case when ψ is small (30) can be written

$$\sin \psi = W^{\frac{1}{2}}\{t(\epsilon - \epsilon_0)/2p_\epsilon\}. \tag{31}$$

This can be transformed into polar co-ordinates, giving

$$r(1 - \cos \theta) = tW^{\frac{1}{2}}(\epsilon - \epsilon_0) \tag{32}$$

which represents a set of confocal parabolas. The focal length of the parabola on which the phase is ϵ is $\frac{1}{2}tW^{\frac{1}{2}}(\epsilon - \epsilon_0)$. According to (28) the equation to the parabola where the phase is ϵ is

$$\frac{y^2}{2xt}W^{-\frac{1}{2}} + \epsilon_0 = \epsilon, \tag{33}$$

which represents a parabola of focal length $\frac{1}{2}tW^{\frac{1}{2}}(\epsilon - \epsilon_0)$ which cuts the axis at $x = 0$ instead of $x = -\frac{1}{2}tW^{\frac{1}{2}}(\epsilon - \epsilon_0)$. The difference is not significant when ψ is small, i.e. when y/x is small.

When y/x is not small (30) is still valid and it is worth while to compare the true line of constant phase represented by (30) with its parabolic approximation. The difference is most marked on the axis at points upstream from the disturbance point. Here $\sin \psi = 1$ so that the distance apart of lines whose phases differ by 2π are the values of λ which are solutions of the equation

$$1 = W\left(\frac{\pi t}{\lambda}\tanh\frac{\pi t}{\lambda}\right);$$

for $W = 0\cdot1$, for instance, $\lambda = 0\cdot1\pi t$. If the parabolic approximations (32) to the phase lines had been assumed to be valid when $y = 0$, $\lambda = 0\cdot31\pi t$. Thus the wave crests upstream of the disturbance are about three times as close together as they would be if the parabolic wave observed downstream were assumed to remain parabolic upstream.

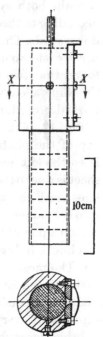

EXPERIMENTAL PRODUCTION OF WAVES IN A UNIFORM SHEET

It is difficult to produce a sheet of uniform thickness as thin as the expanding sheets already described. The apparatus used is shown in fig. 11. Water was forced under pressure through a slit between two curved brass cheeks (section XX, fig. 11) which were screwed to the side of the cylindrical pressure vessel. This vessel was $8\cdot3$ cm. long and $3\cdot8$ cm in diameter and the water entered at one end through a tube $3\cdot8$ cm. in diameter which contained six turbulence-suppressing wire-gauze screens. The pressure head H was maintained constant by a

section XX

Fig. 11. Slit orifice for producing a sheet of uniform thickness.

by-pass weir and was measured by a manometer connected with the tube shown at the top of the chamber in fig. 11. In most of the experiments the slit was 0·0152 cm. wide.

When the water was turned on, a number of jets sprang from the slit. These could be combined into a single sheet by inserting a piece of brass foil into the slit and moving it from end to end. It was thought that the foil acted as a remover of bubbles which had lodged in the converging part of the slit orifice. If no precautions were taken the sheet assumed a triangular form, but by mounting wetable rods extending at right angles to the slit from each end to a distance of 20 cm. a sheet 8·7 × 20 cm. could be produced. Since in the first experiments the slit was vertical the particles in the sheet might be expected to fall under the action of gravity without altering the thickness except in a region close to the upper edge. After measuring the total flux through the slit for a range of values of H, a disturbance point was installed. This consisted of a small air jet from a glass tube drawn down to 0·05 cm. bore and directed on to the sheet at a point near the slit.

ANTISYMMETRICAL WAVES

The waves produced by a point disturbance of this kind would be expected to contain both symmetrical and antisymmetrical components. The latter should according to theory be straight or if gravity is taken into account, be slightly curved downwards. The fact that they are in straight lines rather than in cardioids makes it much easier to photograph them than was the case with the expanding sheet. It was found that two fixed horizontal linear incandescent bulbs placed above and below the camera lens were capable of revealing the whole pattern so that the rather haphazard procedure which produced fig. 5 was not necessary.

Fig. 12, pl. 2, is an example. The orifice slit coincides with the left-hand boundary of the photograph. The glass tube from which the disturbing air jet issued is on the left and the small gap between the end of the tube and the waves in the sheet is due to the fact that the air jet was aimed at 45° to the sheet. The upper pair of lines mark the points on the sheet where the rays from the lower light were reflected into the camera and the lower pair were lighted by the upper light. It seems that the reason a pair of lines appears in each case is that they are situated on opposite sides of a line of inflexion where the slope of the sheet to its undisturbed position is greatest. The most striking feature of fig. 12 is the accuracy with which the two lines of each pair remain parallel to one another. This indicates that the wave is propagated without dispersion. Both pairs are very slightly curved downwards as would be expected if every particle in the sheet is falling under the action of gravity. The wetable glass tubes which keep the sheet intact can be seen at the top and bottom of the photograph. The sloping lines originating from the top and bottom of the slit are antisymmetrical waves starting from these points.

EXPERIMENTS ON SYMMETRICAL WAVES IN A SHEET OF UNIFORM THICKNESS

The method already described for producing antisymmetrical waves by directing a narrow jet of air at a fixed point on to the flowing sheet might also be expected to produce symmetrical waves. Theory indicates that the two sorts of wave should be quite independent and should be propagated at very different speeds so that antisymmetrical waves should lie at a larger angle to the flow than symmetrical waves. When the sheet of which fig. 12 is a reflection photograph taken under nearly normal illumination was looked at very obliquely, a small disturbed area could be seen which is not visible in fig. 12. It was thought that this might be due to symmetrical waves. It has been pointed out that antisymmetrical waves are analogous to those in a stretched string for which the wave amplitude may be much larger than the diameter of the string. Thus the amplitude of an antisymmetrical wave may be many times as great as the thickness of the sheet. Rays of light are therefore reflected through a far greater range of angle than transmitted light which in fact would hardly be deflected at all by waves which leave the thickness of the sheet unchanged. In symmetrical waves the surface cannot be displaced through a distance greater than half the thickness of the sheet, but the thickness of the sheet varies, so that transmitted rays will be deflected through an angle which is greatest where the rate of change in thickness is greatest. This consideration suggested that symmetrical waves should be visible in a schlieren photograph in which antisymmetrical waves might be invisible.

A photographic lens of 7 in. focal length and nominal aperture 2·5 was available. This was mounted at one end of a tube, the other end of which was blocked by a flat sheet containing a wide slit. In the first experiments the tube was mounted horizontally with the wide slit horizontal. The orifice slit was vertical so that the fluid sheet was moving horizontally in a vertical plane. An ordinary electric bulb was placed behind the wide slit and its image fell on a camera lens aiming horizontally at the sheet. Direct light from the slit was cut out by means of a sheet of foil half covering the camera lens. Rays which were deviated in passing through the water sheet escaped round the edge of the foil and so produced a schlieren picture on the sheet. Fig. 13, pl. 2, was taken in this way. It will be seen that the parabolas predicted in equations (28) and (33) are visible, but they are bent slightly downwards owing to the action of gravity. The photograph shows clearly the two lines which mark the position of the otherwise invisible non-dispersive antisymmetrical waves as they cross the system of parabolic symmetrical waves. These also are bent downwards. The photograph also shows that the points at which parabolas cross the antisymmetrical waves are spaced at regular intervals along the radius from the disturbance points. This verifies the assumption used by Rayleigh in the case of waves in deep water that the curves of constant phase form a system of waves which are similar, the centre of similarity being the disturbance point.

The parabolas of fig. 13 are distorted by gravity. It was therefore thought worth while to turn the orifice slit through a right angle so as to produce a horizontal sheet and take a schlieren photograph with the camera pointing vertically downwards. Fig. 14, pl. 3, was taken in this way. It will be seen that the pattern is more symmetrical than that of fig. 13.

MEASUREMENTS OF FIGS. 12, 13, 14

Fig. 12. The angle $2\psi_a$ between the two antisymmetrical waves was 31·4° and the mean rate of discharge was $Q_1 = 5\cdot8$ cm.³/sec. per cm. length of the slit. Since u was not measured, equation (13) must be used in the form

$$2Tt/\rho Q_1^2 = \sin^2 15\cdot7° = 0\cdot073,$$

so that $Tt = \frac{1}{2}(0\cdot073)\,(5\cdot8)^2$. Taking T for water as 73 dyn/cm. gives

$$t = 0\cdot0167\,\text{cm.}$$

As was to be expected this is rather greater than the width of the orifice slit, namely, 0·0152 cm.

Figs. 13 *and* 14. The following measurements were made:

(a) The angle $2\psi_a$ between the two lines along which the parabolas were discontinuous (interpreted as the loci of the antisymmetrical waves). These were 50° in fig. 13, and 40° in fig. 14.

(b) The distances between the discontinuities along these lines. These were 0·47 cm. in fig. 13 and 0·6 cm. in fig. 14.

(c) The y co-ordinates of the centres of the white bands (interpreted as lines of constant phase) were measured on fig. 14 at $x = 4\cdot68$ cm.

The equation to the nth parabolic symmetrical wave is found by writing $2\pi n$ for $\epsilon - \epsilon_0$ in (33) so that

$$y_n^2 = 4\pi n W^{\frac{1}{2}} x t. \tag{34}$$

The equation to the antisymmetrical wave is $y/x = \tan \psi_a$. This cuts the parabola (34) at $y_n = 4\pi n W^{\frac{1}{2}} t \cot \psi_a$ which is at distance $4\pi n W^{\frac{1}{2}} t \cot \psi_a \csc \psi_a$ from the origin. Thus the distance apart of the points of intersection is

$$4\pi W^{\frac{1}{2}} t \cot \psi_a \csc \psi_a.$$

This would be true whatever the value of the angle ψ_a, but when ψ_a is the angle of inclination of the antisymmetrical wave to the stream $\sin \psi_a = W^{\frac{1}{2}}$, so that the distance between intersections is $4\pi t \cot \psi_a$.

The measurements (b) therefore give for fig. 13

$$t = (0\cdot47/4\pi)\tan 25° = 0\cdot0172\,\text{cm.,} \tag{35}$$

and for fig. 14
$$t = (0\cdot61/4\pi)\tan 20° = 0\cdot0177\,\text{cm.} \tag{36}$$

In the measurements (c) of the y co-ordinates at $x = 4\cdot68$ cm. the centres of the white bands were used. Since the phase of $y = 0$ is not known and in a narrow wake downstream of the disturbance point the sheet is certainly thinner than it is elsewhere, (34) was applied in the form

$$\frac{y_{m+n}^2 - y_m^2}{n} = 4\pi^2 W^{\frac{1}{2}} x t, \tag{37}$$

leaving out the two central bands and considering only the differences between the outer bands. Thus when n is positive $m > 1$ and when n and m are negative $|n|$ is greater than 1. The measured values are given in table 1. The average value of $(y_{n+m}^2 - y_m^{2})/n$ is 0·36 cm.². The measured angle ψ_a was 20° and

$$\sin 20° = 0·342 = W^{\frac{1}{2}}$$

so that (37) gives $\qquad t = \dfrac{0·36}{4\pi(4·68)(0·342)} = 0·018 \text{ cm.} \tag{38}$

Comparing (38) and (36), it will be seen that the values of t calculated by measurements of the two types are nearly identical.

Table 1

n	-5	-4	-3	-2	2	3	4
y (cm.)	$-1·34$	$-1·20$	$-1·02$	$-0·84$	$+0·92$	$+1·11$	$+1·25$

SYMMETRICAL WAVES IN AN EXPANDING SHEET

When the thickness of the sheet is not constant, Rayleigh's assumption that the waves from a disturbance point are similar and similarly situated with respect to the disturbance point is no longer valid. On the other hand, if only the part of the field where the lines of constant phase make a small angle with the stream direction is considered, the following argument might be used. The effect of decrease in thickness t with increasing radius is to decrease the velocity of symmetrical waves as r increases. On the other hand, the expansion of the sheet in the tangential direction as it flows outwards causes the wave system to expand by lateral convection. Neither of these causes will affect the *relative* distance of waves from the line $\theta = 0$ which contains the disturbance point at $r = r_0$. In a uniform sheet the wavelength can be expressed in the form $\lambda = f(r) \epsilon^{-\frac{1}{2}}$, where ϵ is the phase, so that a similar expression can be used in the expanding sheet. Hence if (r, θ_ϵ) are the co-ordinates of a point on the line where the phase differs from that of $\theta = 0$ by ϵ

$$r\theta_\epsilon = f(r)\int_0^\epsilon \epsilon^{-\frac{1}{2}} \frac{d\epsilon}{2\pi} = \frac{\epsilon^{\frac{1}{2}} f(r)}{\pi} = \frac{\epsilon\lambda}{\pi}. \tag{39}$$

From (4), $w_s = (2T/\rho t)^{\frac{1}{2}} \pi t/\lambda$ and if t_0 is the thickness of the sheet at $r = R$ where $2T/\rho t_0 u^2 = 1$,

$$\psi = \frac{w_s}{u} = \frac{\pi t_0}{\lambda}\left(\frac{R}{r}\right)^{\frac{1}{2}}. \tag{40}$$

Since on a line of constant phase $\qquad \psi = r\, d\theta_\epsilon/dr \tag{41}$

ψ and λ can be eliminated from (39), (40) and (41) and the resulting equation integrated giving

$$\theta_\epsilon^2 = \tfrac{4}{3}\epsilon t_0 R^{-1}\left\{\left(\frac{r_0}{R}\right)^{-\frac{3}{2}} - \left(\frac{r}{R}\right)^{-\frac{3}{2}}\right\}. \tag{42}$$

MEASUREMENTS OF SYMMETRICAL WAVES IN AN EXPANDING SHEET

Waves were produced in a horizontal expanding sheet by means of a needle point which was carefully introduced into the sheet but did not penetrate it. When the sheet was observed by reflected light two arcs of cardioids, indicating antisymmetrical waves, could be seen. These were photographed with a time exposure of 45 sec. in the manner already described and the result is in fig. 15, pl. 3. The disturbance is confined to a very narrow area close to the cardioid arcs. The photograph demonstrates the remarkable freedom from dispersion of the cardioid wave. The outer edge of the sheet which appears as a blurred edge was in a state of violent agitation, yet in spite of this the cardioids remain sharply defined out to this edge. This indicates that the agitation was confined to vertical motion. No trace of symmetrical waves appears in fig. 15, but when the sheet was observed very obliquely by eye a slight disturbance indicating their possible existence could just be seen. Without disturbing the sheet the schlieren photograph, fig. 16, pl. 4 was then taken. The symmetrical wave system is clearly visible but no trace of the antisymmetrical cardioid waves can be seen. The area covered by the schlieren picture fig. 16 is marked on fig. 15 by a broken line circle. To compare the waves revealed in fig. 16 with those described by equation (42) it was first necessary to measure R. Since no measurement of u was available this was obtained indirectly by measuring the linear dimension of the cardioid in fig. 15. The outer edge of the cardioid, which in the theoretical picture extends to $r = R$, was outside the edge of the continuous part of the sheet so a measurement was made of the radius $r = \tfrac{1}{2}R$, where $\psi = 45°$. To do this two fine lines crossing at 45° were ruled on a transparent sheet. This was moved over the photograph (fig. 15) till one of these lines was tangential to the cardioid at their point of intersection, and at the same time the other line passed through the centre of the sheet. The radius at which this occurred was 4·60 cm. on the photograph which represented the actual sheet with magnification 1/2·29. Thus

$$\tfrac{1}{2}R = 4·60 \times 2·29 \quad \text{and} \quad R = 21·1 \text{ cm.}$$

This radius is shown in fig. 15.

To find $t_0 = 2T/\rho u^2$, it is necessary to know u. This, however, was not easy to measure. It is less than $(2gH)^{\frac{1}{2}}$ owing to loss in energy during the impact of the jet on the target. On the other hand, Q was nearly proportional to $\sqrt{H}$ and the head H was measured at a point just upstream of the point where it converged to form the jet. It was in fact found that with the jet used, 0·298 cm. in diameter,

$$Q = 2·97\sqrt{H} \text{ cm.}^3/\text{sec.}$$

approximately. If R is known as well as Q, t_0 can be found from the expression

$$t_0 = \rho Q^2/8\pi^2 R^2 T, \tag{43}$$

so that with the orifice used $t_0 = 0\cdot1115H/R^2T$ cm. The photographs on figs. 15 and 16 were taken when $H = 151$ cm. so that if

$$T = 73\,\text{c.g.s.}, \quad t_0 = 5\cdot18 \times 10^{-4}\,\text{cm.} \tag{44}$$

To compare the theoretical expression (42) with the waves shown in fig. 16 it is convenient to express (42) in terms of the co-ordinates $y_n = r\theta_n$, of the nth wave from the line $\theta = 0$. Since the phase ϵ_0 of the wave on $\theta = 0$ is not known the appropriate form of (42) is

$$\frac{y_{n+m}^2 - y_m^2}{nR^2} = \tfrac{4}{3}(2\pi)\left(\frac{t}{R}\right)\left(\frac{r}{R}\right)^2\left\{\left(\frac{r_0}{R}\right)^{-\frac{3}{2}} - \left(\frac{r}{R}\right)^{-\frac{3}{2}}\right\}. \tag{45}$$

Values of y_n measured at a distance of $5\cdot9$ cm. from the disturbance point are given in table 2. From these the value found for the left-hand side of (45) is $6\cdot3 \times 10^{-4}$.

Table 2. *Distance (cm.) of centres of dark bands from radial line through the origin of the disturbance (measured at radial distance $5\cdot9$ cm. from it)*

no. of dark band	1	2	3	4	5	6	7	8	9
above centre line in fig. 16	0·418	0·678	0·858	1·018	1·150	1·272	1·385	1·466	1·604
below centre line in fig. 16	0·198	0·571	0·763	0·96	1·085	1·215	—	—	—

The distance of the disturbance point from the centre of the sheet was $3\cdot2$ cm., corresponding to $r_0/R = 0\cdot152$. At the point where the waves were measured, $r = 3\cdot2 + 5\cdot9$ cm. so that $r/R = 0\cdot431$ and

$$\left(\frac{r}{R}\right)^2\left[\left(\frac{r_0}{R}\right)^{-\frac{3}{2}} - \left(\frac{r}{R}\right)^{-\frac{3}{2}}\right] = 2\cdot450.$$

Substituting the values $t_0 = 5\cdot18 \times 10^{-4}$ from (44) and $R = 21\cdot1$ cm., the right-hand side of (45) is $\tfrac{8}{3}\pi(21\cdot1)^{-1}(5\cdot18 \times 10^{-4} \times 2\cdot450) = 5\cdot05 \times 10^{-4}$.

Though this is not very close to the measured value, $6\cdot3 \times 10^{-4}$, of the left-hand side of (45) the nature of the measurements would not lead to the expectation of better agreement.

It is not possible to produce theoretically a diagram which can be compared in detail with fig. 16 because neither the phase of the centre line nor the phase of the edges of the black strips in fig. 16 are known. A diagram suitable for qualitative comparison can be produced by assuming that the phase at the centre is zero and plotting lines of constant phase equal to $2n\pi$, n being a whole number. This has been done in fig. 17, pl. 4, where the lines represent curves

$$\frac{y_n}{R} = n^{\frac{1}{2}}\left(\frac{8\pi t_0}{3R}\right)^{\frac{1}{2}}\left(\frac{r}{R}\right)\left[\left(\frac{r_0}{R}\right)^{-\frac{3}{2}} - \left(\frac{r}{R}\right)^{-\frac{3}{2}}\right]^{\frac{1}{2}} = 1\cdot44 \times 10^{-2}n^{\frac{1}{2}}\frac{r}{R}\left[16\cdot8 - \left(\frac{r}{R}\right)^{-\frac{3}{2}}\right]^{\frac{1}{2}}$$

for the case $t_0 = 5\cdot18 \times 10^{-4}$ cm., $R = 21\cdot1$ cm., $r_0/R = 0\cdot152$. The diagram has been drawn so that the scale is the same as that of fig. 16.

32

THE DYNAMICS OF THIN SHEETS OF FLUID
III. DISINTEGRATION OF FLUID SHEETS

REPRINTED FROM

Proceedings of the Royal Society, A, vol. CCLIII (1959), pp. 313–21

The free edge of a sheet of uniform thickness moves into it at the same speed, $(2T/\rho t)^{\frac{1}{2}}$, as antisymmetrical waves, sweeping the fluid into roughly cylindrical borders. Here T, ρ and t are surface tension, density and thickness of the sheet. In a radially expanding sheet t decreases with increasing radius and beyond a radius R where $(2T/\rho t)^{\frac{1}{2}}$ is greater than u the radial velocity of the sheet, the edge moves inwards faster than it is convected outwards. Photographs show that the edge of an expanding sheet establishes itself near but inside the radius R. The sheet produced by a swirl atomiser expands as a cone but photographs show that its thickness fluctuates very greatly at the point where it emerges from the orifice. The edge of a conical sheet of varying thickness establishes itself at a point well inside the radius at which $(2T/\rho \bar{t})^{\frac{1}{2}} = u$, $\bar{t}$ being the mean thickness.

A moving sheet of uniform thickness can be bounded by a stationary free edge at angle $\sin^{-1}(W^{\frac{1}{2}})$ to the direction of motion. Here W, the Weber number, is $2T/\rho t u^2$. Photographs show free edges at this angle and therefore parallel to antisymmetrical waves. If this remained true in an expanding sheet the edges would coincide with the cardioids discussed in part II, but reasons are given to show that this is not the case. A small obstacle can divide an expanding sheet forming two edges which lie at the same angle to one another as the two cardioids, namely, $2\sin^{-1}(W^{\frac{1}{2}})$ but photographs show that these edges do not subsequently lie on cardioids.

The disintegration of fluid sheets has formed the subject of many researches. Among the more recent Dombrowski & Fraser* showed that flat jets of a homogeneous fluid often break up by the formation of holes which spread till their edges meet leaving a cylindrical thread of fluid which then breaks up owing to the well-known instability of fluid cylinders under the action of surface tension. Most modern workers, however, have been interested in the formation of drops in spraying devices where little precaution was taken to prevent turbulence. The earliest worker in the subject Savart† studied the maximum diameter which a sheet formed by the impact of two co-axial opposing jets can attain. Such an experimental arrangement can produce sheets with little turbulence and the velocity is very close to $(2gH)^{\frac{1}{2}}$, where H is the head of water. Savart did not give a theoretical discussion of these experiments, but they were carefully done and he was able to deduce empirically two laws which are in accordance with what might be expected by a theoretician. Provided H is not too great these are: (1) the diameter of the sheet is proportional to the head of fluid which produces the jets, and (2) the diameter is proportional to the square of the diameter of the two orifices.

---

* *Phil. Trans. Roy. Soc.* A, CCXLVII (1953), 101.

† *Annls. chim. Phys.* LV (1833), 257.

Before discussing Savart's results, it is worth considering theoretically the dynamics of the edge of a sheet. A flat sheet of fluid is stable when disturbed in any way unless two surfaces meet and thus form a hole with a free edge. The surface tension then pulls this edge back, thus concentrating the fluid into a roughly cylindrical mass which constantly increases in volume till instability or other dynamical cause makes it break up into drops. Many cases of this can be seen in the photographs of Dombrowksi & Fraser.

SHEET OF UNIFORM THICKNESS

The simplest case is that of a straight free edge bounding a sheet of uniform thickness. If v is the velocity at any instant and M the mass per unit length of the cylindrical free edge the equation of motion is

$$2T = \frac{\mathrm{d}}{\mathrm{d}\tau}\left(M\frac{\mathrm{d}x}{\mathrm{d}\tau}\right), \tag{1}$$

where τ represents time. If at time $\tau = 0$ the free edge is formed at $x = 0$ and initially $M = 0$, the equation of continuity is

$$M = \rho t x, \tag{2}$$

so that

$$2T = \tfrac{1}{2}\rho t \,\mathrm{d}^2 x/\mathrm{d}\tau^2. \tag{3}$$

The integral of (3) is $x^2 = (2T/\rho t)\,\tau^2 + A\tau + B$, and from the condition that $\mathrm{d}x/\mathrm{d}\tau$ is finite and $x = 0$ at $\tau = 0$,

$$x/\tau = v = (2T/\rho t)^{\frac{1}{2}}. \tag{4}$$

This is the same as the velocity of antisymmetrical waves. Any velocity can be imposed on the whole field of flow without altering the physical quantities concerned and if this velocity is parallel to the edge, both waves and edge remain fixed in space. Thus the angle at which a free edge to a moving sheet of uniform thickness can remain at rest in space is the same as that at which an antisymmetrical wave can remain at rest. In such a sheet the angle at which both waves and the edge could be at rest is $\sin^{-1}(2T/\rho t u^2)^{\frac{1}{2}}$. If $u < (2T/\rho t)^{\frac{1}{2}}$ neither a free edge nor an antisymmetrical wave can remain at rest.

In the course of the experiments on antisymmetrical waves in a uniform sheet which are illustrated in fig. 12 of part II,* an air bubble got into the orifice slit and divided the issuing sheet of water into two parts. The photograph (fig. 1, pl. 1), was then taken. The two turbulent free edges can be seen as thick and rather rough borders. The disturbing air jet was in operation and is marked by an arrow. The two antisymmetrical waves produced by the air jet can be seen and it will be noticed that they are parallel to the free edges.

Fig. 2, pl. 1, is a flash photograph of a horizontal sheet of uniform thickness taken from above after removing the wetable members which form the top and bottom of the sheet in fig. 1. It will be seen that the edges appear straight and

* Paper 31 above.

highly turbulent. This photograph was illuminated by a flash bulb near the camera lens. The large white area is due to approximately normal reflection. The edges of the sheet only appear because the turbulence in them is sufficiently great to cause parts of their surfaces to reflect light back towards the camera.

EXPANDING SHEET

When the sheet is expanding radially, the antisymmetrical waves and free edges no longer necessarily coincide because equation (4) no longer represents the motion of the cylindrical edge except at the limiting radius R, where $t = 2T/\rho u^2$ and the wave crest and free edge are both perpendicular to the stream.

If a radially expanding sheet could be formed so that it extended beyond the radius R, where $W = 2T/\rho t u^2 = 1$ any edge which formed would rapidly run back to this radius, which is the ideal limit that could be reached by a fluid of negligible viscosity when completely free from turbulence. The calculated value of R is given in (4) of part I (paper 30 above).

COMPARISON WITH SAVART'S MEASUREMENTS

Savart gives the measured diameter of the smooth coherent parts of his sheets (nappes) with pairs of orifices 3, 4 and 6 mm. in diameter under heads of water, H, ranging from 10 to 130 cm. His apparatus was so designed that very little loss of head can have occurred in the formation of the jet or in the region of collision. It is therefore legitimate to take u as $(2gH)^{\frac{1}{2}}$. Savart does not seem to have measured the volumetric rate of flow Q. To deduce Q from Savart's data it is necessary to know the properties of the orifices used. These were described as circular orifices in flat plates. The jets from such orifices suffer contraction which in general reduces the area of the stream to a fraction between 0·62 and 1·0 of that of the orifice. This fraction known as the contraction coefficient C, can be inferred at any rate roughly from a remark by Savart,* that the stream from his 6 mm. hole contracted to 4·9 mm. which would make $C = 0·67$. Thus the version of the relation (4) in paper 30 above which could be applied is

$$R = \frac{2\pi a^2 C \rho u^2}{4\pi T} = \frac{a^2 C \rho g H}{T}, \tag{5}$$

where $2\pi a^2$ is the total area of the two orifices. It will be noticed that (5) agrees with Savart's empirical laws for variation of R with a and H. The values of $R/a^2 HCg$ which can be deduced from Savart's table, taking $g = 981$, are 8·4, 8·6 and 8·5 × 10⁻³ for $2a = 0·3$, 0·4 and 0·6 cm.

If $C = 0·67$ and the radius measured by Savart is taken as R this leads to values for T 80, 78 and 79 dyn/cm. On the other hand, the edge of an expanding sheet is always in violent motion and the clear part seldom extends beyond about $0·95R$

* *Annls. chim. Phys.* LV (1833), 266.

even when precautions have been taken to eliminate turbulence. If Savart's measured radius was $0.95R$ the values calculated for T using (5) are 76, 74 and 75 dyn/cm. for the 3, 4 and 6 mm. orifices, respectively. The values obtained from modern measurements of the surface tension of water usually lie in the range 73 to 74 c.g.s., so that Savart's measurements must have been carefully made.

The analysis represented by (4) envisages water moving at high speed up to the edge where the Weber number is 1.0 and being projected into a ring of fluid which then becomes unstable and forms drops which, having no outward velocity, could simply fall off vertically. Fig. 3, pl. 2, is a photograph of such a sheet formed by impact of a 3 mm. vertical jet on a flat impacter. The centre of the camera lens was very slightly above the level of the rim of the sheet which is slightly concave upwards. The horizontal velocity of the drops which are discharged can be estimated roughly by assuming that they fall freely in parabolic paths after being ejected horizontally. Fitting parabolas over the outermost drops shown in fig. 3 the horizontal velocity so calculated was of order 50 cm./sec. Since the horizontal velocity of the fluid in the sheet was above 330 cm./sec. the fraction of kinetic energy remaining in the drops was $(\frac{50}{330})^2 = 0.023$. About 98 % of the kinetic energy therefore disappeared in the region where the drops were formed. A very small fraction of this energy was carried away in the form of surface energy of the drops, and all the rest must have been lost in turbulence within the drops. It will in fact be seen in fig. 3, that the drops are far from spherical and must therefore be in a state of violent agitation.

The loss of velocity at the impacter where the sheet is formed is of the order of 20 % so that the thickness of the sheet at radius R would be

$$t = \frac{1}{0.8}\left(\frac{\pi a^2}{2\pi R}\right) = \frac{0.014}{R}.$$

The radius of the sheet was 9 cm. so that $t \sim 0.0015$ cm. The drops are on the average of order 1 to 3 mm. diameter so that they were of order 100 times the thickness of the sheet.

EDGES FORMED BY OBSTRUCTIONS IN AN EXPANDING SHEET

It has been shown that the edges formed by an obstruction in a uniform sheet moving with velocity u lie in two lines at angle $\sin^{-1}(2T/\rho t u^2)^{\frac{1}{2}}$ to the stream and are parallel to the antisymmetrical waves. In an expanding sheet antisymmetrical waves from an obstruction form two cardioids.

If the obstruction causes the sheet to separate into streams bounded by two edges, these edges will also assume the cardioid form provided that they immediately broke up into drops, which would then pursue paths tangential to the cardioids. Since the surface tension acting at right angles to the edge would have no effect on the component of velocity in the tangential direction, the velocity of a drop formed at a point where the edge was at angle ψ to the direction of the stream would have velocity $u \cos \psi$. If ψ was the same as for the antisymmetrical wave $u \cos \psi = u\{1 - (r/R)\}^{\frac{1}{2}}$. An obstruction placed at radius r would therefore give rise to two cardioid edges from which drops would be discharged tangentially at velocities ranging from $u\{1 - (r/R)\}^{\frac{1}{2}}$ to 0.

An example of this is shown in fig. 10 of paper 31 above, where part of the sheet on the left of the obstruction is bent sharply at a cardioid while the part on the right formed a free edge from which the drops separated tangentially. The existence of the cardioid waves parallel to this edge is indicated by the bright lines close to it. In fig. 9 (paper 31) the edge seen to the right of the obstruction did not break up into drops.

Fig. 4, pl. 2, shows another example of the tangential discharge of drops from an edge which was nearly a cardioid. The obstruction in this case was a clean wire 0·0125 cm. in diameter near the centre of the sheet. The wire did not immediately give rise to two free edges but caused the outer edge to bend inwards to the centre.

At the point where the edge is nearest to the centre it is necessarily at right angles to the stream and at that point the thickness must have been the same as where the unobstructed edge, though rather ragged, had a mean radius of nearly 11·7 cm. The least distance of the edge from the centre, which was downstream of the obstruction, was 4·5 cm. so that the obstruction must have produced a wake in the middle of which the sheet was only $4·5/11·7 = 0·38$ of the thickness of the unobstructed sheet at the same radius. Such a wake could have been produced if the obstructing wire gave the stream a lateral motion, the motion in fact which would also produce the symmetrical waves discussed in part II.

Fig. 5, pl. 2, shows the effect of a waxed, unwetable wire 0·012 cm. in diameter at radius 1·95 cm. in the same sheet as that of fig. 4. Here the sheet parts into two free edges which do not break up into drops but remain coherent till they reach the outer edge of the sheet. It will be noticed that the diameter of the cylindrical free edge increases with distance from the obstruction, as indeed it must, since the total flow in the edge must increase with distance from the obstacle. It will be noticed also that the form of the edge is quite different from a cardioid which would have had greater curvature.

EQUATIONS FOR THE SHAPE OF A FREE EDGE WHEN IT DOES NOT DISINTEGRATE

If it is assumed that the fluid flowing into the free edge mixes so rapidly with the fluid already there that the velocity within it is uniform at any point and equal to qu, the continuity requirement in the edge is

$$\mathrm{d}(mq)/\mathrm{d}s = \rho t \sin \psi, \tag{6}$$

where m is the mass of fluid per unit length. The rate at which momentum parallel to the edge increases along it is $u^2 \mathrm{d}(mq^2)/\mathrm{d}s$ so that the equation for conservation of momentum parallel to the edge is

$$\mathrm{d}(mq^2)/\mathrm{d}s = t\rho \sin \psi \cos \psi. \tag{7}$$

The equation for conservation of momentum perpendicular to the edge is

$$2T - \rho u^2 t \sin^2 \psi = m u^2 q^2 \left[\frac{d\psi}{ds} + \frac{d\theta}{ds} \right]. \tag{8}$$

If t_0 is the thickness of the sheet at $r = R$, where $W = 1$, $tR = t_0 r$ and since

$$\sin \psi \, ds = r \, d\theta,$$

(6) can be integrated giving

$$mq = \rho t_0 R \theta, \tag{9}$$

the constant of integration being chosen so that when $\theta = 0$, $m = 0$. By means of (9), (7) becomes

$$d(q\theta)/d\theta = \cos \psi; \tag{10}$$

since $2T/\rho u^2 = t_0$, (8) becomes

$$1 - \frac{R}{r} \sin^2 \psi = \theta q \frac{R}{r} \left(\frac{d\psi}{d\theta} + 1 \right) \sin \psi. \tag{11}$$

Equations (10) and (11) together with $dr/d\theta = r \cot \psi$ and the conditions at the obstacle $r = r_1$ suffice to determine the shape of the edge, but numerical methods would be required to find the shape for any given value of r_1/R. At the obstacle whose polar co-ordinates are $(r_1, 0)$ it has been assumed that $m = 0$. The initial value of q is then $\cos \psi$ and $\sin \psi = (r_1/R)^{\frac{1}{2}}$.

MEASUREMENTS OF THE ANGLE AT WHICH THE SHEET PARTS
AT AN OBSTACLE

In the above analysis it was assumed that the initial value of m at $r = r_1$ is zero, so that the angle between the two parts of the sheet at the obstacle is $2 \sin^{-1} (r_1/R)^{\frac{1}{2}}$. To test this a horizontal expanding sheet was formed by causing a jet from the orifice previously described to strike an impact disc 1·2 cm. in diameter. The head was maintained constant at 112·8 cm. of water and a wire obstacle of diameter 0·12 cm. was placed at a number of different radii. The resulting edges were photographed. Fig. 6, pl. 3, shows some typical examples. Measurements were made at the obstacle of the angle, $2\psi_0$, between the two edges. When $\sin^2 \psi_0$ was plotted against r, the distance of the wire from the centre, it was found that the resulting curve was nearly straight when r was greater than 6 cm. and that it cut the abscissa $\sin \psi_0 = 1$ at $r_1 = 12·0$ cm., though the highest value of r_1 at which measurements of $2\psi_0$ were practicable was 10·9 cm. It was noticed, however, that $\sin^2 \psi_0/r_1$ increased when the obstacle was placed in the inner part of the sheet. This result is shown in fig. 7. It will be seen that $\sin^2 \psi_0/r_1$ rises considerably as r_1 decreases. It was thought at first that this might indicate a higher surface tension in the part of the sheet where the surface had been newly formed, in fact at the speed 330 cm./sec. the surface had only existed 1·3/330 − 4 msec. when it passed the obstacle in the nearest position to the centre at which ψ_0 was measured. It was realized, however, that another cause might increase ψ_0, namely, the lateral inertia given to the fluid in the edges as it was forced aside by the obstacle. A much smaller obstacle was then substituted in the form of a wire 0·012 cm. in diameter. This reduced the value of $\sin^2 \psi_0/r_1$ though not down to the value 0·083 cm.$^{-1}$ which it had in the outer part of the sheet. The two points on fig. 6 which correspond to the smaller obstacle are shown.

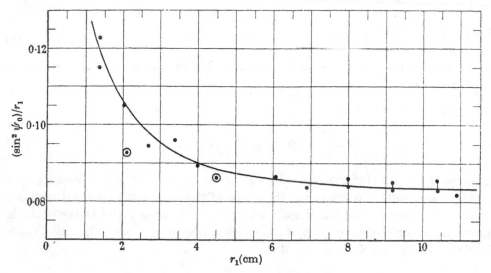

Fig. 7. $(\sin^2 \psi_0)/r_1$ measured on expanding sheet for which $R = 12\cdot0$ cm. ⊙, parting wire 0·012 cm. diameter.

DISINTEGRATION OF SHEET PRODUCED BY A SWIRL ATOMISER

A swirl atomiser produces conical sheets of fluid which break up into drops. It has been shown by Squire* and others that when a sheet of fluid is projected into air, unstable waves can be formed which are analogous to those produced in a flag by the wind. These waves no doubt increase the liability of the sheet to break up, but as Squire pointed out they do not explain why the sheet disintegrates, for sheets projected into air of very low density break up in a manner which seems very similar to that at atmospheric air pressure. The experiments here described lead to the expectation that a non-turbulent sheet would not disintegrate till it reached a point where W was of order 1·0. Squire gave values of W of order 0·01 at the point where break-up was observed. This, however, was found to be due to a numerical error and recalculation gives values of W ranging from 0·25 to 0·40 according to what assumption is made about u. Q was measured as well as H. If no loss of energy occurred in the nozzle of the atomiser, $W = 0\cdot25$, while the maximum loss that would be expected would give the higher value.

It appears therefore that the value of W at the point of disintegration is of the same order though less than what would be expected if the disintegration were due to an

* *Br. J. appl. Phys.* IV (1953), 163.

edge moving back into the conical sheet at the same speed as the fluid. Fig. 8, pl. 4, is a photograph taken at the National Gas Turbine Establishment of a sheet of kerosene issuing from the orifice (0·03 cm. diameter) of a swirl atomiser under pressure of 20 lb./in.^{2} into air at pressure 0·32 atm. The disintegration evidently occurs at an edge which is advancing against the stream. The value of W at the point of disintegration can be estimated by making an assumption about the energy lost in its formation, but a more direct method is available for a rough estimation. The semi-vertical angle of the initially conical sheet shown in fig. 8 is approximately 60°. Comparing its shape with that calculated for a water bell (fig. 1 of part I, paper 30 above) whose semi-vertical angle is 60° it will be seen that they are very similar up to the line where the sheet disintegrates. The calculated shape for $\phi_0 = 57°$ is shown in fig. 9, pl. 4. The scale and ϕ_0 have been chosen so as to fit the photograph as nearly as possible, and the calculated values of W attained at various points are marked on the outline. The position of the edge photographed in fig. 8 is shown roughly in fig. 9. It will be seen that it is situated at the level where $W = 0·33$ approximately.

A possible explanation of the position of the edge at $W = 0·33$ instead of where $W = 1$ may be that the conical sheet is very variable in thickness. That this is the case is indicated by the fine circumferential streaks in fig. 8. These do not seem to be waves of the type called antisymmetrical in part II because they do not appear in the photograph as small projections on the meridian section. They are more likely to be variations in thickness of the sheet and analogous to the symmetrical wave discussed in part II. In that case the edge might be expected to move rapidly towards the orifice as it crosses the thin places and be carried away from it in the thicker parts. The resultant mean velocity of the edge is greater than it would have been if the thickness of the sheet had been uniform and equal to $\bar{l}$, its mean thickness. In the expanding conical sheet the mean thickness $\bar{l}$ is defined in terms of the measurable volumetric rate of discharge Q as

$$\bar{l} = Q/2\pi y u, \tag{12}$$

where y is the distance from the axis of the point where $\bar{l}$ is measured. If the edge rapidly breaks down into drops the velocity with which it moves into a sheet of variable thickness is $(2T/\rho t)^{\frac{1}{2}}$ so that the mean velocity over length X is

$$v = X \Big/ \int_0^X \left(\frac{\rho t}{2T}\right)^{\frac{1}{2}} \mathrm{d}x. \tag{13}$$

If the sheet had been of uniform thickness the velocity would have been

$$v' = (2T/\rho \bar{l})^{\frac{1}{2}},$$

so that

$$\frac{v}{v'} = X \Big/ \int_0^X \left(\frac{t}{\bar{l}}\right)^{\frac{1}{2}} \mathrm{d}x. \tag{14}$$

v'/v is therefore always less than 1.

In an expanding sheet of variable thickness the mean position of the edge will be where $v = u$ so that the mean Weber number at the edge defined as

$$W_E = \frac{2T}{\rho u^2 \bar{t}} = \left(\frac{v'}{v}\right)^2 \qquad (15)$$

is always less than 1 and no lower limit (except 0) is imposed on its value if the only limitation on t is that it is positive.

It is of interest to speculate on causes which might lead to large variations of t in a swirl atomiser. The atomiser consists of a cylindrical box into which fluid is led tangentially through one or more channels. The fluid emerges through an orifice which is concentric with the box. If the flow is *steady* a small spiral variation in thickness of the sheet might be expected in the orifice owing to the want of axial symmetry of the supply: the thicker parts of the sheet would then occupy fixed positions, since the source of the variation would be fixed. The thicker positions would in fact be in streamlines and would therefore be very nearly on generators to the conical sheet.

Another source of unevenness in the conical sheet would arise if the air core which forms within the orifice were oscillating. A comparatively small oscillation would give rise to a large proportional variation in the thickness of the sheet. The distribution of thickness due to this cause would be quite different from that caused by lack of axial symmetry in steady flow. The lines along which the thicker portions of the sheet were distributed would lie nearly in circles round the axis. If, for instance, the core were in circular oscillation and produced a thicker portion which travelled round the orifice in a time τ, the angle of the spiral which contained the thicker portion of the sheet would be of order $\tan^{-1}(\pi\, du/\tau)$ at the point where it emerged from the orifice of diameter d. At a point in the sheet where the distance from the axis is y the angle of the spiral would be $\tan^{-1}(2\pi yu/\tau)$, so that if $\pi\, du/\tau$ is of order 1 the angle of the spiral of the thicker portion rapidly becomes nearly 90° when y/d becomes large. The appearance of the streaks in fig. 8 which are nearly at right angles to the axis suggests that they may be caused in this way in the case there illustrated.

Size of drops produced by a swirl atomiser

The variation in thickness will determine the mean value of W at the point where the sheet breaks up in accordance with (13) and (14). If the mechanism of disintegration is analogous to that described by Dombrowksi & Fraser* for the case when holes form in the sheet, the diameter of the drops produced would be of order 1·9 times the diameter of the cylinders produced at the edge. So far no mechanism has been put forward to describe the separation of the fluid contained in these edges from the continuous part of the sheet. If, however, the sheet parts at its thinnest point the volume contained in unit length of each cylinder after it has

* *Phil. Trans. Roy. Soc.* A, CCXLVII (1953), 101.

PLATE 1

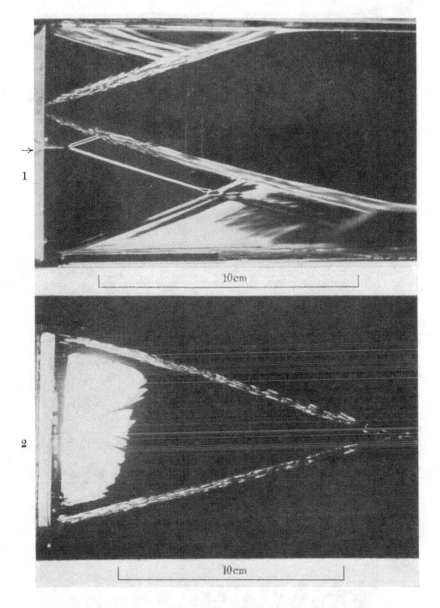

Fig. 1. Two bordered edges and antisymmetrical waves from a disturbance point.
Fig. 2. Bordered edges of unobstructed uniform sheet.

PLATE 2

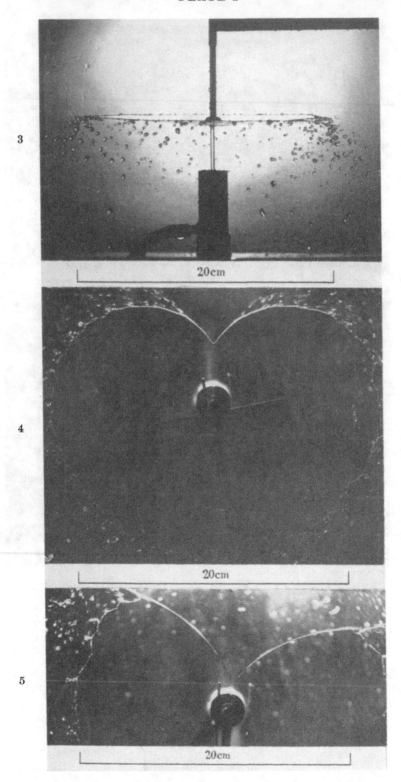

For captions see opposite.

PLATE 3

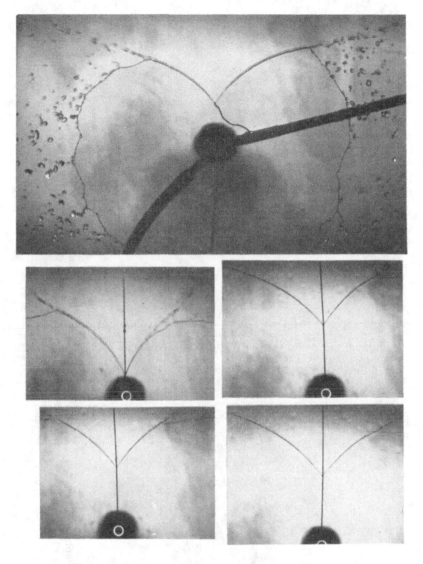

Fig. 6. Edges formed by wire 0·12 cm. in diameter.

CAPTIONS FOR PLATE 2

Fig. 3. Nearly plane horizontal sheet seen from a point nearly in the plane of its rim.
Fig. 4. Edge formed by a wetable wire 0·012 cm. in diameter at $r_1 = 1·9$ cm.
Fig. 5. Edge formed by an unwetable wire 0·012 cm. in diameter at $r_1 = 1·9$ cm. on same sheet as that of fig. 4.

PLATE 4

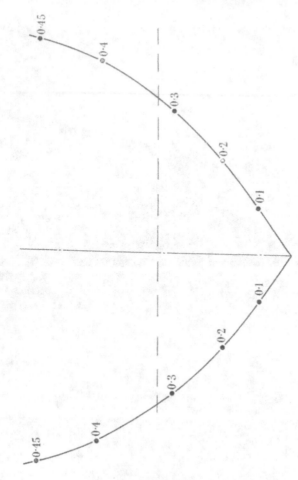

Fig. 9. Calculated shape for continuous part of sheet of fig. 8. The broken line corresponds approximately to the free edge shown in fig. 8. The numbers represent the local value of W.

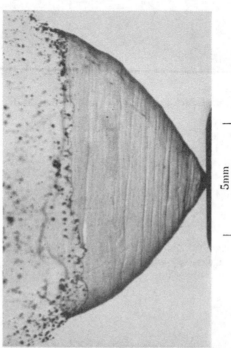

5mm

Fig. 8. Swirl atomiser spray. (Photograph by National Gas Turbine Establishment.)

separated at radius y is $Q\tau/2\pi y$, where τ is the mean period of oscillation of whatever it is in the atomiser box that causes the variation in thickness of the sheet as it leaves the orifice. The diameters of drops produced in this way would therefore be expected to be of order

$$D = 1 \cdot 9 \left(\frac{4}{\pi} \frac{Q\tau}{2\pi y} \right)^{\frac{1}{3}}.$$

If d is the diameter of the orifice of the atomiser, τ might be expected to be of order $\pi d/u$ and Q of order $\frac{1}{4}\pi d^2 u$ so that D is of order $1 \cdot 9(d^3/2y)^{\frac{1}{3}}$. Here

$$y = W_E R = W_E(\rho Q u/4\pi T)$$

and W_E is defined in (15).

33

FORMATION OF THIN FLAT
SHEETS OF WATER

REPRINTED FROM

Proceedings of the Royal Society, A, vol. CCLIX (1960), pp. 1–17

A thin plane sheet of fluid which is limited laterally and converges towards a point transforms itself into a sheet which diverges in a perpendicular plane. When the angle of convergence is small and the distribution of thickness in the converging sheet is elliptic the diverging sheet has the same shape as the converging sheet. Apparatus for producing such a sheet was set up and the shape of the stream and distribution of pressure in the region of the transformation were measured and compared with calculations.

The formation of thin sheets by the oblique impact of two cylindrical jets was studied and the distribution of thickness measured. The shapes of sheets corresponding to the measured distribution of thickness were calculated and compared with photographs.

INTRODUCTION

When two equal cylindrical co-planar jets collide they form an expanding sheet in the plane at right angles to the line containing their axes. If the two jets are coaxial the sheet is symmetrical so that its thickness at any point depends only on

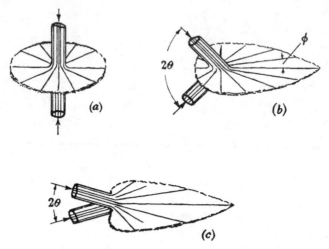

Fig. 1. Sketch of sheets formed by impact of two cylindrical jets.

distance from the axis. This condition is sketched in fig. 1 (a). If the jets are co-planar but not coaxial and meet at an angle 2θ the sheet formed is not symmetrical but it is flat and it bisects the angle between them. The sheet expands radially from the region of the collision and extends furthest in the direction of the component of velocity of the jets in the plane of the sheet. This condition is indicated in fig. 1 (*b*). As θ gets less the extension of the sheet in the opposite direction decreases and at last disappears leaving the sheet in the condition indicated in fig. 1 (*c*).

The object of the present work was to try to present a rational description of the phenomena just described.

THEORETICAL TREATMENTS

In two dimensions the oblique impingement of two equal sheets has been discussed by Mitchell* who showed how to calculate the shape of the free surface in the region of impingement. The impinging sheets, each of thickness t_0, form two sheets of unequal thickness t_1 and t_2 in the plane bisecting the angle between them. Simple momentum considerations show that $t_1 = t_0(1 + \cos\theta)$ and $t_2 = t_0(1 - \cos\theta)$ so that the sheets produced by the impingement can be described exactly without the necessity of thinking about what happens in the region where the jets collide. Similar simple considerations can be applied to describe the conical sheets produced by the impact of two coaxial cylindrical jets which have the same velocity but different diameters. No such simple treatment can be applied to cases where cylindrical jets collide obliquely and the experiments described in the present communication were made in the course of an attempt to produce a description of the way in which a flat sheet is produced under these conditions. Before these experiments are described an approximate theoretical description will be given of an ideal case in which a converging sheet of fluid passes through a point where all the fluid is concentrated into a small cross-section and then diverges into a sheet in a plane perpendicular to its original plane. The first case in which any problem of this kind was described was Dirichlet's† analysis of the motion of an ellipsoidal mass of fluid under its own gravitation. A spheroid under such conditions oscillates from the prolate to the oblate form and back again. The analysis, however, also applies when the kinetic energy is great enough to overcome gravitation and Dirichlet pointed out that it will then tend to one or other of two forms. It extends to infinity either as a line along the axis of symmetry or as a flat sheet in a plane perpendicular to that axis. This applies to the cases where there is no gravitation, however small the energy of the motion is, provided such effects as those of surface tension, and those due to the surrounding air are not considered. It might perhaps be thought that this limiting case of Dirichlet's spheroid in which a column of water converges towards the origin along the axis z and spreads out into a flat sheet in the plane $z = 0$ might be directly related to the formation of a flat sheet by two jets in the manner indicated in fig. 1 (*a*), but the analogy cannot be pressed too far. The maximum pressure, for instance, has been calculated from the version of Dirichlet's analysis given by Lamb.‡ This occurs at the centre of the spheroid when it is

---

* *Phil. Trans. Roy. Soc.* A, CLXXXI (1890), 389.
† *Gött. Abh.* VIII (1860), 3.
‡ *Hydrodynamics*, Cambridge University Press (1932), p. 719.

oblate with ratio of the axes $1 : \sqrt{2}$ and its value is $\frac{3}{8}(\frac{1}{2}\rho U^2)$, where U is the velocity at the end of the axis of symmetry when the spheroid is prolate and very long. The maximum pressure at the point of collision of two steady jets is $\frac{1}{2}\rho U^2$.

TWO-DIMENSIONAL CASE

Limiting the discussion to motion in the plane x, y, only, consider the velocity potential

$$\phi = \tfrac{1}{2}A(x^2 - y^2), \tag{1}$$

where A is a function of time, τ. A particle whose co-ordinates are x_0, y_0 at time $\tau = 0$ will be at x, y at time τ if

$$x = x_0 \exp\left(-\int_0^\tau A\, d\tau\right), \quad y = y_0 \exp\left(\int_0^\tau A\, d\tau\right). \tag{2}$$

Writing

$$\eta = \int_0^\tau A\, d\tau, \quad A = \dot\eta, \quad \dot A = \ddot\eta,$$

the particles which at time $\tau = 0$ lay on the circle $x_0^2 + y_0^2 = r_0^2$ lie at time τ on the ellipse

$$x^2 e^{2\eta} + y^2 e^{-2\eta} = r_0^2. \tag{3}$$

The pressure p corresponding with (1) is given by

$$(p - p_0)/\rho = \tfrac{1}{2}\dot A(x^2 - y^2) - \tfrac{1}{2}A^2(x^2 + y^2), \tag{4}$$

where p_0 is the pressure at the centre of the ellipse, so that the particles which start on $x^2 + y^2 = r_0^2$ will always lie on a surface of constant pressure, which may therefore be a free surface, provided

$$(\dot A - A^2)\, e^{-2\eta} + (\dot A + A^2)\, e^{2\eta} = 0. \tag{5}$$

Hence the equation for η in terms of τ is

$$\ddot\eta/\dot\eta = -\dot\eta \tanh 2\eta. \tag{6}$$

Integrating (6) gives

$$d\eta/d\tau = A_0(\cosh 2\eta)^{-\frac{1}{2}}. \tag{7}$$

A_0 is the value of A when $\tau = 0$

If we write $\xi = e^\eta$, (7) becomes

$$\sqrt{2}A_0\tau = \int_0^\xi (1 + \xi^{-4})^{\frac{1}{2}}\, d\xi. \tag{8}$$

For positive values of η, $\xi > 1$ and the appropriate expansion of (8) is

$$\sqrt{2}A_0\tau = -0{\cdot}8472 + \xi - \frac{1}{2{\cdot}3}\xi^{-3} + \frac{1}{8{\cdot}7}\xi^{-7}\ldots; \tag{9}$$

for negative values of η, $\xi < 1$ and the appropriate solution of (8) is

$$\sqrt{2}A_0\tau = 0{\cdot}8472 - \xi^{-1} + \frac{1}{2{\cdot}3}\xi^3 - \frac{1}{8{\cdot}7}\xi^7\ldots. \tag{10}$$

It will be seen that if τ_ξ represents the value of τ corresponding to ξ, $\tau_\xi = -\tau_{1/\xi}$. Since ξ^2 represents the ratio of the two axes of the ellipse it will be seen that the ratio of the x axis to the y axis at any time before the instant $\tau = 0$ when the ellipse becomes a circle is the same as the ratio of the y axis to the x axis at the same time after $\tau = 0$.

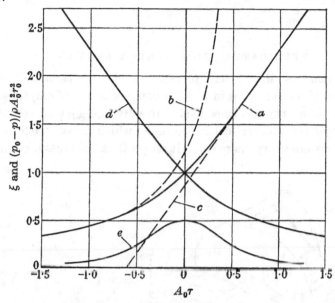

Fig. 2. Changes with time of axes of ellipse. (a) Axis parallel to x; (b) and (c) asymptotic values; (d) axes parallel to y; (e) $(p_0-p)/\rho A_0^2 r_0^2$.

The asymptotic values of $A_0\tau$ are $-0.599+0.7071\xi$ when ξ is large and $+0.599-0.7071\xi^{-1}$ when ξ is small. In figure 2 curve a represents the solution of (8). The ordinates represent the ratio of the axis of the ellipse which is parallel to the axis of x to $2r_0$, the diameter of the circle at line $\tau = 0$, and the broken lines b and c represent the two asymptotic values. It will be seen that it is only in the range $-\frac{1}{2} < A_0\tau < +\frac{1}{2}$ that the eccentricity of the ellipse differs appreciably from the asymptotic values. This might be expected on physical grounds because when the eccentricity is great the fluid velocity in the direction of the major axis is much greater than it is in the perpendicular direction. Energy is conserved and each particle then travels at a constant speed. At the end of the major axis, this speed is $\sqrt{2A_0}r_0$, and is represented in fig. 2 by the slope of the line c. Curve d, fig. 2 represents the ratio of the length of the axis parallel to y to $2r_0$.

PRESSURE IN THE CENTRE OF THE COLLISION REGION

Substituting for A from (7) and writing $x = x_0 e^\eta$, $y = y_0 e^{-\eta}$ in (4)

$$\frac{p_0 - p}{\rho} = \frac{2A_0^2 r_0^2}{\xi^4 + 2 + \xi^{-4}}. \qquad (11)$$

The values of $(p_0-p)/\rho A_0^2 r_0^2$ are shown in fig. 2. The maximum value is $\frac{1}{2}$. The maximum velocity at the end of the major axis is $\sqrt{2}A_0 r_0$ so that $2(\xi^4+2+\xi^{-4})^{-1}$ is the ratio of the pressure at the centre to the Pitot pressure of a stream flowing with the maximum speed in the field of flow. The maximum value of the pressure is half this Pitot pressure.

APPLICATION TO A STEADY STREAM

Though the motion described in the preceding section is essentially unsteady a steady motion could be conceived in which a converging jet of elliptic section with the major axis horizontal transformed itself into a diverging jet with elliptic sections for which the major axis was vertical. If the angles which the stream lines made with the axis of the jet were everywhere small it seems that the cross-sections might be

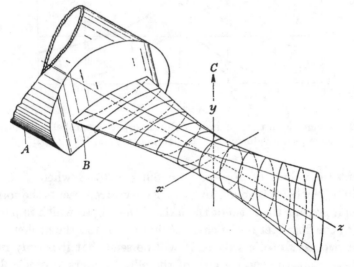

Fig. 3. Projection of jet with elliptic sections on plane perpendicular to diagonal of cube with edges θx, θy, θz. (Position of orifice chamber sketched.)

expected to suffer the same changes of form as those described above. Thus by spacing the successive two-dimensional elliptic sections along an axis z so that $z = U\tau$, an approximate representation of the surface of the jet moving with velocity U might be constructed. The maximum slope of the surface at the end of the major axis of the section would be $\sqrt{2}A_0 r_0/U$ where πr_0^2 is the area of the cross-section. To illustrate this idea the projection shown in fig. 3 was constructed from curve d of fig. 2. The projection is on the plane perpendicular to the diagonal of a cube whose edges are parallel to the co-ordinate axes. The parts of the sections which would be invisible if the fluid were opaque are shown as broken lines. Corresponding points on the successive elliptic sections are stream lines. Some of these are shown on the part of the surface which would be visible.

The reason why a jet issuing from an elongated orifice in a flat sheet transforms itself into a jet which is first flattened in the direction of the long axis of the orifice and then subsequently in the perpendicular direction has long been understood qualitatively, but attention has mostly been directed* to the effect of surface tension, which causes the direction of elongation to be transformed periodically between two perpendicular directions and analysis has been confined to small oscillations about a cylindrical mean surface. In the absence of surface tension the transformation from a section which is flattened in one plane to one which is flattened in a perpendicular plane can only occur once and under many experimental conditions this is what happens.

Experimental

To test whether the theoretical description of the concentration and subsequent expansion of a jet of elliptic section is a good representation of the actual hydro-dynamical situation, apparatus was constructed which it was hoped would enable such a jet to be produced physically. This is indicated in fig. 3. Water flowed through a pipe at constant head through a series of stilling screens into a cylindrical chamber 4·3 cm. in diameter (A, fig. 3). The end of this cylinder B was also cylindrical but with the generators at right angles to the axis of A. In the end B a sharp-edged elliptical orifice was cut so that the minor axis (0·340 cm.) was parallel to the generators of B. The major axis was slightly longer than 2·0 cm. and the chord length was 2·0 cm. The radius of the cylinder from which B was cut was 2·0 cm. so that if the stream lines of the converging stream were at right angles to the surface containing the orifice the angle between the extreme stream lines would be 60°. The apparatus was set up with the cylinder A horizontal and the minor axis of the orifice vertical. A camera C (fig. 3) aiming vertically downwards at the jet took the photograph shown in fig. 4, pl. 1. It will be seen that the outer part of the stream which converges in the horizontal plane appears to slope rather more than the 30° which would be expected if the stream lines from the ends of the major axis were at right angles to the curved surface in which the orifice was cut, but this seems to be an end-effect which is not likely to invalidate the expectation that the stream lines from the greater part of the orifice start at right angles to the surface B when the elliptic orifice is narrow. The surface markings seen in the photograph are waves, not stream lines. In comparing the jet so formed with the theoretical description illustrated in fig. 2 the maximum velocity $\sqrt{2A_0 r_0}$ with which the particles at the ends of the major axes converge towards the axis is taken as $u \sin 30°$ where u is the velocity parallel to the axis.

* Lord Rayleigh, *Proc. Roy Soc.* A, xxix (1879), 71.

PRESSURE MEASUREMENTS

To measure the pressure in the region where the jet changes from being a horizontal to a vertical sheet a very thin flat plate (0·029 cm. thick) containing a pressure hole 0·05 cm. in diameter was supported on a short length of steel hypodermic tube of outside diameter 0·1 cm. This was introduced into the stream so that the sheet was horizontal and on the centre line of the jet. The support was 0·5 cm. behind the leading edge which was curved so that it could move up to make contact with the orifice. It was found that a very small error in alinement of the plane would induce cavitation at the sharp leading edge on one or other of the sides of the sheet and this observation was used as a means of knowing whether the alinement was correct. Simultaneous measurements were made of the head H_1 in the pressure hole and the total head H_0 in the chamber behind the orifice. The position Z of the pressure hole along the axis was measured from an arbitrary origin and the position of the orifice itself $z = 3·36$ cm. The measured values of H_1/H_0 are plotted in fig. 5 with z as ordinate. It will be noticed that they appear to be symmetrically disposed about the position $z = 4·525$ cm.

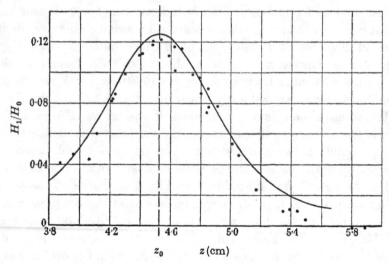

Fig. 5. Distribution of pressure along the axis in region of transition from horizontal to vertical sheet. The full line shows the calculated values from table 1.

To compare the pressure measurements with theory it is necessary to determine the scale on which the theoretical curve of pressure in the middle of the jet which is shown in fig. 2 must be transferred to fig. 5. The assumption that the steady flow is the same as that in the two-dimensional non-steady flow when it is translated with velocity u is represented by taking

$$\sqrt{2A_0}\,r_0 = u\sin\theta \quad \text{and} \quad z - z_0 = u\tau,$$

where z_0 is the position on the arbitrary scale of measurement of the place in the jet where the section is circular. Thus

$$z - z_0 = r_0(\sqrt{2A_0}\tau)\operatorname{cosec}\theta. \tag{12}$$

To find r_0, the rate of discharge Q under head H_0 was measured, and πr_0^2 was taken as equal to $Q/(2gH_0)^{\frac{1}{2}}$. Determinations made in this way yielded values of πr_0^2 ranging from 0·356 to 0·360 cm.². Taking the mean value $r_0 = 0·338$ cm., table 1 has been constructed with ξ as the independent variable. In column 2 of table 1 are the corresponding values of $\sqrt{2A_0}\tau$ calculated from (10). Column 3 contains the corresponding values of $z - z_0$ from (12).

Table 1

1	2	3	4	5	
			$\dfrac{2\sin^2\theta}{\xi^4+2+\xi^{-4}}$	z (cm.)	
ξ	$\sqrt{2A_0}\,\tau$	$z-z_0$			
1·0	0	0	0·1250	4·525	—
0·9	0·151	0·102	0·1197	4·627	4·423
0·8	0·321	0·217	0·1042	4·742	4·308
0·7	0·525	0·356	0·0782	4·881	4·169
0·6	0·783	0·530	0·0507	5·055	3·995
0·5	1·132	0·763	0·0272	5·288	3·762
0·4	1·642	1·115	0·0121	5·64	3·410
0·3	2·44	1·65	—	—	—

To compare the theoretical pressure with that observed, values of the function

$$\frac{H_1}{H_0} = \frac{p_0 - p_1}{\frac{1}{2}\rho u^2} = \frac{2\sin^2\theta}{\xi^4 + 2 + \xi^{-4}} \tag{13}$$

when $\theta = 30°$ are given in column 4 of table 1. The theoretical pressure distribution is symmetrical about z_0, and z_0 could be determined if the value of ξ at the orifice were known, but this could not be determined sufficiently accurately. For this reason z_0 was chosen arbitrarily as 4·525 cm. to coincide with the apparent centre of symmetry of the observed distribution of H_1/H_0. The figures in column 5 of table 1 are 4·525 ± column 3. It will be seen that the agreement between the theoretical model and observation is good. The observations give pressures which are inclined to be lower than those calculated, but the calculations are concerned with the maximum pressure at the centre so that any error in the position of the pressure-measuring device would be expected to decrease the observed pressure below the value at the centre.

COLLIDING JETS

In order that the two colliding jets should be as free from turbulence as possible they were made to issue from sharp-edged orifices each 0·227 cm. in diameter in the flat ends of two tubes each of 2 cm. internal bore. The flow was straightened by means of a number of thin-walled aluminium tubes 0·3 cm. in diameter and 10 cm. long followed by six sheets of wire gauze spaced at intervals of 1 cm. By this means clear cylindrical jets could be obtained.

The jet could be fixed at three angles $2\theta = 60°$, $90°$ and $120°$, and in each case flat fluid sheets were obtained which disintegrated into drops at the edges. Since the flow was very nearly radial from some point within the zone of impact and the velocity is the same as that of the jets since there is no means by which energy could be absorbed, the thickness t must decrease as $1/r$, r being distance from this point.

In any particular experiment tr varied with ϕ, the angular co-ordinate of stream lines in the plane of the sheet, and t was measured by catching the part of the sheet which entered a box between two razor blades the forward edge of which were 0·182 cm. apart. This box was mounted on a turn-table with a vertical spindle which passed as nearly as possible through the point in the region of collision from which the stream spread out. The velocity was taken as $u = \sqrt{(2gH)}$ where H is the head of water in the fluid behind either of the jets above the level of the horizontal sheet formed by their impingement. This last statement seems at first sight to contain an inconsistency because the orifices of the two jets could not be at the same level, but to get a perfect sheet the velocities in the two jets should be the same at the level of impingement, and this condition is obtained by having the head above the level of impingement the same in the two jets so that the head of water at the two orifices is slightly different. For complete equality between the impinging streams the volumetric rate of flow should be the same in the two jets as well as the velocity. This could be attained by having the lower orifice very slightly smaller than the upper one, but the error involved in having them identical is too small to be significant. The supply of water to the jet tubes was through two pieces of large-bore hose which were fitted over the two arms of a T-piece, the third being supplied from a reservoir fitted with a weir and an overflow to ensure a constant head. In each arm of the T-piece there was a valve which could throttle the flow. These valves being in the large-bore part of the hydraulic circuit and upstream of the turbulence reducing screens acted as fine adjustments in producing smooth sheets. The critical feature in the working of the apparatus turned out to be the part of the sheet which was projected backwards. This was much thinner than the forward moving film so that it very rapidly reached the point where $2T/\rho u^2 t = 1$ at which an edge perpendicular to the flow could be established. Here T is the surface tension. If, however, the jets were not accurately set up, or if their speeds were slightly different, the backward-moving part of the sheet could curl over till it came into contact with some part of the apparatus, or even bent back on to the water sheet without being in contact with any solid. In the latter case the flow instantly broke up and no sheet could be formed. By adjusting either of the two valves the backward-moving part could be made to curl upwards or downwards and it was only when an edge could establish itself before the sheet curled back or came into contact with a solid part of the apparatus that a smooth sheet could be established.

Figs. 6(a), (b), (c), pl. 2, show the sheets produced when the angles between the jets were $2\theta = 120°$, $90°$ and $60°$, respectively. It will be seen that the sheet breaks up owing to the formation of rather large drops which fly tangentially off the edges. The order of magnitude of the diameter of the drops can be appreciated by observing that each drop is seen as a pair of short lines. This is because the sheet was illuminated by two flashes operating simultaneously. They were both above the sheet, one on either side of the centre line. The camera was facing downwards from a point vertically above the middle of the sheet. The distance between the images of the two flashes is less than the diameter of the drop but generally of the same order of magnitude.

MEASUREMENTS OF SHEET THICKNESS

The effect of viscosity in the impact region is likely to be very small. The effect of surface tension is also likely to be small because in the region of impact the ratio of surface tension forces to inertia forces is proportional to $2T/\rho u^2 t$ and this is equal to or less than 1 at all points on the edge of the sheet. It is therefore very much less than 1 close to the point of impact. If the forces operating in the impact region are due only to inertia the distribution of thickness in the sheet should be of the form $rt = F(\theta, \phi)$ and independent of u.

The volume of fluid collected per second in the box is $q = tu\,(0.182)$ ml. so that $F(\theta, \phi) = (qr/0.182u)$. $F(\theta, \phi)$ so defined has dimensions L^2 so that to make it non-dimensional it should be divided by the area of the orifices, but since only one pair of orifices was used in the present experiments this has not been done.

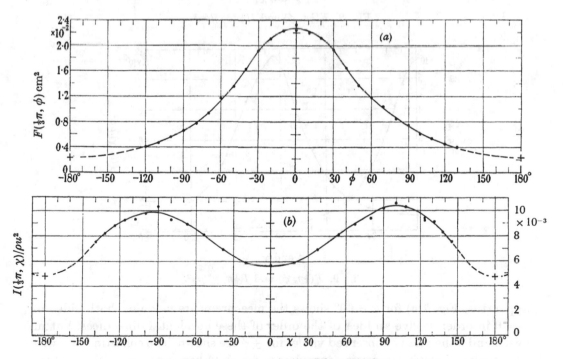

Fig. 7. (a) $F(\tfrac{1}{3}\pi, \phi)$; (b) $I(\tfrac{1}{3}\pi, \chi)/\rho u^2$.

The measured values of $F(\frac{1}{3}\pi, \phi), F(\frac{1}{4}\pi, \phi)$ and $F(\frac{1}{6}\pi, \phi)$ are shown in figs. 7(a), 8, 9. The measured values of $F(\theta, \phi)$ were not always quite symmetrically disposed about $\phi = 0$, but this could only be due to small inaccuracies in setting up the

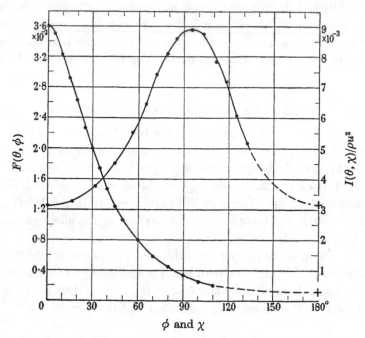

Fig. 8. $F(\frac{1}{4}\pi, \phi)$ and $I(\frac{1}{4}\pi, \chi)/\rho u^2$.

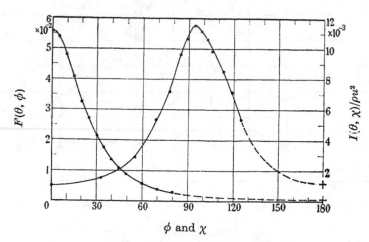

Fig. 9. $F(\frac{1}{6}\pi, \phi)$ and $I(\frac{1}{6}\pi, \chi)/\rho u^2$.

apparatus. Fig. 7(a) for $\theta = 60°$ shows the measurements made on both sides of $\phi = 0$ in order to give an idea of the order of these accidental differences, but in figs. 8 and 9 the mean of observed values for $\pm \phi$ are shown. The collecting box was not capable of making measurements close to $\phi = 180°$ because it came into contact with the jet tubes. The greatest values of ϕ which could be measured are indicated by the outer points in figures. It will be seen in the photographs (fig. 6) that the edge of the sheet near $\phi = 180°$ is perpendicular to the radius from the point of impact. The thickness of the film at that point may be taken as close to $2T/\rho u^2$* so that $F(\theta, \pi)$ may be taken as $2T/\rho u^2$ (distance to edge at $\phi = \pi$ from point of impingement). With T taken as 73 c.g.s. for water the values of $F(\theta, \pi)$ could be estimated. These values are shown by means of crosses at 180° in the figs. 7, 8, 9 and the curve is filled by a purely notional broken line.

LATERAL SPREAD

The lateral spread of the sheet can only be due to a high pressure generated in the area of impact as it is in the converging-diverging jets of elliptic section. This pressure is the agent which destroys the positive and negative components of vertical momentum in the two jets and causes the resultant horizontal flow in the sheet. It causes a vertical reaction force between the two jets equal to $\rho Q u \sin \theta$, where Q is the volume flowing per second in each jet. Since pressure in a Newtonian fluid acts equally in all directions, it might perhaps be expected that the horizontal component of the reaction between the two halves of each jet which causes the lateral spreading would be directly related to the vertical reaction which annihilates the vertical momentum of each jet. The force which changes the direction of each stream line where it passes through the impingement region can be calculated when $F(\theta, \phi)$ is known. The volume of fluid which flows out per second in the range of directions between ϕ and $\phi + d\phi$ is $uF(\theta, \phi) \, d\phi$. To change its direction a force is required whose components are:

$$\rho u^2 F(\theta, \phi) \, (\cos \theta - \cos \phi) \, d\phi \quad \text{in direction } \phi = 0,$$

$$\rho u^2 F(\theta, \phi) \sin \phi \, d\phi \quad \text{in direction } \phi = \tfrac{1}{2}\pi,$$

$$\rho u^2 F(\theta, \phi) \sin \theta \, d\phi \quad \text{vertically.}$$

The horizontal component of this force is therefore

$$\rho u^2 F(\theta, \phi) \, d\phi \{(\cos \phi - \cos \theta)^2 + \sin^2 \theta\}^{\frac{1}{2}} \tag{14}$$

and it acts in the direction χ where

$$\tan \chi = \frac{\sin \phi}{\cos \phi - \cos \theta}. \tag{15}$$

Defining the reaction which covers the angle $d\chi$ as $I(\theta, \chi) \, d\chi$, we have

$$I(\theta, \chi) = \rho u^2 F(\theta, \phi) \{1 - 2 \cos \theta \cos \phi + \cos^2 \theta\}^{\frac{3}{2}} (1 - \cos \theta \cos \phi)^{-1}. \tag{16}$$

It is of interest to insert the measured values of $F(\theta, \phi)$ in (16) in order to study $I(\theta, \chi)$. The values of $I(\theta, \chi)/\rho u^2$ calculated in this way are shown as functions of χ in figs. 7(b), 8 and 9.

* G. I. Taylor, *Proc. Roy. Soc.* A, ccliii (1959), 313 (paper 32 above).

It will be seen that $F(\theta, \phi)$ is much larger when $0 < \phi < \frac{1}{2}\pi$ than when $\phi > \frac{1}{2}\pi$. On the other hand $I(\theta, \chi)$ appears almost symmetrical about $\chi = \frac{1}{2}\pi$ and the lateral reaction is much larger than the longitudinal. In this connection it will be noticed that the internal reactions must be in equilibrium so that

$$\int_0^\pi I(\theta, \chi) \cos \chi \, d\chi = 0. \tag{17}$$

This relationship can in fact be deduced directly from the momentum equation

$$\int_0^\pi F(\theta, \phi) \cos \phi \, d\phi = \cos \theta \int_0^\pi F(\theta, \phi) \, d\phi.$$

The values of $\int_0^\pi I(\theta, \chi) \, d\chi$ obtained by numerical integration using the data given in fig. 7 (b) for $\theta = 60°$ differs from zero only by an amount which could well be due to experimental error. Using the data of fig. 9, for $\theta = 30°$, numerical integration gives values which differ from zero by an amount which is greater than would be expected if the assumption that the water sheet expands from a point is valid.

COMPARISON BETWEEN LATERAL REACTION AND VERTICAL REACTION

The vertical reaction between the upper and lower jets is a force

$$\rho u^2 \sin \theta \int_0^\pi F(\theta, \phi) \, d\phi.$$

The total lateral reaction across the vertical plane containing the impinging jets is

$$\rho u^2 \int_0^\pi F(\theta, \phi) \sin \phi \, d\phi$$

so that this ratio is

$$\frac{\text{lateral reaction}}{\text{vertical reaction}} = \frac{\int_0^\pi F(\theta, \phi) \sin \phi \, d\phi}{\sin \theta \int_0^\pi F(\theta, \phi) \, d\phi} = R.$$

The values of R found by numerical integration of the curves of figs. 7, 8 and 9 were as follows:

θ	30°	45°	60°
R	0·66	0·65	0·66

The value for $\theta = \frac{1}{2}\pi$, i.e. for the horizontal sheet formed by the vertical impingement of two equal jets, is evidently $2/\pi$ or 0·64. This ratio appears to be constant for values of θ over the range which could be measured.

SHAPE OF THE EDGE OF THE SHEET

The form of the antisymmetrical waves which can remain at rest is described*
by the equation

$$\sin^2 \psi = \frac{2T}{\rho u^2 t} = \frac{2Tr}{\rho u^2 F(\theta, \phi)}, \tag{18}$$

where ψ is the angle between the wave front and the radius vector from the point
of impact. In the case where $\theta = \frac{1}{2}\pi$, $F(\theta, \phi)$ is independent of ϕ and the curves which
satisfy (18) are cardioids. With the use of the experimentally determined values of
$F(\theta, \phi)$ it is possible to construct the form of the waves. The edges of the sheet lie
approximately parallel to them* and in figs. 6 (a) (pl. 2) and 11 (pl. 1), portions of such
waves are revealed by reflected light. It is therefore of interest to construct the expected form of antisymmetrical waves from the measured values of $F(\theta, \phi)$ in (18).
The double system of curves so produced, corresponding to equal positive and
negative values of ψ at every point, all represent possible positions where edges
could be maintained if they broke up into drops at the moment of their formation.
If the complete set of curves which represent all the solutions of (18) were constructed it will be noticed that none can exist outside the radius $\rho u^2 F(\theta, \phi)/2T$.
There is one critical curve, namely, that which cuts $\phi = \pi$ at right angles at
$r = \rho u^2 F(\theta, \pi)/2T$, within which antisymmetric waves can exist at all points. An
edge could therefore be established along any of the waves within that unique
curve, by placing an obstacle in the sheet which would cause it to part and form
two edges, but on removing the obstacle the edges which it had formed would be
carried out till they reached the position of the unique wave which is perpendicular
to the radius vector at the point where $F(\theta, \phi)$ is least. This therefore is a position at
which an edge could be established without using an obstruction to hold it in
position.

Unfortunately I was not able to measure $F(\theta, \phi)$ when ϕ exceeded 110° for
$\theta = 68°$, or 80° for $\theta = 30°$ so that I was not able to construct the unique curve
from direct measurements of $F(\theta, \phi)$. The best I could do was to measure on the
photographs, fig. 6, the angle ψ of the edge at the greatest value of θ at which I
could measure $F(\theta, \phi)$, and construct the wave shape using (18) and starting with
these measured values. In this way the forms shown in fig. 10 and in fig. 12, pl. 1,
were constructed.

The curves are continued as broken lines back to $\phi = \pi$, but no calculation was
possible since $F(\theta, \phi)$ was not measured.

It will be noticed that if $F(\theta, \phi)$ is independent of u as it must be, if the
effects of viscosity and surface tension in the region of impingements are
negligible, the shape of the sheets depends only on θ but its linear dimensions are
proportional to u^2, i.e. to the head H of fluid behind the orifices. This was found
to be true within the limited range in which sheets similar to those of fig. 6 were
obtained.

* See G. I. Taylor, *Proc. Roy. Soc.* A, CCLIII (1959), 296 (paper 31 above).

When $\theta = 30°$, for instance, the values of H and L the extreme value of r at the point of the sheet are given below.

Table 2

H (cm.)	160	142·9	140	113	85·3
L (cm.)	24	22·0	21·2	17·7	13·7
L/H	0·15	0·154	0·150	0·156	0·161

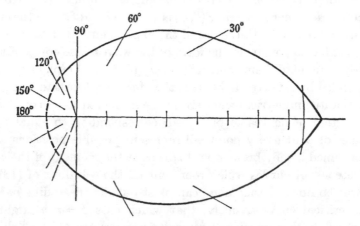

Fig. 10. Shape of critical wave for $\theta = 60°$, calculated using measured values of $F(\theta, \phi)$ in (18).

If H was too low the values of L/H increased considerably because the drops did not detach themselves from the edge as soon as it formed so that the edge established itself outside the critical antisymmetrical wave. In the case $\theta = 30°$ the sheet began to oscillate rather violently when H exceeded 160 cm. so that the position of the pointed end of the sheet could not be determined. At still higher values of H the sheet broke up owing to violent oscillations and so prevented the formation of a pointed end. This no doubt was due to a flag-like instability caused by the reaction of air on the sheet.

INSTABILITY DUE TO NORMAL REACTION OF AIR

The instability of moving sheets, due to the normal reaction of surrounding air, has been discussed by Squire.* Squire's analysis refers to sheets of uniform thickness, but his conclusions may be expected to apply at any rate qualitatively to flat expanding sheets (in fact Squire applied his analysis to this case) and they show that a moving sheet is always unstable provided the wavelength is greater than a critical value. It seems therefore that experiments can be carried out under apparently steady conditions only if the amplification in the observable length of the sheet is small enough.

* *Br. J. appl. Phys.* IV (1953), 163.

PLATE 1

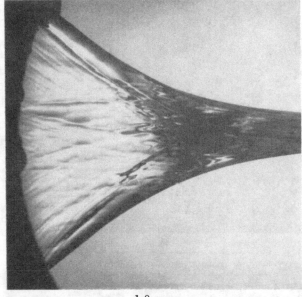

1·0 cm.

Fig. 4. Photograph of converging jet issuing from an elliptic hole
in a cylindrical surface.

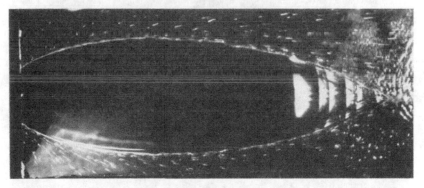

Fig. 11. Unstable waves at the end of sheet formed by two jets, $\theta = 30°$.

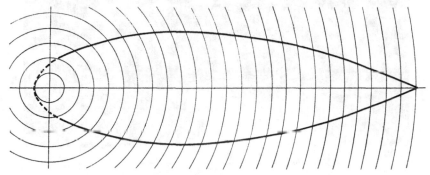

Fig. 12. Shape of critical wave for $\theta = 30°$.

PLATE 2

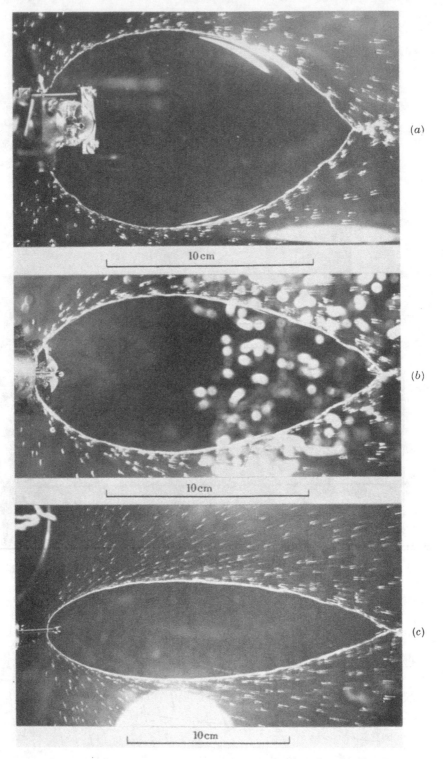

Fig. 6. Sheets produced by impact of cylindrical jets. (a) Jets inclined at $2\theta = 120°$, $H = 67·3$ cm.; (b) jets at $2\theta = 90°$, $H = 67·9$ cm.; (c) jets at $2\theta = 60°$, $H = 139·6$ cm.

In his analysis Squire imagines that a sheet of thickness t and density ρ flows at speed u through stationary air of density ρ_a. Assuming antisymmetric waves whose displacement is proportional to $e^{i(\kappa x - \sigma \tau)}$, Squire's results can be expressed in the form

$$\frac{\sigma}{\kappa u} = \frac{1}{1+\alpha} \{1 + [1 + (1+\alpha)(\tfrac{1}{2}W\kappa t \coth \tfrac{1}{2}\kappa t - 1)]^{\frac{1}{2}}\}, \tag{19}$$

where $W = 2T/\rho u^2$ is the same Weber number as is used by Taylor.* α is written for $(\rho_a/\rho) \coth \tfrac{1}{2}\kappa t$. When the sheet is very thin compared with the wavelength $\coth \tfrac{1}{2}\kappa t = 2/\kappa t$ and (19) can be written

$$\frac{\sigma}{\kappa u} = \frac{1}{1+\alpha} \{1 + [W(1+\alpha) - \alpha]^{\frac{1}{2}}\}, \tag{20}$$

so that the waves are unstable when $W < \alpha/(1+\alpha)$ and the critical value of α below which all waves are stable is

$$\alpha_{\text{crit}} = W/(1 - W).$$

Since

$$k = 2\pi/\lambda \quad \text{and} \quad \alpha = \frac{\rho_a}{\rho}\left(\frac{2}{\kappa t}\right), \quad \frac{\alpha}{W} = \frac{\rho_a}{\rho}\frac{\rho u^2 \lambda}{2\pi T}.$$

When W is small $\alpha_{\text{crit}}/W \sim 1$ so that $\lambda_{\text{crit}} = 2\pi T/\rho_a u^2$. Thus λ_{crit} is independent of W and therefore of the thickness and therefore of position in an expanding sheet. If the jets are formed under head H so that $u^2 = 2gH$

$$\lambda_{\text{crit}} = \frac{\rho}{\rho_a}\left(\frac{\pi T}{\rho g H}\right); \tag{21}$$

for a water sheet in air $\rho/\rho_a = 800$, $T = 73$ dynes/cm. and

$$\lambda_{\text{crit}} = \frac{1\cdot 87 \times 10^2}{H} \text{ cm.}$$

The question which wavelength greater than λ_{crit} will predominate and achieve large amplitude cannot be discussed completely but various hypotheses have been made in similar cases. The commonest is to assume that the wave which increases in amplitude most rapidly will predominate. When W is small this leads to the prediction that the predominating wave occurs when $\alpha = 2W$ and that its magnitude is $\lambda_{\text{max}} = 2\lambda_{\text{crit}}$. This result is equivalent to one given by Squire. For a water sheet projected into air therefore

$$\lambda_{\text{max}} = 3\cdot 74 \times 10^2/H. \tag{22}$$

The fact that λ_{max} does not depend on the thickness of the sheet simplifies calculation of the total amplification of the wave as it passes down it. The amplification per wave of length λ_{max} is

$$\exp\left(\frac{2\pi\{\alpha - W(1+\alpha)\}^{\frac{1}{2}}}{1+\alpha}\right),$$

* Paper 31 above.

or approximately $\exp(2\pi W^{\frac{1}{2}})$, so that the total amplification at distance r from the origin is

$$A = \exp\left(\frac{2\pi}{\lambda_{\max}} \int_0^r W^{\frac{1}{2}} dr\right). \tag{23}$$

The sheets produced by impingement of two jets are not of uniform thickness so that W is not a function of r only but also of ϕ; in an approximate treatment it may be sufficient to consider only the thickness of the centre part of the sheet where $\phi = 0$. In that case

$$W = \frac{2Tr}{\rho u^2 F(\theta, 0)},$$

so that

$$A = \exp\left\{\frac{2\pi}{\lambda_{\max}} \left(\frac{2T}{\rho u^2 F(\theta, 0)}\right)^{\frac{1}{2}} \tfrac{2}{3} r^{\frac{3}{2}}\right\}. \tag{24}$$

For water jets flowing at head H into air this leads to

$$A = \exp\left[\left(\frac{2\pi H}{3\cdot 74 \times 10^2}\right)\left(\frac{146}{2gHF(\theta, 0)}\right)^{\frac{1}{2}} \tfrac{2}{3} r^{\frac{3}{2}}\right].$$

For jets colliding at $60°$ to one another, i.e. $\theta = \tfrac{1}{6}\pi$, with $F(\theta, 0) = 5\cdot 6 \times 10^{-2}$, and $r/H = 0\cdot 15$ (table 2) the value of A at the end of the sheet is

$$A = \exp(8\cdot 07 \times 10^{-4} H^2).$$

At $H = 50, 100$, and $160\,\text{cm}$., $A = 8\cdot 3 \times 10^3, 3 \times 10^6, 4 \times 10^8$.

These are the extreme limits of amplification theoretically possible with the assumption that the amplification in each section of the sheet is the same as in a sheet of uniform thickness equal to the local thickness of the expanding sheet. It is not surprising that sheets made experimentally can be steady when $H = 50\,\text{cm}$. At $H = 100\,\text{cm}$. the sheets also appeared nearly steady. At $H = 140$ the sheets were certainly oscillating at the tip but not sufficiently to prevent the point at the end of the sheet which is shown in fig. 6(c) from forming. At $H = 160$ their motion was so violent that the sheet began to disintegrate at the end before the point could be formed.

Wavelength of unstable waves

By placing a flash bulb above the axis the unstable waves were photographed. Fig. 11, pl. 1, shows the waves for $\theta = 30°$, $H = 155\,\text{cm}$.; for this case according to (22) $\lambda_{\max} = 2\cdot 4\,\text{cm}$.

There are two positions in each wave from which light can be reflected into the camera lens. The wavelength is therefore approximately equal to the spacing of alternate bands. In fig. 11, these are $2\cdot 0\,\text{cm}$. apart. In another case at the same measured head they were $2\cdot 3\,\text{cm}$. apart but the photograph was too dense for good reproduction. Thus the measured lengths of the unstable waves agree quite as well as could be expected with the calculated waves of greatest rate of increase in amplitude. Fig. 12 is the calculated shape of the unique antisymmetrical wave when $\theta = 30°$ for comparison with figs. 11 and 6c.

Appendix

Note on a recent paper by Dombrowski, Hasson & Ward

Since the present paper was written measurements of the thickness of a sheet produced by a special form of nozzle have been published.* These measurements were made with an interferometer, a better technique for this purpose than the cruder method which I used. In analysing their results, however, D., H. & W.* rely on some hydrodynamic ideas which seem to be open to criticism. To compare and contrast their results with those here described it is necessary to quote relevant parts of their paper. On p. 39 of their paper their conception is set forth as follows: 'Although the converging streamlines in the nozzle cause a region of high pressure to be formed behind the orifice it will be assumed in the following analysis that the liquid flows from the nozzle as if there were a line source of high pressure perpendicular to the sheet.'

Since a line source must be a region of high suction rather than high pressure, the meaning of this sentence is obscure and it is only possible to attach a meaning to it by referring to the use which is made of the idea behind it. D., H. & W.'s experiments show that stream lines radiate from a small region in the orifice which can be regarded in the analysis as a point. Their interferometer measurements show that the thickness of the sheet varies along any radius inversely as the distance from this point. These observations are consistent with a radial flow of constant velocity and are therefore consistent with the assumption used in the present paper, but D., H. & W. make a further assumption, namely, that the thickness of the sheet is the same at all points which lie at a given distance from the orifice. They express this mathematically by defining a symbol K which is identical with my $F(\theta, \phi)$ and assuming that K is a constant. In my notation this is to assume that $F(\theta, \phi)$ is independent of ϕ. There seems no theoretical reason to suppose that this is true and my measurements show that when the expanding sheet is produced by the oblique impact of two cylindrical jets it is not true. On the other hand, D., H. & W.'s measurements show that in some cases the flow from the special form of nozzle which they used does approximately correspond with a constant K.

Analysis designed to give the shape of the sheet and based on assuming a constant value of K might therefore be expected to give results in agreement with observation of sheets produced by the particular nozzle used, but even so the shape of the sheet can only be calculated by making further assumptions about the hydrodynamic conditions at the edge. These are described by D., H. & W. in the sentence which follows the quotation given above, namely 'Furthermore, it will be assumed that the contraction of the edges by surface tension does not affect the flow of the sheet, i.e. the liquid corresponding to the "vanished" part of the sheet lying between the theoretical linear and the actual curved boundaries is concentrated in the curved boundary.' This sentence contains two elements, the first up to the abbreviation

* N. Dombrowski, D. Hasson & D. E. Ward, *Chem. Engng Sci.* xx (1960), 35.

'i.e.' seems little more than a restatement of the physics involved in the discussion contained in the previous sentence, but what follows the abbreviation 'i.e.' is an additional physical assumption, namely, that the fluid after entering the curved edge will remain there and pursue a curved path. To maintain the flow in the curved edge a force normal to that edge is required in addition to the force required to neutralise the momentum of the fluid in the sheet. This force is neglected in the treatment given by D., H. & W. The equations for the edge appropriate to D., H. & W.'s assumed edge conditions were given by Taylor as (6) to (11) of paper 32 above.

If instead of assuming that the fluid remains in the curved edge after entering it one goes to the other extreme and assumes that the fluid does not remain in the edge, but leaves it immediately in the form of drops moving in straight tangential lines, the calculated shapes of the edges are cardioids and in fact D., H. & W.'s equation 10 is identical with (16) in paper 31 above by Taylor. Their photograph 3 of the sheet of water produced by their fan spray jet is very similar to part of fig. 6(c) of the present paper, and shows that the edge is surrounded by drops flying off tangentially so that if K is nearly constant the sheet would be bounded approximately by two arcs of cardioids.

34

DEPOSITION OF A VISCOUS FLUID
ON A PLANE SURFACE

REPRINTED FROM

Journal of Fluid Mechanics, vol. IX (1960), pp. 218–24

Two mechanisms by which a viscous fluid can be deposited on a plane surface are described. Measurement of the thickness of the deposit are compared with calculated values. It is found that the two agree within rather wide limits of experimental error provided the effect of surface tension can be neglected, and the conditions under which this is legitimate are discussed.

THE PAINT BRUSH

There are many ways in which a viscous fluid can be deposited on a plane surface. Perhaps the best known is by means of a paint brush. This consists of a number of flexible cylinders or hairs between which the fluid lies. As these hairs are dragged over the surface they lie with their axes parallel to the direction of motion and the paint is dragged through between them by tangential stress. The fluid dragged out by tangential stress acting at one side of the brush on the fluid surrounding the outermost layer of hairs must be replaced by fluid flowing transversely from the interstices between hairs further from the outer layer. This process can be understood qualitatively but to describe it mathematically is difficult. For this reason it seemed worthwhile to describe an ideal structure which would deliver fluid at a calculable rate even though it has no practical use in the art of painting.

The simplest is a plate sliding at height h_0 over a fixed parallel plane. If fluid is supplied at atmospheric pressure between the leading edge of the moving plate and the fixed plane, and if the length of the plane is great compared with h_0, the pressure in the fluid will be constant and a film of depth $\frac{1}{2}h_0$ will be left behind on the fixed plane.

If the moving plate is replaced by a portion of a cylindrical surface of any cross-section moving parallel to the generators it is possible to calculate the volume of fluid left behind. Two examples will be given.

Parallel plates

An arrangement by which plates spaced at distance $2d$ apart could be made to slide on their edges over a horizontal plane is shown in fig. 1. Taking an origin in the surface of the horizontal plane midway between two vertical plates, the co-ordinate x is in the direction of motion, i.e. parallel to the plates and z

is vertical. The velocity u is parallel to the direction of motion at all points and satisfies

$$\frac{\partial^2 u}{\partial y^2} + \frac{\partial^2 u}{\partial z^2} = 0. \tag{1}$$

It is convenient to take u as the velocity relative to the moving plates, so that the boundary conditions are

$$u = 0 \quad \text{at} \quad y = \pm d \quad \text{and} \quad u = U \quad \text{at} \quad z = 0.$$

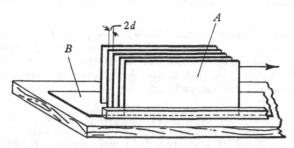

Fig. 1. A, parallel plates; B, sheet on which deposit is collected.

The solution of (1) which satisfies these conditions is

$$u = \frac{4U}{\pi} \sum_0^\infty \frac{(-1)^n}{2n+1} \cos\frac{(2n+1)\pi y}{2d} \exp\left\{\frac{-(2n+1)\pi z}{2d}\right\}. \tag{2}$$

The total volume deposited on the horizontal plane per second from the fluid contained between each pair of plates is

$$Q = \int_{-d}^{+d}\int_0^\infty u\, dz\, dy = \frac{32Ud^2}{\pi^3} \sum_0^\infty \frac{1}{(2n+1)^3} = 1{\cdot}085\,Ud^2. \tag{3}$$

Semicircular grooves

In the above example the way in which the fluid emerges from the spaces between the parallel plates was not considered nor was the manner in which the fluid entered them. The effect of gravity and of possible vertical motion between the plates was also assumed to be negligible. To avoid these uncertainties a 'mathematical paint brush' was designed in which the supply of fluid occurred only at the forward end of the channels which regulated the flow. A number of semicircular grooves were cut in the plane base of a Perspex block. These grooves ended at the supply end in a chamber large enough to ensure that the fluid was supplied at atmospheric pressure. There were 5 grooves each of radius 0·1 cm. and they extend from the rear end of the block through a length of 6·1 cm. to a cavity from which the fluid was supplied. The method of experiment is indicated in fig. 2. The block A was placed on a strip of very thin metal foil B which had been pressed on to a steel surface plate. The block was charged with a viscous fluid, usually glycerine, through the hole C which led into the supply chamber. No fluid could escape except through the grooves. The block rested against a guide so that it could move parallel to the grooves while depositing fluid on the foil. The foil was then peeled off the surface plate and a section cut from the middle using a photographic trimming knife. This section was then weighed with its deposit. After removing the deposit it was weighed again.

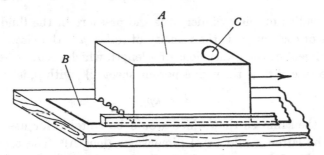

Fig. 2. *A*, Perspex block; *B*, sheet on which fluid is deposited; *C*, filling hole.

In a series of such experiments made with glycerine the weights deposited per cm. of foil varied from 0·0422 to 0·0467 g. The mean of six runs was 0·0437 g. and the corresponding volume was 0·0347 c.c. Thus the experimental value of the volume deposited per cm. from each of the five grooves was 0·00694 c.c. The calculation of the volume deposited from a semicircular groove of radius a was made in the same way as for the parallel plates, giving the result that the volume deposited per unit length of run was $2a^2/\pi$.* When $a = 0\cdot1$ cm., we have $2a^2/\pi = 0\cdot00638$. This is 8 % smaller than the observed value 0·00694. The discrepancy may well be due to small errors in the shaping of the grooves or to slight undetected variations in the flatness of the surface of the foil. The neglect in the calculation of the small pressure gradient along the grooves may give rise to a small error. This error, however, could be calculated and it was much less than the 8 % discrepancy, though its contribution was to increase the deposit above the calculated value $2a^2/\pi$.

POROUS ROLLERS

Another way in which a viscous fluid can be deposited on a plane surface under conditions for which the flow can be described mathematically is by the use of a porous roller. This method is used by printers and sometimes by house-painters. The flow of fluid in this case has been discussed mathematically by Taylor & Miller† though the connection between the thickness of the layer deposited and

* Later Dr F. Ursell pointed out that the calculation could have been much simplified, because the solution of $\nabla^2 u = 0$ which satisfies $u = 0$ on the circumference and $u = U$ on the diameter is

$$u = U\left(\frac{2\theta}{\pi} - 1\right),$$

where θ is the angle subtended at the point where the velocity is u by the two ends of the diameter.

† *Quart. J. Mech. appl. Math.* IX (1956), 129 (paper 25 above).

the hydraulic resistance of the porous cylinder is not contained explicitly in their paper. They express their results in non-dimensional co-ordinates x_1 and y but here p_1 is substituted for their y as a symbol whose meaning is more obvious and

$$x_1 = x(96\kappa R^3)^{-\frac{1}{4}}, \quad p_1 = -\frac{p\kappa R}{\mu U}(96\kappa R^3)^{-\frac{1}{4}}. \tag{4}$$

Here R is the radius of the cylinder, p is the pressure in the fluid at distance x from the point of contact, U is the velocity of roller, μ is the viscosity of the fluid and κ is a coefficient of the dimensions of a length which occurs when Darcy's law connecting the rate of flow through a porous sheet, W, with p, is expressed in the form

$$W = \kappa p/\mu. \tag{5}$$

Thus κ is not the same as that which would be suitable for use with a porous medium, for the latter would be of dimensions (length)^2. The total rate of flow though unit length of the porous cylinder is

$$Q = \int_0^\infty W\,dx, \tag{6}$$

and on substituting from (4) and (5) we find

$$Q = (96\kappa R)^{\frac{1}{2}} U \int_0^\infty p_1\,dx_1. \tag{7}$$

The computed relationship between p_1 and x_1 is given in the form of a curve (fig. 3, p. 134 of paper 25 above). By graphical integration over the main part of the curve and the use of the asymptotic expression for p_1 at large values of x_1, $\int_0^\infty p_1\,dx_1$ was found to be 0·302 so that

$$Q = 2\cdot96 U(\kappa R)^{\frac{1}{2}}. \tag{8}$$

This value for Q is the value which would apply if the whole space between the porous cylinder and the plane on which it rolls were filled with fluid. In fact the fluid only passes through the porous cylindrical surface when there is suction and, at a certain point which my analysis could not determine, the flow separates at a meniscus, some fluid remaining on the flat plate and some adhering to the cylinder. Beyond this meniscus there is no further suction. In the absence of further evidence it seems that the meniscus is likely to divide the fluid equally into two streams as it would if the cylindrical surface had not been porous. Assuming this to be the case the thickness t of the layer deposited would be

$$t = \frac{1}{2}\frac{Q}{U} = 1\cdot48(\kappa R)^{\frac{1}{2}}. \tag{9}$$

Measurements

A perforated cylinder 21 cm. long and 13·5 cm. in diameter was made by wrapping a perforated sheet round some circular discs. These discs had central holes to permit the entry of the fluid to the inner surface of the cylinder. Two layers of flannel were wrapped round the cylinder and the seam pulled tight. The fabric was bent over the ends of the cylinder, sewn through the end perforations and sealed on the inner side to prevent fluid from escaping without passing through the flannel.

Two methods were used to measure t. The first was to roll the cylinder on a sheet of plate glass and then blot up the deposit with weighed sheets of blotting paper which were then rapidly enclosed in a container to prevent evaporation and weighed again with the absorbed fluid. The second, indicated in fig. 3, was

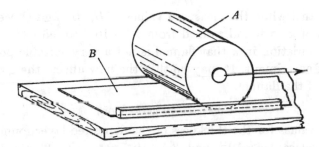

Fig. 3. A, porous roller; B, sheet on which fluid is deposited.

to use the technique employed in the experiments with the 'idealized' paint brushes described earlier, but since the thickness of the film deposited was much less than before larger pieces of foil had to be used. Even so the plotting-paper technique proved to be more satisfactory because it was difficult to prevent the foil from being pulled off the plate on which it had been pressed as the roller passed over it. The porous roller was constructed in the manner described in order to make possible a comparison with the theoretical calculation, but even so the range of fluids for which that comparison could be made was limited.

When water was used the thickness left on the plate was sometimes very small and it was concluded that on these occasions one of the conditions assumed in the theory, that the fluid filled the space between plate and roller, was not valid and cavitation occurred. More consistent results were obtained when more viscous fluids were used, but when fluids as viscous as pure glycerine were used the suction was so great at any but very low speeds that there was danger that the flannel would leave the perforated surface on which it was stretched and thus upset the geometry of the flow.

Some of the measurements are given in table 1.

Table 1. *Measurements of thickness of film deposited by roller*

Liquid	μ (g. cm.$^{-1}$ sec.$^{-1}$)	Film thickness (10^{-3} cm.)	Mean
50 % glycerine	0·085	2·39, 2·12, 1·86, 2·6, 2·44	2·28
30 % glycerine	0·030	2·25, 1·67, 1·95, 1·77	1·91
Water	0·011	0·89, 1·28, 0·64, 1·21, 0·73	0·95

Measurements of κ

It seemed desirable to measure κ with the flannel stretched on the perforated cylinder. The cylinder was therefore set with its axis vertical in a cylindrical vessel from which the outflow could be measured. The height H_2 of the fluid outside the cylinder, as well as the difference in height H_1 between the fluid inside and outside the porous surface, was measured. Water was used in this experiment. The volume Q' flowing through the porous cylinder per second was taken as being given by

$$Q' = \frac{2\pi R \kappa}{\mu} H_1(H_2 + \tfrac{1}{2}H_1). \tag{10}$$

It was found that when the measured values of H_1, H_2, and Q' were inserted in (10) the values of κ so found ranged from $1 \cdot 5 \times 10^{-7}$ to $3 \cdot 5 \times 10^{-7}$ cm. It seems from this large variation in κ that flannel is not a very suitable porous material for this kind of experiment. It may be that the flow affects the geometry of the fibre structure of the flannel.

Comparison with theory

The theoretical values for the thickness of the deposited layer found by inserting these limiting values $1 \cdot 5 \times 10^{-7}$ and $3 \cdot 5 \times 10^{-7}$ for κ in (9) are $1 \cdot 5 \times 10^{-3}$ and $2 \cdot 3 \times 10^{-3}$ cm. Comparison of these with the observed values given in table 1 shows that except in the case of water the agreement with the theoretical analysis is within the limit of experimental error. A possible explanation of the discrepancy in the case of water is given later.

EFFECT OF SURFACE TENSION AT THE AIR-FLUID INTERFACE

In the cases which have so far been discussed it has been tacitly assumed that the surface tension produces negligible effects. This is justifiable if the pressures due to the curvature of the free surface are small compared with those which would produce an appreciable change in the flow. In the case of the 'idealised paint brush', since the curvature of the free surface must be of the order $1/a$ (or $1/d$) for the two cases considered the pressure change on passing through the meniscus is of order T/a, where T is the surface tension. The viscous stresses are of order $\mu U/a$. It might therefore be thought that the analysis is only realistic when $T/\mu U$ is small. This is not the case however. Taking the case illustrated in fig. 2, the change in Q due to change in pressure δp between the ends of the grooves is of order $a^4 \delta p/\mu L$, where L is the length of the groove. Q is of order $a^2 U$, so that the condition that a change δp will make a negligible change in Q is that $a^2 \delta p/\mu U L$ shall be small. If δp is of order T/a, this condition is that $Ta/\mu U L$ shall be small.

When the measurements were made with the apparatus shown in fig. 2, the influence of the surface tension was not fully appreciated so that accurate measurements of U were not made. They were, however, of order $U = 4$ cm./sec. At this speed, and with glycerine, for which $\mu = 9$ g. cm.$^{-1}$ sec.$^{-1}$ and $T = 63$ g. sec.$^{-2}$, $T/\mu U = 1.7$. This is not small, but the length of the grooves was 6 cm. while the radius was 0·1 cm., so that $Ta/\mu UL = 0.03$ and thus was sufficiently small to warrant an expectation that the calculated value of Q (namely, $2a^2/\pi$) might be realized. When water was used instead of glycerine, much thinner layers were deposited.

Similar considerations apply to the porous roller. In making their calculations Taylor & Miller assumed that the whole field of flow was flooded. In fact a meniscus or interface formed itself and divided the fluid into two streams one of which remains on the plane surface and the other is carried round on the outer surface of the roller. The meniscus is not likely to affect the distribution of suction between plane and roller unless it establishes itself within the range where the suction is appreciable. A suitable criterion for estimating whether the meniscus will have appreciable effect is to imagine that the meniscus establishes itself at the point where the calculated suction is equal to the pressure rise on passing through the meniscus due to surface tension. If this point is in the range where the suction is small compared with its maximum value, then it would not be expected to make an appreciable change in the value of Q.

The suction at a distance x from the point of contact of roller and plane is given by the equation (4) and the relationship between the non-dimensional quantities p_1 and x_1 is shown in fig. 3 of the paper 25 above by Taylor and Miller. If the radius of curvature of the meniscus is taken as half of the distance between plane and roller at distance x, the meniscus will establish itself near the point where

$$4TR/x^2 = -p. \tag{11}$$

Substitution for x and p from (4) and (11) gives

$$p_1 x_1^2 = \frac{T}{\mu U} (96)^{-\frac{3}{4}} \left(\frac{\kappa}{R}\right)^{\frac{1}{4}}. \tag{12}$$

When $x_1 > 1$ the asymptotic form of Taylor & Miller's expression may be used. This is

$$p_1 x_1^3 \sim \tfrac{1}{8}. \tag{13}$$

Dividing (13) by (12), an approximate value for the position of the meniscus is

$$x_1 \sim 5\cdot1 \left(\frac{\mu U}{T}\right) \left(\frac{R}{\kappa}\right)^{\frac{1}{4}}, \tag{14}$$

where 5·1 is the approximate value of $(96)^{\frac{3}{4}}(\tfrac{1}{8})$.

When therefore the value of x_1 calculated from (14) is larger than unity, so that the corresponding value of p_1 is small compared with its maximum value 0·38, agreement may be expected between Taylor & Miller's calculation and the measured thickness of the deposited film of fluid. If the value calculated using (14) is less than unity the boundary condition used by Taylor & Miller in their calculation is not valid, so that agreement would not be expected. Taking values appropriate to the apparatus described, namely $R = 6.9$ cm. and $\kappa = 2.5 \times 10^{-7}$ cm. the value of x_1 at the meniscus was, according to (13),

$$x_1 \sim 3.7 \times 10^2 (\mu U/T). \tag{15}$$

In the experiments U was about 3 cm./sec. and $T = 62$ g. sec.$^{-2}$. When 50% glycerine for which $\mu = 0.085$ was used, $\mu U/T = 4.1 \times 10^{-3}$ so that the value of x_1 at the meniscus was $(3.7 \times 10^2)(4.1 \times 10^{-3}) = 1.5$. The corresponding value of p_1, namely, $(1.5)^{-3}/6$, was 0.056. This is well below the maximum value 0.378 which occurs at $x_1 = 0.43$* so that little difference would be expected between the amount deposited when the suction region was curtailed by a meniscus and the amount which would pass through the porous roller if the space between it and the plane were flooded.

When the working fluid is water, for which $\mu = 0.011$ and $T = 73$ in c.g.s. units, the corresponding value of x_1 according to (15) would be 0.17 which is even below the value $x_1 = 0.43$ which corresponds with the maximum value of p_1. Under these conditions where the suction at the meniscus is not small compared with the maximum suction, the boundary condition used by Taylor & Miller is not even approximately valid, so that the lack of agreement between the measured and calculated thickness of the deposit is understandable.

* See paper 25 above.

35

DEPOSITION OF A VISCOUS FLUID
ON THE WALL OF A TUBE

REPRINTED FROM

Journal of Fluid Mechanics, vol. x (1961), pp. 161–5

Measurements of the amount of fluid left behind when a viscous liquid is blown from an open-ended tube are described.

INTRODUCTION

When air is blown into one end of a tube containing a viscous fluid it forms a round-ended column which travels down the tube forcing some of the liquid out at the far end and leaving a fraction m in the form of a layer covering the wall. This fraction has been measured in some cases by Fairbrother & Stubbs,* who were interested in the question whether a bubble of air in a capillary tube through which a fluid is flowing is a true index of the velocity of the flow. If U is the velocity of a bubble and U_m the mean velocity of the fluid ahead of it,

$$m = \frac{U - U_m}{U}. \tag{1}$$

Fairbrother & Stubbs' experiments were limited to the case where m is small, and in that case they found as an empirical relationship, which covered experiments made with fluids in which the velocity U, the surface tension T and the viscosity μ all varied,

$$m = 1 \cdot 0 (\mu U / T)^{\frac{1}{2}}. \tag{2}$$

As these authors point out, it is to be expected that m should be a function of $\mu U / T$ since that is the only non-dimensional combination of these three symbols, but that m should be proportional to $(\mu U / T)^{\frac{1}{2}}$ has not been explained, nor has the coincidence that the empirically determined constant is $1 \cdot 0$. Fairbrother & Stubbs point out that there are theoretical and experimental reasons for believing that the thickness of the film left on a plane sheet when it is pulled vertically out of a fluid is proportional to $U^{\frac{1}{2}}$. This, however, is not a proper analogy since the existence of the film is due to a balance of viscous and gravitational forces, whereas when the fluid is blown out of a horizontal tube gravity plays little part in the phenomenon.

The empirical relationship (2) can only be valid when m is small; in fact if $\mu U / T$ is greater than 1 so that the corresponding value of m is greater than 1, (2) is meaningless. In the experiments to be described it was found that (2) is a good approximation in the range $0 < \mu U / T < 0 \cdot 09$. The maximum value of

* *J. chem. Soc.* I (1935), 527.

$\mu U/T$ obtained in Fairbrother & Stubbs' measurements was 0·014 so that they were well within the range in which (2) holds. When $\mu U/T$ increases beyond 0·09, m is less than would be predicted by (2). The experiments were extended, using very viscous fluids, to $\mu U/T = 1·9$ (i.e. to 135 times Fairbrother & Stubbs' maximum value) when m had reached 0·55. The apparent trend of the curve of experimental points seemed to indicate that m was approaching an asymptotic value at that stage. It seems therefore that if one attempts to blow a very viscous fluid out of a tube more than half may be left behind when the air column breaks through at the far end of the tube.

EXPERIMENTAL DETAILS

The method adopted was similar to that used by Fairbrother & Stubbs, but since only one long interface between air and fluid, rather than separate bubbles, was required, the method of introducing the air column needed careful design.

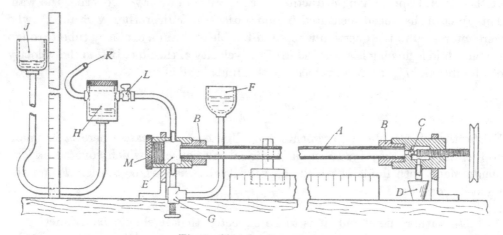

Fig. 1. Sketch of apparatus.

The apparatus is shown diagrammatically in fig. 1. Since the object of the experiment was to obtain large values of $\mu U/T$ without increasing the speed of flow to such an extent that the stresses due to inertia were comparable with those due to viscosity, most of the experiments were carried out with fluids of large viscosity. Glycerine and strong sucrose solution (golden syrup) diluted with water till its viscosity was 28 poise at 20° C. was used. Glass tubes of accurate bore, known as 'Veridia' tubes, of 2 and 3 mm. bore and 4 ft. long were used. The 3 mm. tube was slightly conical so that its cross-section varied uniformly through a range of $1\frac{1}{2}$ % over its length. The 2 mm. tube had only a quarter of that variation. In fig. 1 the tube A has interchangeable brass end pieces B. At the far end these fitted into a fixed ball valve C. The fluid passing the ball passed into a weighed crucible D so that the amount discharged in a measured time could be weighed.

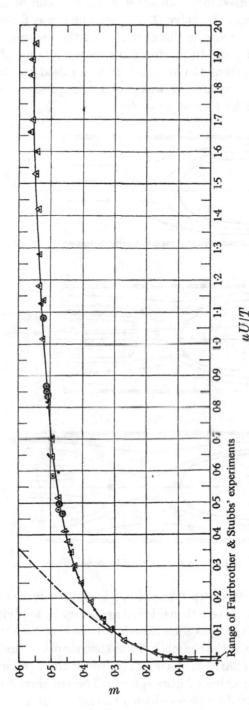

Fig. 2. Experimental results. ●, $\mu = 0.79$ poise at 20° C. (syrup-water mixture); tube diameters 0·15 and 0·2 cm. △, $\mu = 9.3$ poise at 20° C. (glycerine); tube diameters 0·2 and 0·3 cm. ⊙, $\mu = 28$ poise (syrup-water); tube diameter 0·3 cm. ⊚, $\mu = 0.3$ poise (a lubricating oil); tube diameter 0·15 cm. Broken line: Fairbrother & Stubbs' parabola.

At the near end the tube fitted into a chamber E which could be filled with fluid from the container F through the valve G. The air which was forced through the tube was supplied from the Perspex chamber H, and the pressure in H could be raised to 3 atmospheres or more by raising a mercury container J. To maintain the mercury level in H constant the air pressure was raised by means of a subsidiary pump K, while the container J was raised. The pressure vessel could be cut off from the chamber H by turning a tap L.

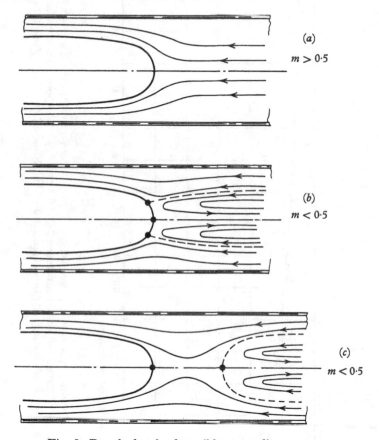

Fig. 3. Rough sketch of possible streamlines.

To perform an experiment the tap L was closed and the valves G and C opened. The chamber E and the tube A were then filled, if necessary by applying air pressure to the container F. The valve C was then closed and the crucible D placed in position. By unscrewing a cover M the chamber E was then emptied leaving a meniscus of fluid in the end of the tube. After replacing the cover M the tap L was opened and the air pressure in E raised to about 3 atmospheres. The apparatus was then ready for the experiment. The ball valve C was suddenly turned to a previously determined position and at the same moment a stop-watch was started. The meniscus then moved at a nearly uniform speed along the tube. After the meniscus had gone some way along the tube the ball valve C was suddenly closed and simultaneously the stop-watch was stopped. The position of the meniscus was read and the length of the air column in the tube determined. The crucible was weighed and the weight w per unit length of the fluid expelled was measured. The weight per unit length of the fluid in the tube when full was $w_0 = \rho \pi a^2$, where ρ is its density and a the radius of the tube. Evidently

$$1 - m = w/w_0.$$

The viscosity of the fluid used was determined by the capillary-tube method and the surface tensions of glycerine and concentrated solutions of sucrose were taken from physical tables.

Results

The results are shown in fig. 2 where m is plotted against $\mu U/T$.

For comparison with the measurements of Fairbrother & Stubbs the parabola $m = (\mu U/T)^{\frac{1}{2}}$ is shown as a broken line, and the range of values of $\mu U/T$ covered by them is also shown. It will be seen that the present experiments are in good agreement with $m = (\mu U/T)^{\frac{1}{2}}$ up to $\mu U/T = 0.09$, but that above this value m increases more slowly than $(\mu U/T)^{\frac{1}{2}}$ till at the highest values reached m approaches 0.55. The trend of the curve suggests that m reaches a limiting value above 0.56, when the stresses due to viscosity are much greater than those due to surface tension. It will be noticed that when $m = 0.5$ the flow velocity at points far from the meniscus is identical with that of the meniscus, so that if $m > 0.5$ the central filament is moving towards the meniscus, whereas if $m < 0.5$ the central filament is moving away from it. If the flow is reduced to steady motion by superposing a velocity $-U$ it is only when $m > 0.5$ that the flow with only one stagnation point is possible. This is shown by the sketch of fig. 3 a. The two simplest possible types of flow when $m < 0.5$ are shown in fig. 3 b and 3 c. In 3 b there is one stagnation point on the vertex and a stagnation ring on the meniscus. In fig. 3 c there are two stagnation points on the axis.

36

ON SCRAPING VISCOUS FLUID
FROM A PLANE SURFACE

REPRINTED FROM

Miszellaneen der Angewandten Mechanik (Festschrift Walter Tollmien),
edited by M. Schäfer, Akademie-Verlag, Berlin (1962), pp. 313–15

The stream function ψ, representing two-dimensional fluid flow, satisfies $\nabla^4\psi = 0$ when the flow is so slow that inertia is negligible. Simple solutions of this equation expressed in polar co-ordinates can be found in the form $\psi = r^n f(\Theta)$ but they seldom have much physical interest except as terms in series expansions. When n is greater than 1 they usually represent flow in a corner produced by agents which can only be described by other types of solution. When n is less than 1 the flow velocity near the origin is infinite. The case when $n = 1$ has physical significance because it can represent flow of a viscous fluid when a flat scraper moves over a flat sheet pushing fluid before it.

The general solution of $\nabla^4\psi = 0$ when $n = 1$ is

$$\psi = r(A \cos \Theta + B \sin \Theta + C\Theta \cos \Theta + D\Theta \sin \Theta). \tag{1}$$

If (1) represents the flow in front of a scraper moving with velocity U over a flat surface (taken as $\Theta = 0$) it is convenient to superpose a velocity $-U$ parallel to $\Theta = 0$ so as to reduce the motion to steady flow. The radial and tangential velocities are

$$u = -\frac{1}{r}\frac{\partial \psi}{\partial \Theta} \quad \text{and} \quad v = \frac{\partial \psi}{\partial r}.$$

If the scraper is inclined at angle α to the flat plate the boundary conditions are

$$\psi = 0, \quad \frac{1}{r}\frac{\partial \psi}{\partial \Theta} = U \quad \text{at} \quad \Theta = 0 \quad \text{and} \quad \psi = \frac{\partial \psi}{\partial \Theta} = 0 \quad \text{at} \quad \Theta = \alpha.$$

Using these conditions (1) becomes

$$\psi = \frac{Ur}{\alpha^2 - \sin^2\alpha}\left\{\alpha^2 \sin \Theta - \sin^2\alpha\, \Theta \cos \Theta \right.$$
$$\left. - \left(\frac{\alpha \sin^3\alpha + \alpha^2 \cos \alpha - \cos \alpha \sin^2\alpha}{\sin \alpha + \alpha \cos \alpha}\right)\Theta \sin \Theta\right\}. \tag{2}*$$

* *Editor's note.* Fig. 1, which shows the streamlines for the case $\alpha = \frac{1}{2}\pi$, is taken from an article by Sir Geoffrey Taylor entitled 'Similarity solutions of hydrodynamic problems' published in *Aeronautics and Astronautics* (Pergamon Press, 1960) and not reprinted in these volumes.

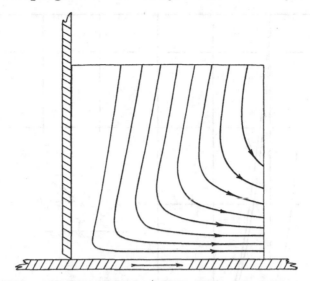

Fig. 1. Streamlines of flow of viscous fluid for $\alpha = \frac{1}{2}\pi$.

If P and S are the stresses normal and tangential to the scraper at distance r from the point of contact and μ is the viscosity

$$P = \frac{2\mu U}{r} \left(\frac{\alpha \sin \alpha}{\alpha^2 - \sin^2 \alpha} \right), \tag{3}$$

$$S = \frac{2\mu U}{r} \frac{\sin \alpha}{\alpha^2 - \sin^2 \alpha}. \tag{4}$$

When α is small these tend to $P = 6\mu U/r\alpha^2$ and $S = 2\mu U/r\alpha$ which are the values that can be found using the approximate equations of lubrication theory.

Resolving the stress in direction L at right angle to the plate and D parallel with it,

$$L = P \cos \alpha + S \sin \alpha = \frac{2\mu U}{r} \left(\frac{\sin^2 \alpha}{\alpha^2 - \sin^2 \alpha} \right),$$

$$D = P \sin \alpha - S \cos \alpha = \frac{2\mu U}{r} \left(\frac{\alpha - \sin \alpha \cos \alpha}{\alpha^2 - \sin^2 \alpha} \right).$$

Values of L, D, P and S divided by $2\mu U/r$ for various values of α are given in table 1, and are displayed in fig. 2. It will be seen that D decreases as α increases, and attains its least value $2\mu U/\pi r$ when $\alpha = \pi$. The most interesting and perhaps unexpected feature of the calculations is that L does not change sign in the range $0 < \alpha < \pi$. In the range $\frac{1}{2}\pi < \alpha < \pi$ the contribution to L due to normal stress is of opposite sign to that due to tangential stress, but the latter is the greater. The palette knives used by artists for removing paint from their palettes are very flexible scrapers. They can therefore only be used at such an angle that P is small and as will be seen in the figure this occurs only when α is nearly 180°. In fact artists instinctively hold their palette knives in this position.

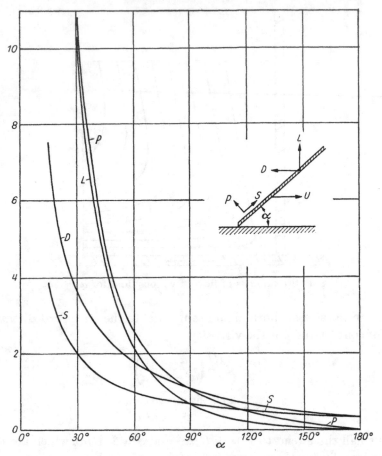

Fig. 2. Values of P, S, L and D divided by $2\mu U/r$.

Table 1

$\alpha°$	$\dfrac{Pr}{2\mu U}$	$\dfrac{Sr}{2\mu U}$	$\dfrac{Lr}{2\mu U}$	$\dfrac{Dr}{2\mu U}$
0	∞	∞	∞	∞
15	43	3·82	42	7·5
30	10·8	2·03	10·3	3·67
45	4·8	1·31	4·30	2·44
60	2·61	0·98	2·15	1·77
75	1·61	0·80	1·19	1·36
90	1·07	0·68	0·68	1·07
105	0·73	0·60	0·38	0·85
120	0·50	0·53	0·21	0·70
135	0·33	0·47	0·10	0·56
150	0·20	0·42	0·04	0·46
165	0·08	0·37	0·014	0·38
180	0	0·32	0	0·32

A plasterer on the other hand holds a smoothing tool so that α is small. In that way he can get the large values of L/D which are needed in forcing plaster from protuberances to hollows.

Though the fluid velocity is everywhere finite the stress becomes infinite at $r = 0$ in this solution. In fact in any real situation continuous contact between scraper and plate along a line will not occur so that infinite stress at $r = 0$ will be relieved over a region comparable with the width of the gap.

37

STANDING WAVES ON A CONTRACTING
OR EXPANDING CURRENT

REPRINTED FROM

Journal of Fluid Mechanics, vol. XIII (1962), pp. 182–92

In a recent work Longuet-Higgins & Stewart* have studied the changes in wavelength and amplitude of progressive waves of constant frequency as they are propagated into regions of surface divergence or convergence. In the work here described complementary conditions are assumed. Standing waves of uniform wavelength, λ, exist in an area of uniform surface divergence. Changes in amplitude and wavelength are studied. These changes depend on the existence of the radiation stress which was discovered by Longuet-Higgins & Stewart but the physical interpretation of this stress is simpler for standing than for progressive waves. Three different ways of obtaining the same rate of strain in the direction of the current caused the amplitude to vary at $\lambda^{-\frac{1}{2}}$, $\lambda^{-\frac{3}{4}}$ and $\lambda^{-\frac{5}{4}}$, respectively.

Experiments in which free-standing waves were generated in a tank one wavelength wide which was then made narrower verified the conclusion that contraction does not alter the periodic character of the waves, even though the ratio of amplitude to wavelength becomes so great that they can no longer be treated mathematically by the usual linearised approximation. The shape of the profile then appears to agree well with calculations of Penney & Price.*

1. Introduction

In a recent paper Longuet-Higgins & Stewart† have investigated the effect of non-uniform currents on short gravity waves. The physical condition assumed to hold was that the waves are generated continuously with a fixed frequency at a fixed point in the current. The frequency observed at all other fixed points is therefore the same as that of the wave generator, but the wave-number varies with position in the current. There is, however, another possible physical wave condition which is worth investigating, namely, the effect of a non-uniform current on a wave train which already exists, or on waves generated by wind action or other cause in the current itself. Most people will have noticed the sudden appearance of smooth areas in the disturbed water downstream of a lock when the sluice gates are opened. These are where rising turbulent currents spread out at the surface with horizontal divergence.

To represent such situations the effect of non-uniform currents with constant horizontal divergence on a train of standing waves whose length is constant in space but variable in time will be discussed. Such a discussion may be expected to be of interest in another connection. In discussing the energy changes which progressive waves of constant frequency suffer when they enter a non-uniform

* *Phil. Trans. Roy. Soc.* A, XXIV (1952), 254.

† *J. Fluid Mech.* X (1961), 529.

current, Longuet-Higgins & Stewart have shown* that it is necessary to take account of an interaction between waves and currents which, in the case of deep water, is equivalent to endowing them with a 'radiation' stress equal to $\frac{1}{2}E_p$, E_p being the wave energy per unit surface area in a progressive wave. The expansion of a progressive wave train as it moves into a region of increasing current velocity gives rise to a transfer of energy from the wave to the current which is identical with that which would occur during expansion at a rate of strain dU/dx under the action of a surface stress $\frac{1}{2}E_p$. Though this stress $S_{px} = \frac{1}{2}E_p$ per unit area must necessarily be present it is rather difficult to understand in detail the mechanism of its action in progressive waves. On the other hand it will be shown that when standing waves are compressed or expanded there exist vertical planes at which there is no horizontal motion due to the waves so that vertical sheets could be inserted there without interfering with the motion. Changes in the energy of waves when they are compressed or expanded could be regarded as being due to the work done by relative motions of the sheets against a 'radiation' stress, S_{sx}, equal to the force acting on a sheet when there are waves only on one side of it. It is therefore possible to calculate S_{sx} by integrating the pressures acting on a vertical sheet at the nodes of a standing wave on a currentless sea.

2. Two-dimensional analysis

A simple irrotational current system in two dimensions which diverges horizontally is that whose velocity potential is $-\frac{1}{2}c(x^2 - z^2)$, z being vertical and positive downwards and x horizontal. The stream lines are rectangular hyperbolas. Consideration of Bernoulli's equation reveals that this flow cannot have a flat free-surface. A small correcting term must be added to the velocity potential to enable the stream function $\psi = 0$ to be a surface of constant pressure. Take as the velocity potential and stream functions of the steady flow

$$\phi_0 = -\tfrac{1}{2}c(x^2 - z^2) + m(z^3 - 3x^2 z), \quad \psi_0 = cxz + m(3z^2 x - x^3), \tag{1}$$

and consider only the part of the field where the slope of the free surface is so small that z/x is small. The equation for the surface, $\psi = 0$, is

$$c\eta_0 = m(x^2 - 3\eta_0^2),$$

or, if η_0/x is negligible, $\eta_0 = mx^2/c$.

The condition that the pressure there is constant is satisfied if $m = c^3/2g$, so that

$$\eta_0 = c^2 x^2 / 2g \tag{2}$$

and the field in which this approximation is useful is that for which

$$|x| < gc^{-2} \times (\text{maximum allowable value of } d\eta_0/dx).$$

To discuss the effect of this horizontal divergence on a train of standing waves,

* *J. Fluid Mech.* VIII (1960), 565 and X (1961), 529.

the time $t = 0$ will be taken as that at which the waves are at their maximum elevation and the initial displacement as

$$\eta - \eta_0 = D \cos kx. \tag{3}$$

Wavelength and amplitude will change with time owing to the divergence c, but the wavelength will remain independent of x so that it is justifiable to assume for η the form

$$\eta = \eta_0 + f(t) \cos kx, \tag{4}$$

where f and k are functions of t only. An appropriate form for the velocity potential is

$$\phi = B e^{-kz} \cos kx - \tfrac{1}{2} c (x^2 - z^2) + (c^3/2g)(z^3 - 3x^2 z), \tag{5}$$

for which the velocity components are

$$u = Bk e^{-kz} \sin kx + cx + (3c^3/g) xz,$$

$$w = Bk e^{-kz} \cos kx - cz - (3c^3/2g)(z^2 - x^2).$$

B is a function of time which must be determined by the condition that ϕ is compatible with the free surface (3) and the condition that the pressure is constant there. The compatibility condition is

$$\left(\frac{\partial \eta}{\partial t}\right)_x = w - u \left(\frac{\partial \eta}{\partial x}\right)_t. \tag{6}$$

When the second-order terms are neglected the terms which are not time-dependent cancel and

$$w - u(\partial \eta / \partial k)_t = Bk \cos kx - cf \cos kx + cx \, kf \sin kx, \tag{7}$$

while

$$(\partial \eta / \partial t)_x = \dot{f} \cos kx - x f \dot{k} \sin kx. \tag{8}$$

(7) and (8) can only be consistent with (6) if

$$ck = -\dot{k}, \tag{9}$$

so that

$$k = k_0 e^{-ct}. \tag{10}$$

Since the distance between vertical planes of particles which are perpendicular to x is proportional to e^{ct} in the undisturbed current, (10) shows that in the disturbed flow particles which at any instant are situated in a nodal plane remain in a nodal plane. The compatibility condition is

$$\dot{f} + cf = Bk. \tag{11}$$

The pressure condition at the free surface is

$$\phi - \tfrac{1}{2}(u^2 + w^2) + g(\eta_0 + f \cos kx) = 0,$$

and, retaining only first-order terms, this reduces to

$$\dot{B} \cos kx - Bx\dot{k} \sin kx - Bxck \sin kx + fg \cos kx = 0. \tag{12}$$

Since $\dot{k} = -ck$, (12) becomes $\quad \dot{B} + fg = 0. \tag{13}$

Eliminating B between (11) and (13),

$$\ddot{f} + 2c\dot{f} + c^2 f + kfg = 0. \tag{14}$$

Writing $ct = \tau - \tau_0$, where $\tau_0 = \ln(c^2/gk_0)$, (14) becomes

$$\frac{d^2 f}{d\tau^2} + 2\frac{df}{d\tau} + f + e^{-\tau} f = 0, \tag{15}$$

and writing $e^{-\tau} = \omega$ so that $\omega = (gk_0/c^2) e^{-ct}$, (15) becomes

$$\frac{d^2 f}{d\omega^2} - \frac{1}{\omega}\frac{df}{d\omega} + \frac{f}{\omega^2} + \frac{f}{\omega} = 0. \tag{16}$$

This equation is a particular case of a more general one which is quoted in tables;* it has a solution

$$f = \omega J_0(2\omega^{\frac{1}{2}}). \tag{17}$$

The equation to the free surface at any time is therefore

$$\eta - \eta_0 = D e^{-ct} \frac{J_0(2\sigma_0 c^{-1} e^{\frac{1}{2}ct})}{J_0(2\sigma c^{-1})} \cos(k_0 e^{-ct} x). \tag{18}$$

Here σ_0 is the frequency of waves of length $2\pi/k_0$ and D is the initial amplitude when $t = 0$. It will be noticed that a single curve in which $\omega J_0(2\omega^{\frac{1}{2}})$ is plotted against $-\ln \omega = \tau$ represents the whole range of values of $(\eta - \eta_0)/D$ for all possible initial values of k and any value of c positive or negative. This curve is shown in fig. 1. On a diverging current the changing amplitude is represented by a point on the curve which moves from left to right, while that on a converging current (c negative) is represented by a point which moves in the opposite direction.

The availability of the curve of fig. 1 for representing the changes in standing waves on a diverging current is limited to the time during which the wavelength is small compared with the dimensions of the area in which the equations are valid approximations; thus if the maximum allowable value of $d\eta_0/dx$ is ϵ, the maximum linear dimension of the wave in which one wavelength covers the whole field is $\lambda = 2g\epsilon/c^2$ so that $2\pi/k$ must be less than $2g\epsilon/c^2$. Since $\omega = gk/c^2 = 2\pi g/c^2\lambda$, the lowest meaningful value of ω in (17) is π/ϵ. Even for a value of ϵ as high as $\frac{1}{10}$, π/ϵ is 10π and the argument of the Bessel function in (17), namely $2\omega^{\frac{1}{2}}$, is about 11. At this value the Bessel function in (17) is close to its asymptotic approximation so that

$$\omega J_0(2\omega^{\frac{1}{2}}) \sim \pi^{-\frac{1}{2}} \omega^{\frac{3}{4}} \cos(2\omega^{\frac{1}{2}} - \tfrac{1}{4}\pi). \tag{19}$$

It appears therefore that a standing wave in an expanding or contracting flow has an amplitude proportional to $k^{\frac{3}{4}}$, i.e.

$$\text{ampl.} \propto \lambda^{-\frac{3}{4}}. \tag{20}$$

This may be compared with an analogous result for progressive waves given in equation (4.9) of the paper by Longuet-Higgins & Stewart.†

---

* See, for instance, *Tables of Functions* by E. Jahnke & E. Emde, Dover (1945).
† *J. Fluid Mech.* x (1961), 529.

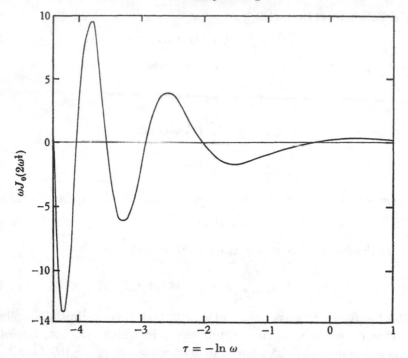

Fig. 1. Changes in amplitude of standing wave on a
contracting or expanding current.

3. Radiation stress in standing waves

Equation (10) shows that the wavelengths expand at the same rate as the distance between particles on wave crests, in other words vertical planes through wave crests always contain the same fluid. The rate at which the energy contained in one wavelength, λ, is transferred to the mean flow can therefore be calculated in two ways.

(i) The result (20) that the amplitude is proportional to $k^{\frac{3}{4}}$ makes it possible to calculate the change in energy contained in one wavelength as λ changes and since the only way in which this change can occur is by means of a force acting on the vertical planes at neighbouring crests which are separating at rate $c\lambda$ this force must be equal to (rate of change of energy per wavelength)$/\lambda c$.

(ii) The second method is to integrate the pressure over vertical planes through the crests of standing waves when there is no diverging current.

The same result is obtained by both methods. The advantage of carrying out both calculations is that a simple conception of the action of radiative stress is obtained. The existence of radiative stress in progressive waves was discovered or at any rate first effectively understood by Longuet-Higgins & Stewart. Their discussion is more general than that of the present work but the conception of radiation stress is simpler for standing than for progressive waves.

The energy of a standing wave of length λ and amplitude a is $E_s\lambda = \frac{1}{4}\rho ga^2\lambda$. If the wavelength changes from λ to $\lambda + d\lambda$ and the amplitude from a to $a + da$,

$$\frac{dE_s}{E_s} = 2\frac{da}{a} + \frac{d\lambda}{\lambda}. \tag{21}$$

The result of the direct calculation of the relationship between λ and a was that $a \propto \lambda^{-\frac{3}{4}}$ so that

$$da/a = -\tfrac{3}{4}d\lambda/\lambda \quad \text{and} \quad dE_s/E_s = -\tfrac{1}{2}d\lambda/\lambda. \tag{22}$$

The loss in energy per wavelength λdE_s during an extension $d\lambda$ must be due to a radiation stress S_{sx} where $S_{sx}d\lambda = -\lambda dE_s$ and using (22),

$$S_{sx} = \tfrac{1}{2}E_s = \tfrac{1}{8}g\rho a^2. \tag{23}$$

The energy per unit area in a region where a progressive wave is reflected at a vertical plane barrier is twice the energy per unit area of the incident wave; (23) shows that the radiative stress in a standing wave is also twice that of the incident progressive wave.

4. AXISYMMETRIC DIVERGING CURRENTS

The smooth mushroom-like areas in the stream below an emptying lock seem to be due to upwelling which spreads out equally in all directions. The same kind of analysis as that used in the two-dimensional case might be used as a model for this situation if the basic current flow is represented by

$$\phi_0 = -\tfrac{1}{2}c(x^2 + y^2 - 2z^2), \tag{24}$$

where y is the co-ordinate parallel to the wave fronts. In order that the free surface may be one of constant pressure, a small corrective term must be added to ϕ_0 which may be taken as $\frac{4}{3}(c^3/g)\{z^3 - \frac{3}{2}z(x^2 + y^2)\}$, and the free surface is then

$$\eta_0 = \tfrac{1}{2}(c^2/g)(x^2 + y^2). \tag{25}$$

A small error will arise with the axisymmetric current which did not affect the two-dimensional case. The correcting term which is necessary to satisfy the pressure condition causes initially straight wave crests to be convected into curves. If this second-order effect can be neglected analysis similar to that for the two-dimensional case can be made. Taking

$$\phi = Be^{-kz}\cos kx - \tfrac{1}{2}c(x^2 + y^2 - 2z^2) + \tfrac{4}{3}(c^3/g)\{z^3 - \tfrac{3}{2}z(x^2 + y^2)\}, \tag{26}$$

and

$$\eta = \eta_0 + f\cos kx, \tag{27}$$

it is found that $ck = -\dot{k}$ as in (9) but the equation of compatibility analogous to (11) is

$$\dot{f} + 2cf = Bk$$

and the surface pressure condition is, as before,

$$\dot{B} + fg = 0.$$

Eliminating B, $$\ddot{f} + 3c\dot{f} + 2c^2 f + kfg = 0. \tag{28}$$

Making the same transformation as before, the equation analogous to (16) is

$$\frac{d^2 f}{d\omega^2} - \frac{2}{\omega}\frac{df}{d\omega} + \frac{2f}{\omega^2} + \frac{f}{\omega} = 0;$$

a solution is $$f = \omega^{\frac{3}{2}} J_1(2\omega^{\frac{1}{2}})$$
so that the asymptotic value is

$$f = \pi^{-\frac{1}{2}}\omega^{\frac{5}{4}}\cos(2\omega^{\frac{1}{2}} - \tfrac{3}{4}\pi). \tag{29}$$

The wave height, a, is therefore proportional to $\lambda^{-\frac{5}{4}}$, or

$$a \propto \lambda^{-\frac{5}{4}}. \tag{30}$$

This result may be used to calculate the radiation stress (if any) acting on vertical planes perpendicular to the wave crests, for, with this type of current, a portion of the surface which was originally square and bounded on two sides by wave crests will remain square so that the energy contained in it is

$$\lambda^2 E_s = \tfrac{1}{4}\lambda^2 g\rho a^2.$$

Expansion from sides of length λ to $\lambda + d\lambda$ changes the energy by an amount $E_s\lambda^2(2da/a + 2d\lambda/\lambda)$. This change must be attributed to stresses S_{sx} and S_{sy} acting on the sides of the square which does work equal to $(S_{sx} + S_{sy})\lambda d\lambda$ during the expansion. And using (30), $da/a = -\tfrac{5}{4}d\lambda/\lambda$ so that

$$(S_{sx} + S_{sy})\lambda d\lambda + E_s\lambda^2(-\tfrac{1}{2}d\lambda/\lambda) = 0, \quad \text{or} \quad S_{sx} + S_{sy} = \tfrac{1}{2}E_s,$$

and since we already know that $S_{xs} = \tfrac{1}{2}E_s$, this argument shows that $S_{sy} = 0$. This result can also be obtained by integrating pressures over a vertical plane perpendicular to the wave crests when there is no expansion retaining terms of the second order of small quantities.

5. Lateral contraction or expansion without upwelling

If the basic flow is one in which the motion at all depths is confined to planes parallel to $z = 0$ the appropriate choice for ϕ_0 is

$$\phi_0 = -\tfrac{1}{2}c(x^2 - y^2) \tag{31}$$

and the undisturbed free surface is again $\eta_0 = (c^2/2g)(x^2 + y^2)$. Using the same symbols as in the two previous cases for the wave disturbance, it is found that the equation of compatibility analogous to (11) is $f = Bk$ and the pressure equation (13), namely $\dot{B} + fg = 0$, is unchanged. Making the same transformations as before the equation for f is

$$\frac{d^2 f}{d\omega^2} + \frac{f}{\omega} = 0 \tag{32}$$

and a solution is $$f = \omega^{\frac{1}{2}} J_1(2\omega^{\frac{1}{2}}); \tag{33}$$

using the asymptotic approximation the amplitude a is proportional to $\omega^{\frac{1}{4}}$, i.e. to $\lambda^{-\frac{1}{4}}$. This result can be compared with §5 of the paper by Longuet-Higgins & Stewart.[*] It can be verified that it also leads to the conclusion that $S_{sx} = \frac{1}{2}E$ and $S_{sy} = 0$.

6. EFFECT OF CURRENTS EXPANDING LATERALLY BUT NOT LONGITUDINALLY

Here the appropriate assumptions for ϕ and η are

$$\phi = Be^{-kz}\cos kx - \frac{1}{2}c(y^2 - z^2) \quad \text{and} \quad \eta - \eta_0 = f(t)\cos kx \tag{34}$$

and the compatibility condition is
$$\dot{f} = Bk - cf,$$

while the pressure condition is
$$\dot{B} + fg = 0,$$

so that
$$\ddot{f} + c\dot{f} + kgf = 0$$

and
$$f = De^{-\frac{1}{2}ct}\cos(gk - \frac{1}{4}c^2)^{\frac{1}{2}}t. \tag{35}$$

A rectangle of length λ and initial breadth b has breadth be^{ct} at time t. The energy contained in it is $\lambda be^{ct}E_s$ and since $E_s = \frac{1}{4}\rho ga^2 = \frac{1}{4}\rho gD^2 e^{-ct}$ the energy in the area remains constant. Since only the barriers at right angles to the wave crests move, the stress S_{sy} must therefore be zero.

7. FORCE EXERTED BY STANDING WAVES ON A VERTICAL BARRIER

Waves of small amplitude exert a fluctuating pressure on a vertical wall and the resultant fluctuating force can be divided into two parts: (i) the steady force due to the hydrostatic pressure of fluid whose surface is at the mean level of the standing waves and (ii) the fluctuating part which is the difference between the total force and the steady part defined in (i). If only the first order of small quantities is considered the mean value of the fluctuating part is zero and the force acting on a vertical barrier extending downwards at a wave crest from the surface to such a depth that the fluctuations in pressure on it are negligible is also zero. To calculate the force on a vertical barrier at one side of which there are standing waves, it is necessary to carry the analysis to the second order of small quantities.

The mechanics of standing waves of finite amplitude has been studied by Penney & Price[†] who showed that standing waves in which all the particles are at rest twice in every period can exist. They showed that there is a single series of such waves depending on the value of a non-dimensional number A and they expressed the velocity potential and displacement of the surface in a Fourier series containing submultiples of the wavelength. The coefficient of each term of this series was a function of time which itself was also expressed as a Fourier series of

[*] *J. Fluid Mech.* x (1961), 529.
[†] *Phil. Trans. Roy. Soc.* A, xxiv (1952), 254.

submultiples of the period. They developed the coefficients of the terms in these functions of time in powers of a single arbitrary number A which determined the amplitude of the waves. The analysis was very complicated and was carried up to terms involving A^5. For the present purpose it is only necessary to include terms up to A^2. Penney & Price showed that there are no terms in ϕ of order A^2. The appropriate expressions for ϕ and η are

$$\phi = \chi(t) + A\sigma k^{-2} e^{-kz} \cos kx \cos \sigma t, \tag{36}$$

$$\eta = Ak^{-1} \cos kx \sin \sigma t - \tfrac{1}{2}A^2 k^{-1} \cos 2kx \sin^2 \sigma t. \tag{37}$$

These equations satisfy the compatibility condition (6) when terms of higher order than A^2 are neglected. The pressure is then given by

$$p/\rho = \dot{\chi}(t) - A\sigma^2 k^{-2} e^{-kz} \cos kx \sin \sigma t - \tfrac{1}{2}A^2 \sigma^2 k^{-2} e^{-2kz} \cos^2 \sigma t + gz. \tag{38}$$

At the surface where $z = \eta$, the condition $p = 0$ is satisfied when terms of higher degree than A^2 are neglected, provided

$$\sigma^2 = gk \quad \text{and} \quad \dot{\chi}(t) = \tfrac{1}{2}A^2 gk^{-1} \cos 2\sigma t. \tag{39}$$

The amplitude, a, of the standing wave is, to the first order, Ak^{-1} so that

$$\rho\dot{\chi}(t) = \tfrac{1}{2}g\rho a^2 k \cos 2\sigma t. \tag{40}$$

This variation in pressure is independent of the depth and is the double-frequency pressure oscillation extending to the bottom of the sea to which several authors have called attention.

The force at any instant exerted by the waves on a vertical barrier of depth D parallel to the wave crests at $x = 0$ is

$$\int_{\eta}^{D} p \, dz - \int_{0}^{D} g\rho z \, dz. \tag{41}$$

The pressure at $x = 0$ is

$$\rho\{\dot{\chi}(t) - a\sigma^2 k^{-1} e^{-kz} \sin \sigma t - \tfrac{1}{2}a^2 \sigma^2 e^{-2kz} \cos^2 \sigma t + gz\}. \tag{42}$$

Integrating (42) the total force to depth D, which is assumed to be much greater than a wavelength, is

$$\rho\{\tfrac{1}{2}Dga^2 \cos 2\sigma t - gak^{-1} \sin \sigma t + ga^2(\sin^2 \sigma t - \tfrac{1}{4}\cos^2 \sigma t - \tfrac{1}{2}\sin^2 \sigma t)\}. \tag{43}$$

The mean value of this force is $\tfrac{1}{8}g\rho a^2$. Comparing this with (23) the mean force of the waves on a vertical wall is equal to $\tfrac{1}{2}E_s$.

The force per wavelength acting on a vertical barrier placed perpendicular to the wave crests can also be calculated, since no fluid crosses these planes. When this is done it is found that the mean value of the force attributable to the term $-A\sigma^2 k^{-2} e^{-kz} \cos kx \sin \sigma t$ in (38) is $\tfrac{1}{4}g\rho a^2 \lambda$ while those due to the last two terms $-\tfrac{1}{2}A^2 \sigma^2 k^{-2} e^{-2kz} \cos^2 \sigma t$ and gz in (38) are each $-\tfrac{1}{8}g\rho a^2 \lambda$ so that the total mean force on barriers perpendicular to the wave fronts is zero.

8. EXPERIMENTS ON STANDING WAVES

The conclusion reached in §2, that slow horizontal contraction alters the wave-length of a simple-harmonic wave of small amplitude but preserves its simple-harmonic character, seemed worth verifying experimentally. A tank 102 cm. deep, sketched in fig. 2, was constructed out of sheet Perspex so that it had two parallel walls and two which converged towards the bottom. The parallel walls were 12 cm. apart while the converging walls were 25·4 cm. apart at the top and 12 cm. at the bottom. In the bottom was a large valve A (fig. 2) which could empty the tank to a mark C in about 5 sec. when raised by the lever B. The waves were produced by causing a narrow wedge D in the middle of the top of the tank to oscillate vertically.

A 16 mm. ciné-camera operated at 69 to 73 frames per second could be placed in two positions so that it could photograph at the top or at levels near the marks C and E.

Since the seatings for the camera were fixed at the same distance from the tank it was possible to take a short length of film covering a few complete periods in the upper position and then move the camera to the lower position and take the surface when it reached the lower position. A disadvantage of this method was that drops were liable to fall off the wave-maker and disturb the wave. This was prevented by installing a trough F (fig. 2) which started to travel down a guide as soon as the valve A was raised. It stopped at position G where it could catch the drops. It can be seen at the right-hand side of the top of the photographs in figs. 3 and 4 (pls. 1 and 2).

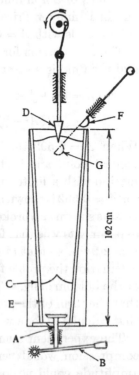

Fig. 2. Converging tank in which the free-surface length was contracted by lowering the water level.

The frames chosen for reproduction in figs. 3 and 4 were those at which the wave crest was at its maximum height during an oscillation. It was found by counting frames between successive maxima that, within the limits of accuracy obtainable by this method, the period was indistinguishable from $(2\pi l/g)^{\frac{1}{2}}$, the period for small oscillations when the length l was taken as that of the free surface at the time the photograph was taken.

The theoretical conclusion described by equation (10) that, when a standing wave is compressed between rigid planes at its nodes, its period will alter so that it remains a simple wave is therefore verified, but the decrement due to viscosity was so great that it was impossible to verify the predicted change in amplitude.

Fig. 3 (pl. 1) shows a case where the initial amplitude is small and the lateral compression of the wave produces a small increase in absolute amplitude in spite of the decrement due to viscosity. The decrement in the 15 or more oscillations which occurred while the water level was falling was never so great that the ratio of wave height to wavelength decreased. It always increased considerably.

If h is the maximum height above the mean level and d the maximum depth below it, the shape of a periodic wave of finite amplitude depends only on $(h+d)/l$. This shape has been calculated approximately by Penney & Price for a series of values of a non-dimensional number A. For small values of A, $(h+d)/l = A/\pi$ and in Penney & Price's approximation the highest wave which has a crest of $90°$ corresponds with $A = 0.592$ and $(h+d)/l = 0.128$.

Their expression for the shape of the wave when its crest is at its highest point and the fluid is instantaneously at rest is

$$2\pi y/l = (A + \tfrac{1}{32}A^3 - \tfrac{47}{1344}A^5)\cos(2\pi x/l) + (\tfrac{1}{2}A^2 - \tfrac{79}{672}A^4)\cos(4\pi x/l)$$
$$+ (\tfrac{3}{8}A^3 - \tfrac{12563}{59136}A^5)\cos(6\pi x/l) + \tfrac{1}{3}A^4\cos(8\pi x/l) + \tfrac{295}{768}A^5\cos(10\pi x/l). \quad (44)$$

When $A = 0.592$ this wave has downward acceleration equal to gravity at its highest point and it should have a pointed crest but the Fourier-series approximation with a finite number of terms cannot represent this. The calculated form for $A = 0.592$ is shown in fig. 4 (pl. 2), and the estimated form at the top of the pointed crest is shown as a broken line.* The highest value of $(h+d)/l$ observed in the present experiments was that for the wave shown in fig. 4. At $l = 14$ cm., the value of $h+d$ was 2.2 cm. so that $(h+d)/l = 0.16$. To compare this wave with Penney & Price's calculation the profile found from (44) for $A = 0.5$ was drawn. This curve is shown at the bottom of fig. 4. The corresponding value of $(h+d)/l$ was 0.177. It will be seen that the form of the profiles revealed by the photographs in fig. 4 is very similar to that calculated for a wave with nearly the same value of $(h+d)/l$.

The experiment seems to indicate that if the viscous decay had been less, slow compression would permit standing waves to remain periodic even when their amplitude could no longer be regarded as small. It would be interesting to use a larger wave tank to find out whether it is possible to compress waves till a pointed crest is attained.

* G. I. Taylor, *Proc. Roy. Soc.* A, CCXVIII (1953), 44 (paper 20 above).

PLATE 1

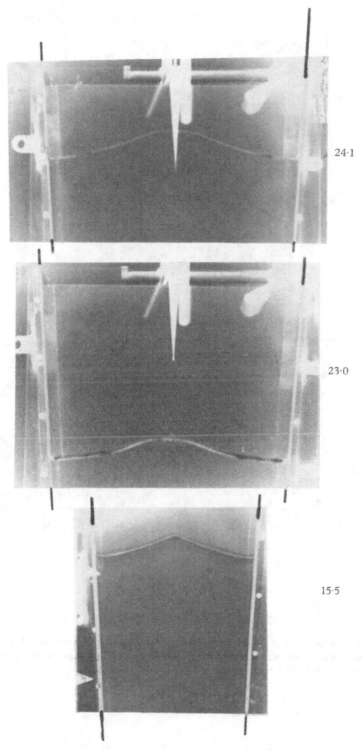

24·1

23·0

15·5

Fig. 3. Top frame: wave generated at length 24·1 cm.; middle frame: wave at maximum amplitude after the length had contracted to 23·0 cm.; lower frame: wave at $l = 15·5$ cm.

PLATE 2

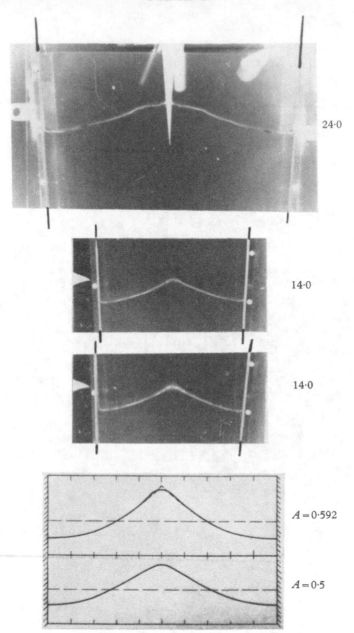

Fig. 4. Top: wave generated at $l = 24 \cdot 0$ cm. Middle: two consecutive frames when $l = 14 \cdot 0$ cm. Below: calculated wave shapes; the shape for $A = 0 \cdot 50$ can be compared with the photographs above; the crests seem rather sharper but otherwise comparable.

38

CAVITATION OF A VISCOUS FLUID
IN NARROW PASSAGES

REPRINTED FROM

Journal of Fluid Mechanics, vol. XVI (1963), pp. 595–619

The conditions which determine the existence and position of cavitation in the narrow passages of hydrodynamically lubricated bearings have been assumed to be the same as those which produce cavitation bubbles, namely a lowering of pressure below that at which gas separates out of fluid. This assumption enables certain predictions to be made which in some cases are verified, but it does not provide a physical description of the interface between oil and air. Theoretical analysis of the situation seems to be beyond our present capacity, and in none of the experiments so far published has it been possible to measure both the most important relevant data, namely the minimum clearance and the oil flow through it.

A method is described here which enables this to be done. It turns out that two physically different kinds of cavitation can occur. One of these is well described by the existing theory and assumption. Surface tension plays no part in it, and in most textbooks on hydrodynamic lubrication is not even mentioned. The other kind, which is akin to hydrodynamic separation rather than bubble cavitation, depends essentially on surface tension. Both kinds appear clearly in published photographs taken through transparent bearings, but the experimenters do not seem to have distinguished between them.

The reason why surface tension, which is only able to supply stresses that are exceedingly small compared with the pressure variation in the fluid itself, may have a large effect on the flow can be understood by considering the flow of a viscous fluid in a tube when blown out by air pressure applied at one end. For any given length of fluid the rate of outflow depends almost entirely on the pressure applied, the surface tension force being negligible; but the amount of fluid left in the tube after the air column has reached the end depends essentially on surface tension.

1. INTRODUCTION

Cavitation or separation in the oil film of hydrodynamic lubrication has long been recognised as an important factor in the design of bearings and engineers have exercised much ingenuity in trying to allow for it, but rather little effort has been devoted to the study of the phenomenon itself. I think that there are three main reasons for this. The first is that a simple journal bearing is essentially a mechanism by which a rotating shaft can be supported against a lateral load so that the geometry of the lubricating fluid depends on the load as well as the peripheral speed and the difference in radii between the shaft and the bearing. This leads to great complication in the analysis which absorbs much of the attention of the workers in the subject. The second is that it is very difficult to make experiments which can reveal the physical conditions that determine the position of the meniscus which separates the fully lubricated from the separated part of a lubrication film, particularly when the geometry of the solid surface is not predetermined, and in nearly all the recorded experiments it is not. A third reason is that in hydrodynamically lubricated bearings the fluid pressures in the high pressure parts are usually great compared with one atmosphere but the negative pressure in the part (if any) of the bearing where cavitation has occurred is usually very much smaller so that it makes little contribution to the total reaction between bearing surface and shaft. It is sufficient for most practical purposes to assume that the pressure is positive throughout.

The interest, even the practical interest, of cavitation or separation of viscous fluids flowing in narrow passages is not limited to its occurrence in lubrication theory and in other cases, where great pressures do not occur, surface tension, which may justifiably be left out of consideration in most problems in lubrication theory, may be of primary importance.

In designing experiments to determine the physical character of the meniscus separating a viscous lubricant from the outside air it seems essential to reduce the number of variables. The first step is obviously to use predetermined geometry and to limit the motion to two dimensions. The simplest practical way to form a meniscus is to fill a cylindrical tube with viscous fluid and blow it out by air pressure applied to one end. The air forms itself into a column with a round end, which may be called the meniscus. This travels down the tube till it reaches the far end. After the meniscus has passed any point in the tube the fluid which is left behind stays practically at rest, because the viscosity of air is so much less than that of the fluid that the pressure in it is nearly constant along its length. The simplest measurement that can be made is the ratio m of the amount of fluid left behind to the internal volume of the tube, and measurements of this kind have been published.* The main result is that, as has been espected on somewhat unsophisticated theoretical grounds, m depends only on the non-dimensional combination $\mu U/T$, where μ is viscosity, U the velocity of the meniscus relative to the wall of the tube and T is surface tension. This may be expressed by the equation

$$m = F_1(\mu U/T). \tag{1}$$

When $\mu U/T$ is small F_1 is small, but as $\mu U/T$ increases F_1 appears to approach an asymptotic limit. In my experiments, which extended up to $\mu U/T = 1\cdot9$, m had risen to $0\cdot56$. In some later experiments by Cox† m nearly reached $0\cdot60$ when $\mu U/T$ was in the range 10–$17\cdot5$. This point is mentioned here to show that the asymptotic value of m is *not* $0\cdot50$ as it would be if the criterion were that the bubble cannot go faster through the tube than the maximum fluid velocity in the Poiseuille flow beyond the bubble which is driven by the air pressure in it. The significance of this fact will appear later.

The other relevant measurement which could be made is the difference in pressure between the fluid on the two sides of the meniscus. It was not possible to do

* F. Fairbrother & A. E. Stubbs, *J. Chem. Soc.* I (1935), 527; G. I. Taylor, *J. Fluid Mech.* x (1961), 161 (paper 35 above); F. P. Bretherton, *J. Fluid Mech.* x (1961), 166.
† *J. Fluid Mech.* xIV (1962), 81.

this in my experiments but it is worth while considering the physical meaning of such measurements if they could be made. The flow far ahead of the meniscus is the Poiseuille flow which is associated with a uniform pressure gradient $8a^{-2}\mu V$, a being the radius of the bore of the tube and V the mean velocity through it. To connect the pressure in the air column with the velocity of the meniscus it would be necessary to know how the uniform pressure gradient in the fluid connects with the practically uniform pressure in the air. The sketch of fig. 1 explains what is needed. AOB is the air bubble, OX is the axis of symmetry of the bubble and OY the radial co-ordinate. The distribution of pressure along the axis is represented on an axial plane OXY by the line CEFOX, and the pressure which would have existed at the vertex, O, of the bubble if the uniform gradient which exists in the fluid far away from it had been continued up to the vertex is represented by D. Though it is a difficult matter to calculate the flow near the vertex, all that is necessary to connect the pressure in the air with the flow in the tube is the pressure difference δp

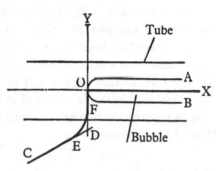

Fig. 1. Distribution of pressure CEFOX in a tube due to bubble AOB. Pressure excess of that in the bubble is represented by the co-ordinate OY.

represented by OD, and since there is only one non-dimensional variable $\mu U/T$ in the problem, it seems that the pressure change δp must be representable by an expression of the form

$$\delta p = -\frac{\mu U}{a} F_2\left(\frac{\mu U}{T}\right). \tag{2}$$

It will be seen that to calculate how fast fluid would be blown out of a tube by a given pressure both F_1 and F_2 must be known, but since F_2 is likely to be comparable with unity when $\mu U/T$ is large and must tend to $2/(\mu U/T)$ when $\mu U/T$ is small, a knowledge of F_2 is not nearly so important as that of F_1 in most cases unless $\mu U/T$ is small.

The problem presented by a bubble in a capillary tube was considered first because it is the simplest definable and easily realisable case of separation in a viscous fluid when the Reynolds number is so small that only the balance of viscous stresses and surface tension need to be taken into account. The analogous problem where a two-dimensional bubble is forced into fluid contained between two parallel plates cannot be materialised because the meniscus is unstable,* though cases with more complicated geometry, such as the flow when a cylinder rolls on a plane covered with viscous fluid, may involve a stable meniscus. There seems, however, to be no reason why the equilibrium configuration should not be calculated except the great difficulty of doing so, even though it is not stable. This

---

* P. G. Saffman & G. I. Taylor, *Proc. Roy. Soc.* A, CCXLV (1958), 312 (paper 28 above).

difficulty is increased by the fact that there appears to be no reason, except physical intuition, to suppose that any numerical solution would be unique. This question is given some speculative consideration in the appendix.

2. Flow in a narrow gap between eccentric rotating cylinders

The flow in a long fully lubricated journal bearing is well understood. To avoid geometrical complexities the simplest case will be considered, namely that of a very eccentric bearing, or more definitely, the narrow space or 'nip' where two cylinders or a cylinder and a plane nearly come into contact, and the equations for the flow in that region will be reproduced in order to develop a method for finding where the meniscus could be located. The physical properties of the meniscus will be assumed to be analogous to those of the bubble in a capillary tube, namely that at the meniscus

$$m = F_1(\mu U/T) \tag{3}$$

and

$$-\delta p = \frac{\mu U}{h} F_2(\mu U/T). \tag{4}$$

Here m is the ratio of the amount of fluid flowing at any section to the amount which would flow if the pressure gradient there were zero, and h is the distance between the surfaces.

The use of (3) and (4) involves the assumption, which experiments to be described later seem to confirm, that the small deviation from exact parallelism of the two surfaces will not appreciably affect m or δp. The convergence or divergence of the passage has a great effect on the stability of the meniscus* and on the distribution of pressure in the passage, and in that way will have a predominating effect on the position at which the meniscus will establish itself, but this is no reason for rejecting the relations (3) and (4) which are concerned only with local conditions close to a two-dimensional meniscus.

Two cases may be considered: (*a*) both surfaces are moving with velocity U, so that $m = q/Uh$ where q is the volume flowing past the nip per unit length; (*b*) only one surface is moving, as in the case of a bearing, so that $m = 2q/Uh$.

If x is the distance along the nip measured from the narrowest point where $h = h_0$, the variation of h with x can be expressed approximately by the equation

$$h = h_0 + x^2/2R. \tag{5}$$

Here

$$R = (R_1^{-1} - R_2^{-1})^{-1}, \tag{6}$$

where R_1 and R_2 are the radii of the cylinders, R_2 being the larger and R_1 and R_2 are taken as both positive if the centre of the smaller cylinder is inside the larger cylinder. When the case considered is that of a cylinder rolling on a plane, R is evidently the radius of that cylinder.

* J. R. A. Pearson, *J. Fluid Mech.* vii (1960), 481; E. Pitts & J. Greiller, *J. Fluid Mech.* xi (1961), 33.

It is convenient to use instead of x a co-ordinate θ defined by

$$\tan\theta = x(2Rh_0)^{-\frac{1}{2}}. \tag{7}$$

The equation for the distribution of pressure according to the Reynolds approximation then takes the non-dimensional form

$$\frac{dp'}{d\theta} = \cos^2\theta - \lambda\cos^4\theta. \tag{8}$$

This equation applies to both cases (*a*) and (*b*), but in case (*a*) the pressure p is related to p' by

$$p = p'\{12\mu U(2R)^{\frac{1}{2}}\,h_0^{-\frac{3}{2}}\}, \tag{9a}$$

and in case (*b*)

$$p = \tfrac{1}{2}p'\{12\mu U(2R)^{\frac{1}{2}}\,h_0^{-\frac{3}{2}}\}. \tag{9b}$$

In both cases (*a*) and (*b*) λ is the ratio of the total flow through the nip to the flow if there had been no pressure gradient at $\theta = 0$, so that in both cases (*a*) and (*b*)

$$\lambda h_0 = mh. \tag{10}$$

The solution of (8) is

$$p' = \tfrac{1}{2}\theta + \tfrac{1}{4}\pi + \tfrac{1}{4}\sin 2\theta - \lambda(\tfrac{3}{8}\theta + \tfrac{3}{16}\pi + \tfrac{1}{4}\sin 2\theta + \tfrac{1}{32}\sin 4\theta), \tag{11}$$

where the constant of integration has been chosen so that $p' = 0$ when $\theta = -\tfrac{1}{2}\pi$, i.e. the pressure is atmospheric far from the nip in the upstream direction, or in other words it is flooded upstream.

When $\lambda = \tfrac{4}{3}$, $p = 0$, at $\theta = +\tfrac{1}{2}\pi$ as well as at $\theta = -\tfrac{1}{2}\pi$ and the distribution of p' is antisymmetrical, being positive when $\theta < 0$ and negative when $\theta > 0$. The distribution of p' in this case is shown in fig. 2.

3. Position of meniscus

To find the position of the meniscus for given values of $\mu U/T$ and h_0/R a diagram like fig. 2 may be used. Lines showing the variation of p' with θ for constant λ can be plotted, but it is not necessary to cover the whole field because, as has been noted many times in the literature of lubrication theory, the flow can only divide at a plane where dp/dx is positive. Fig. 3, which covers a part of the field where a meniscus could occur, has therefore been prepared using the expression (11) and calculating p' for given values of θ and λ.

The diagram of fig. 3 is particularly suitable for finding the locus of the points where condition (3) can be satisfied both in case (*a*) and in case (*b*). The definition of m as the ratio of the amount of fluid passing through the nip to the amount which would flow if there were no pressure gradient leads to

$$m = \lambda h_0/h = \lambda\cos^2\theta. \tag{12}$$

Thus lines of constant m can be drawn and, according to the condition (3), these will be lines of constant $\mu U/T$. Such lines are shown in fig. 3 for values of m from

0·03 to 0·70 and for $m = 1$. $m = 1$ is evidently the line which represents the points where the flow could divide without changing velocity, that is it is the line $dp'/d\theta = 0$ or, using equation (12), $\cos^2\theta = \lambda^{-1}$.

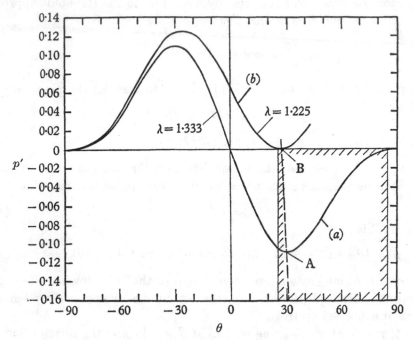

Fig. 2. Distribution of pressure (a) when completely flooded and (b) when the Swift–Stieber condition is satisfied at B.

To fix the point in fig. 3 which represents the position of the meniscus it is necessary to know h_0 as well as $\mu U/T$. Assuming that we have measured $F_2(\mu U/T)$ the corresponding values of p' using (4) and (9) are

$$-p' = \frac{1}{12}\left(\frac{h_0}{2R}\right)^{\frac{1}{2}} F_2\left(\frac{\mu U}{T}\right)\cos^2\theta \quad \text{in case (a)}, \tag{13a}$$

or

$$-p' = \frac{1}{6}\left(\frac{h_0}{2R}\right)^{\frac{1}{2}} F_2\left(\frac{\mu U}{T}\right)\cos^2\theta \quad \text{in case (b)}. \tag{13b}$$

If F_2 were measured experimentally or calculated as a function of $\mu U/T$, the expression (13) would be used to superpose lines of constant h_0/R on fig. 3, and the intersections of these lines with those of constant m and therefore constant $\mu U/T$ would determine the points at which the meniscus would lie for any given value of $\mu U/T$ and h_0/R. It will be noticed that when h_0/R is very small, as it is in most bearings, p' is also small so that unless $F_2(\mu U/T)$ is very large the position of the meniscus is determined simply by the point where the axis $p' = 0$ cuts the appropriate line of constant $\mu U/T$. In other words it is only necessary to know $F_1(\mu U/T)$ in such cases.

To discuss the stability of the meniscus a knowledge of $F_2(\mu U/T)$ is essential. In discussing the origin of the streaks formed when a viscous fluid is spread on a flat sheet by means of a roller, Pearson* assumed that the pressure difference between the two sides of a meniscus was $2T/h$, where h is the distance between the solid surfaces at the position of the meniscus. This is equivalent to assuming that the viscous stress in the neighbourhood of the meniscus which is of order $\mu U/h$ is small compared with $2T/h$, or in other words, that $\mu U/T$ is small. $F_2(\mu U/T)$ is then large, for by the definition in (4), $F_2(\mu U/T) = 2T/\mu U$.

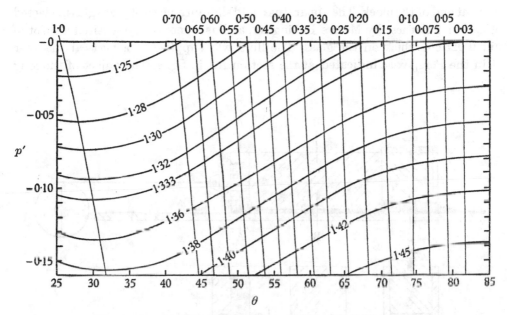

Fig. 3. Contours of p' for constant values of λ (1·25 to 1·45), and of m (0·03 to 0·70 and 1·0).

The investigation can only be extended to cases where $\mu U/T$ is not small either by making an arbitrary assumption that FD in fig. 1 is small compared with OF or by including $F_2(\mu U/T)$ as an arbitrary constant, an assumption which is legitimate when U is constant as it is in cases (*a*) and (*b*).

4. Apparatus for measurements of $F_1(\mu U/T)$

The main difficulty in measuring m is to obtain two-dimensional flow because when the two surfaces are parallel the meniscus is unstable.† It is true that the meniscus may be stable in the diverging region downstream of the nip, but the conditions which lead to such stability are complicated and not very well understood.

* *J. Fluid Mech.* VII (1960), 481.
† See paper 28 above.

Another difficulty is that of measuring m in a journal bearing. To collect and measure q, the volume which passes the nip per second would be difficult. In designing apparatus for determining m as a function of $\mu U/T$ it seemed best to attempt to control m and measure U. If the meniscus had been stable this would perhaps have involved varying U till the meniscus was brought to rest, but the instability prevented the direct use of this method. It was found, however, that the configuration illustrated in fig. 4 stabilised the meniscus at slow speeds.

A Perspex block was cut accurately rectangular and a trough of uniform depth 0·05 cm. and length 12·4 cm. was cut in its lower face. This trough did not extend to the front end of the block. The apparatus could slide on a piece of plate glass, selected optically for flatness by Messrs Pilkington, and as it moved it deposited a sheet of fluid 0·025 cm. thick on the glass. The fluid was supplied at a vertical chamber cut in the Perspex as indicated at the section AA in fig. 4. The object of the long

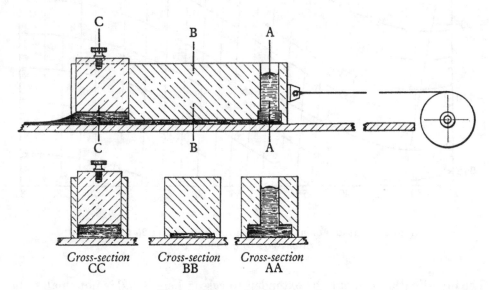

Fig. 4. Perspex block apparatus.

shallow trough hereafter called the regulating chamber was merely to regulate the supply of fluid to an observation chamber between the glass and a second short adjustable block of Perspex which is shown in the section CC. This block was 4·5 cm. long. If h_0 is the depth of the regulating trough and h_1 the height of the channel beneath the movable block, the flow in this chamber can be defined by the ratio

$$m = \frac{\text{actual flow in observation chamber}}{\text{flow which would exist if there were no pressure gradient there}}.$$

If the regulating trough were very long the flow in it would simply be $\frac{1}{2}Uh_0$, where U is the velocity of the block over the glass. The actual flow in the chamber at

C is therefore $\frac{1}{2}Uh_0$, but if there were no pressure gradient in the observation chamber the flow would be $\frac{1}{2}Uh_1$ so that

$$m = h_0/h_1. \tag{14}$$

Owing to the fact that the length of the observation chamber C was small compared with that of the regulating chamber it was first considered justifiable to take m as h_0/h_1. Later a correction (see equation (18)) was applied which took account of the fact that the length of the observation chamber was comparable with that of the regulating chamber, but did not allow for changes in pressure between the two sides of the meniscus due to surface tension.

The experiment consisted in towing the block in a straight line over the glass at gradually increasing speeds. The meniscus was always hollow, but at slow speeds it sloped away from the upper block as indicated in fig. 4. As the speed increased the meniscus became hollower till finally it became apparently horizontal and tangential to the observation chamber at the top. At that stage its horizontal sections were straight lines. A small increase in speed then made the meniscus begin to curve slightly in horizontal sections so that it was no longer two-dimensional. As the speed increased the middle of the meniscus withdrew from the middle of the trailing edge of the observation chamber and the curvature of horizontal sections increased till an unstable condition was reached at which it suddenly began to move forward without any further increase in the speed of the block. The meniscus frequently divided but it always ran forward as far as the rear end of the regulating channel. The value of U which was measured was that at which curvature first appeared in the horizontal sections, because that was the point at which the meniscus was two-dimensional, i.e. cylindrical, and tangential to the top of the observation chamber. The values of U so found were multiplied by μ/T and are plotted against h_0/h in fig. 6.

5. NEW APPARATUS

The curvature in horizontal sections of the meniscus was no doubt due to the fact that the channel was not very wide compared with its depth. A wider block would certainly have reduced the gap between the speed at which the horizontal curvature began and the speed at which the meniscus ran forward unstably to form air fingers. Even so the block had a considerable disadvantage because the glass plate on which it moved was only 30 in. long, and with the larger values of h_0/h_1 the flow pattern had not settled down to its final steady state by the time the block had traversed 30 in.

To overcome these difficulties a new apparatus was designed in which a cylinder could rotate continuously in a trough of fluid and a fixed concentric arc with clearance h_0 was used to regulate the flow into an observation chamber. This apparatus is shown in fig. 5. A cylinder B of radius 7·60 cm. and length 38·0 cm. could rotate in ball bearings held in a strong steel frame. A regulating Perspex block A whose lower surface was a cylinder of radius 7·65 cm. was fixed concentric with the cylinder so that the resulting space between them was 0·05 cm. thick. The angle covered by this block was 78°,* so that the length of the regulating space was 10·4 cm. The width in the direction of the generators of the cylinder was 22·4 cm. The accuracy of the setting was ensured by laying flexible spacers 0·050 cm. thick on the cylinder and then bringing the regulating block on to them.

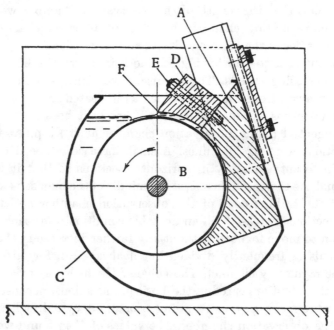

Fig. 5. Cylindrical apparatus.

In this position the block was fixed to a strong frame by suitable adjusting screws. The fluid was contained in a trough C (fig. 5) which was filled up to the level of the top of the cylinder. Leakage at the point where the ½ in. steel axle of the cylinder passes through the ends of the trough C was prevented by means of O-rings set in the steel frame. The trough which was made of tinned steel sheet was slightly flexible so that as the trough filled a small leak began to develop as the weight of fluid increased during filling. This was cured by adjusting supports between the strong frame and the trough.

The observation chamber in which the meniscus was observed was the space under a Perspex block D (fig. 5) which was fixed by bolts E passing through slots in D which allowed for adjustment of the distance h_1 of D from the cylinder. The bottom of the block D was a part of a cylinder of radius 7·65 cm. To set up the

* When fig. 5 was drawn the dimensions of the regulating block A had been mislaid so that the position of its lower corner is incorrectly shown. The angle 78° was found on dismantling the apparatus.

apparatus flexible strips of the required thickness were made and placed on the cylinder. The block D was then laid so that its rear face was in contact with the forward face of the regulating block (which was in an axial plane), and its forward bottom corner F was in contact with the flexible spacers. Since the spacers were lying on the cylinder and the block covered an angle of 36° the thickness of the observation chamber was slightly greater at the rear than at the forward end by an amount $(h_1 - h_0)$ (sec 36° − 1). The cylinder was driven through a train of continuously varying and fixed reduction gears by a constant speed motor so

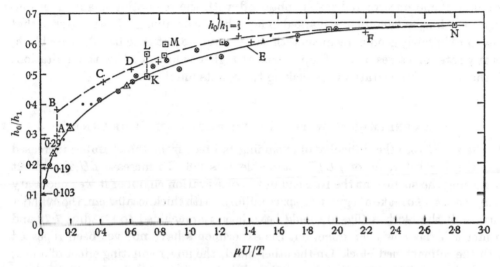

Fig. 6. Measurements of $\mu U/T$ at which flow ceased to be two-dimensional for fixed values of h_0/h_1. Parallel observation chamber: $\triangle$ $h_0 = 0.05$ cm.; $\bullet$ $h_0 = 0.10$ cm.; $\odot$ $h_0 = 0.15$ cm. Rectangular block: $\otimes$ $h_0 = 0.05$ cm. Expanding observation chamber: $+$ angle $= 2.8°$; $\boxdot$ angle $= 1.3°$. Limiting values for $h_0 = 0.05, 0.10, 0.15$ cm. are shown by arrows.

that the peripheral velocity U of the cylinder could be slowly increased till the meniscus withdrew from the forward end F of the observation chamber. As with the Perspex slider shown in fig. 4, the cavitating air-fingers always ran toward to the rear end of the regulation chamber. The speed U at which this occurred could be determined with good repeatability, but to ensure that the flow was two-dimensional it was necessary to fit end-plates lined with sorbo-rubber or felt which could be fitted to the end of regulating block A and to D to prevent air or fluid from being sucked laterally into the rear end of the observation chamber where the lowest pressures occurred. Even this precaution had to be supplemented by guide vanes outside the end-plates to ensure that the outside of the felt or sorbo-rubber lining was flooded with the fluid, for it was found that if any air got through, bubbles in the observation chamber destroyed the two-dimensional character of the flow and upset the conditions leading to the instability of the meniscus.

The fluids used were pure glycerine and glycerine diluted with 5% water. The viscosity of the fluid used was measured at a range of temperatures and the temperature was measured before and after each experiment. Glycerine does not wet Perspex completely and a thin sheet of glycerine on the surface develops dry areas after a time. This fact does not seem to affect the critical value of U, for in the Perspex block experiments both wetting oils and glycerine were used and no difference was found between the results obtained. Lubricating oil was not used in the cylinder apparatus because of the difficulty of cleaning it. The surface tension T was taken as 63 dyn/cm. throughout because the variation with temperature is small. Some uncorrected results obtained with the regulating chamber of uniform depth $h_0 = 0.05$ cm. are marked in fig. 6 by the symbol $\triangle$. It will be seen that these are in fairly good agreement with those obtained with the flat Perspex block, at any rate for values of $\mu U/T$ up to 0.6. At higher values of $\mu U/T$ the limitations of the flat block apparatus were making the results unreliable.

6. Experiments with larger regulating channel

As the speed rose the difficulty of excluding bubbles from the chamber increased and the highest value of $\mu U/T$ obtainable was 0.6. To increase $\mu U/T$ without increasing the suction at the rear end of the observation chamber it was necessary to increase h_0. To make a regulating space uniform with thickness 0.1 cm. the cylinder surface of the block A (fig. 5) should have been re-machined to a radius 7.70 and in fact this has now been done, but the experiments have not yet been repeated with the re-machined block. On the other hand, the flow regulating effect of a long narrow space upstream of the observation chamber does not depend on its having a uniform thickness, but a correction must be applied when it does not.

7. Corrections

If h is known as a function of x the distance along the regulating block, the equivalent thickness of the block of the same length but uniform thickness which would deliver the same volumetric rate of flow can be found by integrating Reynolds's equation. If q is the volumetric rate of flow per cm. of the meniscus Reynolds's approximation is

$$\frac{1}{12\mu}\frac{dp}{dx} = \frac{U}{h^2} - \frac{q}{h^3} \quad \text{in case } (a), \tag{15a}$$

and

$$\frac{1}{12\mu}\frac{dp}{dx} = \frac{U}{2h^2} - \frac{q}{h^3} \quad \text{in case } (b). \tag{14b}$$

When h is uniform and there is no pressure gradient $h = 2q/U$ in case (b), and, if h is variable but there is no difference in pressure at the two ends of the channel,

$$q\int\frac{dx}{h^3} = \tfrac{1}{2}U\int\frac{dx}{h^2}, \tag{16}$$

the integrals extending along the block. Hence

$$m = \frac{2q}{h_1 U} = \int h^{-2} dx \Big/ h_1 \int h^{-3} dx, \tag{17}$$

the integrals extending the total length of the regulating and observation chambers, but it is convenient to express (17) non-dimensionally in the form

$$m = \frac{h_0}{h_1} \int \left(\frac{h_0}{h}\right)^2 dx \Big/ \int \left(\frac{h_0}{h}\right)^3 dx, \tag{18}$$

where h_0 is to be taken as some easily measurable length.

To set the regulating block so that it formed a channel which was wider than 0·05 cm., flexible spacers 0·10 cm. thick were laid on the cylinder and the Perspex block of radius 7·65 cm. was brought down onto them and fixed rigidly. The depth of the channel at all points of the regulating space has first to be found. At its two ends it is h_0 ($h_0 = 0·1$ cm. in the present case). If the radius of the lower surface of the regulating block is $R + \delta$, R being that of the cylinder ($\delta = 0·05$ cm. in the present case), the depth h of the regulating space can be found by using the consideration that the method of setting up the regulating block ensures that h is symmetrical about its mid-point. If ϕ is the angular co-ordinate and the mid-point of the regulating space is $\phi = 0$, it is found that

$$h = h_0 \{\alpha + (1 - \alpha) \sec \phi_0 \cos \phi\}, \tag{19}$$

where $\alpha = \delta/h_0$ and $2\phi_0$ is the angle subtended by the regulating block (in this case 78°). To apply equation (10) it is sufficiently accurate to take the depth in the short observation chamber as uniform and equal to h_1 which is the measured depth of the channel at the forward end F (fig. 5). The contributions of the observation chamber to the integrals in (18) are then $(h_0/h_1)^2 \phi_2$ and $(h_0/h_1)^3 \phi_2$ to the numerator and denominator, respectively, where ϕ_2 is the angle covered by the observation chamber (in the present case $\phi_2 = 36°$). The expression for m is therefore

$$m = \frac{h_0}{h_1} \frac{\left(\frac{h_0}{h_1}\right)^2 \phi_2 + \int_{-\phi_0}^{+\phi_0} \{\alpha + (1 - \alpha) \sec \phi_0 \cos \phi\}^{-2} d\phi}{\left(\frac{h_0}{h_1}\right)^3 \phi_2 + \int_{-\phi_0}^{+\phi_0} \{\alpha + (1 - \alpha) \sec \phi_0 \cos \phi\}^{-3} d\phi}. \tag{20}$$

For the case when $\delta = h_0 = 0·05$ cm., $\alpha = 1$ and equation (20) reduces to

$$m \frac{h_1}{h_0} = K_{0·05} = \frac{(h_0/h_1)^2 + \frac{78}{36}}{(h_0/h_1)^3 + \frac{78}{36}}, \tag{21}$$

where $K_{0·05}$ is a correcting factor to be applied to the approximate equation (14). The correcting factor K_{block} for the Perspex block which had a regulating channel 12·4 cm. long and an observation chamber 4·5 cm. long is the same as (21) except that the friction 78/36 is replaced by 12·4/4·5. The calculated values of $K_{0·1}$, $K_{0·05}$ and K_{block} are given in table 1. The value of $K_{0·1}$ was calculated numerically using (20).

Table 1. *Correction factors to be applied to measured values of* h_0/h_1

h_0/h_1	0·1	0·2	0·3	0·4	0·5	0·6	0·7	0·8
$K_{0·1}$	1·095	1·108	1·125	1·142	1·153	1·156	1·147	1·122
$K_{0·05}$	1·004	1·015	1·029	1·054	1·055	1·060	1·058	1·048
K_{block}	1·006	1·010	1·020	1·035	1·046	1·047	1·050	1·040

8. Effect of gravity

The neglect of the effect of gravity at the meniscus produces little error except at very low speeds. There is a limiting value of h_1 above which the meniscus will leave the leading edge of the observation chamber no matter how small U may be. Since the thickness of the fluid sheet which is carried away is $\frac{1}{2}h_0$ the height of the bottom of the observation chamber above the top of the sheet is $h_1 - \frac{1}{2}h_0$ and the hydrostatic force, which must be balanced by the surface tension force $2T$, is $\frac{1}{2}\rho g(h_1 - \frac{1}{2}h_0)^2$, where ρ is the fluid density. Thus the minimum value of m when $U = 0$ is $h_0/\{\frac{1}{2}h_0 + (4T/\rho g)\}^{\frac{1}{2}}$. For glycerine $(4T/\rho g)^{\frac{1}{2}} = 0·46$ cm. so that the value of m at $U = 0$ is $h_0/(\frac{1}{2}h_0 + 0·46)$. When h_0 is 0·05 cm. this is $m = 0·103$, for $h_0 = 0·10$ cm. it is $m = 0·19$ and for $h_0 = 0·15$ cm. it is $m = 0·29$. These limits are marked on fig. 6. At first sight one might be inclined to think that since the value of m at $U = 0$ is comparable with that at finite values of $\mu U/T$ large errors might arise owing to the effect of gravity. This, however, seems to me unlikely because as soon as U is finite a comparatively large pressure defect at the meniscus can be built up by a very small change in the flow through the regulating channel, and all that gravity does is to change very slightly the rate of flow which is necessary to set up the conditions at the meniscus which lead to its retreat into the observation chamber. The values of m deduced by applying the correcting factors of table 1 are shown in fig. 7.

The principal generalisations that these experiments suggest are as follows.

(1) The value of m at which the meniscus begins to retreat into the channel is a function of $\mu U/T$ only.

(2) As $\mu U/T$ increases m appears to approach an asymptotic value which is certainly above 2/3, the value it has when the flow close to the fixed surface begins to reverse its direction at points in the chamber where the effect of the meniscus in deflecting the streamlines is negligible and the Reynolds approximation holds. This criterion was proposed by Hopkins[*] and used for experiments on flow of type (a) where it predicts $m = \frac{1}{3}$.

(3) At small values of $\mu U/T$ the curve gives the impression of being parabolic. A similar approximation has been noted in the case of a bubble in a capillary tube,[†] though it has been shown[‡] that the approximation ceases to be valid

---

[*] *Br. J. appl. Phys.* VIII (1957), 442.

[†] F. Fairbrother & A. E. Stubbs, *J. Chem. Soc.* I (1935), 527; G. I. Taylor, *J. Fluid Mech.* X (1961), 161 (paper 35 above).

[‡] F. P. Bretherton, *J. Fluid Mech.* X (1961), 166.

when m is less than about 10^{-3}. A rough approximation for the parabolic range of fig. 7 is

$$m = 0\cdot85(\mu U/T)^{\frac{1}{2}}. \tag{22}$$

Fairbrother & Stubbs' empirical formula was $m = 1\cdot0(\mu U/T)^{\frac{1}{2}}$ but there is no reason to expect exact agreement between the two formulae.

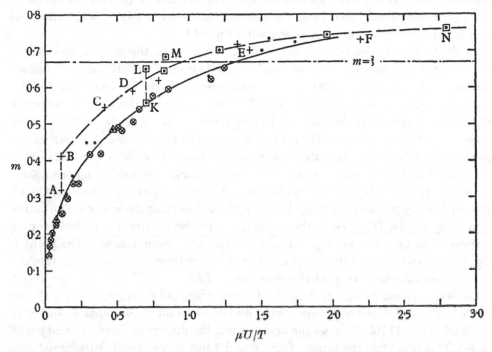

Fig. 7. Corrected values of m. Parallel observation chamber: $\bullet$ $h_0 = 0\cdot10$ cm.; $\triangle$ $h_0 = 0\cdot05$ cm. Expanding observation chamber: $+$ angle $= 2\cdot8°$; $\square$ angle $+1\cdot3°$. Rectangular block: $\otimes$ $h_0 = 0\cdot05$ cm.

9. EXPERIMENTS WITH DIVERGING OBSERVATION CHAMBER

The fact that the meniscus does not seem to be able to travel back when m is greater than some number which is in the neighbourhood of 2/3 suggests that if the observation channel were made to diverge instead of being parallel so that m could vary through the observation chamber from some small number to $1\cdot0$, the air cavity would not be able to reach the regulating block but would stop at some intermediate point. To make the observation chamber diverge some thin Perspex wedges were cut and placed base downwards between the blocks A and D (fig. 5). These are able to give the chamber divergence angles of $1\cdot3°$ and $2\cdot8°$, respectively. They were set with bases at the level of the top of the regulating space so that the maximum values of h_0/h_1 is $1\cdot0$, at the upstream end of the observation channel. As U was gradually increased the meniscus became unstable at the same value of $\mu U/T$ as when the channel was parallel and with the same value of h_1 at the open end. The instability which then occurred was very similar to that observed when air is forced into fluid contained between parallel plates.* It developed into air fingers which were separated by fluid columns whose widths were comparable with that of the fingers. Fig. 8 (pl. 1) is a photograph of the meniscus at that stage using the narrow wedge which made the chamber diverge at $1.3°$. The fingers remained in this condition so long as the speed was kept constant (in fact the exposure of this photograph was 8 sec.) and their ends remained roughly on a generator of the cylinder corresponding with a constant value of h.

The depth of the observation chamber at the level of the ends of the fingers was measured by inserting strips of flexible plastic material (Polythene) of known thickness as feelers. The results of these measurements are shown in figs. 6 and 7 by means of crosses, $+$, marked A, B, C, D, E in the case of the wider divergence, $2.8°$, of the observation chamber, and by squares, $\square$, K, L, M, N for the smaller divergence $1.3°$. The points marked A and K represent the values of h_0/h_1 where the meniscus began to retreat from the forward edge of the observation chamber. Within the accuracy of the measurements they lie on the curve obtained using the nearly parallel but very slightly *converging* channel obtained by bolting the movable block D (fig. 5) directly to the regulating block A without inserting the wedges. The points B and L (figs. 6 and 7) represent the points to which the meniscus retracted keeping the speed constant. The photograph of fig. 8 (pl. 1) corresponds with the point L in fig. 6, but as will be seen this point is not very well determined since the ends of the fingers do not all come to rest at the same value of h.

Figs. 9 (pl. 1) and 10 (pl. 2) show the fingers when the divergence was $2.8°$ and when $\mu U/T$ was 1.4 and 2.2 respectively and correspond with points E and F in figs. 6 and 7. Fig. 11 (pl. 2) shows the fingers when the divergence was $2.8°$ and $\mu U/T$ was 4.0. This is outside the range of figs. 6 and 7 but m was hardly distinguishable from the value it had in the case $\mu U/T = 2.2$.

It will also be noticed in figs. 8–10 that the fluid separating the air fingers extends to the leading edge of the Perspex block, but on fig. 11 (pl. 2) these fluid sheets have become so thin that surface tension pulls them back into the wedge-shaped observation chamber so that they part from the rear edge of the stationary block and the saddle-shaped meniscus appears as a bright spot. Similar effects have been noticed by Pearson† and by Floberg‡ using cylinder and plane corresponding with case (b) and by Pitts & Greiller§ for case (a).

It will be noticed that except for the case when the meniscus first leaves the stabilising edge of the adjustable block the ends of the fingers lie remarkably closely on a line of constant h and that as the speed increases this line, whose movement is represented by the dotted curve in figs. 6 and 7, seems to approach more and more closely to the same asymptote as that of the two-dimensional or

* P. G. Saffman & G. I. Taylor, *Proc. Roy. Soc.* A, ccxlv (1958), 312 (paper 28 above).

† *J. Fluid Mech.* vii (1960), 481.

‡ Dissertation, Chalmers Tech. Univ., Gothenberg (1961) and *Chalmers tek. Högsk Handl.* nos. 234 and 238 (1961).

§ *J. Fluid Mech.* xi (1961), 33.

PLATE 1

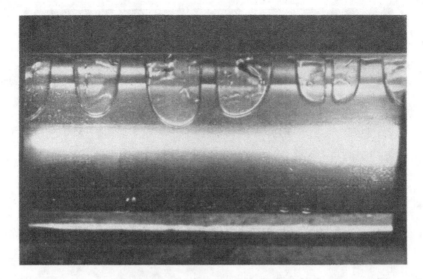

Fig. 8. Air fingers corresponding with $\mu U/T = 0.11$ and divergence $1.3°$, $h_0 = 0.05$.

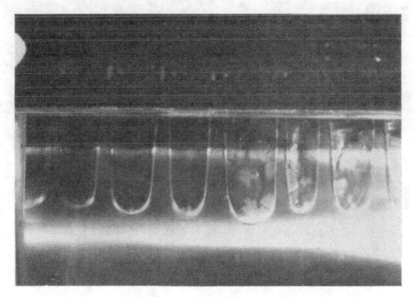

Fig. 9. $\mu U/T = 1.3$, $h_0 = 0.05$, divergence $2.8°$.

PLATE 2

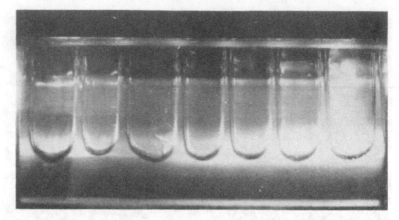

Fig. 10. $\mu U/T = 2\cdot2$, $h_0 = 0\cdot05$, divergence $2\cdot8°$.

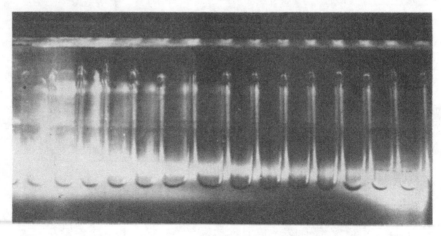

Fig. 11. $\mu U/T = 4\cdot0$, $h_0 = 0\cdot05$, divergence $2\cdot8°$.

PLATE 3

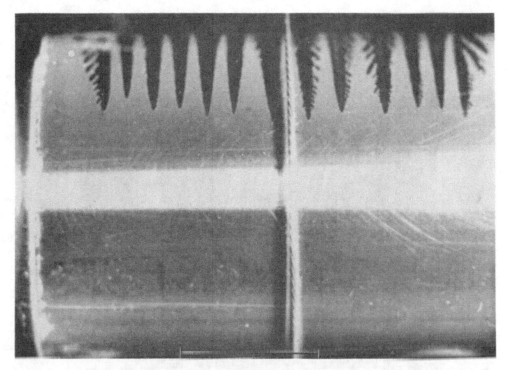

1 cm.

Fig. 12. Cavitation between shaft and transparent bush at $\mu U/T \sim 0\cdot12$, $h_0 \sim 1\cdot0 \times 10^{-3}$ cm.

1 cm.

Fig. 13. Cavitation between shaft and transparent bush at $\mu U/T \sim 0\cdot03$, $h_0 \sim 0\cdot3 \times 10^{-3}$ cm.

PLATE 4

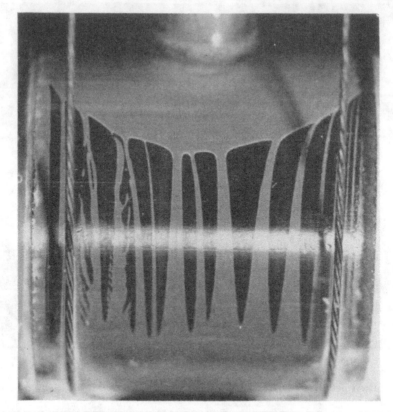

Fig. 14. Photograph showing both the formation of air fingers and the re-formation of the oil film.

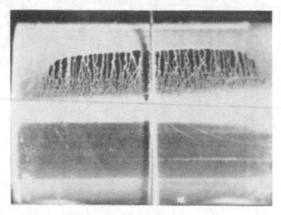

Fig. 15. Photograph showing the re-formation of the oil film for the same value of h_0 as in fig. 13 but with $\mu U/T \sim 0\cdot004$.

cylindrical meniscus which is about to leave the stabilizing edge and become unstable. Thus the asymptotic condition for large $\mu U/T$ seems to be one in which the flow separates, so that most of the fluid which leaves the region of separation is carried away on the cylinder in a sheet of uniform thickness, and only a very little is carried in the form of the thin sheets perpendicular to the cylinder which separate the air fingers. This point is referred to later and seems to be verified in fig. 15 (pl. 4).

10. CAVITATION

The separation just described necessarily involves a pressure gradient upstream of the meniscus which may lead to pressures which are so low that true cavitation must occur. Banks & Mill,[*] for instance, using apparatus in which both surfaces moved (case (a)), showed photographs of cavitation bubbles appearing at the point of lowest pressure in the flooded nip between two rotating cylinders. At speeds where the cavitation pressure is just attained at the point of minimum pressure these bubbles will not grow. They may disappear or they may be carried out as small bubbles. When the speed is higher so that the cavitation pressure extends over a larger area the bubbles will grow and will alter the pressure in the fluid round them. This stage of cavitation has been studied by Floberg. Finally, they may meet and form a continuous air space, or they may extend round the bearing in the form of fingers or streaks which do not join. This kind of cavitation seems to be that visualised by Swift[†] and Stieber.[‡] Curve (a) of fig. 2 shows the non-dimensional pressure curve for a flooded nip for which $\lambda = \frac{4}{3}$. Bubbles will appear first at the point A. As the bubble spreads the minimum pressure will be reduced and λ will therefore also be reduced. This process can proceed till the bubbles extend to the atmosphere and the pressure at the level of their vertices is atmospheric. The point B is then reached on the curve $\lambda = 1.225$, where $p = 0$ and $dp/dx = 0$. The fingers of air cannot penetrate further because that would involve separation in the part of the pressure curve where dp/dx is positive which is impossible. The point B, fig. 2, represents Swift's condition in the case of a very eccentric bearing which is flooded upstream.

The shapes of the air fingers in Swift's type of cavitation must depend on two independent causes both of which tend to make the flooded areas between the fingers get narrower downstream. One cause is geometrical and is the widening of the gap h, and the other is the transfer of fluid beneath the meniscus. The points of the air fingers may be expected to be paraboloidal while they are still narrower than h, since that is the axisymmetrical shape whose cross-section increases linearly. This paraboloidal part would only extend to a length comparable with h and thereafter the shape will be determined by the two causes. If only the first cause were operative, flooded areas between the fingers would occupy a proportion h/h_s of the whole, h_s being the value of h at the points of the fingers. In

---

[*] *Proc. Roy. Soc.* A, CXXIII (1954), 414.   [†] *Proc. Instn civ. Engrs*, CCXXIII (1932), 267.
[‡] *Das Schwimmlager V.D.I.* Berlin (1933).

that case the air spaces would not join together; Cole & Hughes* show some examples of this. The second cause would make the flooded areas get narrower more rapidly than would be expected from purely geometrical considerations.

These speculations resulted from a study of some photographs sent me by Professor J. A. Cole. They show the cavitation of oil through a loaded transparent bush on a rotating shaft. Fig. 12 (pl. 3) shows the air fingers (black) when the shaft 0·9820 in. in diameter is rotating at 138 rev./min. in a bush 0·9840 in. in diameter and 1·63 in. long. Fig. 13 (pl. 3) shows the same bearing with the same load, the shaft rotating at 34·6 rev./min. It was not possible to measure h but it is clear that it must have been greater at 138 than at 34 rev. This may account for the fact that the fingers are a greater distance apart in fig. 12 than in fig. 13. If the geometrical cause for the narrowing of the oil streaks were the only one operating it would have been expected that reduction in their width at a given distance from the beginning of cavitation would be less in fig. 12, than in fig. 13. The fact that this is not true shows that the second cause, namely the transfer of fluid across the meniscus is operative and is probably the principal cause of the much wider angle of the pointed end of the air finger in fig. 12 than in fig. 13.

Professor Cole measured the viscosity of the oil in each of his experiments. He did not measure T but in most oils T lies between 20 and 40 dyn./cm. Using $T = 30$ dyn./cm. the values of $\mu U/T$ in the experiments shown in figs. 12. and 13 are 0·12 and 0·03, respectively. No measurements of h_0 were made but Professor Cole calculated the value of the eccentricity ϵ using the theory of Sassenwald & Walther,† but interpolating between values calculated by these authors in order to make them applicable to his apparatus. The results were $\epsilon = 0·61$ for fig. 12 and $\epsilon = 0·88$ for fig. 13. These correspond with $h_0 = 1·0 \times 10^{-3}$ cm. for fig. 12 and $h_0 = 0·3 \times 10^{-3}$ cm. for fig. 13. The average distance between the fingers in these two cases is 0·22 cm. and 0·065 cm. so that in each case the spacing of the fingers is about 220 times the minimum clearance distance.

11. COMPARISON OF TWO TYPES OF CAVITATION

Comparison of figs. 12 and 13 with figs. 8–11 reveals the physical difference between the two types. In separation cavitation the motion is mainly two-dimensional. The thin partitions between the air fingers carry only a small part of the fluid, the rest is carried in a thin sheet on the moving surface. In true cavitation, starting inside the fluid, much of it is carried in columns filling the space between the two surfaces and separated by air fingers. In bearings these columns may be carried round the shaft unbroken or they may break down leaving a sheet of lubricant of variable thickness on the shaft. When this happens the meniscus which is the boundary of the region where the oil film is re-formed on the far side of the bearing may be expected to reproduce approximately the pattern of the fingers formed in the cavitation region. Fig. 14 (pl. 4) is one of Prof. Cole's photographs showing the re-formation of

* *Proc. Instn mech. Engrs*, CLXX (1956), 499.   † *V.D.I. Forschungsheft* (1954), p. 441.

the oil film under conditions where widely separated air fingers were formed, probably owing to cavitation of the Swift–Stieber type.

It will be noticed that the air fingers grow wider through their whole length till the oil column between them nearly disappears before the oil film is reformed. This can only be because the oil is passing under the menisci which form the edges of the oil columns.

Sometimes, however, Professor Cole obtained quite a different kind of re-formation meniscus. Fig. 15 (pl. 4) is an example. In that case both the speed and the load were each one-eighth of that used in the experiment of fig. 13 so that the eccentricity and, therefore the geometry in the two cases, should be nearly identical. On the other hand, both the maximum negative pressure which would exist if the film were continuous, and the parameter $\mu U/T$ were only one-eighth of that appropriate to fig. 13. It is therefore to be expected that true cavitation would be more likely to occur under the conditions of fig. 13 than those of fig. 15. Unfortunately the illumination failed to show the part of the bearing where cavitation began in fig. 15, but the fact that the re-forming meniscus is smooth suggests that the layer of oil carried found on the shaft was of uniform thickness. The thin lines seen in the cavitated area can only represent a small portion of the oil carried round and may well be the remains of the thin films which separate the air fingers produced by separation rather than by true cavitation.

12. Floberg's experiments

Recently Professor Birkhoff has called my attention to the papers by Floberg cited on p. 440. In the third of these he shows photographs very like Professor Cole's. In the second he shows photographs of cavitation of extra heavy Vactric lubricating oil when a cylinder 8 cm. diameter and 8 cm. long was rotated at distances $h_0 = 0.01$, 0.02, 0.04, and 0.06 cm. below a glass plate. Those taken at $h_0 = 0.02$, 0.04, and 0.06 cm. were very like figs. 8–11 except that they are better photographs, but those at $h_0 = 0.01$ cm. are quite unlike mine. Possibly the former show separation and the latter cavitation. It is therefore of interest to compare the observed positions of the cavitation with those calculated (*a*) using the Swift–Stieber condition, and (*b*) a separation condition based on my experiment.

The experiments of which Floberg published photographs were carried out at speeds of 25 and 100 rev./min. or approximately $U = 10.5$ and 41.9 cm./sec. so that only two values of $\mu U/T$ occurred. Neither the viscosity nor the surface tension of the oil were recorded, but in answer to a letter, Dr Floberg wrote that at $22°$ C. the viscosity was 3.5 poise (i.e. g./cm.sec.). The surface tension was not known but in most lubricating oils it lies in the range 20–40 dyn./cm. so that approximate values of $\mu U/T$ can be estimated by assuming $T = 30$. The experiments were carried out under conditions where the oil space was flooded upstream of the nip so that the curves of fig. 7 should be applicable, using equation (9*b*) for determining the pressure upstream of the meniscus. With $R = 4.0$ cm., $\mu = 3.5$ poise,

$U = 10\cdot5$ and $41\cdot9$ cm./sec. approximate values of p/p' in dyn./cm.[2] are given in table 2. The approximate values of $\mu U/T$ assuming $T = 30$ dyn./cm. are given in the last column of the table.

To find the position at which a two-dimensional meniscus could exist it is theoretically necessary to know δp, the pressure difference between the two sides of the meniscus. This is not known except when $\mu U/T$ is small when it is $2T/h$. When $\mu U/T$ is large the pressure difference must be of order $\mu U/h$. Since $2T/h$ and $\mu U/h$ are less than $2T/h_0$ and $\mu U/h_0$, it is useful to record the latter, when finding out whether it is justifiable to neglect the effect of δp on the position of the meniscus. The relevant figures are given in the last three lines of table 2. It will be seen that they are of order $1/100$ of p/p' in Floberg's experiments.

Table 2. *Data relating to Floberg's experiments with a cylinder rotating under a transparent flat surface*

h_0 (cm.)	0·01	0·02	0·04	0·06	$\mu U/T$
p/p' when $U = 10\cdot5$	$6\cdot2 \times 10^5$	$2\cdot2 \times 10^5$	8×10^4	4×10^4	$1\cdot2$
p/p' when $U = 41\cdot9$	$2\cdot5 \times 10^6$	$8\cdot8 \times 10^5$	3×10^5	$1\cdot3 \times 10^5$	$4\cdot9$
$2T/h_0$ dyn. cm.$^{-2}$	6×10^3	3×10^3	$1\cdot5 \times 10^3$	10^3	—
$\mu U/h_0$ when $U = 10\cdot5$	$3\cdot2 \times 10^3$	$1\cdot6 \times 10^3$	8×10^2	5×10^2	$1\cdot2$
$\mu U/h_0$ when $U = 41\cdot9$	$1\cdot5 \times 10^4$	7×10^3	4×10^3	$2\cdot4 \times 10^3$	$4\cdot9$

In practically all calculations relating to hydrodynamic lubrication it has been assumed that the change in pressure on passing through the meniscus is negligible, so that the pressure distribution when the oil space is flooded upstream depends only on the position of the ends of the cavitation fingers, and is represented in fig. 3 by the curve of constant λ which cuts $p' = 0$ at the position of the meniscus.

To estimate the error in the meniscus position due to the neglect of δp assuming that m is a function of $\mu U/T$ only, note that $m = \lambda \cos^2 \theta$ so that the error $\delta \theta$ is $(\delta \lambda/2\lambda) \tan \theta$. The change in p' is

$$\left(\frac{\partial p'}{\partial \theta}\right)_\lambda \delta \theta + \left(\frac{\partial p'}{\partial \lambda}\right)_\theta \delta \lambda = \delta \theta \left[\left(\frac{\partial p'}{\partial \theta}\right)_\lambda + \left(\frac{\partial p'}{\partial \lambda}\right)_\theta (2\lambda \cot \theta) \right].$$

The values of $(\partial p'/\partial \theta)_\lambda$ and $(\partial p'/\partial \lambda)_\theta$ can be estimated using fig. 3 for any position on $p' = 0$. Thus near $\theta = 45°$, $\lambda = 1\cdot27$ and $(\partial p'/\partial \lambda)_\theta$ is approximately $1\cdot0$ while $(\partial p'/\partial \theta)_\lambda$ is approximately $0\cdot2$, so that $\delta \theta = \delta p'/2\cdot8$ and since $\delta p = (p/p') \delta p'$, it follows that $\delta \theta = (p'/p) \delta p/2\cdot8$.

Comparing the figures in the last two lines in table 2 with those in lines 2 and 3 it will be seen that $p' \delta p/p$ is of order $1/200$ for $h_0 = 0\cdot01$ and about $1/50$ for $h_0 = 0\cdot06$. The error therefore in neglecting δp and using the relationship between m and θ when $p' = 0$ is only a fraction of a degree in θ. This relationship is given by setting $p' = 0$ in (11). It is

$$m = \lambda \cos^2 \theta = \frac{\theta + \tfrac{1}{2}\pi + \tfrac{1}{2}\sin 2\theta}{\tfrac{3}{4}(\theta + \tfrac{1}{2}\pi) + \tfrac{1}{2}\sin 2\theta + \tfrac{1}{16}\sin 4\theta} \tag{23}$$

and it is shown graphically in fig. 16.

If the value of m at which a two-dimensional or cylindrical meniscus can exist is a function of $\mu U/T$ only, as it appears to be in my experiments, it is possible to use the experimental curve of fig. 7 with the theoretical curve of fig. 16 to predict where such a meniscus could exist in Floberg's apparatus. For this purpose fig. 17 was constructed showing the relationship between $\mu U/T$ and $\tan \theta = x(2Rh_0)^{-\frac{1}{2}}$.

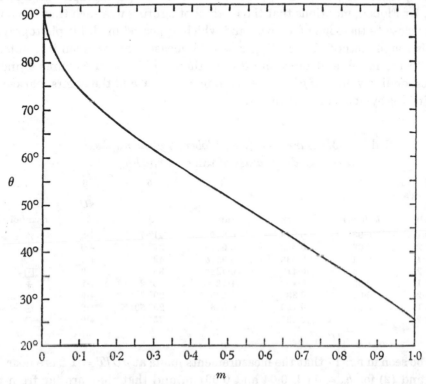

Fig. 16. Relationship between m and θ for points in fig. 3 on $p' = 0$.

Curve (1) shows the relationship so found. In my experiments the two-dimensional meniscus was limited to the range $0 < \mu U/T < 1.75$. On the other hand, measurements of m corresponding with the ends of the air fingers were carried to $\mu U/T = 2.8$ and these are represented in fig. 7 by a broken line. This broken line has been transferred to fig. 17 as curve (2) by the same method as that used to produce curve (1) for the two-dimensional meniscus experiments, but it must be remembered that there is no reason to believe that the broken curve of fig. 7 would represent experiments with other distributions of thickness in the narrow oil passage. Though the two-dimensional meniscus was expected *a priori* to depend only on the local conditions near the meniscus the stability of the meniscus certainly depends on the distribution of thickness of the oil passage* and the positions of the ends of the finger may do so also.

* J. R. A. Pearson, *J. Fluid Mech.* VII (1960), 481; E. Pitts & J. Greiller, *J. Fluid Mech.* XI (1961), 33.

13. COMPARISON WITH MEASUREMENTS OF FLOBERG'S PHOTOGRAPHS

The main difficulty in attempting to measure x, the distance of the ends of the air fingers in Floberg's photograph from the narrowest point in the oil passage is that the position of this point was not marked on the photographs. In a letter, however, Dr Floberg mentions that the method of lighting was such that the point $x = 0$ was close to the edge of a dark band which appeared in all his photographs. Using this rough method for locating $x = 0$ the measurements given in column 3 of table 3 were made, and $\tan\theta$ and θ were then tabulated in columns 4 and 5. The corresponding values of $\mu U/T$ are given in column 6 and the points marked on fig. 17 with the symbols of column 7, table 3.

Table 3. *Measurements from Floberg's photographs and corresponding values of* $\tan\theta = x/(2Rh_0)^{\frac{1}{2}}$

1 Floberg's plate number	2 h_0 (cm.)	3 x (cm.)	4 $\tan\theta$	5 θ	6 $\dfrac{\mu U}{T}$	7 Symbol
30·1	0·06	0·618	0·892	41°·7	1·2	△
30·2	0·06	0·469	0·677	34°·1	4·9	▲
31·1	0·04	0·538	0·9515	43°·6	1·2	⊡
31·2	0·04	0·412	0·728	36°·1	4·9	⊡
32·1	0·02	0·45	1·12	48°·2	1·2	+
32·2	0·02	0·33	0·830	39°·7	4·9	+
33·1	0·01	0·160	0·566	29°·50	1·2	⊙
33·2	0·01	0·137	0·485	25°·9	4·9	⊙

It will be seen in fig. 17 that the measurements made at $\mu U/T = 1\cdot2$ are near the lines (1) and (2) for $h_0 = 0\cdot02$, $0\cdot04$ and $0\cdot06$ cm. and that they are far from the Swift–Stieber line. On the other hand the point corresponding with $h_0 = 0\cdot01$ cm. is close to this line.

Since the highest values of $\mu U/T$ at which I measured the positions of the fingers was $2\cdot8$, it is not possible to make a comparison with Floberg's results at $\mu U/T = 4\cdot9$ but the positions of the points representing his photographs is consistent with a gradual increase in m for $h = 0\cdot02$, $0\cdot04$ and $0\cdot06$ cm. as $\mu U/T$ increases. The point for $h_0 = 0\cdot01$ is very close to the Swift–Stieber line when $\mu U/T = 4\cdot9$.

These results seem to show that Floberg in his experiments got both types of cavitation, though he does not distinguish between them. He noticed that his results for $h_0 = 0\cdot01$ cm. were near the Swift–Stieber point and that the meniscus retreats slightly from the point $x = 0$ as the speed is reduced. This can be seen in fig. 17 where the point for $\mu U/T = 1\cdot2$, $h_0 = 0\cdot01$ cm. corresponds with a slightly greater value of x than that for $\mu U/T = 4\cdot7$. Floberg does not comment on the fact that in his experiments at $h_0 = 0\cdot02$, $0\cdot04$ and $0\cdot06$ cm., x is very much greater than the Swift–Stieber criterion would allow. One interesting feature of fig. 17 is that at each of the values of $\mu U/T$ a decrease in h_0 from 0·06 to 0·02 cm. *increases* $x/(2Rh_0)^{\frac{1}{2}}$ continuously but a further *decrease* in h_0 to 0·01 cm. causes the meniscus to *retreat* to the Swift–Stieber position. This again is evidence that a change in the physical nature of the cavitation took place between $h_0 = 0·02$ and 0·01 cm.

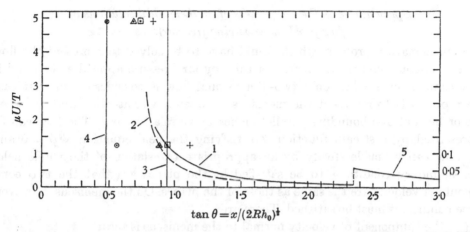

Fig. 17. Points represent observed positions of the ends of air fingers in Floberg's experiments: $\triangle$ $h_0 = 0·06$, $\square$ $h_0 = 0·04$, $+$ $h_0 = 0·02$, $\odot$ $h_0 = 0·01$. Lines represent: (1) position of two-dimensional meniscus using experimental data of fig. 7; (2) positions of ends of fingers using data of fig. 7; (3) Hopkins criterion ($m = \frac{2}{3}$ for case (b)); (4) Swift–Stieber criterion; (5) continuation of line (1) when ordinates are increased in ratio 10:1.

14. POSSIBLE MODE OF TRANSFORMATION OF
ONE TYPE OF CAVITATION TO THE OTHER

It is known that internal cavitation occurs in oils at pressures far exceeding their vapour pressure owing to the existence of gases dissolved in them which are released with a comparatively small drop in pressure. This probably accounts for the fact that the greatest decrease in pressure calculated even for a flooded bearing, Floberg's experiment at $h_0 = 0·01$ cm. and $U = 10·5$ cm./sec., is far less than an atmosphere, yet the Swift–Stieber condition seems to apply fairly closely.

Table 2 shows that the value of p/p' was greater for $\mu U/T = 4·9$, $h_0 = 0·02$ cm. than for $\mu U/T = 1·2$, $h_0 = 0·01$ cm., but fig. 17 suggests that the Swift–Stieber cavitation did not occur in the former case but did in the latter. This may well be due to the fact that separation cavitation necessarily implies a reduction in pressure upstream of the meniscus. If this is sufficiently great internal cavitation may take place, but will be seen from fig. 3 that as the position of the separation meniscus moves back along the axis $p' = 0$ the maximum pressure defect, $-p'$, decreases. Thus though p/p' is greater for $\mu U/T = 4·9$, $h_0 = 0·02$ cm. than for $\mu U/T = 1·2$, $h_0 = 0·01$ cm. the pressure defect at the pressure minimum is probably less in the former case than the latter.

In conclusion I wish to express my thanks to Professor J. A. Cole for his photographs (figs. 12–15), to Dr Leif Floberg for information about his experiments and to Professor Garrett Birkhoff for some useful discussion.

APPENDIX

Speculations on the uniqueness of mathematical solutions of
flow problems involving free surfaces

The mathematical problem which would have to be solved to represent the flow when a viscous fluid is driven from a tube by air pressure applied at one end is very difficult even when only two-dimensional flow is considered. In that case the problem is to represent the meniscus between a viscous fluid and air when one or both of two bounding parallel planes move relative to it. The flow can be represented by a stream function ψ satisfying the field equation $\nabla^4 \psi = 0$ and the flow can be made steady by an appropriate translation of the whole field. The boundary conditions to be satisfied at the planes are that the two components of velocity are the same as those of the planes. On the meniscus, however, three conditions must be satisfied. These are:

(i) the component of velocity normal to the meniscus is zero;

(ii) the component of shear stress parallel to its surface is zero;

(iii) the components of stress normal to its surface is (surface tension)/radius of curvature of the meniscus).

Since only two of these conditions can be satisfied at an arbitrarily chosen surface the possibility of satisfying the third can only be realised by varying the shape of the meniscus. An experimenter would certainly expect a definite shaped meniscus to establish itself, but it may be very difficult to prove uniqueness mathematically.

Some light might be thrown on this subject by recalling some simpler problems relating to the flow of an ideal non-viscous fluid with a free surface. Here the flow is irrotational and at fixed boundaries only one condition can be satisfied, but at a free surface it is possible to satisfy two. For instance, many free-surface problems involving steady flow under the action of gravity have been solved analytically, or by numerical processes and it has been recognized that the solutions are often not unique. The stationary waves produced by an obstacle on the bed of a stream, for instance, is perhaps a trivial example. In that case solutions exist corresponding with cases where waves of arbitrary amplitude and phase are propagated upstream of the obstacle at the speed of the stream. There is a unique case where no such waves exist and it has been pointed out that even the smallest viscosity would tend to make this case the nearest approximation to the real phenomenon exhibited by an obstacle in the bed of a smoothly flowing stream.

In the case just cited the field extends to infinity but cases of non-unique flow with a closed free surface can be imagined. Consider, for instance, free vortex partially filling a rigid circular cylindrical case. In one example the free surface might be a concentric cylinder, but capillary ripples can exist on the free surface and if these are of such a length that they can travel backwards at the speed of the fluid at the free surface the motion is an alternative steady flow satisfying all the necessary boundary conditions. This example is peculiar because for a given amount of fluid in the rigid cylindrical boundary circular free streamlines are possible with all vortex strengths, but the alternative steady motions are only possible for discrete values of this strength, namely for those for which the circular free streamline is an integral number of the critical wavelength which can travel backwards at the speed which makes steady motion possible.

The non-uniqueness of solutions where waves can exist is well understood but recently a more interesting case has been found. Garabedian* has shown that there is a singly infinite set of symmetrical solutions of the two-dimensional version of the problem of emptying water from a vertical tube which is closed at the top. Here both intuition, and experimentation in the axisymmetric case, lead to the expectation that only one of the solutions would represent the actual phenomenon but there seems to be no convincing reason for the choice of any particular solution. Garabedian pointed out that one of them represents the case where the air column rises at a maximum rate, but there seems little justification for choosing that particular solution as the one which would occur if the two-dimensional bubble could be realised experimentally. (It has been pointed out to me by Garrett Birkhoff in a private communication that though one of Garabedian's solutions represents the symmetrical bubble which rises at maximum speed a larger asymmetrical bubble which would rise at a speed $\sqrt{2}$ times Garabedian's maximum is theoretically possible.)

Another case where an infinity of solutions of $\nabla^2 \psi = 0$ can satisfy the relevant boundary conditions is provided by a Hele-Shaw cell.† In that case it was shown experimentally that only one of the motions described by these solutions can be set up. Recently Jacquard & Séquier‡ have shown, by tracing theoretically a method of setting up the motion, that the experimentally observed motion would result from the mode of initiation which they analyse.

Uniqueness of flow when viscous fluid is blown from a tube

Acceptance of equation (1) pre-supposes uniqueness of flow. This is in accordance with experimental measurements made with tubes of varying bore and viscosity. Theoretical justification for (1) based solely on dimensional arguments is not convincing. Dimensional arguments can justifiably be used to state that if m is known all cases of flow are similar for a given value of $\mu U / T$. The justification for taking m as a function of $\mu U / T$ must be based either on detailed analysis of the flow, which at present seems to be outside the range of practical possibility

---

* *Proc. Roy. Soc.* A, CCXLI (1957), 423.
† See paper 28 above.
‡ *Journal de Mécanique* I (1962), 367.

for mathematicians, or on physical intuition. When the tube is flooded upstream from the meniscus (upstream when brought to rest by appropriate translation), as it is when fluid is blown from the tube by pressure applied at one end, μ, T, the radius of the tube a, and the pressure gradient in the flooded portion, dp/dx, are the parameters which can be varied at will. Poiseuille's equation connects $U/(1-m)$, μ and a with dp/dx so that we can regard μ, T, a and $U/(1-m)$ as the factors variable at will. The only non-dimensional combination of these is $\mu U/T(1-m)$, but there is no simple way in which a one-to-one relationship between $\mu U/T$ and $1-m$ can be established.

When the U is reversed so that fluid is blown into a tube which already contains fluid distributed uniformly on the walls, $\mu U/T(1-m)$ is still variable at will but since m is now fixed any value of $\mu U/T$ can occur for any value of m.

It is curious that crude dimensional theory can apply when U is positive but not when it is negative. It seems reasonable therefore to expect that a mathematical solution when found will not be unique, but that the physical situation will be unique when U is positive and $\mu U/T$ is known, whereas when U is negative it is necessary to know m as well as $\mu U/T$ in order to establish a physical situation which is similar for variation in μ, U, T and a.

This paper is an enlarged version of a contribution to the General Motors Symposium on Cavitation held at Detroit in September 1962.

39

DISINTEGRATION OF WATER DROPS IN AN ELECTRIC FIELD

REPRINTED FROM

Proceedings of the Royal Society, A, vol. CCLXXX (1964), pp. 383–97

The disintegration of drops in strong electric fields is believed to play an important part in the formation of thunderstorms, at least in those parts of them where no ice crystals are present. Zeleny showed experimentally that disintegration begins as a hydrodynamical instability, but his ideas about the mechanics of the situation rest on the implicit assumption that instability occurs when the internal pressure is the same as that outside the drop. It is shown that this assumption is false and that instability of an elongated drop would not occur unless a pressure difference existed. When this error is corrected it is found that a drop, elongated by an electric field, becomes unstable when its length is 1·9 times its equatorial diameter, and the calculated critical electric field agrees with laboratory experiments to within 1 %.

When the drop becomes unstable the ends develop obtuse-angled conical points from which axial jets are projected but the stability calculations give no indication of the mechanics of this process. It is shown theoretically that a conical interface between two fluids can exist in equilibrium in an electric field, but only when the cone has a semi vertical angle 40·3°.

Apparatus was constructed for producing the necessary field, and photographs show that conical oil/water interfaces and soap films can be produced at the calculated voltage and that their semi-vertical angles are very close to 10·3°. The photographs give an indication of how the axial jets are produced but no complete analytical description of the process is attempted.

INTRODUCTION

The distortion and bursting of water drops in an electric field has formed the subject of a number of experimental researches. The practical interest of the subject is that it seems to be an important factor in the production of thunderstorms at any rate in those parts of them where it is too warm for ice crystals to exist. Zeleny[*] photographed drops held at the end of capillary tubes and raised to a high potential. He measured the potential at which they disintegrated owing to the formation of a pointed end from which issued a narrow jet. Fig. 1, pl. 1, which is reproduced from Zeleny's paper, shows a jet of glycerine formed in this way. When water was used the end of the drop vibrated violently.

Wilson & Taylor[†] found that a similar phenomenon occurs when an uncharged soap bubble is subjected to a uniform electric field. Macky[‡] and Nolan[§] showed that the same kind of disintegration occurs when a drop of water falls between

[*] *Phys. Rev.* x (1917), 1.
[†] *Proc. Camb. phil. Soc.* XXII (1925), 728; paper 9 above.
[‡] *Proc. Roy. Soc.* A, CXXXIII (1931), 565.
[§] *Proc. Roy. Irish Acad.* XXXVII (1926), 28.

parallel plates when a potential gradient is maintained between them. Zeleny showed experimentally that the disintegration is due to instability rather than the formation of ionic currents and he concluded, on dimensional grounds, that the criterion of instability must be of the form

$$V^2/r_0 T = C, \tag{1}$$

where V is the potential, r_0 the radius of the drop, T surface tension and C a constant. His results were in fact well presented by taking $C = 140$ when V was expressed in electrostatic units.

Wilson & Taylor found that a soap bubble in an electric field F becomes unstable when $Fr_0^{\frac{1}{2}} = 3670 \pm 100$ V. cm.$^{-\frac{1}{2}}$. The variation ± 100 V. represents the standard deviation in eight groups of measurements in which r_0 varied from 0·25 to 1·06 cm. expressed in a form analogous to (1) and using the measured value of the surface tension of the soap solution used ($T = 2 \times 29 = 58$ dyn./cm.). Wilson & Taylor's result, expressed in electrostatic units, is

$$F(r_0/T)^{\frac{1}{2}} = 1·61 \pm 0·04. \tag{2a}$$

Macky's measurements with water drops are represented by

$$F(r_0/T)^{\frac{1}{2}} = 1·51. \tag{2b}$$

The equations representing the equilibrium of a deformed drop under the action of surface tension and an electric field can be set up and the shape of the drop could be determined if a solution could be obtained. This, however, has not yet been done by any worker in this field except when the distortion from the spherical form is small.

In the absence of analysis of this kind Zeleny* assumed that the drop in his experiments became elongated approximately into the form of a prolate spheroid. The intensity of the field at the poles of a spheroid of constant volume can be calculated and the resulting stress normal to its surface found. As the length of the spheroid increases this stress increases and Zeleny tried to use this fact to find a criterion of stability. His method was based on Rayleigh's calculation† of the limiting charge, Q, which an isolated drop of radius r_0 can hold before it becomes unstable. Rayleigh's result was that when $Q^2 > 4\pi r_0^3 T(n+2)$ the drop becomes unstable for a displacement proportional to the Legendre function $P_n(\cos\theta)$, provided $n \geqslant 2$. When $Q^2 < 16\pi r_0^3 T$ or the potential $V < (16\pi r_0 T)^{\frac{1}{2}}$, the drop is stable for all displacements, and when V first exceeds this value the drop becomes unstable only for the disturbance $P_2(\cos\theta)$ for which it becomes slightly ellipsoidal while the displacement is small.

Zeleny's method for adapting Rayleigh's stability criterion for a charged sphere so as to be applicable to a spheroid was to assume that it applies at the polar end when Rayleigh's r_0 is replaced by the polar radius of curvature of the spheroid and the mechanical stress due to the electric field (which can be calculated for a spheroid)

---

* Proc. Camb. phil. Soc. xviii (1915), 71.
† *Phil. Mag.* xiv (1882), 184.

is the same as that on the sphere when it becomes unstable. It seems that this method, though it has been accepted by later workers, is unsound because it takes no account of the fact that the pressure inside the spheroid is not in general the same as that outside when instability occurs. In Zeleny's equations the symbol p, representing this difference, appears in the equation of equilibriun but the pressure is assumed to vanish when the drop becomes unstable, in fact it is really this assumption which is the basis of Zeleny's criterion rather than Rayleigh's criterion for the stability of a charged sphere. The confusion probably arose because when an isolated drop is charged to Rayleigh's critical potential for unstable displacements proportional to $P_n(\cos\theta)$ the internal pressure happens to be equal to the external pressure when $n = 2$. This accidental circumstance only occurs when $n = 2$ and the spherical drop begins to become spheroidal. It does not occur for other values of n, or, as will be shown later, when a spheroid of finite eccentricity becomes unstable in a uniform electric field. It is not surprising therefore that Zeleny's method is not successful as a guide to understanding the mechanics of the instability in spite of the fact that photographs of drops and soap films in a uniform electric field show them to be nearly spheroidal before they disintegrate.

SPHEROIDAL APPROXIMATIONS

Though the equations of equilibrium cannot be completely satisfied over the surface of a freely charged spheroid or a spheroid in a uniform field unless the ellipticity is small, the fact that a soap film in an electric field is nearly spheroidal makes it worthwhile to find out how nearly the equilibrium conditions can be satisfied. Two approximations (I and II) will be considered. In I the equilibrium equations will be satisfied at the poles and the equator. In II they are satisfied at the poles, but at the equatorial section the balance between the internal pressure, surface tension and total force due to the electric field acting over one-half of the spheroid is satisfied. If the equilibrium equations had been satisfied at all points of the spheroidal surface the two approximations I and II would have given identical results. The difference between them is an indication of the error involved in the spheroidal approximation.

Stress on a spheroid due to the electric field

The electric field round an isolated spheroid charged to potential V and also that due to a spheroid in a uniform electric field of strength F can be calculated by methods described in text books.* Taking the equation to the axial section of a prolate spheroid as $x^2/a^2 + y^2/b^2 = 1$, where a is the major axis, the ellipticity is $e = (1 - b^2/a^2)^{\frac{1}{2}}$. For the spheroid charged to potential V the electric field at the surface of any point is

$$\frac{dV}{dn} = \frac{PV}{b^a I_1}.$$ (3)

* See, for example, *Electricity and Magnetism*, by J. H. Jeans, Cambridge University Press (1915), pp. 244–54.

where P is the perpendicular from the centre of the spheroid onto the tangent plane at the point in question and

$$I_1 = \frac{1}{2e} \ln \left(\frac{1+e}{1-e} \right). \tag{4}$$

Since $P^2 = b^2(1 - x^2e^2/a^2)^{-1}$ the normal stress at the point x is

$$\widehat{nn} = \frac{1}{8\pi} \left(\frac{dV}{dn} \right)^2 = \frac{V^2}{8\pi b^2(1 - x^2e^2/a^2)\, I_1^2}. \tag{5}$$

For the uncharged spheroid in a uniform field, F, parallel to the major axis

$$\frac{dV}{dn} = \frac{FxP}{b^2 I_2}, \tag{6}$$

where

$$I_2 = \frac{1}{2e^3} \ln \frac{1+e}{1-e} - \frac{1}{e^2}. \tag{7}$$

The stress is perpendicular to the surface and equal to

$$\widehat{nn} = \frac{1}{8\pi} \frac{F^2x^2}{b^2(1 - x^2e^2/a^2)\, I_2^2}. \tag{8}$$

Approximate equilibrium equations

The discontinuity of normal stress due to surface tension is $T(r_1^{-1} + r_2^{-1})$ where r_1 is the radius of curvature of the ellipse $x^2/a^2 + y^2/b^2 = 1$ and r_2 is the other principal radius of curvature of the spheroid. The analytical expressions for r_1 and r_2 are

$$r_1 = a^2 b^2 \left(\frac{x^2}{a^4} + \frac{y^2}{b^4} \right)^{\frac{3}{2}}, \quad r_2 = b^2 \left(\frac{x^2}{a^4} + \frac{y^2}{b^4} \right)^{\frac{1}{2}}. \tag{9}$$

At the end of the major axis $r_1^{-1} + r_2^{-1} = 2ab^{-2}$, and at the end of the minor axis $r_1^{-1} + r_2^{-1} = ba^{-2} + b^{-1}$. In approximation I the equations of equilibrium are

$$2ab^{-2}T - p = (\widehat{nn})_{x=a} \tag{10}$$

and

$$T(ba^{-2} + b^{-1}) - p = (\widehat{nn})_{x=0}, \tag{11}$$

where p is the difference between the internal and external pressures. In approximation II (10) is replaced by

$$2\pi bT - \pi b^2 p = 2\pi \int_0^b \widehat{nn} y\, dy. \tag{12}$$

Eliminating p, approximation I yields

$$T(2ab^{-2} - ba^{-2} - b^{-1}) = (\widehat{nn})_{x=a} - (\widehat{nn})_{x=0} \tag{13}$$

and II yields

$$T(2ab^{-2} - 2b^{-1}) = (\widehat{nn})_{x=a} - 2b^{-2} \int_0^b \widehat{nn} y\, dy$$

$$= (\widehat{nn})_{x=a} - 2a^{-2} \int_0^a \widehat{nn} x\, dx. \tag{14}$$

Since the drop has constant volume a and b must be expressed in terms of r_0, the original radius of the drop. Writing $\alpha = 1 - e^2$, we have

$$a = r_0 \alpha^{-\frac{1}{3}}, \quad b = r_0 \alpha^{\frac{1}{6}}. \tag{15}$$

For the charged spheroid

$$(\bar{n}\bar{n})_{x=a} = \frac{V^2}{8\pi r_0^2 I_1^2} \alpha^{-\frac{4}{3}}, \quad (\bar{n}\bar{n})_{x=0} = \frac{V^2 \alpha^{-\frac{1}{3}}}{8\pi r_0^2 I_1^2}. \tag{16}$$

Approximation I for the charged spheroid is

$$V(\pi r_0 T)^{-\frac{1}{2}} = N(\alpha) I_1, \tag{17}$$

where

$$[n(\alpha)]^2 = 8\alpha^{\frac{2}{3}}(2 - \alpha^{\frac{1}{2}} - \alpha^{\frac{3}{2}})/(1 - \alpha). \tag{18}$$

For the spheroid in a uniform field F, approximation I is

$$F r_0^{\frac{1}{2}} T^{-\frac{1}{2}} = (8\pi)^{\frac{1}{2}} M(\alpha) I_2, \tag{19}$$

where

$$M(\alpha) = \alpha^{\frac{2}{3}}(2 - \alpha^{\frac{1}{2}} - \alpha^{\frac{3}{2}})^{\frac{1}{2}},$$

and II is

$$F^2 r_0 T^{-1} = 16\pi \alpha^{\frac{5}{6}}(\alpha^{-\frac{1}{2}} - 1) I_2^2/\phi(e)$$

and

$$\phi(e) = \left[\frac{1}{1 - e^2} + \frac{1}{e^4}\{e^2 + \ln(1 - e^2)\}\right]. \tag{20}$$

Accuracy of the spheroidal approximations

The calculated relation for the isolated spheroid between $V(\pi r_0 T)^{-\frac{1}{2}}$ and a/b according to approximation I is shown in fig. 2. For the uncharged spheroid the relationship between $F r_0^{\frac{1}{2}} T^{-\frac{1}{2}}$ and a/b, according to I, is shown in fig. 3. Approximation II is so close to I that it is only in one part of the curve that the difference reaches about 1 % and can be shown in fig. 3. The closeness of the two approximations suggested that the difference between the true form of the interface and the spheroidal approximation must be very small. The difference between the normal force due to pressure and surface tension and that due to electric field was therefore calculated for approximation I. Here the conditions assumed were that the equilibrium equations are satisfied at the equator and the poles. The error of the spheroidal approximation can therefore be appreciated by calculating the difference between the electric and mechanical stress at intermediate points.

If we write $x/a = \xi$, $y/b = (1 - \xi^2)^{\frac{1}{2}}$ and $b^2/a^2 = \alpha$, (9) becomes

$$\frac{1}{r_1} + \frac{1}{r_2} = \frac{1}{a} \left\{ \frac{1 - \alpha^{-1} - \xi^2(\alpha^{-1} - 1)}{\alpha^2[\alpha^{-1} - \xi^2(\alpha^{-1} - 1)]} \right\}. \tag{21}$$

The mechanical stress due to the electric field is proportional to $x^2 p^2$ or to

$$\frac{\xi^2}{\alpha^{-1} - \xi^2(\alpha^{-1} - 1)}. \tag{22}$$

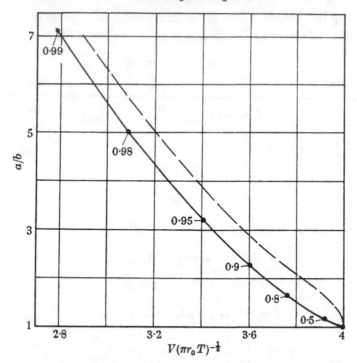

Fig. 2. Charged spheroid. Broken curve shows Zeleny's criterion.

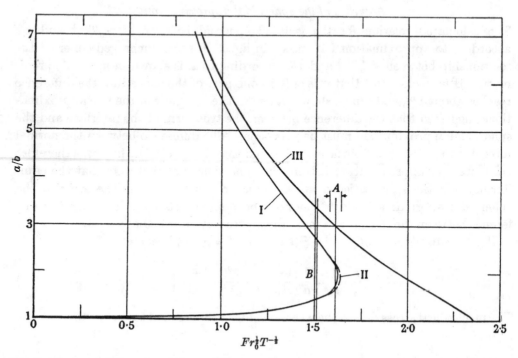

Fig. 3. Uncharged drop in uniform field. I and II according to approximations I and II. *A*, Wilson & Taylor's measurements. *B*, Macky's measurements. III, Zeleny's criterion for stability.

PLATE 1

Fig. 1.

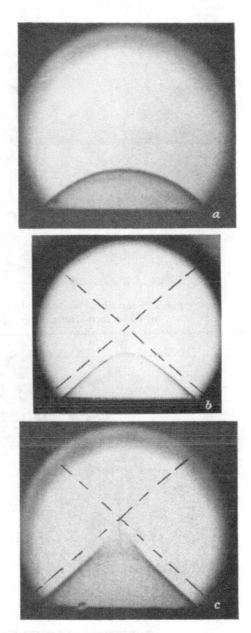

Fig. 6.

Fig. 1. Jet of glycerine from an electrified drop (Zeleny 1917).
Fig. 6. Soap film. Exposure time 1·6 min. Broken lines at angle 98·6°. (*a*) Before oscillation;
(*b*) oscillation beginning; (*c*) exposure covering time of jet formation

PLATE 2

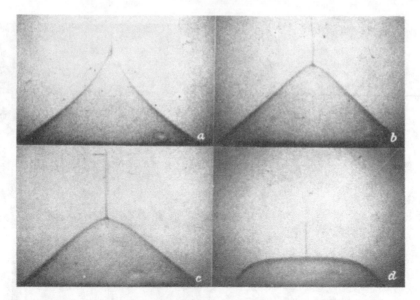

Fig. 7. Soap film, microsecond exposures of successive stages
(a) of jet formation; (b), (c), (d) subsequent collapse.

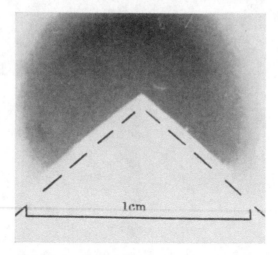

Fig. 9. Pointed summit in oil/water interface.

Fig. 11. Oil/water interface when initial volume was less than
needed for 49·3° conc.

PLATE 3

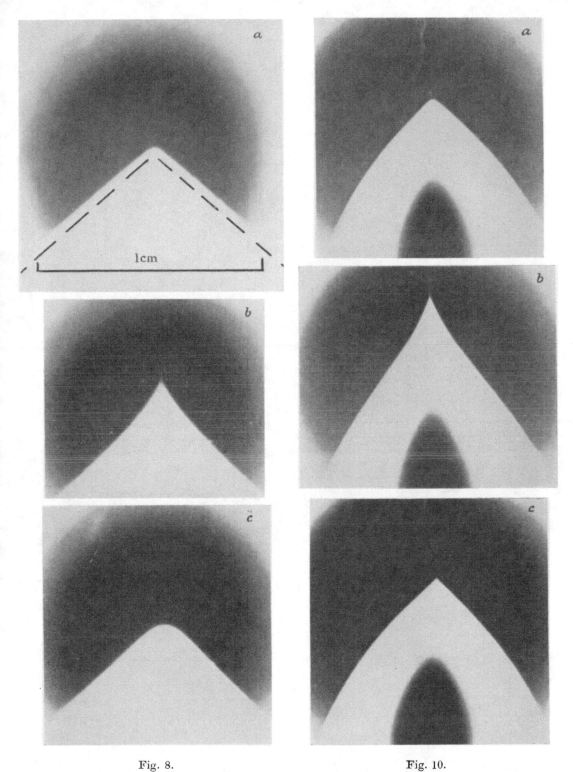

Fig. 8.   Fig. 10.

Fig. 8. Oil/water interface. Three successive frames (1·6 msec. exposures) at intervals of 1/64 sec. (a) Before jet formation; (b) jet forms; (c) subsequent collapse. Broken lines at 98·6° (negative photo).

Fig. 10. Oil/water interface when initial volume was in excess of requirement for 49·3° cone. (a) Before jet forms; (b) jet forms; (c) after jet has become detached and broken up.

For the comparison between the two it is necessary to multiply one of them by a factor to bring the difference between the values at $\xi = 0$ and $\xi = 1$ to equality (this is equivalent to determining $Fr_0^{\frac{3}{2}} T^{-\frac{1}{2}}$ using approximation I).

Numerical comparison

For a numerical comparison the calculation was carried out for the case $a/b = 2$ which is close to the limiting value for stable equilibrium. In this case $\alpha^{-1} = 4$. It is convenient to express the normal stresses as fractions of the value of $\widehat{nn}$ at the poles. In these units (22) gives the stress at ξ due to the electric field as

$$p_E = \frac{\xi^2}{4 - 3\xi^2}, \tag{23}$$

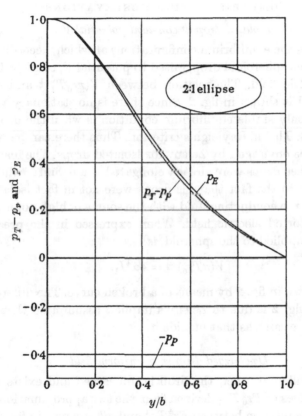

Fig. 4. Distribution of stresses due to electric field p_E, surface tension p_T, internal pressure p_p for case $a/b = 2$.

and since the stress due to the surface tension is proportional to $r_1^{-1} + r_2^{-1}$, (21) becomes

$$p_T = B(5 - 3\xi^2)(4 - 3\xi^2)^{-\frac{3}{2}}, \tag{24}$$

and according to approximation I, B must be chosen so that the difference between the values of p_T at $\xi = 0$ and $\xi = 1$ is $1 \cdot 0$ so that $B = 8/11$.

The equilibrium equation, if it could be satisfied would be

$$p_E = p_T - p_p, \tag{25}$$

where p_p is the pressure expressed in the same units as p_E and p_T. Evidently $p_p = \frac{8}{11}(\frac{5}{8}) = \frac{5}{11}$. The values of p_E and $p_T - p_p$ calculated from (23) and (24) are shown in fig. 4, but the ordinates are $y/b = (1 - \xi^2)^{\frac{1}{2}}$ instead of ξ because this choice reveals the error in (25) more clearly. It will be seen that the error is surprisingly small, amounting to $2 \cdot 0\%$ at its maximum when $y/b \sim 0 \cdot 5$. In fig. 4, $-p_p$ is shown in order to reveal the large part that the internal pressure plays in the equilibrium equation and therefore in the stability condition.

COMPARISON WITH OBSERVATIONS

Isolated drop at constant potential V

Equation (17) gives the equilibrium configurations at which, according to approximation I, an isolated spheroidal drop, raised to potential V, could be in equilibrium with its own electric field. The relation between $V(\pi r_0 T)^{-\frac{1}{2}}$ and a/b according to approximation I is shown in fig. 2. Since there is no stationary value for V as a/b increases, the only stable equilibrium condition is when the drop is spherical and $V(\pi r_0 T)^{-\frac{1}{2}} < 4$. This is Rayleigh's criterion. Thus there can be no criterion of stability of the type envisaged by Zeleny for isolated drops. The instability which Zeleny observed when drops were already elongated at a definite value of $V(r_0 T)^{-\frac{1}{2}}$ must be attributed to the fact that his drops were not in fact isolated, but were connected through the conducting fluid with the source of high potential, and were therefore in a distorted electric field. When expressed in the present notation Zeleny's criterion, applied to the spheroid, is

$$V(\pi T r_0)^{-\frac{1}{2}} = 4\alpha^{\frac{1}{2}} I_1. \tag{26}$$

This relation is shown in fig. 2 by means of a broken curve. The difference between the two curves in fig. 2 is due to Zeleny's implicit assumption that the pressure inside the drop is the same as that outside it.

Uncharged drop in a uniform field

The case is very different when the drop is uncharged and exists in a uniform electric field F. Values of $F r_0^{\frac{1}{2}} T^{-\frac{1}{2}}$ derived from the two approximations I and II are given in table 1. The relation between $F r_0^{\frac{1}{2}} T^{-\frac{1}{2}}$ and a/b is shown in fig. 3 for approximation I, but the two approximations are so close that II is only visible over a small range near $a/b = 2$.

Zeleny's criterion for the stability of drops in a uniform field when expressed in the present notation, is

$$F r_0^{\frac{1}{2}} T^{-\frac{1}{2}} = 4\pi^{\frac{1}{2}} \alpha^{\frac{3}{2}} I_2 \tag{27}$$

and this relation is shown in fig. 3 as the line III. The difference between curves I and III, which is due to Zeleny's implicit assumption that $p = 0$, is striking. It will be seen that as $Fr_0^{\frac{1}{2}}T^{-\frac{1}{2}}$ rises from zero the spheroid becomes increasingly prolate till when $a/b = 1\cdot9$, $Fr_0^{\frac{1}{2}}T^{-\frac{1}{2}} = 1\cdot625$ according to I, or $1\cdot635$ according to II, it reaches a maximum, after which it decreases. Evidently equilibrium configurations corresponding to points on the rising part of the curve are stable while those corresponding with values of $a/b > 1\cdot9$ are unstable.

Table 1. *Values of* $Fr_0^{\frac{1}{2}}T^{-\frac{1}{2}}$

I and II, approximations; III, Zeleny's values.

e	a/b	I	II	III
0·00	1·000	0·000	0·000	2·363
0·20	1·020	0·473	0·46	2·363
0·50	1·154	1·136	1·14	2·308
0·75	1·512	1·564	1·575	2·159
0·80	1·666	1·610	1·622	2·093
0·85	1·898	1·625	1·635	1·992
0·88	2·105	1·608	1·623	1·916
0·90	2·294	1·583	1·588	1·839
0·92	2·550	1·537	1·542	1·747
0·95	3·202	1·406	1·424	1·564
0·98	5·025	1·092	1·086	1·153
0·99	7·092	0·856	0·849	0·889

In curve III, $Fr_0^{\frac{1}{2}}T^{-\frac{1}{2}}$ falls continuously with increase in a/b so that it is only when the pressure difference is taken into account that the spheroidal approximation can predict instability.

Comparison with experiments on the stability of drops in a uniform field

The experimental results of Wilson & Taylor* namely $Fr_0^{\frac{1}{2}}T^{-\frac{1}{2}} = 1\cdot61 \pm 0\cdot04$, marked A in fig. 3, agrees well with the theoretical values $1\cdot625$ according to I and $1\cdot635$ according to II. The experimental result of Macky, marked B in fig. 3, is about 7 % lower. The discrepancy may well be due to the fact that Macky's drops suddenly entered the electric field, dropping into it from an unelectrified region, so that they were not in a static state. They were also acted on by aerodynamic forces which would be expected in any case to create a disturbance which might well displace them from the stable to the unstable part of the curve I in fig. 3 before the field reached the critical value. Wilson & Taylor's experiments on the other hand were conducted under static conditions which permitted the values of both r_0 and F to be measured accurately.

The soap films in Wilson & Taylor's experiments were initially hemispherical and stood on an earthed conducting plate. Their profiles were photographed in some cases just before the instability set in. Fig. 1 of their paper is an example. Measuring the height and horizontal diameter of the bubble in this photograph I find height/ diameter = $4\cdot6/4\cdot2 = 1\cdot1$ so that $a/b = 2\cdot2$. The maximum value of F given in table 1 occurs when $a/b = 1\cdot9$. This is not very far from $2\cdot2$ but it is much less than the value 3 to 4 which Zeleny† quotes from Macky. Since the drops could not

* Paper 9 above. † *J. Franklin Inst.* ccxix (1935), 659.

be observed so closely in the water-dropping experiments of Macky and a close control of drop size was not possible as it was in Wilson & Taylor's experiments, the rough estimates of a/b which Macky gives cannot be regarded as measures of the critical dimensions of a drop at the moment when instability first appears.

Zeleny[*] applied his criterion to drops in a uniform field by assuming that the drop existed with a value of $a/b = 3$ to 4. The ordinate $a/b = 3 \cdot 4$ in fact cuts Zeleny's curve III, fig. 3, close to $Fr_0^{\frac{1}{2}}T^{-\frac{1}{2}} = 1 \cdot 51$ which is Macky's experimental result. Zeleny pointed out that if a/b is assumed to be between 3 and 4 his criterion for stability is satisfied and he regarded this fact as a confirmation of the correctness of his criterion as one which determines the limit of stability. The true interpretation seems to be that as a/b increases the effect of pressure on the equilibrium configuration decreases and when $a/b \sim 3 \cdot 4$ the neglect of pressure in the *equilibrium* calculation raises the calculated value of $Fr_0^{\frac{1}{2}}T^{-\frac{1}{2}}$ from $1 \cdot 38$ up to the value $1 \cdot 51$ which is close to Macky's experimental value for the limit of *stability*.

Further development of the instability

In his discussion of the stability of a charged drop Rayleigh pointed out that no instability occurs till the V reaches the value $(16\pi T r_0)^{\frac{1}{2}}$ when the instability corresponding with the Legendre function P_2 appears. For harmonics of order $n > 2$ Rayleigh shows that instability sets in when $V = (4\pi(n+2)r_0 T)^{\frac{1}{2}}$, so that these instabilities can only appear when V increases beyond the point at which the P_2 instability has begun. Rayleigh evidently knew that jets develop out of the unstable ends of drops for he remarked[†] that for great values of the charge (i.e. great values of V) the drop is unstable 'for all values of n below a certain limit, the maximum instability corresponding to a great, but still finite, value of n. Under these circumstances the liquid is thrown out in fine jets, whose fineness, however, has a limit'. It seems that Rayleigh's conception of the mechanics of the formation of fine jets was linked with the appearance of unstable disturbances corresponding with higher values of n for which higher values of V are required than those needed for the slightly ellipsoidal P_2 form. This appears to be contrary to the experimental evidence for the jets seem to form shortly after the instability sets in without increases in the electric field. On the other hand, the ellipsoidal analysis suggests that a series of unstable ellipsoidal equilibrium states are possible in which the length increases indefinitely as the field is reduced. It seems therefore that theory might lead one to guess that when instability sets in the ellipsoidal form might persist, the drop becoming very long before it disintegrates. In fact experiments show that though it elongates to a limited extent it quickly develops an apparently conical end (fig. 1) which usually oscillates and a narrow jet appears at the vertex. At this stage the ellipsoidal approximation is evidently useless and a different type of analysis is necessary.

[*] *J. Franklin Inst.* ccxix (1935), 659.
[†] *Theory of Sound*, ii, 2nd ed. (1896), 374.

Analysis of conditions at the point of a deformed drop

The failure of the spheroidal analysis to provide a useful model for thinking about the mechanics of the development of jets led me to approach the problem from another point of view and try to find the conditions under which a conical point could exist in equilibrium. Since the curvature of a conical surface is inversely proportional to the distance from the vertex the stress normal to it which can balance the surface tension must also be inversely proportional to distance from the vertex. Since the fluid will be assumed to be conducting the conical surface must be an equipotential, and to balance the surface tension the potential gradient there must be proportional to $R^{-\frac{1}{2}}$ where R is the radial co-ordinate.

The electric field expressed in spherical polar co-ordinates which satisfies the stress condition has potential

$$V = V_0 + A R^{\frac{1}{2}} P_{\frac{1}{2}}(\cos \theta), \tag{28}$$

where the line $\theta = 0$ or $\theta = \pi$ is the axis of the cone and $P_{\frac{1}{2}}(\cos \theta)$ is the Legendre function of order $\frac{1}{2}$. If $\theta = \theta_0$ is the conical equipotential surface where $V = V_0$

$$P_{\frac{1}{2}}(\cos \theta_0) = 0. \tag{29}$$

The function $P_{\frac{1}{2}}(\cos \theta)$ has been tabulated.* It has only one zero in the range $0 < \theta < \pi$ at $\theta = \theta_0 = 130 \cdot 7099°$. Since $P_{\frac{1}{2}}(\cos \theta)$ is finite and positive in the range $0 < \theta < \theta_0$ but is infinite at $\theta = \pi$ the only possible electric field which can exist in equilibrium with a conical fluid is that external to a cone of semi-vertical angle $\alpha = \pi - \theta_0$ or $49 \cdot 3°$.

Electric field required for conical point

The curvature of a cone of semi-vertical angle α is $\cot \alpha / R$ so that the equation of equilibrium is

$$\frac{T \cot \alpha}{R} = \frac{1}{8\pi} \left(\frac{dV}{R \, d\theta} \right)^2 \tag{30}$$

and from (28)

$$\frac{1}{R} \frac{dV}{d\theta} = A R^{-\frac{1}{2}} \left[\frac{d}{d\theta} P_{\frac{1}{2}}(\cos \theta) \right]_{\theta = \theta_0}. \tag{31}$$

I am indebted to Dr J. C. P. Miller for computing the value of $dP_{\frac{1}{2}}(\cos \theta)/d\theta$ at $\theta = \theta_0$. It is $-0 \cdot 974$, so that

$$0 \cdot 974 A = 300 \, [8\pi T \cot \alpha]^{\frac{1}{2}}. \tag{32}$$

The factor 300 is inserted so that electric potential is expressed in volts instead of electrostatic units. The field necessary for equilibrium with a fluid cone is therefore that represented by (28) when

$$A = 1 \cdot 432 \times 10^3 T^{\frac{1}{2}} \text{ V. cm.}^{-\frac{1}{2}}. \tag{33}$$

---

* M. C. Gray, *Quart. appl. Math.* XI (1953), 311.

Experiments designed to produce a conical point under controlled conditions

It has already been pointed out that Zeleny's photographs of a disintegrating liquid surface show a thin jet emerging from an axisymmetric surface ending in a cone whose vertical angle appears to be about a right angle. In general the outline of the fluid surface does not consist of two straight lines so that there is some uncertainty about the angle at the vertex. To make a satisfactory experimental

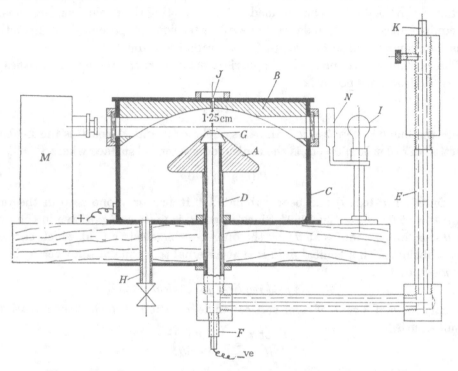

Fig. 5. Chamber for producing field necessary for conical interface.

test of the usefulness of the analysis apparatus was constructed in which it might be possible to produce the field represented by (28) between two metallic surfaces. In the apparatus shown diagrammatically in fig. 5, one surface, A, was an aluminium cone of semi-vertical angle $49 \cdot 3°$ while the other, B, was an aluminium disc hollowed out so that its lower surface was represented by

$$R = R_0[P_{\frac{1}{2}}(\cos \theta)]^{-2}. \tag{34}$$

The value of R_0 chosen was $1 \cdot 25$ cm. The disc B was supported in a horizontal position with its centre $1 \cdot 25$ cm. above the point of the cone by the cylindrical wall of a brass box, C, which was connected to earth. The cone A was supported on a non-conducting tubular pillar, D, which passed through the bottom of the box and was connected with a non-conducting tubular reservoir, E. A short metal rod, F, projected through the non-conducting pillar, D, so that the cone B could be connected with a source of supply of high voltage. The object of the aluminium cone B was to ensure that the electric field near the cone could have the calculated distribution when, and if, the fluid surface assumed a conical form. The cone B was truncated so that its upper edge was a horizontal circle 1 cm. diameter which could form the lower edge of a conical liquid surface if such a surface could in fact be formed. Two types of experiment were performed.

(i) A horizontal soap film was stretched across the edge G, and in order to ensure that this film was initially flat it was found necessary to undercut the aluminium so that the inner surface of the hollowed cone was a cone of semivertical angle smaller than 49·3°. For this experiment the inside of the cone was connected with the terminal F by a wire passing through the pillar D, and since the reservoir E was empty the pressure was equal on the two sides of the soap film.

(ii) The reservoir was filled with water or other conducting fluid up to the level of the edge G. A cork in the top of the tube E then prevented the level of this fluid from altering while the outer box C was filled with the non-conducting fluid, which entered through the tube H. To ensure that the interface at G was not disturbed the box was left open till the level of the non-conducting fluid rose just above G. The upper conductor B was then lowered into position the air escaping through a central vent J. The box was then filled through the tube H till the fluid appeared at J. The level of the interface could be adjusted by a non-conducting rod K moving vertically in E.

In order to observe the interface or soap film it was necessary to bore horizontal holes in the disc B. The axes of these holes passed through the point where the vertex of the aluminium cone A would have been if it had not been truncated. The holes were sealed by glass discs and photographs were taken of the profiles of the interfaces or soap films by means of a cinema camera, M. The background was a translucent screen N which was illuminated either by a steady light, I, or by a stroboscopic flashing light.

The electric field was supplied by apparatus which had been used in an experimental electron microscope and was kindly lent by Professor C. W. Oatley. The surface tension of the soap solution and that of a transformer oil/water interface were determined by the hanging drop method and the use of the calculations of Fordham.* In two determinations for the soap solution T was found to be 29·0 and 28·8 dyn./cm. and for the oil/water interface it was 37·2.

RESULTS

Soap film

The calculated voltage for conical equilibrium was

$$V = 1·432 \times 10^3 (2 \times 29·0)^{\frac{1}{2}} (1·25)^{\frac{1}{2}} = 12·2 \times 10^3 \,\text{V}. \tag{35}$$

* *Proc. Roy. Soc.* A, CXCIV (1948), 1.

In view of (28) and (33) the film was stretched over the truncated top of the cone and the potential gradually raised. It showed little sign of convexity till the potential was 8000 V., it then rose with increasing voltage forming an apparently spherical cap, till V reached a value which recorded as $11\frac{1}{2}$ kV. it suddenly began to oscillate and at the top of each oscillation it ejected a narrow jet of soap solution. Fig. 6 (*a*), pl. 1, shows the film just before oscillation began. During the early unstable stages before the formation of the jet, the film rose sufficiently slowly for the camera operating at 64 frames a second to obtain clear images. The film appeared to rise at increasing speed into an imaginary conical envelope, the round top contracting as it approached the centre of the field. The semi-vertical angle of the conical part of the film was very close to the calculated equilibrium value 49·3°. Fig. 6 (*b*) shows the film approaching the vertex and two straight lines at an angle 98·6° have been drawn in ink on the photograph to show how very close to the calculated equilibrium position most of the film was at that stage. Fig. 6 (*c*) was taken with an exposure time of the order of 1·6 msec. and during that time the film had reached its highest position, thrown out its jet and was beginning to retreat.

To obtain better definition of the formation of the jet a stroboscopic flash with duration of the order of microseconds was used. This was set to flash 100 times a second because the camera was recording at 64 frames a second and more frequent flashes would have made possible more than one exposure on the same frame. The film was in a state of violent oscillation. Fig. 7 (*a*), plate 2, shows the only occasion when the jet was caught at the top of its motion in 50 ft. of film. The subsequent positions are shown in figs. 7 (*b*), (*c*) and (*d*), but since these frames were taken from different parts of the film it is not certain that they can be taken as representing successive states in one oscillation. Indeed it seems likely that 7 (*b*) was very close to the top of an oscillation of smaller amplitude than those shown in 7 (*a*) and 7 (*d*). An interesting feature of the photographs is that the jet either continues to discharge when the top of the film is nearly flat at the bottom of the stroke (fig. 7 (*d*)) or is pulled down by the retreating film. The latter explanation is the more likely.

The mechanics of the process seems clear. The film does not become unstable till the potential is as great or slightly greater than would be required for equilibrium if it assumed the conical form. The rapid acceleration of the top towards the point in space where the vertex of the equilibrium cone could exist indicates that the electric field is locally greater than that necessary for equilibrium, but the fact that the greater part of the film is nearly conical until the jet appears shows that away from the region close to the accelerating top of the film the field is of the form calculated for conical equilibrium. The extremely rapid acceleration which occurs just before the jet appears must be due to the fact that the extent of the region where the field differs appreciably from that of equation (28) is rapidly diminishing so that the maximum field strength at the top of the film is rapidly increasing. If similarity is preserved it must be of the same order as that at the same distance from the vertex in the field represented by (28).

The rate at which the top of the film approaches the vertex may perhaps be controlled by the inertia of the film and the air which moves with it. If h is its thickness and ρ its density, similarity would be preserved during the final stages if

$$\rho h \frac{\mathrm{d}^2 R}{\mathrm{d}t^2} = -K\left(\frac{T}{R}\right), \tag{36}$$

where K is a constant depending on the applied potential. The velocity $\mathrm{d}R/\mathrm{d}t$ found by integrating is

$$\frac{\mathrm{d}R}{\mathrm{d}t} = -\left(\frac{2KT}{\rho h}\right)^{\frac{1}{2}}\left(\ln\frac{R_0}{R}\right)^{\frac{1}{2}}, \tag{37}$$

where R_0 is a constant of integration. It will be seen that it is only when R/R_0 becomes very small indeed that $\mathrm{d}R/\mathrm{d}t$ becomes large. The limiting velocity cannot be infinite as $R \to 0$ and it may well depend on the viscosity of the film. The fact that the jet, once formed, seems to remain coherent and attached to the top of the film as the latter descends, suggests that viscosity is the agent which can exert a downward force on the material of the jets and perhaps the jet is prevented from disintegrating because the electric field normal to its surface has a stabilizing influence on the varicose instability which would otherwise make it break up into small drops. On the other hand, the instability for lateral motions caused by the mutual repulsion of the charges on neighbouring parts of the jet may be prevented from developing by the stabilizing effects of the viscous tension in the jet which must exist while it is attached to the descending film. It is noticeable that the jet has always disappeared before the film begins to rise again.

It seems likely that the prime cause of the oscillation is the reduction of the field strength near the vertex which must occur when the jet, which is an equipotential surface, reduces the field strength in its neighbourhood.

The violence of the oscillation of a soap film made it desirable to experiment with interfaces where inertia and viscosity might damp the unsteady motions. In fact Zeleny's photograph, fig. 1, pl. 1, shows that viscosity does have that effect on jet formation.

Water/oil interface

When the interface is between a non-conducting fluid and a conducting one, gravity and viscosity as well as inertia are likely to affect the form of the interface. Attempts were made to use two fluids of the same density but practical difficulties made them unsuccessful. Experiments with transformer oil of density 0·882 and water were successful. The level of the interface was adjusted by the plunger K (fig. 5). In order to give the fluid as great a chance as possible of forming a conical interface this level was adjusted in some experiments so that the volume of fluid in the space between the initially nearly spherical interface and the plane through the edge G (fig. 5) was equal to that of a cone of semivertical angle 49·3° standing on the same edge.

Fig. 8, pl. 3, shows the interface (a) just before reaching the vertex, (b) while the jet was in action, (c) subsiding after the jet had stopped. These three photographs were spaced at intervals of 1/64 sec. and the exposure time was 1·6 msec. The oscillation was much less than for the soap film. Fig. 9, pl. 2, shows the summit of another oscillation just before the jet formed. The accuracy with which the semi-vertical angle 49·3° is attained is again remarkable. It will be noticed in fig. 8(c) that the jet does not follow the descending interface.

When the amount of conducting fluid above the truncated metal cone is greater than that necessary for the formation of the 49·3° cone the field necessary for conical formation cannot occur and the interface becomes ogival. Fig. 10, pl. 3, contains three photographs at successive intervals of 1/64 sec., (a) just before jet formation, (b) while the jet is in action, (c) after the jet had stopped.

It will be seen in fig. 10(a) that the semi-vertical angle is rather less than 49·3° though still greater than 45°, but at the moment when the jet forms and the top of the conducting fluid is in rapid motion (fig. 10(b)) the angle is considerably less.

When the volume of fluid above the truncated top of the metal cone is less than that of the 49·3° cone the motion tends to become asymmetrical. Fig. 11, pl. 2, shows the jet which has formed under this condition and it will be seen that the interface below the jet is conical with semi-vertical angle rather greater than 45°.

Critical potential for oil/water interface

The specific inductive capacity of the transformer oil used was 2·2, and the interfacial surface tension was found to be 37 using the hanging drop method. The calculated potential for the formation of a point in the apparatus shown in fig. 5 was therefore

$$V = 1·432 \times 10^3 (37)^{\frac{1}{2}} (1·25)^{\frac{1}{2}} (2·2)^{-\frac{1}{4}} = 6·5 \times 10^3 \, \text{V}.$$

The measured potential was $7·2 \times 10^3$ V. for the case shown in figs. 8 and 9, and $7·6 \times 10^3$ V. for that in fig. 11. These are rather greater than the calculated values but imperfect insulation of molecular surface effects introduced errors into the measurements.

In conclusion I express my thanks to Professor C. W. Oatley for lending apparatus and to Mr A. D. McEwan for help in the design of the apparatus and in carrying out the experiments.

40

THE STABILITY OF A HORIZONTAL FLUID INTERFACE IN A VERTICAL ELECTRIC FIELD*

REPRINTED FROM

Journal of Fluid Mechanics, vol. XXII (1965), pp. 1–15

The stability of the horizontal interface between conducting and non-conducting fluids under the influence of an initially uniform vertical electric field is discussed. To produce such a field when the conducting fluid is the heavier it is imagined that a large horizontal electrode is immersed in the non-conducting fluid. As the field increases the part of the interface below the electrode rises till at a voltage V, which depends on the interfacial tension, the height of the electrode above the interface and the density difference, the interface becomes unstable for vertical displacements Z which satisfy the equation
$$\left(\frac{\partial^2}{\partial x^2} + \frac{\partial^2}{\partial y^2} + k^2\right) Z = 0.$$

The value of k consistent with the lowest value of V is found. When the electrode is situated above the interface at less than a certain distance the lowest value of V is attained when $k = 0$ so that the horizontal extent of an unstable crest is likely to be great. As the electrode height increases above this critical value k increases and the unstable crests become more closely spaced till an upper limiting value of k is obtained.

Experiments made with several pairs of fluids verify these theoretical conclusions. In some cases sparking occurs before the potential V is reached, but in others, air at atmospheric pressure over water, for instance, the instability occurs first and the jet of water which results permits the passage of a spark which may inhibit further development of the instability. The physical condition which determines the sparking voltage to a fluid may therefore be very different from that which is operative between solid electrodes. This consideration might be relevant to the performance of power-line insulators in wet weather.

1. Introduction

In discussing the disintegration of water drops in a strong electric field it was shown† that the internal pressure, though constant inside the drop, has an important effect on its stability. An analogous effect must occur at the horizontal interface between a conducting fluid such as water and a lighter non-conducting fluid when a strong vertical electric field is applied. A local vertical displacement of the interface will concentrate the lines of force and so increase the vertical force on the interface. If this increase is great enough to counterbalance the pressure drop due to gravity and the surface tension, the interface will be unstable. The physical conditions under which this instability may occur can be realised by supporting a

---

* With A. D. McEwan.

† G. I. Taylor, *Proc. Roy. Soc.* A, CCLXXX (1964), 383 (paper 39 above).

horizontal conducting plate at height H above the interface and raising it to potential V_0. The value of V_0 at which the interface is in neutral equilibrium can be calculated by straightforward methods.

Suppose that the potential of the conducting plate is gradually raised from zero. The water surface under it will rise and, if the vessel containing it is of finite extent, the parts of the surface which are outside the influence of the plate will fall. The ratio of the rise under the plate to the fall at distant points will depend on the ratio of the area of the plate to that of the container in which the experiment is performed. If h is the height of the plate above the raised interface when neutral conditions are established and h_0 is its height above the level of the interface at points outside the range of the electric field the decrease in pressure under the raised interface due to gravity is $(\rho_1 - \rho_2)(h_0 - h)g$, where ρ_2 and ρ_1 are the densities of the upper and lower fluids respectively. In the following analysis it is assumed that the upper electrode radius is much greater than its height above the conducting fluid.

If the interface is displaced vertically through a small height

$$Z = Bf(x, y) \tag{1}$$

from the level to which it had been raised by the field under the upper electrode, the equation for neutral equilibrium is satisfied when the surface stress due to the electric field is

$$g(\rho_1 - \rho_2)\{h_0 - h + Bf(x, y)\} + T\left(\frac{1}{r_1} + \frac{1}{r_2}\right). \tag{2}$$

Here r_1 and r_2 are the principal radii of curvature of the displaced interface and T is the interfacial tension. Since B will be assumed small compared with the lateral extent of disturbances to be considered

$$\frac{1}{r_1} + \frac{1}{r_2} = -B\left(\frac{\partial^2}{\partial x^2} + \frac{\partial^2}{\partial y^2}\right)f(x, y). \tag{3}$$

If the potential of the interface is zero, that of the upper electrode is V_0 and the vertical co-ordinate z is measured from the level to which the lower fluid is raised when the neutral equilibrium is attained, a possible potential field which satisfies the necessary electrical boundary condition $V = V_0$ at $z = h$ is

$$V = V_0\left\{\frac{z}{h} + \frac{\sinh k(z-h)}{\sinh kh} Bf(x, y)\right\}, \tag{4}$$

and V is a potential function provided

$$\frac{\partial^2 f}{\partial x^2} + \frac{\partial^2 f}{\partial y^2} + k^2 f = 0. \tag{5}$$

The component of stress normal to the interface is

$$\frac{K}{8\pi}\left\{\left(\frac{\partial V}{\partial x}\right)^2 + \left(\frac{\partial V}{\partial y}\right)^2 + \left(\frac{\partial V}{\partial z}\right)^2\right\},$$

where K is the dielectric constant of the non-conducting fluid. Since the slope of the interface is assumed to be small, only $K(\partial V/\partial z)^2/8\pi$ need be taken into account in the linearized equation of neutral equilibrium, which is

$$\frac{K}{8\pi}\frac{V_0^2}{h^2}\{1+2kB\,(\coth kh)f(x,y)\} = g(\rho_1-\rho_2)\{h_0-h+Bf(x,y)\}+BTk^2f(x,y),\quad (6)$$

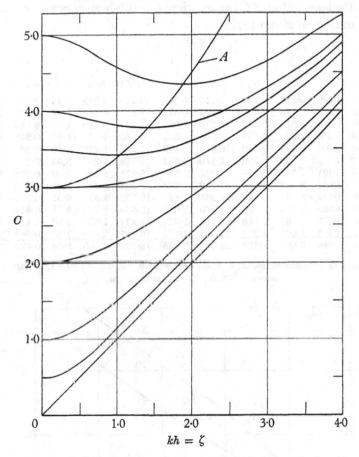

Fig. 1. Variation of C with ζ for constant h'^2. $h'^2 = C$ when $\zeta = 0$.
Curve A is $C = 4\zeta\sinh^2\zeta/(\sinh 2\zeta - 2\zeta)$.

and (6) is satisfied provided

$$\frac{KV_0^2}{8\pi h^2} = g(\rho_1-\rho_2)\,(h_0-h), \tag{7}$$

and

$$\frac{KV_0^2}{4\pi h^2} = \tanh kh\left\{Tk+g\frac{(\rho_1-\rho_2)}{k}\right\}. \tag{8}$$

These can be expressed in non-dimensional form. Writing

$$h' = hg^{\frac{1}{2}}(\rho_1-\rho_2)^{\frac{1}{2}}T^{-\frac{1}{2}},\quad k' = kT^{\frac{1}{2}}(\rho_1-\rho_2)^{-\frac{1}{2}}g^{-\frac{1}{2}},\atop C = KV_0^2/4\pi Th,\quad \zeta = kh = k'h', \tag{9}$$

(7) becomes
$$h_0/h = 1 + C/2h'^2,$$
(10)

and (8) becomes
$$C = \frac{\tanh \zeta}{\zeta}(\zeta^2 + h'^2).$$
(11)

In particular, when $\zeta = 0$, $C = h'^2$. Fig. 1 shows how C varies with ζ when h' is constant. It will be noticed that when $h' > \sqrt{3}$ each curve has a minimum but when $h' < \sqrt{3}$ the least value of C occurs when $\zeta = 0$. These curves may be regarded as the curves of neutral stability.

Table 1

1	2	3	4	5	6	7	8	9	10	11	12	13	14
					h_0	V_0	h_0	V_0	h_0	V_0	h	h_0	V_0
Kh	h'	C	h/h_0	$(h'C)^{\frac{1}{2}}$	(cm.)	(kV.)	(cm.)	(kV.)	(cm.)	(kV.)	(cm.)	(cm.)	(kV.)
0	0·4	0·160	1·500	0·213	0·164	1·125	0·340	0·780	0·283	0·36	0·062	0·93	1·16
0	0·8	0·640	1·500	0·715	0·3274	3·393	0·681	2·354	0·566	1·07	0·124	1·86	3·50
0	1·2	1·440	1·500	1·314	0·491	6·236	1·020	4·327	0·850	1·99	0·186	0·279	6·40
0	1·6	2·560	1·500	2·024	0·655	9·606	1·360	6·665	1·13	3·07	0·248	0·372	9·90
0	1·732	3·000	1·500	2·279	0·709	10·816	1·473	7·505	1·226	3·46	0·268	0·403	11·16
0·6	1·775	3·143	1·499	2·362	0·726	11·210	1·501	7·778	1·257	3·58	0·276	0·414	11·58
1·0	1·860	3·395	1·491	2·513	0·756	11·927	1·571	8·275	1·309	3·82	0·288	0·430	12·30
1·5	2·043	3·875	1·464	2·813	0·816	13·350	1·695	9·263	1·412	4·27	0·316	0·461	13·75
2·0	2·319	4·518	1·420	3·237	0·898	15·363	1·867	10·659	1·551	4·91	0·359	0·510	15·8
2·5	2·675	5·289	1·369	3·761	0·999	17·850	2·077	12·385	1·728	5·71	0·414	0·566	18·4
3·0	3·096	6·154	1·321	4·365	1·116	20·716	2·326	14·374	1·930	6·62	0·480	0·634	21·4
3·5	3·544	7·078	1·282	5·008	1·239	23·768	2·575	16·491	2·144	7·60	0·548	0·700	24·5
4·0	4·021	8·040	1·248	5·686	1·369	26·896	2·845	18·724	2·369	8·63	0·622	0·774	27·8

Columns 1–5, non-dimensional; 6–7, air–water; 8–9, oil–water; 10–11, oil–alcohol water mixture; 12–14, mercury–oil.

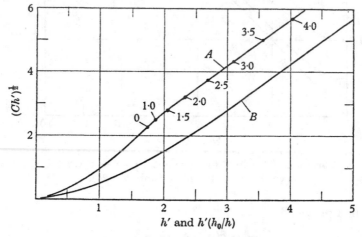

Fig. 2. Minimum values of $(Ch')^{\frac{1}{2}}$ and corresponding values of h' (curve A), and $h'(h_0/h)$ (curve B). Numbers are values of kh.

For a given value of h', the value of V_0 at which instability first occurs when V_0 increases is that for which C is a minimum as ζ varies. This condition gives

$$C = \frac{4\zeta \sinh^2 \zeta}{\sinh 2\zeta - 2\zeta} \quad \text{and} \quad h'^2 = \zeta^2 \frac{\sinh 2\zeta + 2\zeta}{\sinh 2\zeta - 2\zeta}.$$
(12)

The data which can conveniently be measured experimentally are V_0 and h_0, and from (9)

$$h = h' \left[\frac{T}{g(\rho_1 - \rho_2)} \right]^{\frac{1}{2}}, \tag{13}$$

$$V_0 = C^{\frac{1}{2}} h'^{\frac{1}{2}} \left(\frac{4\pi}{K} \right)^{\frac{1}{2}} \left(\frac{T}{g(\rho_1 - \rho_2)} \right)^{\frac{1}{4}} T^{\frac{1}{2}}, \tag{14}$$

$$\left. \begin{array}{l} h_0 = h \left(1 + \dfrac{C}{2h'^2} \right) \quad \text{when} \quad h' > \sqrt{3}, \\[2mm] h_0 = \tfrac{3}{2} h \quad \text{when} \quad h' < \sqrt{3}. \end{array} \right\} \tag{15}$$

For experiments in which V_0 is raised till instability occurs it is therefore convenient to tabulate the value of $C^{\frac{1}{2}} h'^{\frac{1}{2}}$ corresponding with the minimum value of C when h' is fixed. When $h' > \sqrt{3}$ this can be done by calculating C and h' for arbitrarily chosen values of ζ using (12). When $h' < \sqrt{3}$, $C^{\frac{1}{2}} h'^{\frac{1}{2}} = h'^{\frac{1}{2}}$. Some values are given in column 5 of table 1 and are shown in fig. 2.

2. PHYSICAL CHARACTER OF DISTURBANCES

The small disturbances which occur when V_0 reaches the critical value are characterized only by the value of k corresponding with the minimum V_0. Any disturbance which satisfies (5) is then possible. To obtain a unique solution which would define the disturbances more exactly it would be necessary to find boundary conditions to be satisfied at the edge of the disturbed area. It has been assumed, however, that h_0 is small compared with the linear dimensions of the charged plate and under these conditions the distant boundary must have little influence in determining the form of the disturbances. The only limitation to the geometrical character of the disturbed surface is contained in (5). The practical effect of this limitation seems to be to define roughly the distance of a crest or summit from its nearest neighbours. Thus, for instance, in the regularly spaced disturbances represented by

$$f = \sin kx \quad \text{or} \quad f = \sin \frac{kx}{\sqrt{2}} \sin \frac{ky}{\sqrt{2}},$$

neighbouring summits are at distance $2\pi/k$ apart, while for Christopherson's hexagonal distribution,*

$$f = 2 \cos \tfrac{1}{2}\sqrt{3}\, kx \cos \tfrac{1}{2} ky + \cos ky,$$

this distance is $4\pi/\sqrt{3}\,k$. The linear theory does not seem capable of distinguishing between them. Analogous conditions occur in the cases of instability due to heating a fluid from below and of the instability of a horizontal free surface subjected to vertical oscillations.

* *Quart. J. Math.* XI (1940), 63.

Limiting conditions

When $h' < \sqrt{3}, kh = 0, C = h'^2$, or

$$V_0 = \left[\frac{4\pi g(\rho_1 - \rho_2)}{K}\right]^{\frac{1}{2}} h^{\frac{3}{2}}, \tag{16}$$

and (15) becomes

$$h_0 = \tfrac{3}{2}h. \tag{17}$$

The significance of this expression can be appreciated by assuming from the outset that the interface remains plane and is disturbed only by a vertical displacement. The equation of equilibrium is then

$$KV_0^2/8\pi h^2 = g(\rho_1 - \rho_2)(h_0 - h). \tag{18}$$

The maximum value of V_0 for which (18) can be satisfied corresponds with the maximum value of $h^2(h_0 - h)$ as h varies while h_0 remains constant. This maximum is $\tfrac{4}{27}h_0^3$ or $\tfrac{1}{2}h^3$ when $h = \tfrac{2}{3}h_0$. When this value is inserted in (18), (16) is recovered.

When kh is large, $\tanh kh \sim 1\cdot0$ so that (8) becomes

$$\frac{V_0}{h} = \left(\frac{4\pi}{K}\right)^{\frac{1}{2}} \left\{Tk + \frac{g(\rho_1 - \rho_2)}{k}\right\}^{\frac{1}{2}}. \tag{19}$$

The minimum value occurs when $k^2 = g(\rho_1 - \rho_2)/T$ and is

$$\frac{V_0}{h} = \left(\frac{8\pi}{K}\right)^{\frac{1}{2}} \{gT(\rho_1 - \rho_2)\}^{\frac{1}{4}}. \tag{20}$$

If the disturbance is in the form of plane waves, their length is

$$\lambda = 2\pi(T/[g(\rho_1 - \rho_2)])^{\frac{1}{2}}. \tag{21}$$

It is of interest to note that when $\rho_2 = 0$ this is the length of the surface wave which travels with least velocity on deep water.

3. Comparison with observation

Experiments were made in which a horizontal charged metal disc $12\cdot7$ cm. in diameter was supported at various heights H above the conducting fluid which was water, dilute alcohol or mercury. The conducting fluid was contained in a tank with transparent walls so that its surface could be photographed. Fig. 3 is a sketch of the apparatus. The height H of the electrode E_1 above the water surface was measured before charging it to potential V_0. When the potential gradient was established the level of the water under the electrode rose and that at distant points fell by an amount δh which was measured.

The object of the experiment was to observe the form of the instability, and to obtain data for a comparison with the calculated relation between V_0 and h, the distance between the electrode E_1 and the liquid interface at the onset of instability.

PLATE 1

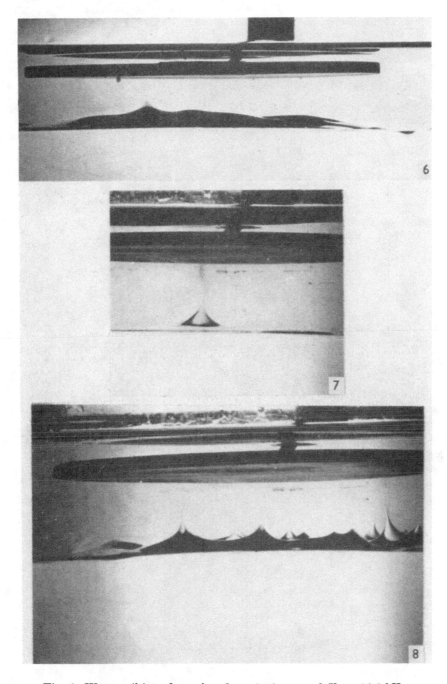

Fig. 6. Water–oil interface when $h_0 = 1\cdot79$ cm. and $V_0 = 10\cdot1$ kV.
Fig. 7. Water–oil interface when $h_0 = 2\cdot62$ cm. and $V_0 = 15$ kV.
Fig. 8. Water–oil interface when $h_0 = 2\cdot62$ cm. and $V_0 = 17\cdot7$ kV.

It was found that when the liquid interface became unstable, agitation of the surface made a direct measurement of h impossible. Instead the fall, δh, in level of the interface at a point remote from E_1, was measured, and by measuring the height H of E_1 above the interface in the absence of an applied voltage, the value of h_0 could be estimated closely, as $H + \delta h$.

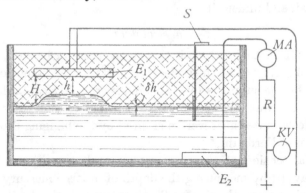

Fig. 3. Apparatus.

The source of high potential, kindly lent to us by Professor C. W. Oatley, was capable of generating up to 25 kV. d.c. Readings on a kilovoltmeter KV and micro-ammeter MA in series with a resistance of 2×10^9 ohm (fig. 3) enabled the mean voltage between E_1 and E_2 to be measured. The current through the micro-ammeter became appreciable as the interface became unstable owing to the formation of conduction paths in the non-conductor by the conducting liquid which were of appreciable conductivity. The period of time for which E_1 and E_2 were actually short circuited in this way was estimated using a cathode-ray oscilloscope during experiments with the water–air interface, and was found to form a minute propor-tion of the time between successive short circuits. This is in agreement with the observation* that in a 50 ft. length of a cinema record of this instability only one frame showed a jet in process of formation.

Though the time of existence of short circuiting paths was very small it was found that the mean conductivity produced in the non-conducting fluid as a result of the instability raised the potential recorded on KV (fig. 3), but when V was corrected using the appropriate formula,

$$V_0 = \text{reading on } KV - 2 \times (\text{reading on } MA), \tag{22}$$

the resulting values of V_0 were remarkably constant, and agreed with the value observed when the instability first made its appearance.

When the experiments were made with air and water it was noticed that directly the instability started sparks occurred and it was thought that possibly the dis-turbed state of the water surface was due to the occurrence of cascading ionization by collision rather than to instability. Experiments were therefore performed with

* Paper 39 above.

oil as the non-conducting fluid. This had the advantage that electrical breakdown of the oil was unlikely and also that the dielectric constant of oil is more than twice as great as that of air so that the voltage necessary for instability is correspondingly reduced. Later, experiments were made in which the lower fluid was a mixture of equal volumes of alcohol and water. Experiments were also made with mercury under air and under oil.

Physical data

The oil used was Shell Diala which is known as 'transformer' oil. Its density was found to be 0.882 g. cm.$^{-3}$ and the dielectric constant was taken to be 2.2 which is the value given by the suppliers (and also in several physical tables). The density of the dilute alcohol was 0.924 g. cm.$^{-3}$. The surface tension of the oil–water and the oil–dilute alcohol interfaces were measured by the hanging drop method using the computations of Fordham.*

The interfacial tension of the mercury–air and mercury–oil interfaces were found in two ways (i) by measuring the depth of a large mercury drop on a horizontal plane (Quincke's method); and (ii) by measuring the depth of the interface in a 1 mm. diameter glass tube below a large interface into which the tube was thrust. The second method involved inserting a fine iron wire into the upper part of the capillary tube. Two other iron wires were mounted on a Perspex carrier which could slide on the tube. These two wires were connected together and when either of them touched the mercury they completed a circuit which showed the instant of contact. The wire inside the capillary tube was also connected with another voltmeter which showed when the mercury inside the tube had risen to the lower end of the inner wire. To make a measurement the position of the Perspex carrier on the capillary tube was adjusted till on inserting the tube into the mercury through the air or oil which lay above it the contacts inside and outside the tube were made simultaneously. The vertical distance between the iron wires inside and outside the tube was then the hydrostatic head which was balanced against the interfacial tension.

Table 2. *Instability data for different interfaces. Surface tensions (column 3) are as given (i) by Quincke's method and (ii) by glass capillary tube method. In column (6), the limiting gradient is* $0.3\{gT(\rho_1 - \rho_2)\}^{\frac{1}{4}}(8\pi/K)^{\frac{1}{2}}$

1	2	3	4	5	6
	$\rho_1 - \rho_2$ (g. cm.$^{-3}$)	T (dyn. cm.$^{-1}$)	h/h' (cm.)	$V_0/C^{\frac{1}{2}}h'^{\frac{1}{2}}$ (kV.)	limiting gradient (kV. cm.$^{-1}$)
Water–air	1·0	73	0·273	4·75	24·6
Water–oil	0·118	37·2	0·567	3·29	8·2
50/50 water alcohol–oil	0·042	9·3	0·469	1·40	4·5
Mercury–air	13·6	(i) 356 (ii) 350	0·162	5·4	70·2
Mercury–oil	12·7	(i) 295 (ii) 303	0·155	4·89	44·6

* *Proc. Roy. Soc.* A, CXCIV (1948), 11.

To make accurate measurements by either of the methods (i) or (ii) it would be necessary to know the angle of contact, but with mercury and Perspex or glass this angle was near enough to 180° to make the error due to assuming it to be 180° tolerable in view of other uncertainties in the experiment.

The data used in comparing the observed with the calculated conditions for instability are given in columns 2 and 3 of table 2 and the factors which enable the corresponding values of h and V_0 to be found are given in columns 4 and 5. Column 6 gives the limiting electric intensity in kilovolts per cm. at which a horizontal interface becomes unstable as the gap between plate and interface increases. It should be noticed that potentials in the theoretical formula are in electrostatic units and to convert them to kilovolts these must be multiplied by a factor 0·3.

Experimental results

As the critical voltage for instability was approached the interface became rather agitated and occasional isolated points would rise from the lower (conducting) fluid. These sometimes appeared to be connected with bubbles of the water in the oil or detached drops of water on the water surface when the non-conducting fluid was air. Sometimes they were connected with visible specks of dust.

At this stage a small increase in V gave rise to violent agitation of the interface. Pointed conical disturbances rose and fell in many places over the interface and jets of water penetrated into the upper fluid. Although it was not always easy to associate exactly a definite visible amount of disturbance with the theoretical neutral stability the scatter of readings of V_0 was not large when the correction formula (22) was used, even when there was a large scatter in the readings of the voltmeter kV. of fig. 3. When the upper fluid was oil it was found that after the apparatus had operated for a short time the surface of the lower fluid appeared to become slightly clouded. A scraper S (fig. 3) similar to that used in Langmuir's experiments on surface films was therefore arranged so that it could be pulled the length of the trough. It was found that a surface film was produced whenever a number of jet-like disturbances had passed through the oil. The conductivity of oil rose and a current of several micro-amperes might occur before neutral equilibrium was attained. The normal non-conducting character of the oil was restored by a passage of the scraper along the trough, and before each recorded observation this was done.

Air–water interfaces

The voltage at which the surface first became disturbed and sparking began was very clearly defined and repeatable, but immediately this happened the voltage between the electrodes dropped owing apparently to the formation of an unbroken jet of water which immediately lowered the resistance between the electrodes. When this occurred the potential drop between the electrodes could not usually be raised to the value at which the jet first appeared even though the voltage of the supply was much greater. In fact the drop in potential at the resistance $R = 2 \times 10^9$ ohm (shown in fig. 3) increased faster than the rise in voltage applied by the generator. The measurements of V_0 and h_0 are given in table 3. The calculated values using the data of table 2 are given in columns 6 and 7 of table 1, and are plotted in the full curve of fig. 4. It will be seen that the agreement is extremely good. The conclusion which must be drawn from this is that when sparking occurs in these experiments at a water surface the physical cause of the phenomenon lies in the behaviour of the water surface rather than in ionization by collision or other processes of breakdown in the air. To test this conclusion, experiments were made in which the water was replaced by a metal (brass) and the sparking potential measured. Since the general level of the water under the upper electrode had been raised by the electric field before instability and sparking occurred the lower electrode was made in the form of a disc of the same size as the upper one. In the first of these experiments neither the upper nor the lower electrode had well-rounded edges and sparks always occurred between the two edges. When the upper electrode which was 0·37 cm thick was rounded so that its edge section was a semi-circle and the lower electrode was rounded on the upper side at the edge to a larger radius, the sparks appeared at random all over the flat central parts of the lower electrode. The results of these measurements are also shown in fig. 4. Apart from a small scatter they show a linear relationship between sparking voltage and h_B, the distance between the brass plates corresponding to 34·0 kV./cm.

Table 3. *Observed voltages for instability at a water–air interface*

h_0 (cm.)	0·226	0·336	0·440	0·556	0·660	0·762	0·862	0·962	1·062	1·166	1·270
V_0 (kV.)	1·6	3·3	5·3	7·2	9·8	12·1	14·6	17·3	19·6	20·5	25·1

The sparking voltages to water and to brass could be compared on the basis that either h_0 or h is comparable with h_B. Since h_0 is the quantity which could be measured fig. 4 is constructed to reveal the comparison between h_0 and h_B. It will be seen that h_B is always less than h_0 for a given sparking voltage. It was found that if the values of h corresponding to any given value of h_0 were calculated from equation (15) the water–air–brass sparking curve is still below the brass–air–brass curve. The difference between h and h_0 is only significant at small values of h' and for sparks from water through air the difference between h_0 and h is never greater than 0·273 cm. At distances greater than 1 cm. the potential gradient for instability at the water surface is

$$V_0/h_0 = 300(8\pi)^{\frac{1}{2}}\{(981)(73)\}^{\frac{1}{4}} = 24\cdot 6 \text{ kV./cm.},\qquad(23)$$

which is less than 34·0 kV./cm. This explains why in our experiments with air over water instability preceded sparking. However, the potential gradient for very large spark gaps in air at atmospheric pressure is less than 34 kV./cm. so that it cannot be stated as a general proposition that instability could always occur before dielectric breakdown.

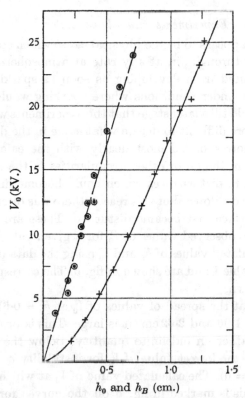

Fig 4. Sparking potentials in air between horizontal electrodes. Experimental points: + lower electrode water, ⊙ lower electrode brass. Full curve calculated instability for water; calculated limiting slope 24·6 kV./cm. Broken line 34·0 kV./cm.

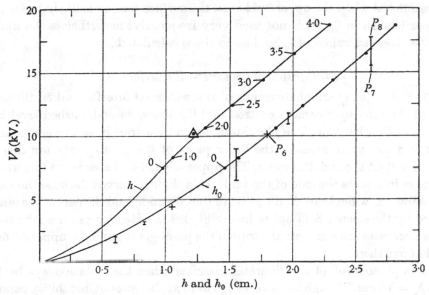

Fig. 5. Voltage necessary for instability of oil–water interface. Full lines h and h_0 are calculated. Vertical lines and crosses represent experimental results. P_6, P_7 and P_8 represent values of h_0 at which photographs fig. 6, 7 and 8 (pl. 1) were taken. △ is mean value of h derived from fig. 6.

Experiments with oil over water

Though the instability appears to be the physical cause which starts sparking from a metal plate to water through air, at any rate at atmospheric pressure, the instability may be inhibited from developing as soon as sparking occurs. It was to study this instability under conditions where sparking would not be expected to modify the electric field till a later stage that the experiments with oil were undertaken. It was found more difficult to decide what stage in the development of the unstable disturbance corresponded most nearly with the calculated instability, because water bubbles in the oil or other irregularities in the oil–water interface were liable to give rise to disturbances which could become finite even when infinitesimal disturbances die down. For this reason there was often a range of values of V_0 at which the interface first became disturbed. These are indicated in fig. 5 where the spread of the observed values of V_0 for a given value of h_0 is shown by vertical lines. The calculated values of h_0 and V_0 using the data of table 2 are given in columns 8 and 9 of table 1 and are shown in fig. 5. The corresponding calculated values of h are also shown.

It will be noticed that the spread of values of V_0 at $h_0 = 0.61$, 0.86, 1.05 cm. is small but at $h_0 = 1.55$, 1.96 and 2.62 cm. it is large. This is probably because the interfacial tension is rather an indefinite quantity. Below the value of h_0 corresponding with $h' = \sqrt{3}$, the lowest value of V_0 for instability corresponds with infinitely long waves ($kh = 0$). The calculated value of V_0 at which $h' = \sqrt{3}$ is 7.5 kV. (column 9, table 1). This is marked in fig. 5 on the curves for h and h_0 by the symbol 0. Other values of kh are marked 1.0, 2.0, 2.5, 3.0, 3.5, 4.0 in fig. 5. Below $V_0 = 7.5$ kV. the instability depends only on $\rho_1 - \rho_2$, but above that value the surface tension begins to be important. With water under air the surface tension has a definite value but when the upper fluid is oil the surface tension is likely to vary if the scraper (shown in fig. 3) is not used very frequently: nevertheless the upper limit of the observed values of V_0 lie close to those calculated.

Photographs of the water–oil interface

Figs. 6, 7, and 8 (pl. 1) are photographs of the water–oil interface taken through the wall of the tank by a camera centred slightly above its undisturbed level but below that of the electrode. The illumination was by a translucent screen behind the tank and the white areas of the lower parts of the photographs are due to unobstructed light through the water. The upper edge of this area is the line where the interface intersects the side of the tank. It is slightly curved because the tank was only a few cm. wider than the diameter of the electrode, and the part of the interface raised by the electric field before instability set in extended as far as the tank wall. The electrodes are seen near the tops of the photographs and the upper surface of the oil above them.

Fig. 6 is a photograph of the oil–water interface when the agitation was beginning and $h_0 = 1.9$ cm. Though the level of the fluid at the onset of instability cannot be measured accurately it will be seen that it has risen about $\frac{1}{3}H$ above the undisturbed level, as it should when $kh = 0$. The value of h obtained by drawing a horizontal line which lies between the crests and troughs on the central part of the field gave h as 1·22 cm. This is shown in fig. 5 as a triangle. The observed value of V_0 corresponding with fig. 6 was 10·1 kV. which is very close to the calculated value.

Fig. 7 shows what sometimes occurred at values of V_0 below that corresponding with violent agitation and the point P_7 in fig. 5 represents this condition. A single cone, possibly caused by some accidental disturbance would move about over the interface. In this case h_0 was 2·62 cm. and V_0 was about 15 kV. On raising the voltage to 17·7 kV., P_8 in fig. 5, the surface became violently agitated and the photograph shown in fig. 8 was taken. It will be noticed that there are a large number of pointed crests when $h_0 = 2\cdot62$ cm. There were never more than one or two when $h_0 = 1\cdot79$ cm. This seems to be in qualitative agreement with the fact that the calculated value of kh for the photograph of fig. 6 is 1·8 while that corresponding with fig. 7 is 3·5.

Oil–dilute alcohol interface

The reduction in density difference between the two fluids from 0·118 to 0·042 made the instability occur at lower potential gradients but the reduction in surface tension did not make it possible to attain higher values of kh than were attained with the oil–water interface. This was due to the limitation imposed by the size of the apparatus. At the highest value of kh used, V_0 was under 9 kV. but the ratio of h_0 to the diameter of the electrode was not small enough to permit the use of the theoretical model which virtually assumes this ratio to be negligible. Fig. 9 shows the experimental results and the theoretical relationship calculated using the data of table 2. Corresponding calculated values of h_0 and V_0 are given in columns 10 and 11, table 1. These are shown in the curve of fig. 9. It will be seen that except at the low values of h_0 the experimental points are not far below the calculated curve. The reason for the large errors below $h_0 = 1$ cm. was not clear but it was noticed that the readings of the ammeter were often large and varying so that the correction given in (22) was rapidly varying and often nearly as large as the reading of the voltmeter. This large current might have been due to a residue left in the oil after the collapse of a jet. Contamination of the oil in the presence of alcohol no doubt contributed to residue formation.

Experiments with mercury as the conducting fluid

No unstable disturbance was produced when a shallow circular trough 8·1 cm. diameter filled with mercury was placed under the upper electrode (12·7 cm. diameter) and the potential raised till sparking through the air occurred. The sparks were always on the mercury meniscus at the rim of the trough, no doubt because of the concentration of electric field there. The tank was then filled with oil and the potential of the upper electrode raised. The mercury became agitated at the rim and jets of mercury came off it. To make a comparison with the theory it was necessary to avoid electric stress concentration at the curved edge of the

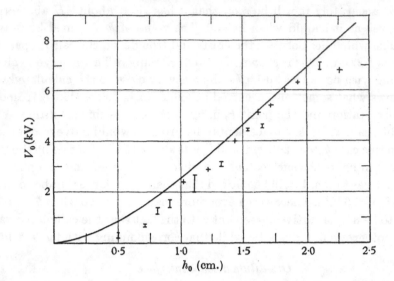

Fig. 9. Interface between transformer oil and a mixture of equal volumes of water and alcohol. Line is theoretical value of h_0 using $\rho_1 - \rho_2 = 0.042$, $T = 9.3$ dyn/cm.

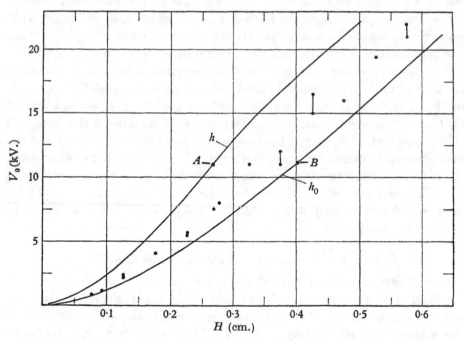

Fig. 10. Experimental values of H, transformer oil–mercury interface. A and B calculated values of h and h_0 for $h' = \sqrt{3}$.

mercury. A steel disc 6·35 cm. diameter and 6·5 mm. thick with rounded edges was attached to the lower side of the electrode E_1 (fig. 3). This diameter being smaller than that of the mercury made it possible for the maximum electric stress at the mercury surface to be confined to the central region of its surface and in fact when unstable points of mercury rose they were equally likely to appear at any point in the central region. After each experiment the mercury surface was scraped so that any film caused by the previous experiment would be removed.

It was found that as H the height of the steel electrode above the original height of the mercury increased definite and repeatable values of V_0 at which mercury jets rose from the surface were found so long as H was less than 0·31 cm. The experimental points are plotted in fig. 10.

The values of V_0 and corresponding values of h and h_0 were calculated from the data of table 2, and are given in columns 12, 13 and 14 of table 1. These are plotted in fig. 10. It will be seen that the experimental values of H lie between them. Since the experimental arrangements made it impossible to measure δh, the drop in level at points outside the influence of the electric field experimental measures of h and h_0 were impossible. It is satisfactory, however, that the experimental points lie between the calculated h and h_0 curves. The value of V_0 at which $h' = \sqrt{3}$ is 11 kV. Repeatable experimental results were obtained when H was less than 0·3 cm. and $V_0 < 11$ kV., but when $H > 0·3$ cm. the measured critical values of V_0 were variable. This is indicated in figure 10 by the spread of points at $H = 0·34$, 0·41 and 0·54 cm.

4. CONCLUSIONS

It has been shown theoretically that horizontal interfaces between conducting and non-conducting fluids become unstable under the action of a sufficiently great electric field. Experiments to verify the theory have revealed that electric failure of the non-conductor can arise in two manners. In the first, the non-conductor suffers the normal form of breakdown (occurring between solid electrodes, for example) at a voltage gradient lower than that which will cause instability of the interface. Such a situation arises in air over mercury. In the second, instability precedes dielectric breakdown, and the conducting jets formed from the unstable disturbances precipitate the failure of the non-conductor.

The theory for the stability of the interface does not involve a detailed knowledge of the form of the unstable disturbance. Nevertheless, in experiments the onset of instability was predicted generally with good accuracy.

41

CONICAL FREE SURFACES AND FLUID INTERFACES

REPRINTED FROM

Proceedings of the 11th International Congress of Applied Mechanics
(*Munich, 1964*), Springer-Verlag (1966), pp. 790–6

INTRODUCTION

The bursting of drops by forces applied over their surface has been the subject of many experimental researches and slight deviations from the spherical form have been investigated by linearized equations. In two cases experiments have shown that in the final stage the drop appears to develop conical ends, but no rational description of this phase of the disruptive process seems to have been published. These two cases are (1) when the drop is torn apart by an electric field and (2) when a drop of fluid of small viscosity exists in a fluid of much greater viscosity which is extending in one direction with a uniform rate of strain. Analytical difficulties have prevented me from presenting more than a partial treatment in either case, but even so they may lead to a better understanding of the mechanics and physics of the breakdown of free surfaces.

1. BURSTING OF DROPS IN AN ELECTRIC FIELD*

2. CONICAL INTERFACES BETWEEN VISCOUS FLUIDS

The disintegration of drops of a viscous fluid owing to stresses imposed by a surrounding fluid has been studied experimentally. Small deviations from the spherical form have been analysed and the results compared with observation.† The way in which the drop disintegrates depends on the ratio β of the viscosity of the drop to that of the surrounding fluid. For all values of β the drop becomes unstable in a flow which is extending steadily in one direction when a number $F_1 = 2cK_0r_0/T$ exceeds a critical value depending on β. Here K_0 is the viscosity of the outer fluid and c, which has dimensions t^{-1}, represents the rate of strain in the outer fluid. When β is large the instability results in an indefinite lengthening of the drop, the ends remaining rounded, but when β is small the drop develops pointed ends which remain pointed and fixed in space while F_1 lies within a certain range of values. At a certain critical value of F_1 the drop begins to extend and ultimately becomes a

---

* *Editor's note.* There followed here a summary of work which is described fully in paper 39 above on 'Disintegration of water drops in an electric field'.

† G. I. Taylor, *Proc. Roy. Soc.* A, CXLVI (1934), 501 (paper 11 above).

thin thread, which in time breaks up into very small drops. It is not certain whether, during the apparently stable pointed state, a very thin thread is emitted from the end, but photographs (see fig. 4, pl. 1, of paper 11 above) do not reveal one.

An approximate method for discussing the mechanics of the pointed end assuming it to be a narrow cone of semi-vertical angle α, has been developed on lines analogous to those which have proved successful in considering the aerodynamics of spindle-shaped bodies. The outer very viscous fluid will be assumed to move bodily with velocity U and perpendicular to the axis with a small velocity v which is equal to $U\alpha$ at the surface of the cone. In this approximation the radial velocity at distance r from the axis is $U = \alpha U r_1/r$ where r_1 is the local radius of the narrow cone. If x is the distance from the vertex, $U = \alpha^2 x U/r$. The radial stress at the conical surface is $\bar{rr} = -2K_0 U\alpha/r_1 = -2K_0 U/x$.

The approximation to be used for the flow of the much less viscous fluid inside the conical interface is that of Poiseuille, the flow being determined by the condition that the velocity at the interface when β is very small is equal to U. When the flow is steady the total amount of fluid flowing over any section is zero. Under these circumstances the pressure, p, inside the cone is given by

$$\frac{dp}{dx} = \frac{8K_1 U}{r_1^2} = \frac{8K_1 U}{\alpha^2 x^2}, \tag{1}$$

where $K_1 = \beta K_0$ is the viscosity in the inner fluid. Integrating (1)

$$p = -\frac{8K_1 U}{\alpha^2 x}. \tag{2}$$

The approximate equation of continuity of stress at the interface is

$$\frac{T}{\alpha x} = -\frac{2K_0 U}{x} - \frac{8\beta K_0 U}{\alpha^2 x} \tag{3}$$

and since the factor $1/x$ occurs in all the terms, equilibrium at all points of the cone is possible. Writing

$$\xi = -\frac{T}{4K_0 U}\beta^{-\frac{1}{2}}, \quad \eta = \alpha\beta^{-\frac{1}{2}} \tag{4}$$

the solution of (3) is

$$\eta = \xi \pm (\xi^2 - 4)^{\frac{1}{2}}. \tag{5}$$

Fig. 1 shows η as a function of $1/\xi$ instead of ξ so that the effect of increasing $-U$ can be seen directly. Evidently U must be negative, i.e. the flow outside the drop must be directed towards the point. For any given value of ξ there are two values of η or

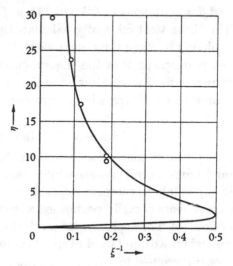

Fig. 1. Theoretical relation between semi-vertical angle of conical end and velocity of outer fluid. o Observed angles.

none, but it can be shown that $\eta = \xi - (\xi^2 - 4)^{\frac{1}{2}}$ represents an unstable equilibrium. $\eta = \xi + (\xi^2 - 4)^{\frac{1}{2}}$ represents physically possible stable conditions. As $-U$ increases η decreases till when $-(4K_0 U/T)\beta^{\frac{1}{2}} = 0 \cdot 5$ the minimum possible value of $\alpha = 2\beta^{\frac{1}{2}}$ is reached. If $-U$ is increased beyond this value the conical interface cannot be maintained and the cone will continue to extend indefinitely.

Comparison with experiment

The only experiments available for comparison seem to be my own experiments made 30 years ago, and since I did not then realise the possibility of making the kind of approximation here described I did not make any attempt to obtain the data necessary for a comparison over a range of the significant variables. The experiments were made in a glass-walled box filled with the more viscous fluid. The flow was produced by rotating four rollers situated at the corners of a square. These gave a flow which was shown to be representable approximately in the portion of the field containing the drop by
$$u = cx, \quad v = -cy. \tag{6}$$

The drops were photographed and the necessary measurements for calculating $F_1 = 2cK_0 r_0/T$ in each case were made. These data are given in paper 11 above. For the comparison it was necessary to take $-U$ in the formulae for the cone as $\frac{1}{2}cL$ where L is the length to which the drop had been extended, ξ^{-1} is then $\beta^{\frac{1}{2}}(L/r_0) F_1$. Fig. 2* shows a drop which was being maintained in the centre of the field as a steady motion by hand control of small differences between the velocity of rotation of the right- and left-hand pairs of rollers. These rollers can be seen out of focus in the four corners of the photograph. The viscosities were $K_0 = 110$ and $K_1 = 0 \cdot 034$ poise, so that $\beta = 3 \cdot 1 \times 10^{-4}$. In this case F_1 was $1 \cdot 40$, $L = 1 \cdot 90$ cm., $r_0 = 0 \cdot 25$ cm. The measured values of 2α at the two ends or the drop were $18°$ and $20°$, and $\beta^{-\frac{1}{2}} = 57$ so that $\xi^{-1} = 0 \cdot 187$ or $\xi = 5 \cdot 35$, and $\eta = 10$ and 9 at the two ends. The theoretical value of η calculated from (5) is $5 \cdot 35 + (5 \cdot 35^2 - 4)^{\frac{1}{2}} = 10 \cdot 3$ which is probably in closer agreement with the value derived from direct measurement on the photograph than the theory warrants. The points representing the condition shown in fig. 2 are the lowest of the experimental points shown in fig. 1. The others represent conditions which were also reproduced in pl. 1 of paper 11 above.

Extension of analysis to cover the whole field of flow

The physical conception which led to equation (3) was that a conical interface might exist at rest in space while a viscous fluid flows past it with velocity U. Though it is possible to conceive experiments in which this situation could be realized the experiments actually performed were with drops elongated in the extending field of flow (6). The approximate method which led to (3) can be extended to cover the case where a drop of fluid of small viscosity is immersed in the field of axisymmetric flow represented by
$$u = cx, \quad v = w = -\tfrac{1}{2}cr. \tag{7}$$

* *Editor's note.* Fig. 2 is not reproduced here, as it is identical with fig. 4 on pl. 1 of paper 11 above.

The space required to give this analysis is greater than that allotted for communication to the Congress, but the result may be set down.

The shape attained by the drop is

$$\frac{r}{a} = 1 - \gamma \frac{x^2}{a^2}, \tag{8}$$

where

$$\frac{T}{K_0 ca} = 4 + \frac{4\beta}{\gamma}$$

and a is the radius of the drop at its mid-point. a is related to the original radius r_0 by the equation

$$\frac{L}{a} = \frac{5}{4}\left(\frac{r_0}{a}\right)^3 = \gamma^{-\frac{1}{2}}$$

where $2L$ is the length of the drop and the equation analogous to (3) which relates

$$B' = \frac{T}{K_0 cr_0} \quad \text{with} \quad \gamma' = \frac{\gamma}{\beta}$$

is

$$\tfrac{1}{4}(\tfrac{4}{5})^{\frac{1}{3}}\beta^{-\frac{5}{6}}B' = (\gamma')^{\frac{1}{6}} + (\gamma')^{-\frac{5}{6}}. \tag{9}$$

The right-hand side of (9) has a minimum value $1 \cdot 569$. Thus the drop cannot exist in a stationary condition when B' is less than a certain value, i.e. when $K_0 cr_0/T$ exceeds a certain value. Thus there is a limit to the value of c at which a spindle shaped drop of form (8) can exist and in fact at this limit the conical point has not yet attained the fineness represented by $\alpha = 2\beta^{\frac{1}{2}}$ but is more than twice as large for the case when $\beta = 3 \times 10^{-4}$.

Finally I must point out that only an approximate analysis has been presented here. No solutions exist in which the conical interface extends to infinity except those which involve singularities on the axis on *both* sides of the vertex of the cone. The flow inside the interface is probably sufficiently well represented by the Poiseuille equation, but when the tangential drag on the outer fluid is not neglected the outer flow must be represented by a distribution of the singularities which have been called 'Stokeslets' as well as the distribution of sources and sinks which can give rise to the spindle shaped interface. There are indications that it may not be possible for the point to remain truly conical when these are taken into account, but this is not certain.

42

THE FORCE EXERTED
BY AN ELECTRIC FIELD ON A
LONG CYLINDRICAL CONDUCTOR

REPRINTED FROM

Proceedings of the Royal Society, A, vol. CCXCI (1966), 145–58

The calculations and experiments here described were undertaken in the hope of giving a rational description of the fine jets which a strong electric field can drag from the surface of a conducting fluid. The force on a long axisymmetric conductor in contact with a conducting plane and subjected to an electric field parallel to its length is found by replacing the conductor by an axial distribution of charge. This distribution can be determined by means of an integral equation. The solutions for some particular cases were found by means of a computer, and Professor Van Dyke, in an appendix, gives a more analytical method of solution. The equivalent distribution of charge is found for half a spheroid standing on a plane. The force acting on this distribution is compared with the known force on the curved surface of a hemispheroid and the result used to show that the error involved in taking them as equal is small.

Experiments are described in which cylinders and hemispheroids standing on a horizontal earthed plate were lifted by a vertical field. Agreement between these experiments and calculation when the conductors are sufficiently light indicates that the space charge in the intense field at their upper ends is not large enough to invalidate the calculation, but when the conductors are heavy enough they oscillate instead of rising, an effect which must be due to electric breakdown of the air producing space charges which upset the field.

INTRODUCTION

It was first recorded I believe about 2500 years ago that light objects lying on a surface will jump up to a piece of rubbed amber when it is placed sufficiently close above them. In spite of the immense amount of work on electrostatics that has been done since then I cannot find that this experiment has been repeated under conditions where measurements of the lifting force could be made. My interest in the matter was roused by observing the fine threads which rise from the surface of a conducting fluid when it is exposed to a high electric field. To understand the mechanism of this situation it is necessary to know the mechanical force which can act on the end of a thread-like conductor where the field strength is highly concentrated. The only axisymmetric case of this kind for which the distribution of stress is known seems to be that in which the thread-like conductor is a very long ellipsoid in a uniform field or what amounts to the same thing, a half ellipsoid standing on a conducting plane. It cannot be assumed that the longitudinal stress in the fluid jets is the same as that in a long ellipsoid because the particular distribution of cross-section characteristic of a long ellipsoid is not likely to be the same as that of the continuous part of a jet. For this reason a study has been made of the force which the electric field can exert on a long solid conducting cylinder of uniform section when the field is uniform apart from the disturbance produced by the cylinder. If the cylinder is assumed to be standing on a conducting plane the effect of the plane can be treated by the method of images. The advantage of considering a cylinder of uniform section is that the stress over the cylindrical surface makes no contribution to the longitudinal equilibrium so that if the equivalent distribution of charge along the centre line which would make the cylindrical surface an equipotential can be found, the difference between the forces on the two ends of the cylinder can be calculated. This method, if it is valid, enables the end force to be calculated independently of the details of the shape of the end beyond the point where it is cylindrical, though the distribution of stress over the end must depend greatly on this shape. In the experiments which will be described the long cylinders used had roughly hemispherical ends in order to reduce the chance of occurrence of spots of high local electric field, but no measurements of their actual shape were made. The agreement of the measured force with that calculated by the method indicated above seems to show that such details are irrelevant so far as the total force is concerned, at any rate so long as the field is not distorted by space charges due to breakdown of the insulating property of air in high electric fields.

CALCULATION OF THE EQUIVALENT AXIAL DISTRIBUTION OF CHARGE

The cylinder of length L and radius R will be assumed to be in contact with a plane at zero potential and the axial charge distribution $\sigma(x)$ which gives rise to zero potential along the cylindrical surface is to be calculated. The relevant integral equation for determining $\sigma(x)$ in the presence of a uniform field F is

$$xF + \int_0^L \sigma(\xi)\left[\{(x-\xi)^2 + R^2\}^{-\frac{1}{2}} - \{(x+\xi)^2 + R^2\}^{-\frac{1}{2}}\right]d\xi = 0 \qquad (1)$$

for all values of x. The second term in the integral represents the effect of the charge distribution in the reflected system. No way to solve this equation could be seen but since there was available a computer capable of handling 20 simultaneous linear equations, x/L was divided into 20 equal intervals in each of which $\sigma(x)$ was taken as constant but unknown. The value of the integrals in (1) were calculated for the mid-points of each interval (i.e. at $x = \frac{1}{40}L$, $x = \frac{3}{40}L$, etc.) and the resulting 20 equations solved numerically for $L/R = 10^4$, $\frac{1}{3} \times 10^4$, 10^3, $\frac{1}{3} \times 10^3$, 10^2, $\frac{1}{3} \times 10^2$. These results, for which I am indebted to A. D. McEwan, are shown in fig. 1 where values of $\sigma(x)/FL$ are displayed. It will be seen that the distribution is roughly linear except near $x = L$ and indeed it was realised that the diagonal terms in the matrix corresponding with the contribution of the axial charge in the section nearest to the corresponding point on the surface of the cylinder, are much larger than the other terms. If all the other terms are neglected the result is simply

$$\sigma(x) = -\frac{Fx}{2\ln(L/R)}. \qquad (2)$$

On the other hand, the charge density increases at an increasing rate as the end $x = L$ is approached and when the rise in $\sigma(x)$ between neighbouring intervals is comparable with $\sigma(x)$ the approximation involved in taking $\sigma(x)$ constant over an interval is clearly unjustifiable. The points shown on fig. 1 suggest that 20 points may not be enough to give a reasonable description of the charge in the last one-tenth of the range when $L/R < 330$.

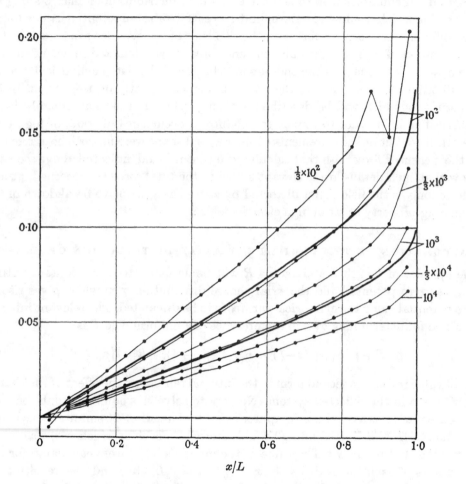

Fig. 1. Computed values of $\sigma(x)/FL$ for six values of L/R. Thick lines, Van Dyke's calculation for $L/R = 10^2$ and 10^3.

In analogous hydrodynamic problems relating to the fluid flow round axisymmetric slender bodies moving parallel to the axis, methods have been developed for determining the axial distribution of sources which produces the same external flow as that round the body. These methods are not directly applicable to the present problem but it seemed likely that analogous ones might be developed in connection with it. Accordingly Professor Van Dyke, whose work on this subject is well known, was consulted and he kindly consented to write the note which accompanies this paper. In (A 8) of his appendix he finds that the equivalent distribution of charge for a long conducting cylinder in contact with a conducting plane and exposed to a uniform field F is approximately

$$\sigma(x) = \frac{-Fx}{2\ln L/R}\left[1 - \frac{1}{\ln(L/R)}\{\ln 2\sqrt{(1-x^2/L^2)} - 1\}\right]. \tag{3}$$

In the limit when L/R is very large $\sigma(x)$ tends, as had been expected, to $Fx(2\ln L/R)^{-1}$, but very close to the end it rapidly becomes infinite. The values of $\sigma(x)$ were calculated from (3) for two cases, $L/R = 100$ and 1000, and are shown in fig. 1. as full curves. It will be seen that except in the last one-tenth of the range the agreement between the two methods is good.

Calculation of force on axisymmetric conductor in contact with a plane

In calculating the total force on the cylinder one might be tempted to assume that it is the product of the total equivalent electric charge on the axis and the local field. For a long cylinder this is approximately true but it is not obviously so and to illustrate the kind of difficulty which may be encountered it is worth considering how the force on a hemisphere of radius a in contact with a conducting plane, and in a uniform field of strength F, can be deduced from the equivalent doublet which represents the external field. The doublet is of strength $2Qd = Fa^3$ and by integration over the surface it is found that the total force on the surface of the hemisphere is

$$P_1 = \tfrac{9}{16}a^2F^2. \tag{4}$$

The axial charge $-Q$ and its image may be regarded as being situated at $x = \pm d$. The force which the external field exerts on the charge $-Q$ is $+FQ$ while the force which the image exerts is $-Q^2/4d^2$; both are infinite when $d \to 0$. To form an equilibrium equation the force exerted by the electric field over a closed surface containing a charge must be found and equated to the force which the electric field exerts on that charge. The appropriate closed surface is the whole hemisphere including the base. The total force on this surface is $P_1 - P_0$ where P_0 refers to the base and P_1 to the hemispherical surface, so that the equation of equilibrium is

$$P_1 = P_0 + FQ - Q^2/4d^2 \tag{5}$$

and

$$P_0 = \frac{1}{8\pi}\int_0^a \left\{\frac{2Qd}{(y^2+d^2)^{\frac{3}{2}}} - F\right\}^2 2\pi y\,dy. \tag{6}$$

Integrating (6) it is found that when $d \to 0$ while $Qd = \tfrac{1}{2}Fa^3$

$$P_0 = \frac{Q^2}{4d^2} - \frac{Q^2d^2}{4a^4} + \frac{QdF}{a} - QF + \frac{F^2a^2}{8}$$

so that (5) becomes

$$P_1 = -\frac{Q^2d^2}{4a^4} + \frac{QdF}{a} + \tfrac{1}{8}F^2a^2 \quad \text{or} \quad \tfrac{9}{16}a^2F^2, \tag{7}$$

which is in agreement with the result (4) of direct integration. The same method can be applied to any axisymmetric conductor in contact with a plane if the equivalent distribution of charge along the axis is known. The geometry is indicated in fig. 2. P_1 is the force exerted on the curved surface and P_0 that which would be exerted on the plane $x = 0$ over the circle of contact with the

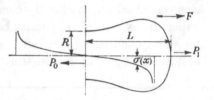

Fig. 2. Sketch explaining symbols.

plane by the equivalent source and image system. The equation of equilibrium is

$$P_1 - P_0 = -\int_0^L \sigma(x) \left(F + \int_0^L \frac{\sigma(\xi)}{(x+\xi)^2} \, d\xi \right) dx \tag{8}$$

and

$$P_0 = \frac{1}{8\pi} \int_0^R \left(\frac{dV}{dx} \right)_0^2 (2\pi y) \, dy, \tag{9}$$

where

$$\left(\frac{dV}{dx} \right)_0 = F + 2 \int_0^L \frac{\sigma(x) \, x \, dx}{(x^2 + y^2)^{\frac{3}{2}}}. \tag{10}$$

Spheroid

The equivalent source distribution is known for few axisymmetric bodies, but it can be found for a spheroid.

The potential in the electric field surrounding an uncharged conducting spheroid in a uniform field F is*

$$V = Fx \int_0^\lambda \frac{d\lambda}{(a^2 + \lambda) \Delta_\lambda} \Big/ \int_0^\infty \frac{d\lambda}{(a^2 + \lambda) \Delta_\lambda}, \tag{11}$$

where $2a$ and $2R$ are the major and minor axes and

$$\Delta_\lambda = (a^2 + \lambda)^{\frac{1}{2}} (R^2 + \lambda). \tag{12}$$

The stress normal to the surface of the spheroid

$$\frac{x^2}{a^2} + \frac{y^2}{R^2} = 1 \quad \text{is} \quad \frac{1}{8\pi} \left(\frac{\partial V}{\partial n} \right)^2,$$

where $\partial V / \partial n$ is the normal component of potential gradient when $\lambda = 0$. Its value†
is

$$\widehat{nn} = \frac{F^2 x^2}{8\pi R^2 I_2^2 (1 - x^2 e^2 / a^2)}. \tag{13}$$

Integrating this over the half surface corresponding with $0 < x < a$ we find that the force is

$$P_1 = \frac{F^2 a^2 I_1}{8 I_2^2}, \tag{14}$$

where

$$I_1 = \frac{1}{e^4} \{ -\ln(1 - e^2) - e^2 \}, \quad I_2 = \frac{1}{e^2} \left\{ \frac{1}{2e} \ln \frac{1+e}{1-e} - 1 \right\} \tag{15}$$

and e is the eccentricity $(1 - R^2/a^2)^{\frac{1}{2}}$.

* See *Electricity and Magnetism*, by J. H. Jeans, Cambridge University Press (1915).
† G. I. Taylor, *Proc. Roy. Soc.* A, CCLXXX (1964), 383 (paper 39 above).

When $a/R \gg 1$ so that $e \to 1$ the approximate expressions for I_1 and I_2 are $2\ln(L/R) - 1$ and $\ln(L/R) - 1 + \ln 2$ so that

$$P_1 \approx \frac{F^2 a^2}{4\ln a/R} \frac{\left(1 - \dfrac{1}{2\ln(a/R)}\right)}{\left(1 - \dfrac{1 - \ln 2}{\ln(a/R)}\right)^2}. \tag{16}$$

Equivalent axial charge. The confocal spheroids

$$\frac{x^2}{a^2 + \lambda} + \frac{y^2}{R^2 + \lambda} = 1$$

degenerate into the line between the foci when $\lambda = -R^2$.

Close to an axial distribution of sources the value of dV/dy is $-2\sigma(x)/y$ so that to find $\sigma(x)$ it is necessary to find the limiting value $-\frac{1}{2} y \, dV/dy$ over the surface of the spheroid

$$\frac{x^2}{a^2 - R^2 + \beta a^2} + \frac{y^2}{\beta a^2} = 1$$

as $\beta \to 0$. Under these conditions $(dV/dy) \sim (dV/dn)$ and after some reduction it is found that

$$\sigma(x) = \frac{-Fx}{2e^3 I_2} \tag{17}$$

It appears therefore that for the spheroid the distribution is truly linear and extends over the line between the foci. When $e \sim 1$ so that a/R is large, (17) reduces to

$$\sigma(x) = \frac{-Fx}{2\ln(a/R) + \ln 2 - 2}. \tag{18}$$

It will be noticed that when a/R is very large the distribution tends slowly towards the same limit as that for a long cylinder.

Force on hemispheroid. Taking the distribution of equivalent axial charge as

$$\sigma(x) = \gamma x \tag{19}$$

the integrals in (8) can be evaluated. The result is

$$P_1 - P_0 = -\tfrac{1}{2} L^2 \gamma F - (\tfrac{3}{4} - \ln 2) L^2 \gamma^2. \tag{20}$$

For the spheroid $\gamma = -F/2e^3 I_2$ so that $(P_1 - P_0)/L^2 F^2$ can be evaluated in terms of e, and since $L = ae$ this result can be compared with the direct calculation of (14) namely $P_1/F^2 a^2 = I_1/8 I_2^2$ by multiplying the latter by $L^2/a^2 = e^2$.

The comparison is shown in table 1. Even when a/R is as low as 3, P_0 is only $3\frac{1}{2}\%$ of P_1, while for $a/R = 10$ it is only 0.1% of P_1.

The object of the calculations recorded in table 1 was to provide justification for neglecting P_0 in calculations of the force on long cylinders.

Table 1. *Comparison between* $P_1 - P_0$ *computed from the equivalent source distribution for a spheroid and* P_1 *computed directly*

a/R	$P_1 - P_0$	P_1
2	0·239	0·294
3	0·208	0·215
5	0·1560	0·1573
10	0·1119	0·1120

Calculations of the force on long slender bodies

It has been pointed out that the first approximation to the charge distribution is $\sigma(x) = -Fx/2\ln(L/R)$. The first approximation to the force which the external field exerts on the charge distribution is therefore

$$\frac{F^2 L^2}{4\ln(L/R)}. \tag{21}$$

The effect of the image of the body on this force manifests itself in two ways. First the image system must be taken into account in determining $\sigma(x)$. This was done in both Van Dyke's calculation where it is given in (A10) and (A11), and in the 20-step machine solution exhibited in fig. 1.

The second effect is due to the modification of the field which acts on $\sigma(x)$, or in other words the attraction between the body and its image owing to the charges induced on them. This is represented by the term.

$$-\int_0^L \sigma(x) \int_0^L \frac{\sigma(\xi)\,d\xi}{(x+\xi)^2}\,dx$$

in (8). For slender bodies it is much smaller than the direct effect represented by

$$-\int_0^L F\sigma(x)\,dx$$

and can be calculated by using the approximate value $-F\xi/2\ln(L/R)$ for $\sigma(\xi)$ in the integral. Thus

$$-\int_0^L \frac{\sigma(\xi)\,d\xi}{(x+\xi)^2} = \frac{F}{2\ln(L/R)}\left[\ln\left(\frac{x+L}{L}\right) - \frac{L}{x+L}\right] \left.\vphantom{\int}\right\}$$

and

$$\int_0^L \sigma(x)\int_0^L \frac{\sigma(\xi)\,d\xi}{(x+\xi)^2}\,dx = -\frac{F^2 L^2}{\{2\ln(L/R)\}^2}(-\ln 2 + \tfrac{3}{4}), \tag{22}$$

so that (8) becomes

$$P_1 - P_0 = -F\int_0^L \sigma(x)\,dx - \frac{0\cdot0569 F^2 L^2}{\{2\ln(L/R)\}^2}. \tag{23}$$

If we use Van Dyke's expression (A11) the force acting on the cylinder is

$$P_1 = P_0 + F^2 L^2\left\{\frac{1}{4(\ln L/R - \tfrac{3}{2} + \ln 2)} - \frac{0\cdot0569}{(2\ln L/R)^2}\right\}. \tag{24}$$

The integral P_0 can be calculated, but it is very small and may justifiably be neglected. The last term in (24) is also very small, amounting to less than 2 % when $L/R = 4$ and rapidly decreasing as L/R increases.

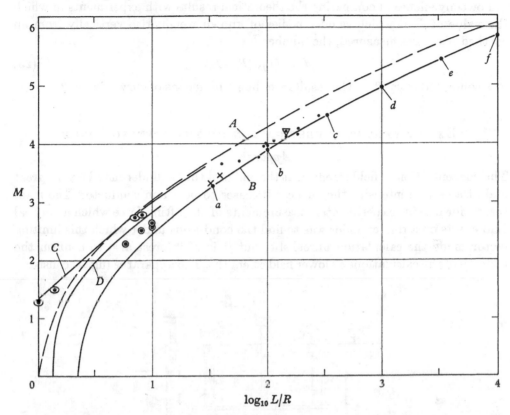

Fig. 3. Comparison of observations with theoretical curves. Experimental, $M = FL/\sqrt{(mg)}$: ⊙, spheroids; ◉, light plastic cylinders; •, aluminium wire; ×, wood cylinders; ▽, painted glass tube. Theoretical: a to f, calculations by 20-variable computer; A, $M = 2\sqrt{\{\ln(L/R)\}}$ (rough approximation); B, $M = 2\sqrt{\{\ln(L/R) - \frac{3}{2} + \ln 2\}}$ (Van Dyke's (A 11)); C, $M = I_2\sqrt{(8/I_1)}$ spheroid (accurate); D, Van Dyke's approximation for spheroid.

Van Dyke's expressions (A 10) and (A 11) do not include the part of the total force which is due to the attraction of the image system so that to compare them with the computer solutions the 'total force' of his table A 1 must be taken as

$$F \int_0^L \sigma(x)\,\mathrm{d}x.$$

The value of
$$F \int_0^L \sigma(x)\,\mathrm{d}x$$

given by the computer is $F^2 L^2$ times $\frac{1}{20}$ of the sum of the 20 numbers displayed in fig. 1 for each value of L/R. The last column of Van Dyke's table A 1 was derived by multiplying the figures derived in the manner just described by $4\ln L/R$. The agreement of the 20-step computer solution with Van Dyke's formula (A 11) is remarkable, but it must be remembered that there is no theoretical reason for preferring (A 11) to (A 10).

For convenience in comparing the theoretical results with experiments in which the value of F at which slender bodies of known weight rise vertically from an earthed plate was measured, the number

$$M = FL/\sqrt{(P_1 - P_0)} \tag{25}$$

was computed from (24). The results are shown by means of curves in fig. 3.

EXPERIMENTAL INVESTIGATION OF THE LIFTING FORCE

Apparatus

The concentration of field strength at the end of a thin cylinder may be very great and at a certain intensity the air near it ceases to be a non-conductor. The main reason for making experimental measurements of the lifting force which a vertical field exerts on a thin cylinder was to find the conditions under which this limiting factor made the calculation unrealistic, but it is of interest also to confirm the validity of the calculations at lower field strengths. The apparatus (fig. 4) consisted

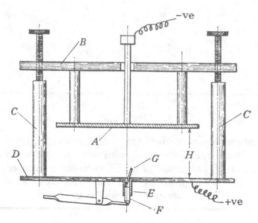

Fig. 4. Sketch of apparatus.

of a thick, circular, round edged, horizontal brass plate A, 30 cm. in diameter, which was supported horizontally under a Perspex plate B. This plate stood on four adjustable legs, C, and held plate A at height H above an earthed plate, D. The potential of the plate A was raised gradually till at a measured value V_0 a test object placed on the earthed plate D jumped up. When this object had a wide enough base to stand stably it was erected on the centre of D. For experiments with very narrow cylinders (in fact cylinders of length up to 288 times their diameter were used) it was not possible to erect them in this way. A vertical tube E 1·5 cm. long and 0·2 cm. bore was therefore fixed below the centre of the earthed plate D. This tube contained a piston F which could be raised till its top was level with the surface of D. The cylinder G being tested was dropped into the tube E so that its base was on the piston F and its top was above the level of D. The bore of E was greater than the diameter of the cylinder D (fig. 4) which therefore lay at a slight angle to the vertical. The potential of the upper plate was first raised to such a value that the cylinder G stood upright in the tube E but the lifting force was not great enough to overcome gravity. A vertical rod can stand stably on a horizontal plane if it is supported at the upper end by a vertical force greater than half its weight. The piston F was then gradually raised, sometimes reducing V_0 during this operation but keeping the specimen vertical till the top of F was level with D. The potential V_0 was then gradually raised till the weight of the specimen was taken by the electric field acting on the top of the cylinder.

General description of results

When the weight of the cylinder was small enough it jumped vertically till it discharged itself on the upper plate and then rapidly oscillated between the plates. When the weight of the cylinder increased the jumps became smaller till they were so small that they could not be appreciated as separate events. The cylinders could then move about freely on the earthed plate while standing vertical. When heavier cylinders were used this stage could not be reached. At a field strength below that which could provide enough vertical force to lift them they began to oscillate and as the field strength was raised to higher values they remained in contact with the plane even though the field was much stronger than that which would have lifted them if they had been vertical.

Specimens tested

Only very light specimens could be used. Their dimensions and weights are given in table 2. Four hemispheroids, one a hemisphere, were constructed by mounting pieces of a very light plastic packing material of density $0 \cdot 020$ g./cm.^{3} on a lathe. They were then rubbed lightly to the required shape with fine sand paper. They were then made conducting by painting them with 'aquadag', a conducting paint. Cylindrical specimens with roughly hemispherical ends were also constructed of the same material. A few solid wood specimens were made by rubbing down matches and treating them with 'aquadag'.

To obtain really long cylinders aluminium wire was drawn down to $0 \cdot 0208$ cm. diameter and in this way a specimen was produced for which L/R was as great as 288.

COMPARISON OF THEORY WITH OBSERVATION

The potential V_k at which the specimen either jumped or in some cases started to move freely over the plate was measured and is recorded in kilovolts in column 5 of table 2. The experimental result may be compared with the theoretical by defining a number

$$M = \frac{FL}{\sqrt{(mg)}} = \frac{V_k L}{0 \cdot 3H \sqrt{(mg)}}.$$

Table 2. *Measured values of H, L, R, m and V_k. Values of M and $\overline{F}_m$ computed from them*

H (cm.)	L (cm.)	R (cm.)	m (mg.)	V_k (kV.)	M	L/R	$\overline{F}_m$ (kV/cm.)
			Spheroids				
		light plastic with conducting coat, in fig. 3					
7·66	2·10	0·30	22·0	13·75	2·74	7·0	13
7·66	2·54	0·305	34	14·75	2·78	8·3	16
4·70	1·48	1·07	273	23·5	1·50	1·38	13
4·20	1·025	1·025	217	23·25	1·30	1·00	12
			Round topped cylinders				
		light plastic with conducting coat					
7·66	1·80	0·307	31·3	16·25	2·30	5·85	15
7·66	2·46	0·243	31·7	13·4	2·57	10·14	19
7·66	2·57	0·317	47·5	15·3	2·51	8·10	18
			wood with conducting coat				
7·66	2·38	0·0625	17·5	13·75	3·45	38·1	56
			glass with conducting coat				
7·66	1·27	0·0185	4·36	11·0	3·98	93	95
			aluminium wire				
7·66	3·0	0·0245	15·1	oscillated		122·4	[133]
7·66	2·05	0·0245	10·3	13·5	3·77	83·7	110
7·66	0·99	0·0245	5·0	19·0	3·68	40·4	77
7·66	3·0	0·0175	7·4	oscillated		171·5	[131]
7·66	2·0	0·0175	4·95	10·25	4·05	114·3	107
7·66	0·99	0·0175	2·45	13·5	3·75	56·6	75
			aluminium wire, 0·0208 cm. diameter				
7·66	3·0	0·0104	2·73	5·6	4·45	288	134
7·66	1·92	0·0104	1·75	6·5	4·14	184·5	107
7·66	1·52	0·0104	1·386	7·25	4·10	144·3	95
7·66	0·99	0·0104	0·902	—	3·22	95·1	76

The theoretical value of M for a hemispheroid of length L and diameter $2R$ (curve C fig. 3) is found by writing $F = V_k/0\cdot3H$, $P_1 = mg$ and $e^2 = 1 - R^2/L^2$ in (24) and

$$M = I_2\sqrt{(8/I_1)}. \tag{26}$$

The limiting value of M for a hemisphere ($e = 0$) is $\frac{4}{3}$. Van Dyke's formula (A 11) for the cylinder (curve B, fig. 3) is

$$M = 2\{\ln(L/R) + \ln 2 - \tfrac{3}{2}\}^{\frac{1}{2}} \tag{27}$$

and the computer solution (points a to f, fig. 3) is hardly distinguishable from Van Dyke's (A 11) over the range it appeared to be convergent, namely $L/R > 33$ (see table 3).

The experimental values for M are plotted in fig. 3 in which the abscissae are $\log_{10}(L/R)$. It will be seen that the agreement when L/R is greater than 30 is good, but that when L/R is less than 10 the observed values are greater by about 10 to

15 % than those on Van Dyke's curve. To estimate M for $L/R < 10$ it might be necessary to estimate the value of P_0 (equation (9)) as well as to calculate coefficients for higher powers of $\ln (L/R)$. It will be noticed that the experiments with spheroids including the hemisphere agree well with the theoretical formula (26) which was not derived from slender-body theory.

The rough approximation to M namely $2(\ln L/R)^{\frac{1}{2}}$ is shown as a broken line A in fig. 3. The values of M for the ellipse cross the curve $M = 2(\ln L/R)^{\frac{1}{2}}$ at about $\log_{10}(L/R) = 0\cdot 8$ and at higher values the two curves lie very close together.

Conclusion

The object in view when these experiments were made was to find out whether ionization, which is likely to occur locally at the field strengths necessary to lift the specimens, alters the distribution of potential enough to invalidate calculations which take no account of this effect.

Though the distribution of field strength over the ends of the cylinders cannot be calculated, an average value can be defined by assuming a uniform stress over their convex ends. If $\overline{F}_m$ is the average field strength at the ends corresponding with the condition when the specimen rises from the plate

$$\pi R^2 \left(\frac{1}{8\pi} \overline{F}_m^2 \right) = mg. \tag{28}$$

Here $\overline{F}_m$ is expressed in electrostatic units. In kilovolts per centimetre,

$$\overline{F}_m = \frac{0\cdot 3}{R} \sqrt{(8mg)} \tag{29}$$

and values are given in the last column of table 2. These are so high that ionization and brush discharges are probable, but this does not prevent the cylinders from being lifted at the calculated voltage unless they begin to oscillate at lower voltages. When this occurs they go on oscillating without breaking contact with the plate on which they are standing even when the vertical field is much stronger than that which could lift them if they had been constrained to remain vertical. It seems likely that the cause of the instability is the repulsion between space charge above the top of the cylinder and the induced charge in it.

My thanks are due to Professor Van Dyke for writing the Appendix and to Mr A. D. McEwan for help in the experiments and computations.

Appendix by M. D. Van Dyke

An asymptotic approximation for slender bodies

A body whose meridian curve is sufficiently smooth can be represented by a continuous distribution of charges along its axis of strength $\sigma(x)$, and images of equal but opposite strength. Together with the uniform field, this gives the potential as

$$\phi = Fx + \int_0^L \sigma(\xi) \left[\frac{1}{\sqrt{\{(x-\xi)^2 + r^2\}}} - \frac{1}{\sqrt{\{(x+\xi)^2 + r^2\}}} \right] d\xi. \tag{A1}$$

We can approximate to this in the vicinity of a slender body by methods familiar to aerodynamicists.* We first integrate by parts, using the fact that $\sigma(0) = 0$ by continuity and antisymmetry, which gives

$$\phi = Fx + \sigma(L) \left[\sinh^{-1} \frac{L-x}{r} - \sinh^{-1} \frac{L+x}{r} \right] + \int_0^L \sigma'(\xi) \left[\sinh^{-1} \frac{x-\xi}{r} + \sinh^{-1} \frac{x+\xi}{r} \right] d\xi. \tag{A2}$$

Now we can approximate for small r, using the asymptotic form of $\sinh^{-1}$ for large value of its argument. With a relative error of order r^2, which is of order $(R/L)^2$ except near too abrupt tips or other discontinuities, this gives

$$\phi = Fx + 2\sigma(x) \ln \frac{2}{r} - \sigma(L) \ln \frac{L+x}{L-x} + \int_0^x \sigma'(\xi) \ln (x-\xi) \, d\xi$$

$$- \int_x^L \sigma'(\xi) \ln (\xi - x) \, d\xi + \int_0^L \sigma'(\xi) \ln (x+\xi) \, d\xi. \tag{A3}$$

Requiring this to vanish at the body, which is conveniently described in terms of the small parameter $\epsilon = R_0/L$ by $r = \epsilon L f(x)$, yields an integral equation for $\sigma(x)$:

$$Fx + 2\sigma(x) \left[\ln \frac{1}{\epsilon} + \ln \frac{2}{Lf(x)} \right] - \sigma(L) \ln \frac{L+x}{L-x} + \int_0^x \sigma'(\xi) \ln (x-\xi) \, d\xi$$

$$- \int_0^L \sigma'(\xi) \ln (\xi - x) \, d\xi + \int_0^L \sigma'(\xi) \ln (x+\xi) \, d\xi = 0. \tag{A4}$$

Solving this integral equation would give the charge distribution with an error of only order ϵ^2 almost everywhere. However, the solution can be found only in special or inverse problems. For example, trying a linear variation of $\sigma(x)$ over the length L yields the slender-body solution for an ellipsoid:

$$\sigma(x) = \frac{-Fx}{2\{\ln (2/\epsilon) - 1\}}. \tag{A5}$$

This differs by order ϵ^2 from the exact result. (Of course the charge distribution should actually extend only between the foci, but except near the end—where we could provide a correction if necessary—this causes an error of only order ϵ^2.) The total force on the half ellipsoid is given to this approximation by

$$-F \int_0^L \sigma(x) \, dx = \frac{F^2 L^2}{4\{\ln (2/\epsilon) - 1\}}. \tag{A6}$$

* See 'The calculation of pressure on slender airplanes in subsonic and supersonic flow' by M. A. Heaslet and H. Lomax, *Nat. Advis. Comm. Aero.*, Washington, *Tech. Note* no. 2900 (1953).

For a general shape we are forced to accept in the first approximation a much larger error, of order $\{\ln (1/\epsilon)\}^{-1}$, by neglecting all but the first two terms in (A 4) to obtain

$$\sigma_1(x) = \frac{-Fx}{2 \ln (1/\epsilon)}. \tag{A 7}$$

This can be improved by iteration, the second approximation being

$$\sigma_2(x) = \frac{-Fx}{2 \ln (1/\epsilon)} \left[1 + \frac{1}{\ln (1/\epsilon)} \left(1 - \ln \frac{2\sqrt{(L^2 - x^2)}}{Lf(x)} \right) \right]. \tag{A 8}$$

Here the relative error has been reduced to order $\{\ln (1/\epsilon)\}^{-2}$, and it would be $\{\ln (1/\epsilon)\}^{-n}$ in the nth approximation. (However, further approximations are difficult to compute.) This is a problem, like viscous flow past a circle at low Reynolds number, where perturbation theory creeps toward the solution with an infinite series of inverse powers of a logarithm.

For a cylindrical body the second approximation is

$$\sigma_2(x) = \frac{-Fx}{2 \ln (L/R)} \left[1 + \frac{1}{\ln (L/R)} \{1 - \ln 2 \sqrt{(1 - x^2/L^2)}\} \right]. \tag{A 9}$$

Hence the total force on the body is

$$\frac{F^2 L^2}{4 \ln (L/R)} \left[1 + \frac{1}{\ln (L/R)} (\tfrac{3}{2} - \ln 2) + O \left(\frac{1}{\ln^2 (L/R)} \right) \right]. \tag{A 10}$$

For a spheroid the exact value for P_1 is given in (14). The approximation (16) can be written in the alternative forms

$$P_1 = \frac{F^2 a^2}{4 \ln (a/R)} \left\{ 1 + \frac{2 \ln 2 - \tfrac{3}{2}}{\ln (a/R)} + O \left(\frac{1}{\ln (a/R)} \right)^2 \right\},$$

$$P_1 = \frac{F^2 a^2}{4 \ln (a/R)} \left\{ \frac{1}{1 - \dfrac{2 \ln 2 - \tfrac{3}{2}}{\ln (a/R)} + O \left(\dfrac{1}{\ln (a/R)} \right)^2} \right\},$$

if terms of order $\{\ln (a/R)\}^{-2}$ are ignored. Comparing these two expressions numerically with the exact values it is found that the second is rather better than the first when $\{\ln (a/R)\}^{-2}$ is no longer negligible. It seemed worth while, as an empirical experiment, to make the same change in the approximate expression (A 10). (A 11) is produced in this way

$$P_1 = \frac{F^2 L^2}{4\{\ln 2(L/R) - \tfrac{3}{2}\}}. \tag{A 11}$$

Since no exact expression analogous to (14) was available in the cylindrical case for comparison with approximation (A 11), this expression was compared with Taylor's numerical solution. The results are shown in table A 1. It will be seen that the agreement is close but this cannot be taken as evidence that there is any theoretical reason for preferring (A 11) to (A 10).

Table A 1. $\dfrac{total\ force}{F^2 L^2/\{4\ln{(L/R)}\}}$

			numerical solution
L/R	(A 10)	(A 11)	(Taylor)
10^4	1·087	1·096	1·094
$1/3 \times 10^4$	1·099	1·110	1·109
10^3	1·116	1·132	1·131
$1/3 \times 10^3$	1·139	1·161	1·159
10^2	1·175	1·212	1·213
$1/3 \times 10^2$	1·230	1·299	1·303
10	1·348	1·539	—

43

THE CIRCULATION PRODUCED IN A DROP BY AN ELECTRIC FIELD

REPRINTED FROM

Proceedings of the Royal Society, A, vol. CCXCI (1966), pp. 159–66

The elongation of a drop of one dielectric fluid in another owing to the imposition of an electric field has previously been studied assuming that the interface is uncharged and the fluids at rest. For a steady field this is unrealistic, because however small the conductivity of either fluid the charge associated with steady currents must accumulate at the interface till the steady state is established.

It is shown that equilibrium can only be established in a drop when circulations are set up both in the drop and its surroundings. A relation is found between the ratios of the conductivity, viscosity and dielectric constant for the drop and surrounding fluid which permits the drop to remain spherical when subjected to a uniform field. The streamlines of the circulation for this case are shown and criteria are given for distinguishing between circulations which carry the surface of the drop towards or away from the poles and for predicting whether the drop will become prolate or oblate.

Experiments by S. G. Mason and his co-workers are compared with the theoretical predictions and agreement is found in all cases for which the necessary data are given.

INTRODUCTION

A uniform electric field elongates a conducting drop immersed in a non-conducting fluid into a shape which is very nearly a prolate spheroid.* When both the drop and its surroundings are non-conducting dielectrics it is also distorted into a spheroidal shape and it has been stated by O'Konski & Thacker† and by Garton & Krasucki‡ that in all such cases the spheroid is prolate. Some cases have been reported in which drops have been observed to become oblate§ and it has been suggested by O'Konski & Harris‖ that such an effect could be due to the existence of some conductivity in the fluids. Indeed they have put forward the following expression for small values of the ellipticity, e, produced by a small uniform electric field, F, when both fluids are conducting:

$$\tfrac{3}{4}F\left(\frac{a\kappa_1}{\pi\gamma}\right)^{\frac{1}{2}}\left[\frac{(R-1)(R^2+7R-2-6\kappa_2/\kappa_1)}{(R+2)^2}\right]^{\frac{1}{2}}. \tag{1}$$

Here κ_1 and κ_2 are the dielectric constants of the surrounding fluid and drop respectivity, γ is interfacial tension, $R = \sigma_1/\sigma_2$ where σ_1 and σ_2 are the resistivities

---

* G. I. Taylor, *Proc. Roy. Soc.* A, CCLXXX (1964), 383 (paper 39 above).
† *J. phys. Chem.* LVII (1953), 955.
‡ *Proc. Roy. Soc.* A, CCLXXX (1964), 211.
§ R. S. Allan & S. G. Mason, *Proc. Roy. Soc.* A, CCLXVII (1962), 45.
‖ *J. phys. Chem.* LXI (1957), 1172.

of the two media expressed in electrostatic units, and a is the original radius of the drop.

Expression (1) was derived by minimising the difference between the energy of the electric field before and after the introduction of the drop. The energy was calculated by taking the known distribution of potential for a steady current and using the usual electrostatic expression for the energy of electric fields in dielectrics. O'Konski & Harris attempted to justify the neglect of energy which might arise owing to separation of charges on the interface during the introduction of the drop by assuming that this event occurred too quickly to permit surface charges to be generated. While admitting that this assumption would simplify the mathematical problem it seems to me to be definitely wrong, at any rate in its application to steady state problems, because while the surface charge is being built up the potential is not that which was used in the calculation, and when the field has reached a steady state after the introduction of the drop there must be a surface charge. One effect of this charge is to produce circulatory currents which are ignored in (1).

Though (1) does not seem to be correct, its form is suggestive because it envisages a state in which the drop might remain spherical provided

$$R^2 + 7R - 2 - 6\kappa_2/\kappa_1 = 0. \tag{2}$$

It does not seem possible for a spherical drop to exist at rest in equilibrium in an electric field because neither the surface tension nor constant internal and external pressures can balance the stress due to the electric field, which varies over the surface.

If it is possible for a drop to remain spherical in an electric field, equilibrium can only be maintained through a balance between the electric stress and a variable pressure difference between the inside and outside of the drop. Such a pressure difference can only exist if the fluids inside and outside the drop are in motion. It is the purpose of this communication to show that this equilibrium can exist and to find the conditons which make this possible.

STRESS DUE TO THE ELECTRIC FIELD

It will be assumed that the distribution of potential round a conducting drop in a conducting fluid is the same as that which would exist in a static state. This assumption cannot be quite accurate when currents of fluid flow exist which convect charge, but in most cases the effect is likely to be small. This assumption was made by O'Konski & Harris who did not consider the existence of fluid circulation in the drop. The solution of the problem presented by a spherical conductor immersed in another conductor is known. If σ_1 and σ_2 are the resistivities of the outer conductor and the sphere respectively, a the radius of the sphere and F the uniform electric field far from the sphere, the potentials V_1 and V_2 outside and inside the sphere are, in polar co-ordinates,

$$V_1 = F \cos \theta \left(r + \frac{1-R}{2+R} \frac{a^3}{r^2} \right), \tag{3}$$

$$V_2 = \frac{3Fr \cos \theta}{2+R}, \tag{4}$$

where $R = \sigma_1 / \sigma_2$.

The normal and tangential components of the electric field at the surface $r = a$ are

$$E_{1n} = \frac{\partial V_1}{\partial r} = \frac{3RF \cos \theta}{2+R}, \quad E_{1t} = \frac{\partial V_1}{a \, \partial \theta} = -\frac{3F \sin \theta}{2+R}, \tag{5}$$

$$E_{2n} = \frac{\partial V_2}{\partial r} = \frac{3F \cos \theta}{2+R}, \quad E_{2t} = \frac{\partial V_2}{a \, \partial \theta} = -\frac{3F \sin \theta}{2+R}. \tag{6}$$

In order that the radial components (5) and (6) of the potential gradients inside and outside the interface between the dielectrics may exist there must be a surface charge ρ such that

$$\kappa_1 \frac{\partial V_1}{\partial r} - \kappa_2 \frac{\partial V_2}{\partial r} = -4\pi\rho, \tag{7}$$

so that

$$\rho = \frac{3F \cos \theta}{4\pi} \left(\frac{\kappa_1 R - \kappa_2}{2+R} \right). \tag{8}$$

The stress that the electric field exerts on the interface has radial and tangential components $(\widehat{rr})_E$ and $(\widehat{r\theta})_E$ where

$$(\widehat{rr})_E = \frac{\kappa_1}{8\pi} (E_{1n}^2 - E_{1t}^2) - \frac{\kappa_2}{8\pi} (E_{2n}^2 - E_{2t}^2)$$

$$= \frac{9F^2}{8\pi(2+R)^2} [\kappa_2 - \kappa_1 + \{\kappa_1(R_2+1) - 2\kappa_2\} \cos^2 \theta] \tag{9}$$

and

$$(\widehat{r\theta})_E = \frac{E_{1t}}{4\pi} (\kappa_1 E_{1n} - \kappa_2 E_{2n}) = \frac{-9}{4\pi(2+R)^2} (\kappa_1 R - \kappa_2) \cos \theta \sin \theta. \tag{10}^*$$

STRESS DUE TO FLUID FLOW

The tangential stress $(\widehat{r\theta})_E$ can only be balanced by stresses associated with hydrodynamic currents in the drop and its surroundings, and such stresses can only exist in viscous fluids. In order to ensure that the drop can be spherical it is necessary to find motions in both fluids for which the variable part of the difference between the normal components is $(\widehat{rr})_E$, and the difference between the tangential components is $(\widehat{r\theta})_E$. Only the case in which the stress due to inertia is small compared with that due to viscosity will be considered, so that it is legitimate to use the linear (Stokes) equation of motion.

* Equation (10) could also be derived by multiplying the surface charge ρ by $-E_{1t}$.

Using spherical polar co-ordinates, the Stokes stream function ψ is related to the velocity components, u, v by the relations

$$u = \frac{1}{r^2 \sin\theta}\frac{\partial\psi}{\partial\theta}, \quad v = \frac{-1}{r\sin\theta}\frac{\partial\psi}{\partial r}, \tag{11}$$

and ψ must satisfy the equation*

$$\mathrm{D}^4\psi = 0, \tag{12}$$

where the operator D^2 is

$$\frac{\partial^2}{\partial r^2} + \frac{\sin\theta}{r^2}\frac{\partial}{\partial\theta}\left(\frac{1}{\sin\theta}\frac{\partial}{\partial\theta}\right). \tag{13}$$

Some solutions of (12) which would correspond with appropriate angular distributions of stress can be found by assuming the form

$$\psi = r^n \sin^2\theta \cos\theta. \tag{14}$$

It is found that (12) and (14) are consistent when

$$n = -2, 0, 3 \text{ or } 5.$$

The appropriate solutions of (12) are therefore

$$\psi_1 = (Aa^4 r^{-2} + Ba^2)\sin^2\theta\cos\theta \tag{15}$$

for the flow outside and

$$\psi_2 = (Ca^{-1}r^3 + Da^{-3}r^5)\sin^2\theta\cos\theta \tag{16}$$

for the flow inside the drop, the factors being chosen so that A, B, C, D have the dimensions of velocity.

Table 1. *Velocity components and stresses for four independent solutions of* $\mathrm{D}^4\psi = 0$

1	2	3	4	5	6	7
ψ	u	v	p	$\widehat{rr}$	$\widehat{r\theta}$	
$\sin^2\theta\cos\theta$	$1-3\cos^2\theta$	$\sin\theta\cos\theta$	$\mu(1-3\cos^2\theta)$	$\mu(1-3\cos^2\theta)$	$\mu(\cos\theta\sin\theta)$	
r^{-2}	$-r^{-4}$	$2r^{-4}$	0	$8r^{-5}$	$-16r^{-5}$	$+Aa^4$
1	$-r^{-2}$	0	$-2r^{-3}$	$6r^{-3}$	$-6r^{-3}$	$-Aa^2$
r^3	$-r$	$-3r$	0	-2	-6	$+Aa^{-1}$
r^5	$-r^3$	$-5r^3$	$-7r^2$	$+r^2$	$-16r^2$	$-Aa^{-3}$

The velocity components can be calculated separately for each of the four solutions using (11). They are tabulated in columns 2 and 3 of table 1 for the stream functions listed in column 1. The pressure divided by viscosity can be calculated by inserting the values of u and v in the equations of motion and integrating. The results are given in column 4. In spherical polar co-ordinates the normal stress component is $\widehat{rr} = -p + 2\mu\,\partial u/\partial r$ and the tangential component is

$$\widehat{r\theta} = \mu\left\{r\frac{\partial}{\partial r}\left(\frac{v}{r}\right) + \frac{1}{r}\frac{\partial u}{\partial\theta}\right\}.$$

The values of $\widehat{rr}/\mu$ and $\widehat{r\theta}/\mu$ are listed in columns 5 and 6 of table 1.

* See *Modern Developments in Fluid Dynamics*, edited by S. Goldstein, Oxford University Press (1938), p. 115.

PLATE 1

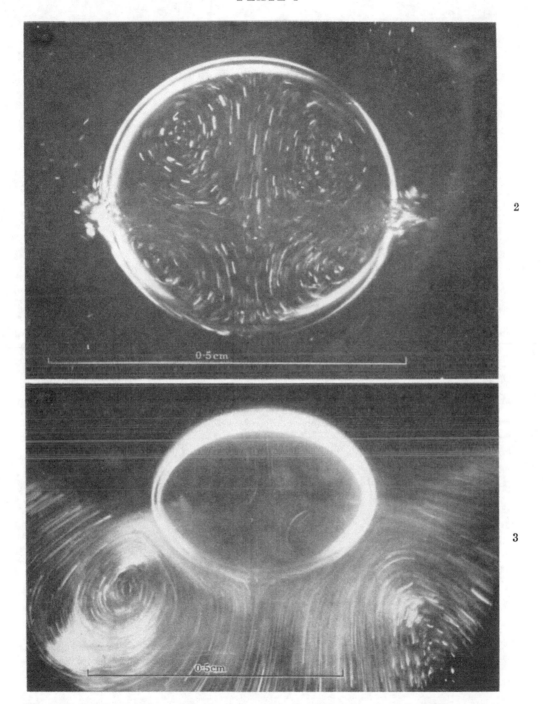

Fig. 2 Circulation inside a silicone oil drop in a mixture of castor oil and corn oil in d.c. field. Distance between electrodes, 4·62 cm. Applied voltage, 3·0 kV. The field is directed vertically. Exposure, 2 sec.

Fig. 3 Circulation outside a silicone oil drop in a mixture of castor oil and corn oil in d.c. field. Distance between electrodes, 4·62 cm. Applied voltage, 7·5 kV. The field is directed vertically. Exposure, 5 sec.

The continuity equations to be satisfied at $r = a$ are

$$u_1 = u_2 = 0 \quad \text{and} \quad v_1 = v_2,$$
$$A + B = C + D = 0 \quad \text{and} \quad -2A = 3C + 5D,$$
$$\left. \right\} \tag{17}$$

so that
$$A = -B = C = -D. \tag{18}$$

The constants by which the relevant terms of table 1 must be multiplied are given in column 7.

The stress components at $r = a$ in the two fluids are

$$(\widehat{rr})_1 = Aa^{-1}\mu_1(8-3)(1-3\cos^2\theta), \quad (\widehat{rr})_2 = Aa^{-1}\mu_2(-2-1)(1-3\cos^2\theta),$$
$$(\widehat{r\theta})_1 = Aa^{-1}\mu_1(-16+6)\cos\theta\sin\theta, \quad (\widehat{r\theta})_2 = Aa^{-1}\mu_2(-6+16)\cos\theta\sin\theta. \tag{19}$$

The equilibrium equations to be satisfied at $r = a$ are

$$(\widehat{rr})_E + (\widehat{rr})_1 - (\widehat{rr})_2 = C, \tag{20}$$

$$(\widehat{r\theta})_E + (\widehat{r\theta})_1 - (\widehat{r\theta})_2 = 0, \tag{21}$$

or

$$\frac{9F^2}{8\pi(2+R)^2}[\kappa_2 - \kappa_1 + \{\kappa_1(R^2+1) - 2\kappa_2\}\cos^2\theta] + A(2\mu_1 + 3\mu_2)(1 - 3\cos^2\theta) = C, \tag{22}$$

$$\frac{-9F^2}{8\pi(2+R)^2}(\kappa_1 R - \kappa_2)(2\cos\theta\sin\theta) - 10A(\mu_1 + \mu_2)\cos\theta\sin\theta = 0. \tag{23}$$

The constant C determines the difference between the internal pressure and the normal component of stress due to surface tension acting over the spherical surface. If we equate to zero the coefficients of $\cos^2\theta$ in (22) and $\cos\theta\sin\theta$ in (23), $(9F^2\kappa_2)/8\pi\mu_2(2+R)^2$ can be eliminated leaving an equation between the ratios $R = \sigma_1/\sigma_2$, $M = \mu_1/\pi_2$, $S = \kappa_1/\kappa_2$ namely

$$\frac{S(R^2+1)-2}{SR-1} + \frac{3}{5}\left(\frac{2M+3}{M+1}\right) = 0. \tag{24}$$

Equation (24) expresses the relation between the ratios R, M, S, which would permit a spherical drop to remain spherical when subjected to a uniform electric field.

CONDITION DISCRIMINATING BETWEEN PROLATE AND OBLATE DROPS

The foregoing analysis is confined to spherical drops. To find out whether under conditions differing slightly from those described by (24) the drop would be oblate or prolate suppose that a stress $C'\cos^2\theta$ applied normally to the surface of the drop is necessary to keep it spherical. The only change in the equilibrium equations will be to add a term $C'\cos^2\theta$ to (22), C' being positive if the added stress is directed inwards at the poles. Eliminating A between (22) and (23), we have

$$C' = \frac{9F^2\kappa_2}{8\pi(2+R)^2}\left\{S(R^2+1) - 2 + 3(SR-1)\left(\frac{2M+3}{5M+5}\right)\right\}. \tag{25}$$

Since all the symbols in (25) refer to positive quantities the condition that the normal stress $C' \cos^2 \theta$ is of the kind which would be exerted by surface tension if the sphere were strained into a prolate form is that C' is positive. The expression

$$\phi(S, R, M) = S(R^2 + 1) - 2 + 3(SR - 1)\left(\frac{2M + 3}{5M + 5}\right) \tag{26}$$

is therefore a discriminating function. If $\phi(S, R, M)$ is positive the drop will be prolate. If it is negative the drop will be oblate. It will be noticed that $\phi(S, R, M)$ discriminates between prolate and oblate forms independently of F.

CIRCULATION OF FLUID IN AND ROUND THE DROP

On any radius the maximum velocity occurs at the interface $r = a$ where its value (column 3 of table 1) is $2A \cos \theta \sin \theta$. The maximum velocity in the whole field occurs at $\theta = \frac{1}{4}\pi$ and is therefore A; which, according to (23) is

$$A = -\frac{9F^2 a \kappa_2}{8\pi(2 + R)^2}\left[\frac{SR - 1}{5(\mu_1 + \mu_2)}\right]. \tag{27}$$

The direction of circulation is from pole to equator at the interface when A is positive, that is when $SR - 1$ is negative.

The streamlines are shown in fig. 1 where the direction of motion for the case $SR < 1$ is indicated by arrows. The numbers shown in fig. 1 are the calculated values of ψ/Aa^2. It will be noticed that the velocity of circulation depends on the sum of the viscosities but not on their ratios.

In any comparison with observation the assumption that the distribution of potential depends only on the conductivities of the two fluids should be borne in mind. If the conductivities are too small to supply charge fast enough to replace the charge which is removed by the circulation at the interface the distribution of potential will not be that which was assumed.

COMPARISON WITH OBSERVATION

The only experiments in which all the relevant physical properties of drop and surrounding medium seem to have been recorded are those of Allen & Mason.[*] In table 1 on p. 51 of that paper the physical properties of 13 pairs of fluids, described as systems 1 to 13 are given. These data were transformed to the notation used here and are recorded in table 2. At the head of that table the correspondence between the two is recorded. Column 1 of table 2 gives Allen & Mason's number specifying the pair of fluids used. Columns 2 to 6 give the measured values of σ_1, σ_2, S, M, R. Column 7 gives the values of $SR - 1$ which according to theory should discriminate between the two possible directions of circulation. Allen & Mason did not report the existence of circulation so no comparison can be made.[†]

[*] *Proc. Roy. Soc.* A, CCLXVII (1962), 45.

[†] In reply to a letter describing my calculations Dr S. G. Mason wrote that he had in fact observed circulations both inside and outside both prolate and oblate drops.

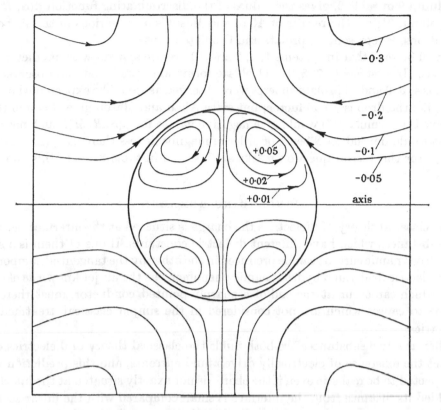

Fig. 1. Streamlines when SR is less than $1\cdot0$. The numbers are values of ψ/Aa^2. a is the radius of the drop and A is the maximum velocity given in (37).

Table 2. *Comparison between Allen & Mason's observations (column 9)*
and the sign of $\phi(S, R, M)$

In column 9, P indicates that a prolate form was observed; O, an oblate form.

1 A. & M. T. system	2 $1/\kappa_2$ σ_1	3 $1/\kappa_1$ σ_2	4 $1/q$ S	5 $1/p$ M	6 R R	7 — $SR-1$	8 — $\phi(S, R, M)$	9 obs. —
1	$> 3 \times 10^{12}$	$2\cdot2 \times 10^{11}$	$0\cdot46$	$0\cdot26$	> 13	> 5	> 80	P
2	$> 3 \times 10^{12}$	10^{11}	$0\cdot44$	$1\cdot0$	> 30	> 12	> 400	P
3	$> 3 \times 10^{12}$	$2\cdot8 \times 10^{10}$	$0\cdot52$	$5\cdot3$	> 107	> 55	$> 5 \times 10^3$	P
4	$> 3 \times 10^{12}$	$0\cdot9 \times 10^3$	0	5×10^3	$> 3 \times 10^9$	?	?	P
5	$> 3 \times 10^{12}$	$0\cdot9 \times 10^3$	0	5×10^3	$> 3 \times 10^9$	?	?	P
6	10^{11}	$> 3 \times 10^{12}$	$2\cdot27$	$0\cdot08$	$< 3 \times 10^{-2}$	$-0\cdot93$	$-1\cdot7$	O
7	10^{11}	$> 3 \times 10^{12}$	$2\cdot27$	$1\cdot00$	$< 3 \times 10^{-2}$	$-0\cdot93$	$-2\cdot1$	O
8	10^{11}	$> 3 \times 10^{12}$	$2\cdot27$	$5\cdot2$	$< 3 \times 10^{-2}$	$-0\cdot93$	$-2\cdot6$	O
9	10^{11}	$0\cdot9 \times 10^3$	0	5×10^3	$1\cdot1 \times 10^8$	?	?	P
10	$2\cdot8 \times 10^{10}$	$> 3 \times 10^{12}$	$1\cdot92$	$0\cdot19$	$< 10^{-2}$	$< -0\cdot98$	< -2	O
11	$2\cdot8 \times 10^{10}$	$> 3 \times 10^{12}$	$1\cdot92$	$1\cdot0$	$< 10^{-2}$	$< -0\cdot98$	$< -2\cdot4$	O
12	$2\cdot8 \times 10^{10}$	$0\cdot83 \times 10^6$	$0\cdot14$	50	$3\cdot4 \times 10^4$	$4\cdot8 \times 10^3$	$1\cdot6 \times 10^8$	P
13	$2\cdot8 \times 10^{10}$	$0\cdot9 \times 10^3$	0	10^3	3×10^7	?	?	P

Column 8 of table 2, gives the values of the discriminating function $\phi(S, R, M)$ and column 9 gives the results of Allan and Mason's observations of the shapes of their drops, P representing prolate and O oblate drops.

It will be seen that in systems 1, 2, 3 and 12 prolate spheres were predicted and observed. In systems 6, 7, 8, 10, 11 oblate forms were predicted and observed. In systems 4, 5, 9 and 13 prolate spheres were observed as would be expected when R is great, i.e. the drop is a conductor; but since Allan and Mason give $S = 0$ in these systems the formula (26) would give negative values for $\phi(S, R, M)$. Such negative values would depend on S being of order of magnitude less than $1/R^2$, (i.e. of order 10^{-14}). For this reason question marks are inserted in table 2 where this anomaly exists.

CONCLUDING REMARKS

In the classical theory of dielectrics the charge is situated at the interface between two substances which have different dielectric constants. If one of them is a solid no hydrodynamic current can be produced in the other by the tangential component of the electric field. Such hydrodynamic phenomena as the air jet known as electric wind which can occur at the end of a sharply pointed conductor, must therefore be due to causes which are not considered in the simple classical treatment of dielectrics.

When the two substances are both fluids the classical theory of dielectrics does predict the existence of electrically driven fluid currents, and this prediction may be expected to be realistic even if the charge is not exactly situated at the interface, provided its distance from the interface is small compared with the linear scale of the situation.

ADDENDUM

By A. D. McEWAN AND L. N. J. de JONG

Experimental verification

One difficulty in trying to observe the predicted circulation is that a drop is liable to become charged and to migrate at a rate sufficient to distort substantially the field of flow. To avoid this difficulty two vegetable oils, castor oil (sp.gr. 0·963) and corn oil (sp.gr. 0·919) were partially mixed to form a stratified fluid with small density gradient ($0·43 \times 10^{-2}$ g. cm.$^{-4}$). Within the fluid a drop of a silicone fluid I.C.I. 535/10 of intermediate density (sp.gr. 0·942) could be suspended. The conductivity of the mixture was much greater than that of the silicone drop and $SR < 1$ ($O(10^{-2})$), so that the predicted surface circulation would be from pole to equator. The stratified fluid was contained in a glass-walled box between two horizontal brass electrodes ($10·3 \times 10·3$ cm.) separated by 4·62 cm. To reveal the circulation, particles of Pearlite dry powder were introduced in the silicone drop and a vertical plane normal to the line of vision was illuminated in a narrow band.

The circulation was clearly visible and was in the direction predicted. Photographs were taken by a camera outside the cell containing the fluid. Examples are given in figs. 2 and 3 (pl. 1). The particles were seen to leave the drop on the equatorial diameter and thus revealed the lower half of the external circulation, which is shown in fig. 2. The external motion is also revealed by a few particles in fig. 2, the shortness of their tracks shows that the external circulation is slower than the internal in most parts of the field. In comparing this with fig. 1 it will be noticed that the external circulation is in closed streamlines. This may well be due to the stratification of the external fluid and the finite extent of the cell.

44

OBLIQUE IMPACT OF A JET
ON A PLANE SURFACE

REPRINTED FROM

Philosophical Transactions of the Royal Society, A, vol. CCLX (1966), pp. 96–100

The dynamics of the region where a jet, striking a plane surface obliquely, is transformed into a thin sheet will be discussed. The maximum (stagnation) pressure is the same for all angles of incidence but the area over which the high pressure acts is much reduced as the angle of incidence, θ, becomes small. The main transformation from a jet to a sheet is due to a pressure of order $\sin^2 \theta \times$ the stagnation pressure acting over an area of order $\operatorname{cosec} \theta \times$ the jet cross section. The pressure is due to the destruction of the component of velocity normal to the impact surface, but since pressure acts equally in all directions it imparts lateral velocity to streams which are not in the plane of symmetry. Each element of the stream can be regarded as passing through a small region where an impulsive body force changes its direction without changing its velocity, and some properties of this impulse will be described.

One case will be given where the transformations from a jet to a thin flat sheet can be described completely and the calculated distribution of pressure in the region where it occurs compared with experimental measurements.

Though a jet cannot produce a pressure greater than the stagnation pressure as a steady state, it seems theoretically possible to attain a much higher pressure for a short time when a very oblique jet is moved sideways.

When a jet strikes a flat plate it is transformed into a sheet which flows outwards radially over the plate from a region of impact whose dimensions are of the order of the cross section of the jet. A complete discussion of the mechanics of the situation would include friction effects at the solid surface, but in most cases this is relatively unimportant and it is sufficiently realistic to discuss the flow as though it were produced by the impact of two equal jets. In that case the plane of symmetry replaces the impact plate and there is no friction on that plane.

The physical event which occurs in the impact region is that a pressure is built up which serves to deflect streamlines from the direction of the jet to lines in the impact plane which spread out radially from the impact region. The stream velocity varies in passing through this region but, if the fluid is inviscid, regains its original value as it passes out of it. The overall effect on a stream tube of fluid passing through the impact region is that which would be produced by a steady force applied in a direction that bisects the angle between its original and final directions.

TWO-DIMENSIONAL JETS

In general it has not been possible to calculate the distribution of pressure in the impact region but a complete solution was given by Michell* for the impact of a flat jet in two-dimensional flow. Without knowing the details of what happens in the impact region, one can show that a jet impinging obliquely on a plane at angle θ will divide itself into two, a fraction $\cos^2 \tfrac{1}{2}\theta$ going forward and $\sin^2 \tfrac{1}{2}\theta$ going backwards. This is a consequence of the conservation of momentum parallel to the plate. To find how the pressure is distributed over the plate it is necessary to use Michell's analysis. He did not give the necessary expressions in explicit form but I find that u_1 the non-dimensional velocity on the plate, i.e. velocity$/U$ (U being the velocity of the jet) is related to a non-dimensional co-ordinate x_1 on the plate though an auxiliary variable q by the relations

$$u_1 = \frac{-1 - q\cos\theta + \sqrt{(1-q^2)}\sin\theta}{q - \cos\theta}, \tag{1}$$

$$x_1 = \tfrac{1}{2}\{(1+\cos\theta)\ln(1+q) - (1-\cos\theta)\ln(1-q)\} + \sin\theta\sin^{-1}q + \text{const.}, \tag{2}$$

and the pressure p at x_1 is
$$p = \tfrac{1}{2}\rho U^2(1 - u_1^2). \tag{3}$$

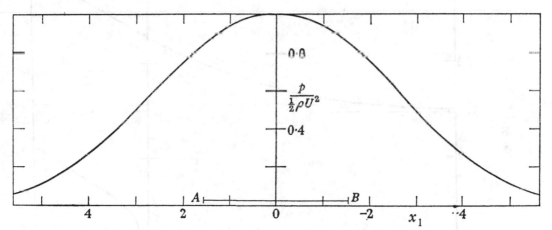

Fig. 1. Pressure on plate, $\theta = 90°$. AB is jet width.

Figs. 1 and 2 show the distribution of $p/\tfrac{1}{2}\rho U^2$ as a function of $x_1 = \pi x/$(width of jet) for $\theta = 90°$ and $\theta = 30°$. The areas of the curves in fig. 1 and the upper part of fig. 2 represent the force that the jet exerts on the plate, which is $\rho U^2 \sin\theta$ (width of jet). Since $\sin 30° = \tfrac{1}{2}$, the area of the pressure curve in fig. 1 is twice that of fig. 2.

In fig. 1 and the upper part of fig. 2 the width of the jet is marked and in the lower half of fig. 2 the shape of the jet calculated from Michell's equations is also shown. It will be seen that when $\theta = 30°$ the jet divides into two portions so that only a fraction $\sin^2 15°$ or 6·7 % of it goes backwards. It will be noticed that though the

* *Phil. Trans. Roy. Soc.* A, CLXXXI (1890), 389.

downward force exerted on the plane decreases as θ decreases the maximum value of the pressure, namely $\frac{1}{2}\rho U^2$, remains unchanged but when θ becomes very small the area over which the pressure is near $\frac{1}{2}\rho U^2$ becomes very small.

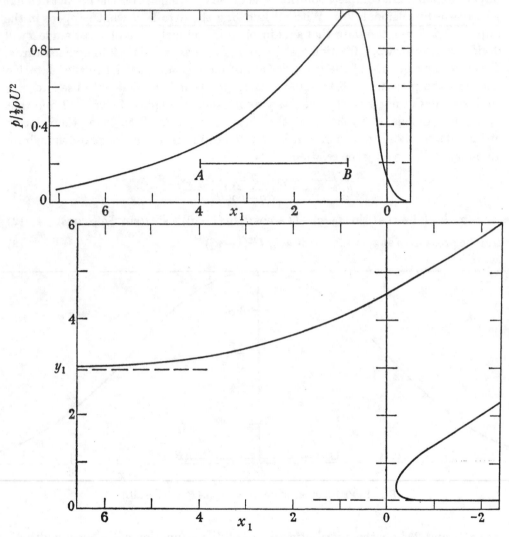

Fig. 2. The upper half: pressure on plate $\theta = 30°$, AB is jet width.
The lower half: shape of jet when $\theta = 30°$.

IMPACT OF A JET WHEN FLOW IS NOT
CONFINED TO TWO DIMENSIONS

No complete description of the impact region analogous to Michell's has been given for any three-dimensional case and it can be proved that considerations of conservation of momentum alone are incapable of giving the angular distribution of the thickness of the outflowing sheet formed by the impact. If we define the angular co-ordinate of a radius vector from the impact region on the impact plane as ϕ, and take $\phi = 0$ as the line of the projection of the inflowing jet on the plane, the proportion of the fluid which flows out between ϕ and $\phi + d\phi$ is $F(\theta, \phi)$ and by definition

$$\int_0^{2\pi} F(\theta, \phi)\, d\phi = 1. \tag{4}$$

$F(\theta, \phi)$ can be represented by a Fourier series

$$F(\theta, \phi) = \sum_0^\infty A_n \cos n\phi + \sum_1^\infty B_n \sin n\phi, \tag{5}$$

and since the jet is symmetrical in section, $B_n = 0$. The inward rate of flow of inertia parallel to any direction ϕ_0 is $\rho U^2 \cos \theta \cos \phi_0$ per unit area of cross section. The rate of outward flow in this direction is

$$\rho U^2 \int_0^{2\pi} F(\theta, \phi) \cos(\phi - \phi_0)\, d\phi. \tag{6}$$

Since the pressure of the plate on the stream cannot contribute to this,

$$\cos \theta \cos \phi_0 = \int_0^{2\pi} \left(A_0 + A_1 \cos \phi + \sum_2^\infty A_n \cos n\phi \right)(\cos \phi \cos \phi_0 + \sin \phi \sin \phi_0)\, d\phi. \tag{7}$$

This equation is satisfied for all values of ϕ_0 provided $A_1 = (1/\pi) \cos \theta$. The only other condition is that to satisfy (6), $A_0 = \frac{1}{2}\pi$. Thus for all values of $n \geqslant 2$ the coefficient A_n can have any values so far as the equations for inertia and continuity are concerned .

Though there seems to be no way in which $F(\theta, \phi)$ can be determined without making a complete analysis of the flow in the impact region it is worthwhile to look at the inverse problem and see what can be learned about the forces in the impact region if the value of $F(\theta, \phi)$ which can be measured, were known. Some years ago I measured $F(\theta, \phi)$ for the sheet produced by pairs of equal jets of circular cross section impinging at angles of $2\theta = 60$, 90 and 120°. In the paper* which described the measurements I gave the equations from which the component of force parallel to the impact plane which was needed to deflect each element of the jet into its observed angular position could be calculated. If $\rho U^2 I(\theta, \chi)\, d\chi$ is the force required, I gave the relation between ϕ and χ, namely

$$\tan \chi = \frac{\sin \phi}{\cos \phi - \cos \theta}, \tag{8}$$

and also the formulae relating $F(\theta, \phi)$ to $I(\theta, \chi)$. The distribution of $I(\theta, \chi)$ round the plane must be such that all the forces are in equilibrium because every element in the set is balanced by the reaction of the rest. In other words, if a set of masses proportional to $I(\theta, \chi)$ were distributed on a circle their centre of gravity would be at the centre.

* *Proc. Roy. Soc.* A, CCLIV (1960), 1 (paper 33 above).

The measured values of $F(\theta, \phi)$ as a function of ϕ and the values of $I(\theta, \chi)$ calculated from them were given in figs. 7 to 9 of my paper.* Among the things I noticed but could not find any theoretical reason for was the fact that the $I(\theta, \chi)$ curve appeared to be very nearly symmetrical about $\chi = \frac{1}{2}\pi$. This was particularly remarkable because of the very great asymmetry of the $F(\theta, \phi)$ curves. Indeed this was so great that $I(\theta, 0)$ was almost exactly equal to $I(\theta, \pi)$, although $F(\theta, 0)$ for $\theta = 30°$ was 100 times as great as $F(\theta, \pi)$. This may be compared to the two dimensional case where for $\theta = 30°$ the thickness of the forward-going stream (analogous to $F(\theta, 0)$) was about 14 times as great as that of the backward-going stream analogous to $F(\theta, \pi)$. If the distribution of reaction force is symmetrical about $\chi = \frac{1}{2}\pi$ as well as $\chi = 0$ the force in every angular sector $d\chi$ is exactly balanced by an equal reaction in the same interval $d\chi$ but oppositely directed.

Another point I noticed was that the total reaction force between the two halves of the stream across the plane of symmetry (i.e. the force which gives rise to the lateral spread) was 0·64, 0·65 and 0·66 times the vertical force $\rho U^2 \sin \theta$ when θ was 30, 45 and 60°. The reaction across any axial plane when a round jet strikes a plane at right angles ($\theta = 90°$) can be shown theoretically to be $2/\pi = 0·64$. The significance of this ratio (approximately $2/\pi$) probably lies in the fact that the jet was circular in section. It might be interesting to measure $F(\frac{1}{2}\pi, \phi)$, i.e. the angular distribution of thickness of the sheet produced by two jets of elliptic cross section aimed directly at one another.

Transformation of a converging into a diverging jet

The only case where I have been able to calculate the flow and distribution of pressure in a region where a diverging sheet is formed is that of a converging jet of elliptical cross section. The analysis for this case is given in paper 33 above and the experimental arrangements for producing the jet and for measuring the pressure in the region where the converging jet is transformed into a diverging jet are also described. The agreement between calculated and measured distribution of pressure along the axis is striking. Unfortunately there is a misprint in the description of the photograph of the jet (fig. 4 of the paper) where the scale length is given as 10 cm. instead of 1·0 cm.*

When this jet is converging the flow is somewhat analogous to two jets converging but the small high pressure region, which always occurs when the angle of convergence of two separate jets is small, does not occur in that case.

Theoretical possibility of the existence of a region where the maximum pressure can be much greater than $\frac{1}{2}\rho U^2$

Consider what happens when a jet of velocity U is directed downwards at a small angle θ on to a horizontal plane. Suppose that the horizontal plane now moves upwards with constant velocity V, the orifice remaining fixed in space. The impact

* The error has been corrected in this volume. *Ed.*

plane can be brought to rest by moving the whole frame of reference downwards with velocity V. This of course makes the orifice also move downwards with velocity V. If now the whole frame of reference is given a velocity $V \operatorname{cosec} \theta$ parallel to the plane, the flow is reduced to a steady state in which, relative to the new frame of reference, the orifice approaches the impact point with velocity $V \operatorname{cosec} \theta$. Thus in this frame of reference a steady stream approaches the impact point with velocity $U + V \operatorname{cosec} \theta$ and the pressure distribution is the same as it was before except that its magnitude is increased in the ratio $\{1 + (V/U) \operatorname{cosec} \theta\}^2$ and the maximum pressure is $\frac{1}{2}\rho(U + V \operatorname{cosec} \theta)^2$. Suppose for instance that a jet at a head of 1 m. of water is directed vertically downwards at a plate the maximum pressure is that of 100 cm. of water and the jet velocity is 4·4 m./sec. Suppose now that the plate is moved upwards with velocity 50 cm./sec. the maximum head of water is increased to $100(1 + 1/8\cdot8)^2 = 124$ cm. of water. If now the jet is at 5° to the plane and it is moved upwards at 50 cm./sec. the maximum head will be $100(1 + \operatorname{cosec} \theta/8\cdot8)^2 = 730$ cm. This high pressure spot will of course move very rapidly across the impact plane.

45

THE PEELING OF A FLEXIBLE STRIP ATTACHED BY A VISCOUS ADHESIVE*

REPRINTED FROM

Journal of Fluid Mechanics, vol. XXVI (1966), pp. 1–15

The peeling of a flexible strip from a rigid surface to which it is attached by a thin layer of adhesive is discussed, treating the adhesive as a Newtonian viscous fluid. This makes it possible to examine the flow and stress distributions ahead of the point where separation occurs. The conditions at this point are taken to be the same as those observed in other cases where a stream of viscous fluid separates into two. In particular, the effect of surface tension at the separating meniscus on the speed of peeling is predicted.

Experiments are described in which a sheet of 'Melinex' 4 μm. thick was laid on a sheet of fluid covering a piece of plate glass. The apparatus was designed to ensure that this was peeled off at a constant angle, and the speed of the separation meniscus, as well as the load on the sheet, was measured. The experimental results are analysed in the light of the theory and shown to be consistent with it.

An interesting feature is the prediction that at low peeling speeds there is a great reduction in the thickness of the adhesive layer immediately ahead of the line of separation. Although the initial thickness of the layer dictates the scale of the shape adopted by the strip ahead of this line, it exerts no effect upon the relation between the external variables.

It is noted that, when the adhesive layer remains intact ahead of separation, the physical appearance of commercially available tapes in slow peeling can resemble that of simple viscous adhesives.

1. INTRODUCTION

When a flexible sheet is peeled from a rigid surface to which it has been attached by a layer of adhesive, the stress within the adhesive resisting the peeling is confined to a region near to the line of separation or rupture. Such a situation commonly arises in the stripping of what are technically known as 'pressure-sensitive' adhesive tapes. In this application, numerous experimental studies have been made of peeling adhesion, but a satisfactory rationalisation of the observed behaviour is impeded by the difficulty in taking full account of the complicated rheological properties which most good adhesives are known to possess.

A simplification that has been the basis of several theoretical models is the assumption of purely Hookean elastic adhesive properties, with each connecting adhesive element acting independently of its neighbours and with failure occurring when a limiting stress condition is reached.† With such a model, the strain energy in the

---

* With A. D. McEWAN.

† See, for example, J. J. Bickermann, *J. appl. Phys.* XXVIII (1957), 1484, and D. H. Kaelble, *Trans. Soc. Rheol.* IV (1960), 45.

adhesive when it fails (W) is related to its breaking strength and its elastic properties in such a way that W could be determined by appropriate experiments using other means than peeling a strip from a rigid surface. In this model, the tension T needed to peel unit width of strip when applied at an angle θ to the rigid surface is directly connected with W through the equation

$$W = T(1 - \cos\theta),\tag{1}$$

so that T could be calculated using values of W determined independently.

However, models assuming elastic deformation to failure cannot account for the observation that many adhesives peel at a rate which depends upon T. Then although (1) remains true it is of less value in relating T with θ. These quantities might still be connected using approximations to the visco-elastic properties of the adhesive determined perhaps by simple tension-extension-time experiments as in the work of Chang,* but it is questionable whether the distribution of stress, strain and rate of strain within much of the critical region would be the same as that in a simple one-dimensional experiment.

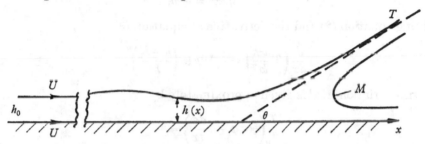

Fig. 1. Peeling model.

For some adhesives, the rate of propagation U of the separation region is nearly proportional to T over a considerable range of values of T. For such cases, (1) shows that W is proportional to U. This is characteristic of Newtonian viscosity.

It is perhaps surprising that few theories of peeling have been proposed on the assumption that the adhesive possesses no elastic strength and is purely viscous. This neglect is probably due, in part, to the difficulty of defining suitable criteria for finding the position of the line of separation or rupture in an adhesive layer between diverging surfaces, if no specific 'failure' condition can be given.

In this note, use is made of the similarity between the separation of a viscous layer in the present context, and the previously studied cases in which a liquid layer confined within a narrow space between rigid boundaries is forced to divide about an advancing free surface, to enable the rupture position to be defined and a full description found. The model considered is a completely flexible sheet separated from a plane rigid surface by a Newtonian liquid layer of viscosity μ and initial thickness h_0. Reference is made to fig. 1. The co-ordinate system is fixed with respect to

* *Trans. Soc. Rheol.* IV (1960), 74.

the moving tear by imposing the propagation velocity U. The motion is taken to be steady and two-dimensional. When the liquid separates, part of it adheres to the surface and part to the sheet. The total amount per unit area adhering to both surfaces is equal to that far upstream of the separation point, but the amount of fluid occupying the diverging space immediately before the separating meniscus M must depend upon how easily it is able to flow in the region of this space.

We confine our attention to small peeling angles θ, for which the Reynolds approximations greatly simplify the analysis; furthermore, we assume that the Reynolds number $\rho U h_0/\mu$ is negligibly small in cases of practical interest. Then the depth of the fluid h at position x will depend upon μ, and the longitudinal pressure gradient dp/dx is related to h by

$$(h - h_0)\, U = (h^3/12\mu)\, dp/dx, \tag{2}$$

p being the pressure excess over the atmospheric pressure. When the strip is inextensible but possesses no flexural rigidity, then, for small slopes,

$$T\, d^2h/dx^2 = -p. \tag{3}$$

Combining equation (2) and the derivative of equation (3),

$$\frac{d}{dx}\left(T\frac{d^2h}{dx^2}\right) = -12\mu U\left(\frac{h - h_0}{h^3}\right). \tag{4}$$

The x-wise variation in T is given approximately by

$$\frac{T}{T_0}\left(\cos\phi + h\frac{d\phi}{dx}\right) = 0, \tag{5}$$

where T_0 is the tension at points where the local slope ϕ of the strip is zero, and internal pressure is $-T\, d\phi/dx$. Since ϕ is always small, however, the tension T can be taken as constant, and, with new variables $\zeta = h/h_0$, $\xi = x\alpha^{-\frac{1}{3}}/h_0$, $\alpha = T/12\mu U$, equation (4) becomes

$$\zeta''' = (1 - \zeta)/\zeta^3, \tag{6}$$

which is valid when $\alpha^{-\frac{1}{3}}\zeta' = \tan\phi$ is small. Primes denote differentiation with respect to ξ.

The intial boundary condition for all solutions of interest in the present case is $\zeta \to 1$ as $\xi \to -\infty$, and in the vicinity of $\zeta = 1$ the equation is approximated by

$$\zeta''' = 1 - \zeta \tag{7}$$

for which the solution is

$$\zeta - 1 = A\, e^{-\xi} + B\, e^{\frac{1}{2}\xi}\sin\left(\tfrac{1}{2}\sqrt{3}\xi + \epsilon\right). \tag{8}$$

A must be taken as zero to ensure that as $\xi \to -\infty$ the fluid is stationary with respect to the bounding surfaces. Thus

$$\zeta' = \zeta'' = \tfrac{1}{3}B\sqrt{3} \quad \text{as} \quad \zeta \to 1. \tag{9}$$

This provides a convenient starting-point for the numerical integration of equation (6). For reasons explained later it is necessary to obtain a set of solutions represented by different values of B. Since $\zeta - 1$ is periodic, with period $4\pi/\sqrt{3} = 7\cdot26$, solutions which start with a given value of B are repeated when B increases in ratio $e^{2\pi\sqrt{3}} = 37\cdot622$ times. This provides a convenient check as to whether any particular choice of B is small enough to warrant the use of equations (8) and (9) for initial values in the numerical solution of equation (6).

1.1. *Physical conditions at the meniscus*

The physical conditions at the meniscus are unknown but there is no reason to suppose that they differ from those at the meniscus between rigid diverging surfaces. The case bearing the closest similarity to the present one would be that in which the rigid surfaces each travelled at the same velocity with respect to the minimum gap position. Experiments have been performed by Pitts & Greiller* in which two partly immersed rollers were driven in opposite directions of rotation. The liquid drawn through the narrow gap between the rollers formed a meniscus ahead of this gap, the position of which could be measured. Unfortunately, the data are of limited extent and it is necessary to consider the analogous case dealt with by Taylor† of motion in which one surface is fixed and the other moves away from the meniscus. Such motion arises in a partly filled journal bearing. Taylor pointed out that two conditions must be fulfilled at the meniscus and that they both depend upon $\mu U/\sigma$, where σ is the surface tension. The conditions are: (i) that for determining the pressure drop‡ between the atmosphere and some point upstream of the meniscus in the liquid; and (ii) that which gives m, the ratio of the thickness of the layers of fluid adhering to the moving surface (surfaces) to the width of the gap at the meniscus.

The experimental determination of these conditions as a function of $\mu U/\sigma$ is a matter of some difficulty. No measurements have yet been made of pressure drop across the meniscus (condition (i)), for either case. For the second condition, results were obtained in both the above-mentioned studies. However, a complication arises in the definition of m from these results. The value of m can be found simply by determining both the meniscus position (which defines h_m, the width of the gap at the meniscus) and the amount of liquid adhering to the moving surfaces after passing the meniscus. This method was used by Pitts & Greiller in an attempt to find $\lambda = q/Uh_0$, a parameter defining the amount of liquid passing through h_0, the maximum gap width upstream of the meniscus. Here q is the volume flux per unit width, and U is the surface velocity. Since the determination of q was insensitive they confine their published experimental results to measured values of h_0/h_m. λ is a number which depends upon the geometry of the surfaces and the fluid

---

* *J. Fluid Mech.* xi (1961), 33.

† *J. Fluid Mech.* xvi (1963), 595 (paper 38 above).

‡ The quantity here described as 'pressure drop' and represented later by the symbol δp in equation (13) is the difference between the pressure in the air outside the meniscus and the pressure which would exist according to Reynolds approximation if the flow had continued undisturbed up to the position of the meniscus.

conditions on both sides of the nip h_0, but, if λ could be defined, then m would be given as

$$m = \lambda h_0/h_m. \tag{10}$$

The value of λ, when both sides of the nip are flooded, is $\frac{4}{3}$. If h_m is substantially larger than h_0 the effect of the meniscus in altering the value of λ should be small.

Adopting this value for λ, the results given in fig. 3 of Pitts & Greiller's paper, together with some extra data kindly supplied by Dr Pitts, were used to prepare fig. 2. The results are seen to lie fairly close to the line $m = 0.63(\mu U/\sigma)^{\frac{1}{2}}$ but the

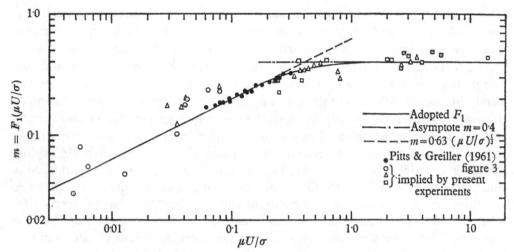

Fig. 2. $m = 1/\zeta = F_1(\mu U/\sigma)$ as a function of $\mu U/\sigma$.

exponent of the curve of best fit seems to be slightly lower. Although there is no theoretical reason for supposing that the points asymptotically approach a parabolic form for low $\mu U/\sigma$, this form occurs in two other analogous situations, namely in the driving of a viscous liquid from a cylindrical tube by a long bubble* (Bretherton† theoretically predicts a $(\mu U/\sigma)^{\frac{2}{3}}$ relationship at very low driving speeds, but his experimental results are inclined to suggest a somewhat lower exponent), and in the journal-bearing case of Taylor.‡ In each case the multiplicative factor is different, as might be expected from the different physical configurations. A second characteristic common to these analogous situations is that the value of m apparently approaches asymptotically a value less than 1 as $\mu U/\sigma$ becomes large,§ and, from the appearance of Pitts & Greiller's results on fig. 2, such levelling off of m may also be occurring in their experiments using two rollers. There is no direct experimental evidence which would enable us to give an exact value for m_∞, the asymptotic limit of m, and for the analysis it was assumed to be 0.4. Subsequent experimental results, given later, confirmed that this assumption was reasonably accurate.

* F. Fairbrother & A. E. Stubbs, *J. chem. Soc.* I (1935), 527; G. I. Taylor, *J. Fluid Mech.* x (1961), 161 (paper 35 above).

† *J. Fluid Mech.* x (1961), 166.   ‡ Paper 38 above.

§ G. I. Taylor, papers 35 and 38 above; B. G. Cox, *J. Fluid Mech.* xiv (1962), 81.

PLATE 1

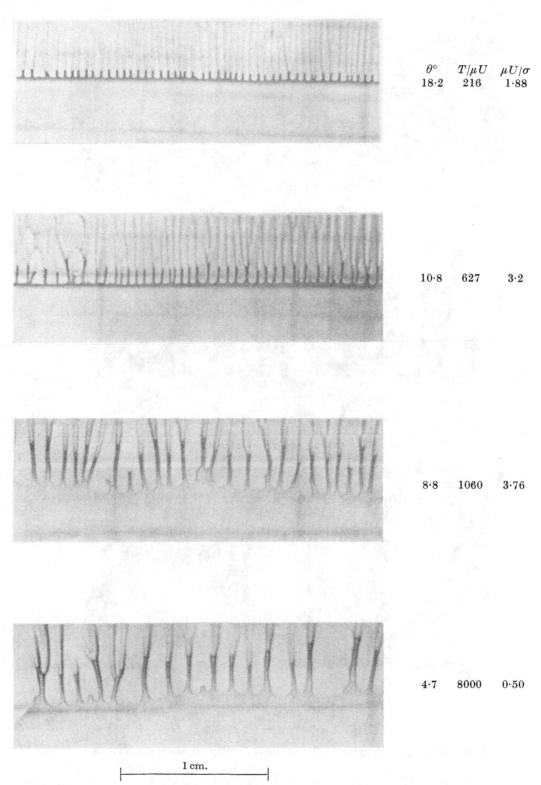

$\theta°$	$T/\mu U$	$\mu U/\sigma$
18·2	216	1·88
10·8	627	3·2
8·8	1060	3·76
4·7	8000	0·50

1 cm.

Fig. 8. Meniscus appearance—viscous liquid. Peeling advances downwards.
$\mu = 73$ poise, $\sigma = 33·6$ dyne/cm.

PLATE 2

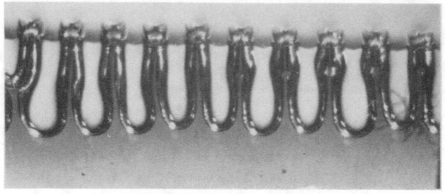

(I) $T = 0.48$ kg./cm. $2\theta = 38°$

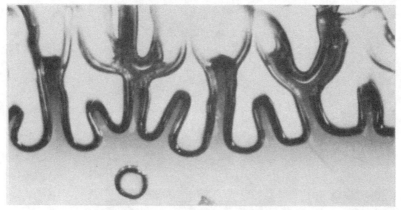

(II) $T = 0.48$ kg./cm. $2\theta = 55°$

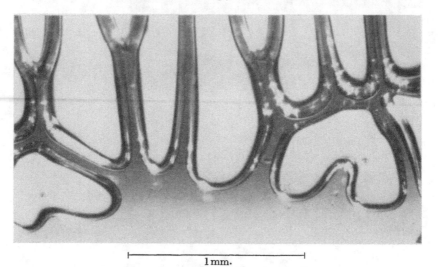

1 mm.

(III) $T = 0.87$ kg./cm. $2\theta = 38°$.

Fig. 9. Peeling of 'Acrylate' adhesive tape. Peeling advances downwards.

1.2. *Application of meniscus conditions to the peeling-strip solution*

Details of the solutions of equation (6) describing the gap width as a function of position are given in the next subsection. It is useful to consider first what information is required from these solutions.

Far upstream of the meniscus, the liquid film is undisturbed and is stationary with respect to the plane and the strip, occupying a gap of width h_0. For this case therefore, $\lambda = 1$ and m is simply

$$m = h_0/h_m = \zeta_m^{-1}. \tag{11}$$

In this case we assume that with small peeling angles

$$m = F_1(\mu U/\sigma). \tag{12}$$

In the complete absence of data on the pressure drop through the meniscus one can only speculate upon its probable magnitude, but dimensional analysis suggests that its form is

$$-\delta p = \frac{\mu U}{h_0 \zeta} F_2\left(\frac{\mu U}{\sigma}\right). \tag{13}$$

When $\mu U/\sigma$ is very small, surface tension will dominate and the pressure drop will be approximately
$$-\delta p = 2\sigma/h_0\zeta,$$

so that $$F_2(\mu U/\sigma) = 2\sigma/\mu U. \tag{14}$$

For $\mu U/\sigma$ large, the shape of the meniscus approaches one defined by viscous stress alone and dimensional arguments suggest that

$$-\delta p = C\mu U/h_0\zeta,$$

where C is a constant, almost certainly positive and probably less than $1\cdot0$. Thus, when $\mu U/\sigma$ is large, the change in pressure across the meniscus is of the same order as that occurring over an x-wise distance of h_0.

Immediately ahead of the meniscus the pressure condition, using equation (3) and the terms defined before equation (6), is

$$\alpha^{-\frac{2}{3}}T/h_0 = -\delta p, \tag{15}$$

or $$\zeta'' = \tfrac{1}{12}\alpha^{-\frac{1}{3}}F_1F_2. \tag{16}$$

Since the slope of the strip does not change after the meniscus,

$$[\zeta']_{\zeta=m} = \alpha^{\frac{1}{3}}\tan\theta. \tag{17}$$

Thus, the end boundary conditions describing the shape of the strip rely upon a knowledge of F_1 and F_2. For F_1, the solid line of fig. 2 is taken, but for F_2 equation (14) can only be justifiably adopted for low values of $\mu U/\sigma$.

From equations (11), (16) and (17) the solution of equation (6) is related to the externally measurable quantities in a peeling experiment, namely $\mu U/\sigma$, α and θ. Solutions of the equation must therefore yield the functional relationship between ζ, $d\zeta/d\xi$ and $d^2\zeta/d\xi^2$. The value of ξ is immaterial.

1.3. *Solutions of equation* (6) *and their conversion to a usable form*

Equation (6) was solved by numerical integration using the Cambridge University Mathematical Laboratory's 'Edsac II' digital computer. A Runge–Kutta method was adopted, and the interval of integration was made variable depending upon the value $d^2\zeta/d\xi^2$ in order that accuracy could be retained when quantities were varying rapidly. A test of accuracy could be obtained by performing the integration with various basic step sizes.

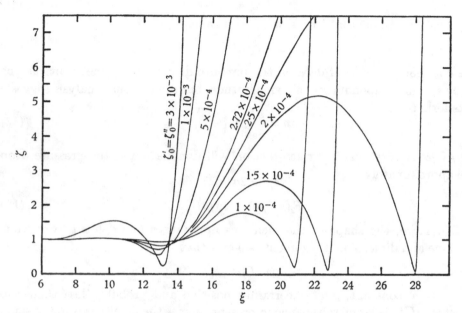

Fig. 3. Dimensionless form of peeling strip.

Integration was commenced from $\zeta = 1$ with ζ' and ζ'' equal according to equation (9) and with values of B between $8\cdot66 \times 10^{-5}$ and $2\cdot60 \times 10^{-3}$ chosen from trial solutions performed by Dr R. Herczynsky, to cover adequately the resultant range of variables. Some representative results are presented in fig. 3. The pattern of results repeated itself in ξ with a period of $4\pi/\sqrt{3}$. Integration was commenced from $\xi = 0$, but for all initial values of ζ' and ζ'' covered by the calculations the variation in ζ is insignificant before $\xi = 6$. It will be seen that for all initial values ζ undergoes one or two cycles of oscillation of rapidly increasing amplitude before departing monotonically from the ξ-axis. The number of cycles of oscillation before departure is fixed by the initial values of ζ' and ζ'', but it can be demonstrated that all solutions must finally approach infinity.

There seems to be no analytical way of finding the initial conditions for monotonic departure in a given cycle. This occurs in the nth cycle, where n is the least integer for which $B > (37\cdot622)^{-n} B_0$, and B_0, determined from computer solutions of (6), lies between $2 \cdot 365 \times 10^{-4}$ and $2 \cdot 375 \times 10^{-4}$. Below the latter value, ζ will execute a further half cycle before departure, and in this case the minimum value of ζ after this half cycle varies inversely with its amplitude, and the gradient of departure is subsequently steeper. Fig. 4 plots ζ' and ζ'' as a function of ζ for those parts of the curves of fig. 3 which exhibit positive values of these derivatives. The abscissa is written, for convenience, ζ^{-1}, which by equation (11) corresponds to m.

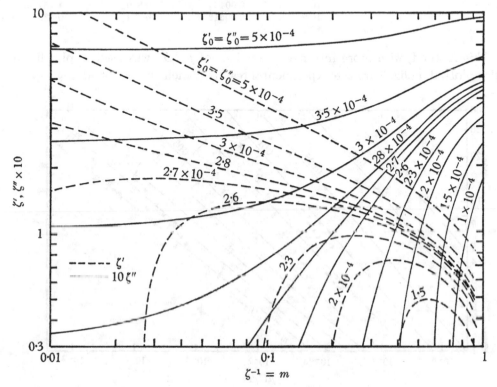

Fig. 4. ζ' and ζ'' vs. $1/\zeta$.

From fig. 4, a plot of ζ' against ζ'' can be made for constant values of m. Now, by equation (16), ζ'' is given as a function of $F_1 F_2$ for constant values of α. Since F_1 and F_2 have been taken as functionally related to $\mu U/\sigma$ with values given by fig. 2 and equation (14), respectively, ζ'' can be expressed in terms of $m, \mu U/\sigma$ and α:

$$\zeta'' = m(\mu U/\sigma)\,\sigma/6\alpha^{\frac{1}{3}}\mu U, \tag{18}$$

which enables m, α and θ to be related. As a typical calculation, take $m = 0 \cdot 3$, $\zeta = 3 \cdot 333$ and, from fig. 2, $\mu U/\sigma = 0 \cdot 249$. Try $\frac{1}{12}\alpha = \mu U/T = 10^{-2}$. Then, by equation (18), the required value of ζ'' is $0 \cdot 0002$. From the plot of ζ' against ζ'' this corresponds to $\zeta' = 1 \cdot 075$, and by equation (17) $\theta = 28 \cdot 0°$. Table 1 gives a set of such calculations.

Table 1

m	Appropriate $\mu U/\sigma$	Selected $\mu U/T$	Corresponding		
			ζ''	ζ'	$\theta°$
0·3	0·249	10^{-2}	0·0992	1·075	28·0
		10^{-3}	0·0459	0·969	14·0
		10^{-4}	0·0214	0·917	5·58
		10^{-5}	0·00992	0·892	2·53
		10^{-6}	0·00460	0·880	1·157
		10^{-7}	0·00212	0·874	0·532

This method, with more intermediate values of $\mu U/T$, was used to plot fig. 5. Also plotted on fig. 5 are the experimental results, which are discussed in §3.

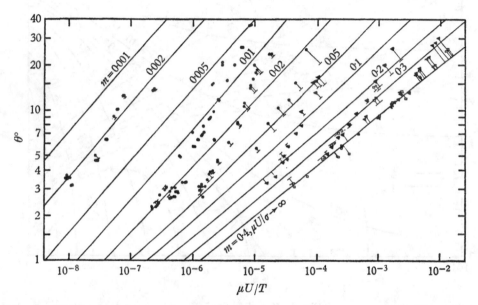

Fig. 5. $\mu U/T$ as a function of θ; theory and experiment. The flags on symbols represent the comparison between theory and experiment. The line at the base of the flag marks the m curve upon which the point should lie if fig. 2 were correct. For points without flags the curve coincides to within the symbol radius. ●, ethyl alcohol; ▲, castor oil; □, Limea oil.

2. EXPERIMENTS

A systematic control of at least one of the relevant parameters, θ, $\mu U/\sigma$ or $T/U\mu$, was desirable for data correlation, but in fact could not be achieved easily in an experiment. A thin flexible sheet peeling at a small angle from a plane surface appears to be torsionally unstable when the peeling rate is a function of the applied force. Hence some form of lateral constraint was necessary. Furthermore, in order to avoid the conflicting limitations introduced by gravitational force and sheet stiffness, the plane surface had to be nearly horizontal.

The arrangement adopted, described below and sketched in fig. 6, controlled none of the above variables directly, but permitted wide variations in them to be produced with ease, and enabled a steady state to be maintained for the major part of each test. It comprised a horizontal plate-glass surface (*a*) over which was laid a thin transparent plastic sheet, 30 cm. in width and over 100 cm. long (*b*).* A film of liquid (*c*) was sandwiched between the sheet and the glass surface. The

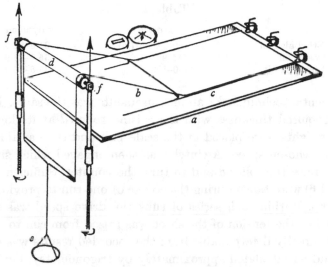

Fig. 6. Experimental arrangement.

sheet, of 'Melinex', approximately $4\,\mu$m. thick, was clamped at one end to the glass surface, and at the other end passed over a freely moving roller (*d*) the axis of which was parallel to the glass surface. The sheet at this end had attached along its edge a rigid metal strip, from which were suspended weights in a scale pan (*e*). The roller could be raised vertically at a constant pre-set rate by screws at the axle trunnion (*f*) driven by an infinitely variable speed drive. The rate of raising of the roller and the weight in the scale pan dictated the rate of peeling of the sheet from the glass, and the peeling angle. The peeling velocity and angle remained virtually constant for each run, except for the short starting period affected by inertia and the formation of a steady meniscus. A camera recorded time, meniscus position, and the counter which measured roller elevation.

Although precise control of liquid–film thickness was not theoretically necessary, the computation of corrections to the measured quantities required that it be known. Accordingly, films of uniform thickness were formed by use of a height-adjustable roller, manually operated, which was passed over the sheet as the liquid was being laid for each experimental run. For most tests the film was made 0·0025 cm. thick, but variations in film thickness exerted no systematic effect upon the results.

* For some tests at low peeling angles, a sheet 10·1 cm. wide and 25 μm. thick was used.

Three liquids were used to form the adhesive layer; these were ethyl alcohol, caster oil, and a heavy lubricating oil, Shell 'Limea 81'. The viscosity of each of the oils was determined over a range of temperature by the falling-ball method, with appropriate corrections, and surface tension was found by the pendant drop method with Fordham's calculations,* and by capillarity in a small tube. Alcohol properties were taken from physical tables. The range of properties so determined is listed below in table 2.

Table 2

	μ poise	σ dyne/cm.
Ethyl alcohol	0·0112–0·0126	21·6–22·7
Castor oil	7·8–9·2	35·0–35·9
Limea 81 oil	76·0–100	33·6–36·0

The experimental technique for all measurements was the same. The layer was rolled to the required thickness with the raising roller d at its lowest position. The requisite weights were placed in the scale pan and the variable-speed drive was pre-set at a chosen speed. A clutch was then engaged which simultaneously commenced to raise the roller d and to turn the counter. Sufficient photographs (between 3 and 6) were taken during the course of one run to provide at least two reducible results. Within each series of runs the drive speed was held more or less constant while the tension of the sheet was raised from run to run. The only parameter not directly determinable from the recorded results was θ, the peeling angle; this could be calculated approximately by trigonometry, but errors due to the weight of the liquid-laden strip and (in the tests at low $\mu U/\sigma$ and low θ) the thickness of the separating meniscus caused θ to be consistently underestimated. This first error could be accounted for by the correction

$$\Delta\theta = -wL/2T,$$

in which w is the loading per unit area on the strip and L is the peeled length. For the second error use is made of the observation that the height h is limited by gravity independently of $\mu U/\sigma$. For small peeling angles this height is

$$h_{\lim} = (4T/\rho g)^{\frac{1}{2}}$$

(where ρ is the liquid density), which for alcohol is 0·335 cm. Below this height the meniscus rises to h_0/m, being 0·0025 cm. in most tests. As argued by Taylor,† gravity is unlikely to affect the value of m materially, but, to correct θ, h_0/m or $h_{\lim}$, whichever was the smaller, was subtracted from the recorded roller height.

3. Results and Discussion

All the experimental results are plotted on fig. 5, and are compared with their theoretical location. On this figure each experimental result is represented by its appropriate symbol, and the difference between its value, and the value of m it

---

* *Proc. Roy. Soc.* A, cxciv (1948), 1.   † Paper 38 above.

should possess as given by fig. 2, is shown as a flag. Each flag terminates on a short line which represents the line of constant m to which it belongs. It is not possible to locate the point on this line because the source of the discrepancy could lie in any one of the three variables.

Symbols on fig. 5 which have no flags are those for which the theoretical m-line coincides with the experimental points to within the radius of the symbol.

The number of symbols not possessing flags represents a large proportion of the total number of experiments, but before remarking on the quality of the agreement between theory and experiment it is desirable to see what a plot of this nature signifies.

When m is small, ζ is large, and the solutions of equation (6) corresponding to this large ζ are of the 'non-returning' variety. These solutions are characterised by a value of ζ'' which is nearly independent of ξ, so that the shape adopted by the peeling strip is parabolic:

$$\zeta = \tfrac{1}{2}c_0 x^2 + c_1 x + c_2,$$

from which

$$\zeta \zeta'' = \tfrac{1}{2}\zeta'^2 + c_0 c_2 - \tfrac{1}{2}c_1^2.$$

If the quadratic is a satisfactory fit to the exact solution above $\zeta = \zeta_1$, then $(\zeta - \zeta_1)\zeta_1'' = \tfrac{1}{2}(\zeta'^2 - \zeta_1'^2)$ for any greater value of ζ. Therefore, the solutions asymptote to

$$\zeta \zeta'' = \tfrac{1}{2}\zeta'^2. \tag{19}$$

In this case, by equations (11), (12), (16) and (17), the motion becomes independent of μU, and

$$\tan^2 \theta = 4\sigma/T,$$

or for small peeling angles

$$\tan^2 \tfrac{1}{2}\theta = \sigma/T. \tag{20}$$

But this equation is the same as that for the static balance between σ and T. This implies that, at low values of $\mu U/\sigma$, all experimental points will correspond to the theoretical lines regardless of the value of μU. The function for m assumed in fig. 2 could be very much in error, but its influence upon the relationship between T and θ would be small.

It is noted that the disagreement between experiment and theory becomes more marked as m and θ increase. Discrepancies at high θ are to be expected because of the approximations contained in the analysis, but for those results below $\theta = 10°$, say, the disagreement probably arises largely from the inaccuracy of the assumed forms of F_1 and F_2.

With the computed solutions of equation (6) at hand it is possible to make an approximate estimate of F_1 at high values of $\mu U/\sigma$, from the experimental results. A plot is prepared of $\zeta \zeta''$ against ζ' and lines of constant ζ are marked upon it. Fig. 7 shows part of such a plot. Now, if F_2 is assumed to be given by equation (14), and $F_1 = 1/\zeta$, then the experimental value of $\zeta \zeta''$ is given using equation (16). The experimental value of ζ' is calculated by equation (17), and each experimental point can be located upon fig. 7. From this figure, ζ can be interpolated and the 'true' value of m calculated. It must be remembered, however, that

F_2 has been *assumed* so that, quite apart from the crowding of ζ-lines at high ζ, errors must arise due to the inaccuracy of this function. Note that F_2 only appears in $\zeta\zeta''$, and that the experimental value of this product is small as $\mu U/\sigma$ becomes of order unity (the experimental points group near the ζ'-axis). Therefore, the error in determining ζ is small for reasonably high values of $\mu U/\sigma$.

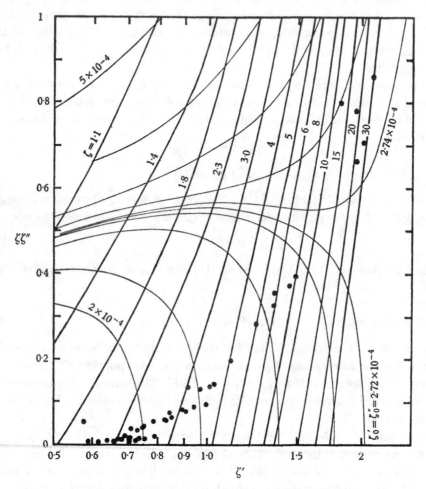

Fig. 7. Determination of m from experiment.

In fig. 2 the experimental values of m, so determined, are compared with the assumed function. Only points for $\theta < 10°$ are plotted. At high $\mu U/\sigma$, the results seem to display asymptotic behaviour, but the asymptote is possibly somewhat greater than 0·4. F_1 is nevertheless a good approximation. At lower values of $\mu U/\sigma$ the scatter of results becomes more pronounced, because the experimental results become less and less sensitive to F_1. Within the limitations of the measured quantities, the agreement can be considered as quite acceptable.

4. VISUAL COMPARISON WITH REAL ADHESIVES
IN PEELING

A characteristic feature of the meniscus formed when a viscous fluid separates between two diverging surfaces is the formation of 'webs' of fluid extending behind the leading edge of the meniscus 'fingers'. The appearance and form of these webs is observed to be dependent upon the angle of divergence of the surfaces and also upon $\mu U/\sigma$. Although their existence was known previously, analysis by Pearson* to find the most unstable wavelength of lateral disturbances to the meniscus, and to relate it to experiments on spreading (one surface only in motion) of viscous liquids, was the first attempt at a rational description of the webs. Taylor† publishes some photographs of webs produced in a similar physical situation, and Pitts & Greiller‡ also comment on the ribbed appearance of liquid films rolled between the nip of adjacent counter-rotating cylinders. The same webbed appearance of the meniscus was noted in the present experiments on separation peeling. Fig. 8 (pl. 1) shows some typical examples of the form taken by the meniscus at values of $\mu U/\sigma$ corresponding with the approach of m to its asymptotic limit.

It is of interest to compare the present experimental observations with the appearance of the line of rupture in the peeling of ordinary commercially available impermeable pressure-sensitive tapes.

Fig. 9 (pl. 2) reproduces photographs of the menicus formed using tape kindly provided by the Minnesota Mining and Manufacturing Co. Ltd., having what was described as an 'acrylate adhesive, adhesion level 30 units/in.'. The two sticky surfaces were rolled together, care being taken wherever possible to avoid the trapping of air bubbles. They were then peeled apart in a small spring jig. In the first photograph, (I), peeling had just commenced and fingering is seen to be quite regular. The increase in peeling angle resulted in the shortening of the finger pattern (II). As the peeling progressed the meniscus approached small defects in the layer which resulted in the growth of cavities ahead of the fingers so that the regular pattern was lost (III). The occasional forking of webs is similar to that observed with the viscous-liquid experiments, and arises when the advancement of one air finger is inhibited by the more rapid advancement of the fingers adjacent to it. The behaviour is similar to that observed by Saffman & Taylor§ in a Hele-Shaw cell. As might be expected, the forking is inhibited by an increase in peeling angle.

Two other kinds of adhesive tape displayed a similar rupture-line behaviour: because of the superior uniformity of the adhesive layer, the former tape was preferred for photographic purposes.

Notwithstanding the apparent close similarity between the zone of rupture

---

* *J. Fluid Mech.* VII (1960), 481.
† Paper 38 above.
‡ *J. Fluid Mech.* XI (1961), 33.
§ *Proc. Roy. Soc.* A, CCXLV (1958), 312 (paper 28 above).

with Newtonian liquids and with practical adhesives, there is little ground for supposing that the real adhesives examined behave in all relevant respects as normal liquids. While some types of adhesive backing are known to exhibit an almost linear relationship between peeling rate and peeling force per unit width, which on the basis of the models presented would be reasonable evidence of 'Newtonian'-viscosity-dominated behaviour, such behaviour is the exception rather than the rule. Kaelble* and Busse, Lamberg & Verdery,† for example, give experimental results of peeling tests at high peeling angles in which some of the adhesives tested showed almost linear force/rate, but in most cases the relationship is found to approximate to a power law, of exponent greater than unity. However, such tests cannot be compared meaningfully with the present ones since they involve very high peeling angles, for which not only are the approximations of the present analysis no longer valid, but the effects of strip stiffness would almost certainly have influenced the results. Furthermore, these tests did not specifically distinguish the states in which the adhesive separated cleanly from one of the surfaces, and those in which the adhesive remained attached to both after rupture. The present analysis can justifiably be related only to the second case.

5. Concluding remarks

It appears that a Newtonian viscous adhesive layer, in peeling, can be described successfully by means of a simple Poiseuille model of behaviour, provided that the layer remains intact ahead of the line of rupture and that the peeling angle is small. The description relies upon a prediction of the significance of the parameter $\mu U/\sigma$ in defining the ratio of the thickness of the layer far upstream of rupture to the thickness at the rupture line. Within the experimental limitations, the prediction is confirmed, demonstrating that at low values of the peeling velocity U the relationship between peeling force T per unit width and peeling angle θ depends upon surface tension σ, but that a well-defined limiting relationship between $\mu U/\sigma$ and θ is approached; as σ becomes insignificant U is proportional to T for a given θ.

This result is analogous to the experimentally observed existence of an asymptotic limit to the proportion of liquid remaining after the evacuation of a two- or three-dimensional viscous-liquid-filled passage by a penetrating cavity or bubble, a situation which finds application in other cases of practical interest, as in cavitating lubrication and in applying surface coatings.

It is noted that the physical appearance of the zone of rupture of commercially available impermeable pressure-sensitive tapes in peeling is similar to that arising in the present experiments, provided that the adhesive remains attached to the tape and is unflawed ahead of the rupture line. The characteristic instability of the

---

* *Trans. Soc. Rheol.* IV (1960), 45.

† *J. appl. Phys.* XVII (1946), 376.

free surface gives rise to the formation of webs of adhesive separated by penetrating air fingers. This similarity cannot however be considered as evidence of any detailed applicability of the present results to real adhesives in peeling.

One of us (A. D. McE.) expresses his gratitude for assistance given by an Australian C.S.I.R.O. Studentship and later by an Australian Public Service Board Scholarship.

46

THE COALESCENCE OF CLOSELY SPACED DROPS WHEN THEY ARE AT DIFFERENT ELECTRIC POTENTIALS

REPRINTED FROM

Proceedings of the Royal Society, A, vol. CCCVI (1968), pp. 423–34

To discuss the coalescence of two neighbouring drops at different electric potentials as a problem of stability it is necessary to assume that they are held on fixed supports, for otherwise there is no equilibrium state to become unstable. In this work these supports are two equal rings. Experiments at first using soap films instead of drops, are in agreement with the analysis when the geometry of the rings conforms to the requirements of the theory.

When the distance between the drops is small compared with that between the supporting rings the critical potential difference does not depend on the size or position of the rings but only on the radius of the two drops and the distance between them. Under these circumstances a simple expression derived from computer solutions of the relevant equation has been found. This limiting formula is in agreement with some unpublished experiments with water drops carried out by J. Latham, and also with experiments in which instability was observed when neighbouring glycerine drops were subjected to potential differences as low as 9·8 V.

The reaction between two neighbouring drops in an electric field, tending to make them coalesce, has been discussed as a problem of stability by Latham & Roxburgh.* Each drop in their analysis is supposed to become unstable in the same way as an isolated drop in a uniform field† and the effect of the neighbouring drop is merely to increase that field. This approach, though no doubt valid when the drops are widely separated, cannot be regarded as satisfactory when they are close together because there is no equilibrium state to become unstable. To discuss the stability of two neighbouring drops at different potentials it is necessary to introduce into the analysis physical constraints which will enable them to be at rest till, without moving the constraints, the drop surfaces spring together at a critical potential difference. In the present work a particular form of constraint is considered which would permit the drops to remain axi-symmetric and be amenable to mathematical and experimental treatment. The drops, each of radius R_0, are suspended on two small equal coaxial rings of radius a. To simplify the analysis the planes of the rings are separated by distance $2h$ which is small compared with a so that

$$h \ll a \ll R_0. \tag{1}$$

Fig. 1 shows the geometry. The initial position of the portions of the drops which lie within the rings are shown by a broken line which is part of a sphere of radius R_0.

---

* *Proc. Roy. Soc.* A, CCXCV (1968), 84.

† G. I. Taylor, *Proc. Roy. Soc.* A, CCLXXX (1940), 383 (paper 39 above).

The sections of the supporting rings by an axial plane are shown at A, B, C, D and $AD = BC = 2h$, while $AB = CD = 2a$.

The rings, which are conducting, are maintained at potentials $\pm V$ and the distorted positions of the surfaces under the influence of the electric field is shown in fig. 1 by the full curves joining AB and CD. The parts of the drop outside the ring will be slightly distorted by the field but it will be assumed that this distortion does not affect the shape and volume of the drop enough to alter the internal pressure by a significant amount, so that it will remain $2T/R_0$ when the field is applied. In the first experiments to be described the 'drop' was a soap bubble so that T was twice the surface tension of the interface and equal to $54\cdot8 \times 10^{-3}$ in M.K.S. units. Taking cylindrical co-ordinates (r, Z) with $Z = 0$ as the plane midway between the two equal drops the assumption that h/a is small permits the approximation $V^2\epsilon_0/2Z^2$ for the stress normal to the surface, and the approximate equilibrium equation for the surface in M.K.S. units is

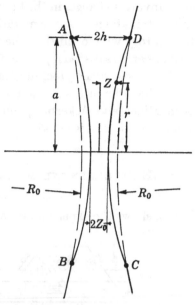

Fig. 1. Portions of two drops suspended on two rings whose sections are at A, B and C, D respectively.

$$T\left(\frac{\mathrm{d}^2Z}{\mathrm{d}r^2} + \frac{1}{r}\frac{\mathrm{d}Z}{\mathrm{d}r}\right) = \frac{2T}{R_0} + \frac{V^2\epsilon_0}{2Z^2}, \tag{2}$$

where ϵ_0 is the factor $8\cdot854 \times 10^{-12}$.

Writing $Z = hy$, $r = ax$ (2) becomes

$$\frac{\mathrm{d}^2y}{\mathrm{d}x^2} + \frac{1}{x}\frac{\mathrm{d}y}{\mathrm{d}x} = \alpha + \frac{\beta}{y^2}, \tag{3}$$

where

$$\alpha = \frac{2a^2}{hR_0} \quad \text{and} \quad \beta = \frac{V^2\epsilon_0\alpha^2}{2h^3T}. \tag{4}$$

The boundary conditions are

$$y = 1 \quad \text{at} \quad x = 1 \quad \text{and} \quad \mathrm{d}y/\mathrm{d}x = 0 \quad \text{at} \quad x = 0. \tag{5}$$

It will be noticed that when $\beta = 0$ the solution of (3) which satisfies the boundary conditions is

$$y = 1 + \tfrac{1}{4}\alpha(x^2 - 1), \tag{6}$$

and since y must be positive at $x = 0$, $1 - \tfrac{1}{4}\alpha > 0$. In other words, to find stability limits it is necessary to find the greatest value of β for which the boundary conditions can be satisfied for every value of α in the range $0 < \alpha < 4$.

The case $\alpha = 0$ is specially simple because by appropriate scaling of y, equation (3) can be reduced to a single equation with $\alpha = 0$, $\beta = 1$. This simplification, however, involves changes in the boundary conditions which now involve β. Preliminary calculations for $\alpha = 0$ were made by Dr J. C. P. Miller of the Cambridge Mathematical Laboratory who found that the maximum value of β for which the boundary conditions can be satisfied is $\beta = 0.7912$. Later calculations made by Dr R. C. Ackerberg of the Brooklyn Polytechnic Institute using a computer gave $\beta = 0.7892$ for this case. The difference is negligible so far as my experiments are concerned. For $\alpha = 0$ both Miller and Ackerberg found that $y(0)$, the value of Z/h at the axis when instability occurs, is 0.555.

EXPERIMENTAL MEASUREMENTS OF CRITICAL POTENTIALS WHEN $\alpha = 0$

Fig. 2. shows the experimental arrangement.

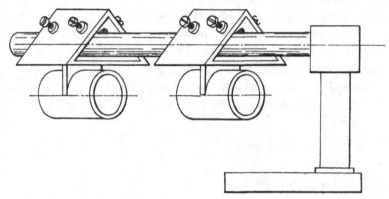

Fig. 2. Apparatus for holding two circular soap films in position.

A horizontal insulating rod supports two tubes of triangular section to each of which a short copper tube of circular section is rigidly fixed. The triangular tubes are each pierced by four levelling screws whose lower ends can slide on the insulating Bakelite rod. One end of each copper tube is turned so that the inner surface converges to a sharp-edged circular hole whose diameter $2a$ is a little less than the inner diameter of the tube. Thus a soap film stretched over the end of the tube will be flat and in the plane of the end of the tube. With this arrangement the end of each tube can be dipped into a shallow dish of soap solution and rapidly pushed on to the supporting rod without risking wetting the insulation. If the adjusting screws have been set correctly the two ends of the copper tubes and the soap films on them will be parallel and coaxial.

The experiment consisted in gradually raising the potential difference $2V$ between the cylinders till the films became unstable and ran together. Usually when this happened the film did not break but formed a flat circular disc held at the edge by short sections of minimal surfaces which were very nearly frustrums of 60° semivertical angle cones attached to the rings.

The calculated potential difference $2V$ according to (3) and using Miller's value $\beta = 0.791$ is

$$2V = 2\sqrt{\frac{(0.791)(2h^3T)}{\epsilon_0 a^2}},$$

and since a was 0·02 m. and the surface tension of the soap solution used was

$$27.4 \times 10^{-3}\,\mathrm{Nm^{-1}}, \quad T = 54.8 \times 10^{-3}\,\mathrm{Nm^{-1}}, \quad 2V = 9.9h^{\frac{3}{2}} \times 10^6, \tag{7}$$

$2h$ being the distance in metres between the two soap films before the potential difference $2V$ was applied.

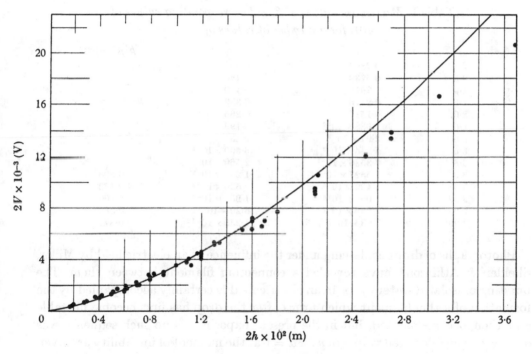

Fig. 3. Critical difference in potential, $2V$ in volts, between two circular soap films of radius 0·02 m. and $2h$ part. (Curve, equation (7).)

The first experiments with the apparatus shown in fig. 2 gave results which were not in very good agreement with (7) and it was thought that this might be due to the fact that the electric field near the edge of the film might be greater than that assumed in the calculation. Copper guard rings 0·10 m. in diameter were then fitted over the ends of the tube so as to make the flat soap films parts of conducting discs 0·1 m. in diameter. The results of the experiments are shown in fig. 3 and the curve representing (7) is also shown. The agreement between theory and observation is good when $h/a < 0.8$.

Experiments when α ≠ 0

Calculations of the solutions of (3) for a range of values of α were made for me by Dr Ackerberg. Table 1 gives in column 2 the computed maximum values of β corresponding with a range of values of α. The corresponding values, $y(0)$, of y at the axis are given in column 3. Since one of the objects of the calculation was to estimate how closely drops at different potentials could be brought together before they become unstable and spring into contact, some values of α close to 4 were chosen for more accurate evaluations. One of the questions which has been asked is whether when instability occurs, the drops tend to become pointed at the positions of maximum electric stress as isolated drops do in some circumstances.

Table 1. *Maximum values of β and corresponding values of*
y(0) for a number of values of α

α	$\beta_{\max}$	$y(0)$	$\beta^{\frac{1}{2}}/(1-\frac{1}{4}\alpha)$
0	0·7892	0·5556	
0·5	0·5739	0·483	
1·0	0·3976	0·410	
1·5	0·2576	0·339	
2·0	0·1517	0·266	
2·5	$7·688 \times 10^{-2}$	0·199	
3·0	$2·97 \times 10^{-2}$	0·131	
3·5	$5·944 \times 10^{-3}$	$6·500 \times 10^{-2}$	
3·8	$7·3536 \times 10^{-4}$	$2·560 \times 10^{-2}$	
3·9	$1·5527 \times 10^{-4}$	$1·280 \times 10^{-2}$	0·498
3·98	$4·499 \times 10^{-6}$	$2·530 \times 10^{-3}$	0·422
3·99	$1·002 \times 10^{-6}$	$1·254 \times 10^{-3}$	0·400
3·995	$2·258 \times 10^{-7}$	$6·29 \times 10^{-4}$	0·380
3·999	$7·33 \times 10^{-9}$	$1·255 \times 10^{-4}$	0·344

Photographs of drops coalescing under the influence of an electric field by Miller, Shelden & Atkinson* have revealed a connecting filament between them. The bursting of isolated water drops in an electric field is certainly accompanied by the formation of a thin filament which forms after the drop has first become unstable and then become pointed, but in the present experiments no such sequence was observed. The calculated values of y for $\alpha - 0$ at the moment of instability are given in table 2, line 2, and for $\alpha = 3·8$ in table 2, line 3. To give an idea of the amount by which the surface is displaced by the electric field the values of y for $\alpha = 3·8$, $\beta = 0$ (equation (6)) are given in line 4 and the difference between lines 3 and 4 shown in line 5, gives this displacement due to the electric field. The curvature of the film is a maximum at $x = 0$, but it is not a different order of magnitude from that of the rest of the film when instability occurs.

The theoretical treatment which led to equation (2) assumed that the pressure behind the film or drop surface remains constant as the surface within the supporting ring is displaced by the electric stress. This can only be an approximation based on the smallness of the ratio a/R_0, because in that case the small change in volume

* *Physics of Fluids* VIII (1965), 1921.

Table. 2. *Calculated values of y for* $\alpha = 0$ *and for* $\alpha = 3\cdot8$

	x	0	0·1	0·2	0·3	0·4
$\alpha = 0$	y	0·555	0·562	0·581	0·610	0·650
$\alpha = 3\cdot8$	$y\,(\beta_{max})$	0·0256	0·0373	0·0700	0·1211	0·1908
$\alpha = 3\cdot8$	$y\,(\beta = 0)$	0·0500	0·0595	0·0800	0·1350	0·2020
$\alpha = 3\cdot8$	displacement	0·0244	0·0222	0·0180	0·0139	0·0112

	x	0·5	0·6	0·7	0·8	0·9	1·0
$\alpha = 0$	y	0·697	0·750	0·808	0·870	0·943	1·0
$\alpha = 3\cdot8$	$y\,(\beta_{max})$	0·2788	0·3855	0·5110	0·6551	0·8181	1·0
$\alpha = 3\cdot8$	$y\,(\beta = 0)$	0·2875	0·3920	0·5150	0·6580	0·8200	1·0
$\alpha = 3\cdot8$	displacement	0·0087	0·0065	0·0040	0·0029	0·0019	0

of the segment cut off by the plane of the ring due to displacement by the electric field will only give rise to a small change in the volume of the much larger portion of the drop which lies behind the plane of the ring. Thus it seems justifiable to assume constant pressure $2T/R_0$ when the equations refer to drops supported in the way assumed. In the first experiments designed to produce a non-zero pressure behind a soap film a complete soap bubble was blown and attached to a copper wire ring. This method turned out to be difficult because the ring and its support had to be sufficiently strong to ensure that no appreciable deflexion occurred when the field was applied and this made it necessary to have a ring made of wire whose thickness might be comparable with h. Another difficulty with this method was that to obtain values of α in the range $0 < \alpha < 1\cdot8$ the soap bubbles were too large to be supported on rings in a vertical plane. Some results however were obtained by this method. These are shown in fig. 5. To overcome these difficulties the apparatus shown in fig. 4 was constructed. Films were stretched across opposite ends of a short length of copper tube A. Opposite one was a metal plate C (fig. 4) which could be raised to potential V and would be represented in the analysis by $Z = 0$. The pressure of the air between the two films was kept constant by supplying air to the middle of the copper tube A when the potential of the plate C was raised, so as to keep the curvature of the film at the opposite end constant. This objective was attained by fixing a small light (B, fig. 4) to a swinging arm so that the light was reflected from the edge of film back to B. For a given value of R_0 the arm holding B was fixed at an angle to θ to the axis of the copper cylinder so that

$$a/R_0 = \sin \theta. \tag{8}$$

As the potential of the plate C increased air was supplied to the space between the films so as to keep the image of the light B at the edge of the film. Values of α and β at the moment of instability were calculated from (4) and (8) by using the measured values of V, a, Z_0, θ and T. They are plotted in fig. 5. The first set of measurements were made using the method of controlling the pressure in the chamber A shown in fig. 4. The results are shown as crosses in fig. 5. The method was made more

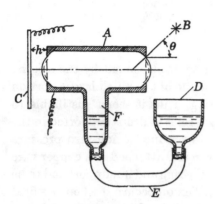

Fig. 4. Apparatus for ensuring constant pressure behind a soap film.

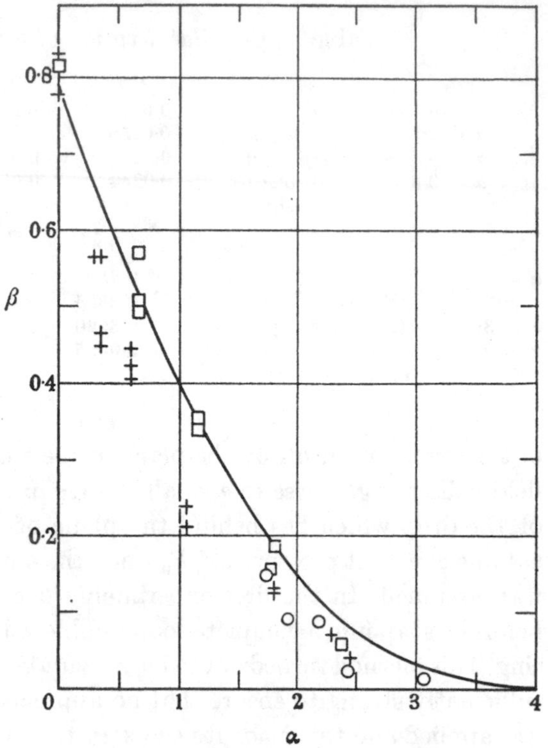

Fig. 5. Calculated β (full line). Observed β: $+$, with arrangement as in fig. 4; $\square$, with improved pressure control; $\bigcirc$, complete soap bubble held on wire ring near a vertical plate.

controllable by replacing the reservoir D and its rubber connecting tube by an open topped beaker standing on a shelf below A so that as the shelf was raised the water from the beaker passed straight into the chamber (fig. 4) without having to pass through a rubber tube. The results obtained with this improved arrangement are shown in fig. 5, by squares. It will be seen that these are close to the full curve which represents the theoretical results given in table 1.

DISPLACEMENT OF SURFACE WHEN INSTABILITY OCCURS

The computed values of $y(0)$ when β is a maximum were given in table 1. They are plotted in fig. 6. It will be seen that they are slightly above the line

$$y(0) = 0 \cdot 5 - \tfrac{1}{8}\alpha.$$

(which is shown in fig. 6), but appear to be tangential to it at $\alpha = 4$. The distance of the drop surface at $x = 0$ from the plane $y = 0$ is $\tfrac{1}{4}\alpha h$ when $\beta = 0$, so that the displacement of the drop at $x = 0$ due to the electric field is $h(1 - \tfrac{1}{4}\alpha) - y_0 h$. The displacement is less than half the initial distance from drop to plate but seems in fig. 6 to approach this value asymptotically as α approaches 4. No asymptotic solution has so far been found.

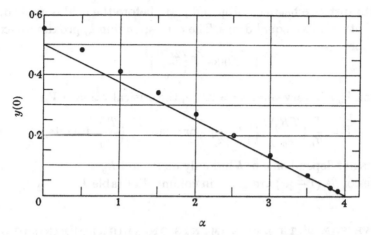

Fig. 6. Computed values of $y(0)$ and line $y = 0 \cdot 5 - \frac{1}{8}\alpha$.

APPROXIMATE MAXIMUM POTENTIAL DIFFERENCE BETWEEN CLOSELY SPACED DROPS

To find how β varies as α approaches 4 the logarithmic plot of β against $4-\alpha$ which is shown in fig. 7 was made. The line $\log_{10}\beta = 2\log_{10}(4-\alpha) - 2 \cdot 0$ nearly passes through the points corresponding with $\alpha = 3 \cdot 995$, $\alpha = 3 \cdot 99$ and $\alpha = 3 \cdot 98$ so

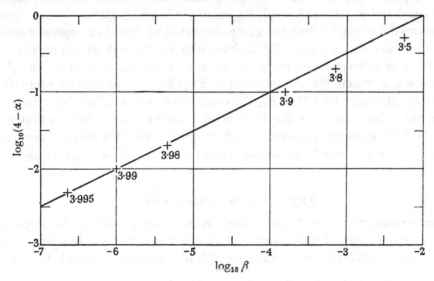

Fig. 7. Logarithmic plot of computed values of β for values of α close to 4.
The figures are the values of α.

that when $4 - \alpha$ is small $\beta \sim 0\cdot01(4-\alpha)^2$ or $\sqrt{\beta} \sim 0\cdot1(4-\alpha)$ or $0\cdot4\,Z_0/h$. If the error when $\alpha = 3\cdot995$ is taken as negligible the computed results of table 1 lead to

$$\sqrt{\beta} = 0\cdot095\,(4-\alpha) = 0\cdot38Z_0/h, \tag{9}$$

where Z_0 is the distance between drop and plate before the field is applied, i.e. half the distance between two equal drops. The corresponding approximate expression for V is therefore

$$V = 0\cdot38\,\frac{Z_0}{a}\left(\frac{2hT}{\epsilon_0}\right)^{\frac{1}{2}}, \tag{10}$$

and since when Z_0/h is very small $a^2 = 2R_0h$, the critical value of V is

$$V = \frac{Z_0}{R_0}\left(\frac{TR_0}{\epsilon_0}\right)^{\frac{1}{2}}\left(\frac{\beta^{\frac{1}{2}}}{1-\frac{1}{4}\alpha}\right) \quad \text{or} \quad 0\cdot38\,\frac{Z_0}{R_0}\left(\frac{TR_0}{\epsilon_0}\right)^{\frac{1}{2}} \text{volts}, \tag{11}$$

so that V does not depend on a or h but only on Z_0 and R_0.

Some values of $\beta^{\frac{1}{2}}/(1-\frac{1}{4}\alpha)$ are given in column 4 of table 1.

COMPARISON WITH EXPERIMENTS ON NEIGHBOURING DROPS

The experiments of Latham & Roxburgh[*] were made with water drops supported on Teflon rods in a uniform field. In their theoretical discussion the drops remain uncharged and it is assumed that this is also true in their experiments. The potential differences between the drops are those calculated by Davies[†] for neighbouring uncharged spheres in a uniform field. Since the displacements of the drop surfaces when they are very close together must depend on the difference in potential between them, experiments in which this difference in potential is measured directly can be interpreted directly without the uncertainty which lies in assuming them to be uncharged and exactly spherical. Experiments in which the critical potentials were measured directly for drops of water supported so that their centres were in a vertical line were also made by Mr Latham who kindly sent me his results. The values of V were expressed non-dimensionally in the c.g.s. system as $A = V(\pi R_0 T)^{-\frac{1}{2}}$ and in the m.k.s. system this is equivalent to $2V(TR_0/\epsilon_0)^{-\frac{1}{2}}$ so that for comparison with (11), $0\cdot5\,A$ should be $0\cdot38Z_0/R_0$. The comparison is shown in table 3, in which a broken line is drawn above which Z_0/R_0 is less than $0\cdot12$. It will be seen that when $Z_0/R_0 < 0\cdot12$ the agreement is good enough to confirm the idea that coalescence of closely spaced drops is due to instability when they are at different potentials.

FURTHER EXPERIMENTS

The approximate expression (11) was based on the assumption that the drops were supported on two rings of radius a distant $2h$ apart and it was only when h/a and also a/R_0 are small that the analysis applies. When however Z_0/h is small, the symbol

---

[*] *Proc. Roy. Soc.* A, ccxcv (1966), 84.
[†] *Quart. J. Mech. appl. Math.* xvii (1964), 499.

Table 3. *Latham's measurements* $(0.5\,A)$, *compared with* $0.38Z_0/R_0$.
Above the broken line $Z_0/R_0 < 0.12$

$2Z_0$ (mm.)	$R_0 = 1.06$ mm.		$R_0 = 1.34$ mm.		$R_0 = 1.68$ mm.	
	$0.5\,A$	$0.38Z_0/R_0$	$0.5\,A$	$0.38Z_0/R_0$	$0.5\,A$	$0.38Z_0/R_0$
0.1	0.020	0.018	0.017	0.014	0.0008	0.011
0.2	0.036	0.036	0.031	0.028	0.023	0.023
0.4	0.115	0.072	0.075	0.057	0.048	0.045
0.6	0.175	0.107	0.125	0.085	0.075	0.068

a drops out of the expression for the critical value of V. The supports in Latham's experiments did not satisfy the condition that a/R_0 is small, nevertheless as table 3 shows the approximate expression (11) for the critical potential difference agrees well with experiments when Z_0/R_0 is small and the supports are not separated by a distance which is small compared with the radius. It seemed worthwhile therefore

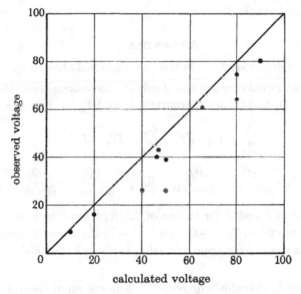

Fig. 8. Observed potential difference for coalescence of glycerine drops compared with that calculated using (11).

to make experiments in which equal drops were supported in a different manner to see whether (11) still applies. Two hypodermic steel tubes 1 mm. external diameter were supported on the stage of a travelling microscope. The ends were bent down through an angle of 45° and fluid could be forced through them till drops formed at the ends and remained hanging there. The drops could be brought close together and a potential difference ranging from 0 to 110 V. could be applied. The ends of the two hypodermic tubes were at the same level so that when the profiles of both drops appeared sharply in the field of the microscope they had the same radius. The drop diameters were measured by a transverse motion of the stage of the microscope.

Experiments with water showed that oscillations usually preceded the instability which occurred usually at considerably less potential difference than that calculated from (11). This seemed to be due to the fact that water could flow through the tubes and alter the volumes of the drops when the field was applied. For this reason glycerine was used instead of water because the fluid resistance in the hypodermic tube was great enough to prevent any appreciable change in volume while the potential was being raised. The range of values of Z_0 was from 0·01 to 0·11 mm. while the range of R_0 was from 0·8 to 1·2 mm. and the critical potential differences $2V$ ranged from 9·8 to 90 V. The results are shown in fig. 8 and though the experimental critical potential was still rather less than that calculated, the scatter due to experimental error is great enough to mask any systematic difference which might exist.

In conclusion I wish to express my thanks to Dr J. C. P. Miller and Dr R. C. Ackerberg for their help in computation and to Mr J. Latham for sending me the experimental results given in table 3.

APPENDIX

Potential difference between two equal isolated charged spheres

The potentials of two equal charged equal spheres have been calculated by Kelvin.[*] If their charges are Q_1 and Q_2 and potentials V_1 and V_2, in the C.G.S. units used by Kelvin

$$\frac{Q_1}{R_4} = IV_1 - JV_2, \quad \frac{Q_2}{R_0} = IV_2 - JV_1,$$

so that

$$R_0 V_1 = \frac{IQ_1}{I^2 - J^2} + \frac{JQ_2}{I^2 - J^2}, \quad R_0 V_2 = \frac{IQ_2}{I^2 - J^2} + \frac{JQ_1}{I^2 - J^2}.$$

Kelvin gives values of I and J for values of $2Z/R_0$ from 0 to 2 at intervals of 0·1. For the particular case when $Q_1 = -Q_2$ and $V_1 = -V_2$ the range in which the present analysis might be a reasonable approximation is given in table 4.

Table 4. *Potential difference* $2V$ *between equal spheres*
containing charge $\pm V$

$2Z_0/R_0$	0	0·1	0·2	0·3	0·4	∞
VR_0/Q	0	0·4055	0·4640	0·5040	0·7397	1·0

It will be seen that when the spheres are brought near to contact the difference in potential between them decreases so that analysis which applies to experiments with drops at known voltages cannot be applied directly to find the stability conditions for charged insulated drops.

[*] *Phil. Mag.* (1853), 287.

47

INSTABILITY OF JETS, THREADS, AND SHEETS OF VISCOUS FLUID

REPRINTED FROM

Proceedings of the 12th International Congress of Applied Mechanics (*Stanford, 1968*), Springer-Verlag (1969), pp. 382–8

The instability which I propose to discuss is well-known in the form it assumes when a very viscous fluid like thick oil or honey falls in a stream onto a plate. A short distance above the plate the stream begins to wave or rotate in spiral form. It is possible to calculate the diameter at points in a *steady* falling stream of viscous fluid, and it is found that it thickens up very rapidly a short distance above the plate which stops the motion. The reason for the instability is clear. If the stream is very thin the longitudinal compression acts in the same way as end compression in a thin elastic rod. The rod becomes unstable at a certain load, and less force is needed to move the ends towards one another when the rod is bent than when it is straight. This is called the Euler instability. If however the stream is thick it may be that the work required to produce a given amount of longitudinal compressive strain is less if it remains straight than if it bends. Figs. 1 and 2 (pl. 1) show thin and thick streams of glycerine falling onto a plate. In fig. 2 the stream spreads out so steadily that a reflection of the upper part of the stream can be seen in the fluid spreading on the plate.

ELECTRICAL VERSUS MECHANICAL INSTABILITY

My attention was drawn to this kind of instability by observing the behaviour of a thin jet of viscous fluid drawn from the end of a metal capillary tube by electric forces. That thin fluid jets produced in this way are very unstable is well known. Fig. 3 (pl. 1) shows two photographs* of an electrically driven jet emerging from the tip of a tube 0·22 mm. in diameter; on the left is a time exposure and on the right a spark exposure of the same jet. It has been assumed that the violent instability is due to the mutual repulsion of charges carried in different parts of the jet, and indeed when the jet breaks into drops this seems a reasonable explanation. Zeleny† found that similar effects occur with viscous jets, but in that case the length of the straight part, which in fig. 3 is only 0·3 mm. long, could extend to nearly 1 cm. before instability appeared. It is true that a charged thread at the axis of a conducting cylinder can be proved to be unstable but the spark photograph of fig. 3 does not suggest that the jet is unstable when first formed.

* From an article by W. R. Lane & H. L. Green in *Surveys in Mechanics*, edited by G. K. Batchelor and R. M. Davies, Cambridge University Press (1956).

† *Phys. Rev.* x (1917), 1.

In Zeleny's and Lane's apparatus the end of the tube from which the jet emerged was placed near a charged plate, but no precautions were taken to control the distant field, on the implicit assumption that when the distance between tube and plate is small the potential field far from this gap has little effect on the jet. In apparatus which I described some years ago* the distant field was much more uniform, and no instability or scatter was observed even after the jet broke into drops. Fig. 4 (pl. 1) is an example. The drops into which the jet broke owing to the Rayleigh varicose instability moved along the axis of the 1 mm. tube from which the jet was pulled by the electric field.

While experimenting with glycerine streams produced electrically, I found that the stream could sometimes be completely steady for its whole length (up to 5·5 cm. in my apparatus), but by altering some of the physical conditions I could photograph with an exposure of a millisecond a stream which seemed to vanish at a definite point (fig. 5). Of course it did not in fact disappear; it must have started waving at the point indicated by the arrow. I therefore took a photograph of the same jet under the same conditions using a flash of 2 μsec. duration (fig. 6, pl. 1). It will be seen that oscillations begin at the point marked by the arrow in fig. 6. They do not increase rapidly as would be expected if there were an instability due to reaction between the electric field and the jet. The jet becomes crooked to a limited extent. It was this that made me think the instability was purely mechanical and due to longitudinal compression of a viscous stream. This would explain the extreme straightness of the jet before it began to oscillate, as well as the limited bending afterwards, provided the instability began in a steady jet at or near the point where the diameter reached its minimum, because that is the place where the longitudinal stress due to viscosity changes sign. I therefore measured the diameters along a downward pointing jet similar to that of figs. 5 and 6 and found that the instability did not occur till just after the jet had reached a minimum diameter. The reason for using jets directed downwards is that the only force which would contribute to a change in sign of the viscous stress is then the aerodynamic drag on the moving column.

Gravity driven jets of viscous fluid

The difference between the instability of glycerine falling onto a plate (fig. 1) and the instability of an electrically driven jet seems to be that the deceleration which produces the instability is much more sudden in the former case and the fluid piles up immediately after the compression sets in. To obtain a small deceleration in a falling stream many methods are available, but the simplest is to let the fluid fall into a fluid of less density than itself and then, at the point where the instability is to be established, to alter the density of the surrounding fluid so that the gravitational driving force decreases. Fig. 7 (pl. 2) shows coloured glycerine (specific gravity 1·255) falling in water onto a plate. The jet is quite straight till it reaches a point a few jet diameters above the plate where it begins to oscillate like the jet of fig. 1.

* *Proc. Roy. Soc.* A, ccxci (1966), 145 (paper 42 above).

PLATE 1

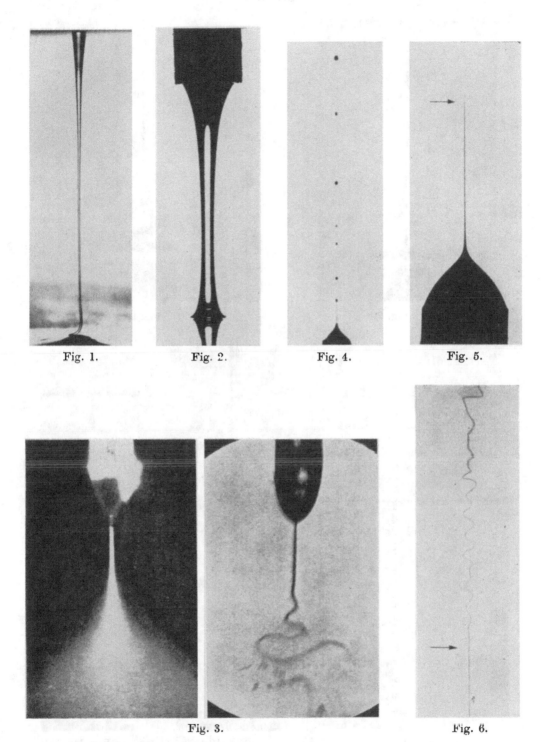

Fig. 1.   Fig. 2.   Fig. 4.   Fig. 5.

Fig. 3.   Fig. 6.

Fig. 1. Thin stream of glycerine falling on a plate.
Fig. 2. Thick stream of glycerine falling on a plate.
Fig. 3. Water pulled from a tube by electric force (Lane & Green).
Fig. 4. Tellus oil pulled from tube in apparatus designed for constant electric field.
Fig. 5. Electrically driven glycerine jet, 1 msec. exposure.
Fig. 6. Electrically driven glycerine jet, 2 μsec. exposure.

PLATE 2

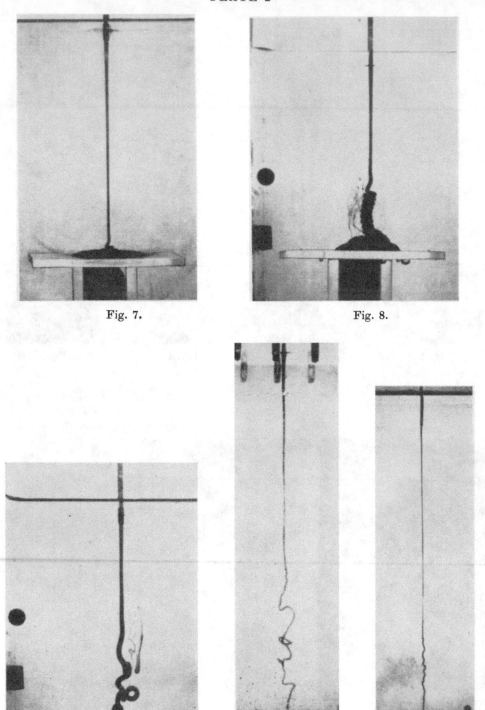

Fig. 7.

Fig. 8.

Fig. 9.

Fig. 10.

Fig. 11.

Fig. 7. Coloured glycerine falling through water.

Fig. 8. Glycerine falling into water above round mark, salt solution of specific gravity 1·237 below.

Fig. 9. Glycerine falling into fluid of specific gravity 1·225 above round mark, 1·237 below.

Fig. 10. ½ mm. jet in specific gravity 1·225 above, 1·237 below.

Fig. 11. ½ mm. jet in specific gravity 1·225 above, 1·230 below.

PLATE 3

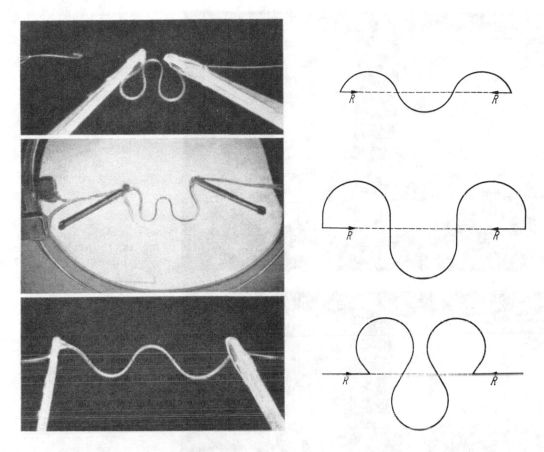

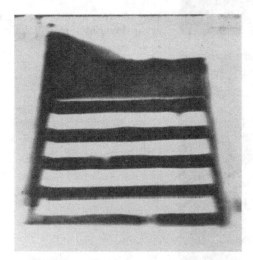

Fig. 12. Threads of SAIB floating on mercury compared with Love's calculated elasticas.

Fig. 13. Reflections of striped card in layer of fluid (40 poise, 1 cm. thick) when $-\epsilon \approx 0.1$ sec.$^{-1}$.

Fig. 14. Reflection when $-\epsilon \approx 0.4$ sec.$^{-1}$

PLATE 4

Fig. 15. 15 cm. disc rotating in
syrup layer at 1 rev. in 2·7 sec.

Fig. 16. 15 cm. disc rotating in
syrup layer at 1 rev. in 2·4 sec.

Fig. 17. 15 cm. disc rotating in
syrup layer at 1 rev. in 1·4 sec.

On filling the tank up to the level of the round mark on the left (fig. 8, pl. 2) with fluid of specific gravity 1·237 and the remainder carefully with water, it will be seen that the glycerine begins to oscillate in a spiral manner near the level of the change in density. The downward velocity has in fact changed in the ratio $d_1/\pi d_2$, where d_1 is the diameter of the jet and d_2 that of the spiral. In fig. 8 this reduction is about 12:1 and the change in the driving force is from $1·255\,g. - 1·000\,g. = 0·255\,g.$ to $1·255\,g. - 1·237\,g. = 0·018\,g.$, i.e. the driving force has been reduced in the ratio 14:1. On repeating the experiment with a surrounding specific gravity of 1·225 above and 1·237 below, the driving force is changed from $1·255\,g. - 1·225\,g. = 0·030\,g.$ to $0·018\,g.$, i.e. the driving force is reduced in the ratio of only 1·67:1. Fig. 9 (pl. 2) shows this case. The change in jet velocity at the level of density change is much less than in fig. 8; in fact the length of the coiled portion is about twice that contained in the same vertical distance in the uncoiled portion. A finer jet, $\frac{1}{2}$ mm. diameter, is shown in figs. 10 and 11 (pl. 2). In fig. 10 the glycerine is falling into the same density distribution as in fig. 9. Fig. 11 shows glycerine falling through a fluid for which the specific gravity is 1·225 at the top and 1·230 at the bottom. Thus the driving force at the level of density change is only reduced from $1·255\,g. - 1·225\,g. = 0·030\,g.$ to $1·255\,g. - 1·230\,g. = 0·025\,g.$ Thus the force changes in the ratio 1·2:1. The crumpling in fig. 11 is much less than in fig. 10.

CRUMPLING OF THREADS OF VISCOUS FLUIDS IN THE ABSENCE OF GRAVITY

The analogy between the equations of motion of a viscous fluid and that of an incompressible elastic solid is well-known. It was pointed out to me by Brooke Benjamin that when a thin rod is subjected to end compression it forms an elastica and that Love's *Theory of Elasticity* contains a picture of the set of elasticas. I therefore found a very viscous fluid known as SAIB (Sucrose Acetate Isoburate) and stretched a thread of it between two sticks. I then laid this thread on the surface of a dish of mercury and either pushed the sticks towards one another or laid two matches on the thread and pushed them towards one another, and photographed the thread. Fig. 12 (pl. 3) shows some of these photographs together with some of Love's calculated elasticas. The comparison would not be expected to be at all exact because this method of producing the threads does not make them with uniform diameter.

INSTABILITY OF SHEETS OF VISCOUS FLUID FLOATING ON AN IMMISCIBLE NON-VISCOUS FLUID

In the absence of gravity a sheet of viscous fluid of thickness h under end load due to rate of extension ϵ is unstable for all wavelengths when the total viscous compressive stress per unit width $-4\mu\epsilon h - T$ is positive. Here T is the sum of the surface tensions of both sides of the sheet. As $-\epsilon$ increases, instability of wavelength λ is possible as soon as $-4\mu\epsilon h - T$ becomes positive; but the rate of growth is a maximum when the wavelength is

$$\lambda = 2\pi\{(-4\mu\epsilon h - T)/\rho g\}^{\frac{1}{2}}, \qquad (1)$$

ρ being the density of the lower fluid (Benjamin, private communication).

Though it has been difficult to realise the physical conditions under which this formula is legitimately applicable because of the difficulty of obtaining a thin viscous sheet floating on an immisible fluid, some experiments were made in which SAIB was floated on concentrated brine and syrup on carbon tetrachloride. The thickness of these sheets could not be less than about 0·6 to 1·0 cm. because the equilibrium depth of the lower surface of a large bubble of the upper liquid in the lower was of order 1 cm., and if a hole was pierced through it, it would not close up again. Two methods were used in these qualitative experiments, and in each case no instability was observed till the rate of compressive strain reached a certain value and then waves suddenly appeared. In the first method the layer was spread between two vertical walls which penetrated a short distance into the lower liquid. The sides of the viscous sheet were contained by means of vertical rubber sheets joining the ends of the vertical walls so that when the walls approached one another at speed U the rate of strain was unidirectional and equal to $-U/L = \epsilon$, where L was the distance apart at the instant considered. To observe the waves a horizontally striped card was set up and photographed obliquely by light reflected from the surface on the viscous layer. Figs. 13 and 14 (pl. 3) are photographs of the stripes taken when $-\epsilon$ was roughly 0·1 sec.$^{-1}$ and 0·4 sec.$^{-1}$ respectively. The viscosity was about 40 poise and the surface tensions at the upper and lower surfaces about 30 and 15 dyne cm.$^{-1}$ respectively. The thickness of the layer was about 1 cm. so that in fig. 13, $-4\mu\epsilon h$ was about 16 dyne cm.$^{-1}$ while in fig. 14 it was roughly 64 dyne cm.$^{-1}$. Thus it seems that waves appeared between the conditions in which $-4\mu\epsilon h$ was 16 and 64, and since T was roughly $30 + 15 = 45$ dyne cm.$^{-1}$ it seems that they appeared in the range of ϵ where they would be expected.

The other qualitative experiment was to immerse a horizontal disc of 5·5 cm. diameter into some syrup floating on carbon tetrachloride contained in a glass vessel of 15 cm. diameter. The depth of the syrup was 1 cm. On rotating the disc the surface remained flat till the angular speed was 1 revolution in 2·7 sec. Fig. 15 (pl. 4) shows it in this condition. Fig. 16 (pl. 4) shows the situation at 1 revolution in 2·4 sec. Short waves can be seen at about 45° to the periphery, but most of the surface appears smooth. Fig. 17 (pl. 4) shows the surface at 1 revolution in 1·4 sec. When this experiment was made it was thought that a skin with some rigidity might have formed, but the surface flattened leaving no visible trace of the wave pattern in a few seconds after stopping the rotation. It may well be that some surface viscosity is present, but the complete disappearance of the pattern suggests that there was no element of rigidity. Wave crests of 45° would be perpendicular to the direction of maximum compressive strain so that the expected phenomenon seems to be like that shown in fig. 16. The suddenness with which the waves appear gives the impression that the rotating disc presents a true instability problem.

48

MOTION OF AXISYMMETRIC BODIES IN VISCOUS FLUIDS

REPRINTED FROM

Problems of Hydrodynamics and Continuum Mechanics (1969), pp. 718–2
Copyright 1969 by Society for Industrial and Applied Mathmatics

The resistance to slow motion of a long axisymmetric body moving in a viscous fluid is twice as great when the direction of motion is perpendicular to the axis as when moving along it.
This is known when the body is a prolate spheroid, but is true more generally.

It gives me great pleasure to make a small contribution to the birthday volume dedicated to Professor Sedov, though I fear its subject is not among those to which he has devoted special attention.

1. Introduction

It is known* that the resistance of a very long prolate spheroid moving in a viscous fluid at very low Reynolds numbers is twice as great when the direction of motion is perpendicular to the axis of symmetry as it is when moving along the axis. Recently I wrote a film† for students illustrating the mechanics of Stokes, or creeping flow. In one experiment I showed that a metal cylinder, or straight piece of wire, falls through a viscous fluid (corn syrup in the film) twice as fast when its long axis is vertical as when it is horizontal, and I asserted that this is a property of all long axisymmetric bodies, no matter what the distribution of cross-section may be, provided the centre of gravity is in such a position on the axis that the body can remain horizontal while falling. The truth of this proposition is not at once obvious but it is borne out accurately under experimental conditions, and when the film was shown I was asked whether my assertion can be proved mathematically. Though the discussion given below could no doubt be improved it gives the reasoning which led me to make the assertion.

2. Motion along the axis

The motion of the fluid could be regarded as being due to an axial distribution of sources and Stokeslets, and the problem is to determine whether any such disturbances can make the velocity zero over the body when the flow at infinity

---

* See *Low Reynolds Number Hydrodynamics*, by J. Happel & H. Brenner, Prentice-Hall (1905), p. 226.

† *Low Reynolds Number Flows*. Educational Services Inc. (1967).

is parallel to the axis and equal to $-U_L$. A Stokeslet of strength S on the axis of x and pointing along it is a singularity which gives rise to a flow whose components in spherical polar coordinates (r, θ) are $u' = 2(S/r) \cos \theta$, $v' = -(S/r) \sin \theta$.

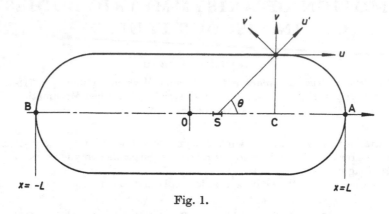

Fig. 1.

The components parallel and perpendicular to Ox are (fig. 1)

$$u = \frac{S(1 + \cos^2 \theta)}{r}, \quad v = \frac{S \cos \theta \sin \theta}{r}.$$

We may suppose that the surface of the body is described by $R(x)$, the distance of a point P on it from the axis Ox. The contribution to u at P due to a Stokeslet $S(\xi)\, d\xi$ on the axis at $x = \xi$ is $r^{-1}(1 + \cos^2 \theta)\, S(\xi)\, d\xi$, where

$$\tan \theta = \frac{R(x)}{\xi - x} \quad \text{and} \quad r^2 = R^2 + (\xi - x)^2. \tag{2.1}$$

The integral equation for determining $S(x)$ so that $u = 0$ at $(x, R(x))$ is therefore

$$0 = -U_L + \int_{-L-x}^{L-x} \frac{1 + \cos^2 \theta}{r} S(\xi)\, d\xi, \tag{2.2}$$

where $x = \pm L$ are the two ends of the body. Over the greater part of the range $1 + \cos^2 \theta$ may be taken without appreciable error as 2, but in a range from $x - e$ to $x + e$, θ varies from a small value, say α, to $\pi - \alpha$. In this small range we take $S(\xi)$ as constant and equal to $S(x)$ so that the equation for $S(x)$ is

$$0 = -U_L + 2 \int_{-L-x}^{L-x} \frac{S(\xi)}{r}\, d\xi + S(x) \int_{x-e}^{x+e} \frac{\cos^2 \theta - 1}{r}\, d\xi; \tag{2.3}$$

writing $\cot \alpha = e/R(x)$, the second of these integrals is $2S(x) \cos \alpha$, and since $R(x)/L$ is small, the variation in $S(x)$ is small in the distance over which α varies from near 0 to near π, so that no appreciable error is committed if $\cos \alpha$ is taken as 1, and hence the integral equation for $S(x)$ is

$$0 = -U_L + 2 \int_{-L-x}^{L-x} \frac{S(\xi)}{r}\, d\xi - 2S(x). \tag{2.4}$$

So far the velocity which the Stokeslets produce in the direction perpendicular to the body has not been considered. This is small but can be evaluated once $S(x)$ has been determined from (2·4) and

$$v(x) = \int_{-L-x}^{L-x} \frac{S(\xi)}{r} \cos\theta \sin\theta \, d\xi. \tag{2·5}$$

$v(x)$ can be neutralized by adding a distribution of sources of strength

$$m(x) = -2\pi R(x) v(x). \tag{2·6}$$

It will be noticed that if, for instance, the body is symmetrical about $x = 0$ and $S(x)$ is constant, the body will resemble a long prolate spheroid and $v(x)$ will be positive in the forward half and negative in the rear half. In that case $m(x)$ will be negative in the forward half and positive in the rear half which is the opposite situation to the distribution of sources which would represent the motion if the fluid were non-viscous.

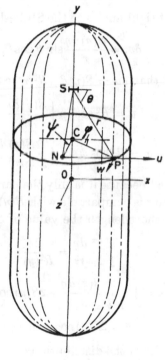

Fig. 2.

3. Motion perpendicular to the axis

Consider the motion produced by Stokeslets of intensity $S(y)\,dy$ distributed along the axis Oy, but directed parallel to Ox. The geometry is now no longer axisymmetrical and the relevant lengths and angles used for calculating the velocity at points on the axisymmetrical body are shown in fig. 2; P is a point on the surface, S is a point on the axis, and C is the centre of the section at y so that $SC = \eta - y$, $SP = r$, $CP = R(y)$. PN is the line through P in the section at x which is parallel to Ox, ϕ is the angular coordinate of P. θ is the same angle as in fig. 1 and ψ is the angle SNC which is needed in calculating the component of velocity perpendicular to the direction of motion Ox. The geometry of fig. 2 shows that

$$r^2 = R^2 + (\eta - y)^2, \quad \cos\theta = R\frac{\cos\phi}{r}, \quad \sin\theta = \frac{[(\eta - y)^2 + R^2\sin^2\phi]^{\frac{1}{2}}}{r},$$

$$\cos\psi = \frac{R\sin\phi}{[(\eta - y)^2 + R^2\sin^2\phi]^{\frac{1}{2}}}.$$

The contributions of the Stokeslet of strength $S(\eta)\,d\eta$ at S to the velocity at P are

$$\delta u = \frac{S(\eta)\,d\eta}{r}\left[1 + \frac{R^2\cos^2\phi}{r^2}\right],$$

$$\delta v = \sin\psi\,\delta v_1, \quad \delta w = \cos\psi\,\delta v_1,$$

where δv_1 is the velocity at right angles to the Stokeslet direction, i.e.

$$\delta v_1 = \frac{S(\eta)\,d\eta}{r}\sin\theta\cos\theta.$$

From these it is found that $\delta w = S(\eta)\,d\eta(R^2/r^3)\cos\phi\sin\phi$ so that the variable parts of u and w are

$$u = R^2\cos^2\phi\int\frac{S(\eta)\,d\eta}{r^3} \tag{3.1}$$

and

$$w = R^2\cos\phi\sin\phi\int\frac{S(\eta)\,d\eta}{r^3}. \tag{3.2}$$

$\int S(\eta)/r^3\,d\eta$ can be evaluated because it is only the part of the Stokeslet distribution close to the point P, where r is comparable with $R(y)$, that makes any appreciable contribution, and in that short length the value of $S(\eta)$ can be taken as constant and equal to $S(y)$. Also

$$\int_{-\infty}^{+\infty}\frac{d\eta}{r^3} = \frac{2}{R^2(y)}$$

so that

$$u = \int_{Y_2 - y}^{Y_1 - y}\frac{S(\eta)\,d\eta}{r} + 2S(y)\cos^2\phi, \tag{3.3}$$

$$w = 2S(y)\cos\phi\sin\phi, \tag{3.4}$$

where Y_1 and Y_2 are the limits of the distribution of $S(y)$.

Now consider the effect of adding a distribution $N(y)\,dy$ of doublets along the axis Oy. The components u_d and w_d at P are due almost entirely to the doublets close to P and in this range $N(y)$ may be regarded as constant. The velocity components at P are

$$u = \frac{N(y)}{R^2}(-2 + 4\cos^2\phi), \tag{3.5}$$

$$w = \frac{N(y)}{R^2}4\cos\phi\sin\phi, \tag{3.6}$$

so that if $N(y)/R^2$ is taken as $-\frac{1}{8}S(y)$, the variable parts of u and w will be neutralised leaving a u-component $-S(y)$.* Thus, adding (3·5) and (3·6) to (3·3) and (3·4), the u-component at P due to both Stokeslets and doublets and superposed velocity $-U_P$ is

$$-U_P + \int_{Y_2-v}^{Y_1-v} \frac{S(\eta)\,d\eta}{r} - S(y) = 0 \quad \text{and} \quad w = 0. \tag{3·7}$$

Comparing (3·7) with (2·4), it will be seen that if the value of $S(x)$ has been found by solving the integral equation (2·4), the same distribution of Stokeslets along the y-axis will satisfy the conditions round the body moving transversely provided

$$U_P = \tfrac{1}{2}U_L.\dagger \tag{3·8}$$

4. Force on axisymmetric body

When the flow round one or more bodies moving in a fluid of viscosity μ can be represented by Stokeslets, sources and doublets, the force on them in any direction is $4\pi\mu$ times the sum of the components of all the Stokeslets resolved in that direction. This can be proved by integrating the stress over a very large spherical surface containing them. The doublets and sources make no contribution to the total force though they are necessary to satisfy the boundary conditions at the surface of the bodies. For this reason the force

$$4\mu \int_{-L}^{+L} S(x)\,dx$$

moves the body at speed U_L when applied longitudinally and $\frac{1}{2}U_L$ when applied at right angles to the axis. To verify this experimentally by measuring the rate of fall of such a body in the two positions, it is necessary to ensure that the centre of gravity is in such a position that the viscous forces do not exert a couple about it whatever the orientation of the body to the force. This is most simply secured by using an axisymmetric body of uniform density which is itself symmetrical about a plane at right angles to its axis. Such a body will preserve its orientation while falling. In the film which led to this discussion this condition was attained by using a solid metal cylinder with rounded ends sinking in corn syrup.

---

* *Editor's note.* There is an (uncorrected) error of sign here; $-S(y)$ should be $+S(y)$, and likewise in (3·7). The two integral equations (2·4) and (3·7) are now not identical in form, and the result represented by (3·8) cannot be deduced from them as they stand. However, it may be shown that, for large values of the ratio of length to breadth of the body, the integral terms in (2·4) and (3·7) are dominant, so that (3·8) is a valid asymptotic result. A mathematical investigation of the problem has been made by J. P. K. Tillett, *J. Fluid Mech.* **44** (1970), 419.

† It has been pointed out to me that in the transverse case a third distribution of singularities can eliminate the surface velocity parallel to the axis of symmetry, and this makes no contribution to the total force on the body.

5. DIRECTION OF MOTION OF INCLINED AXISYMMETRIC BODY

The inclination to the vertical of the path of an inclined axisymmetric body falling in a viscous fluid can be found even though the actual velocity of fall is unknown. This can be seen at once in fig. 3 which shows a body of approximately the same proportions as those of the body used in my film when falling at angle β to the vertical. The components of velocity parallel and perpendicular to the axis are each proportional to the corresponding components of force. In fig. 3 the triangle of force is

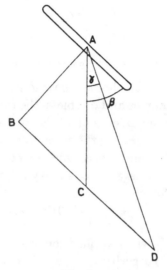

Fig. 3.

ABC, where AB represents the component perpendicular to the axis and BC that parallel to it. AC is vertical. The triangle of velocities is found by producing BC to the point D so that BD = 2BC. AD is then the direction of motion. As β varies the point B moves in a semicircle, and the angle $\gamma = $ CAD, which is the deviation of the path from the vertical, is given by $\tan \gamma = \tan \beta / (2 + \tan^2 \beta)$. γ has a maximum value of 19° 30′ when $\cos^2 \beta = \frac{1}{3}$, so that a long axisymmetric body can never fall along a path which is inclined to the vertical at more than 19·5°, and this occurs when the axis of the body lies at $\cos^{-1} (1\sqrt{3})$ or 54° 44′ to the vertical. This result for a long ellipsoid or cylinder is given in Happel and Brenner's book.

49

ELECTRICALLY DRIVEN JETS

REPRINTED FROM

Proceedings of the Royal Society, A, vol. CCCXIII (1969), pp. 453–75

Fine jets of slightly conducting viscous fluids and thicker jets or drops of less viscous ones can be drawn from conducting tubes by electric forces. As the potential of the tube relative to a neighbouring plate rises, viscous fluids become nearly conical and fine jets come from the vertices. The potentials at which these jets or drops first appear was measured and compared with calculations.

The stability of viscous jets depends on the geometry of the electrodes. Jets as small as 20 μm. in diameter and 5 cm. long were produced which were quite steady up to a millimetre from their ends. Attempts to describe them mathematically failed. Their stability seems to be due to mechanical rather than electrical causes, like that of a stretched string, which is straight when pulled but bent when pushed.

Experiments on the stability of water jets in a parallel electric field reveal two critical fields, one at which jets that are breaking into drops become steady and another at which these steady jets become unsteady again, without breaking into drops.

Experiments are described in which a cylindrical soap film becomes unstable under a radial electric field. The results are compared with calculations by A. B. Basset and after a mistake in his analysis is corrected, agreement is found over the range where experiments are possible. Basset's calculations for axisymmetrical disturbances are extended to those in which the jet moves laterally. Though this is the form in which the instability appears, calculations about uniform jets do not seem to be relevant.

In an appendix M. D. Van Dyke calculates the attraction between a long cylinder and a perpendicular plate at a different potential.

It was pointed out 369 years ago by William Gilbert* that a spherical drop of water on a dry surface is drawn up into a cone when a piece of rubbed amber is held at a suitable distance above it. This phenomenon has recently been subjected to theoretical treatment† and it was shown that a conducting fluid can exist in equilibrium in the form of a cone under the action of an electric field but only when the semivertical angle is 49·3°. Apparatus was constructed in which the necessary field could be set up and very nearly conical fluid surfaces or interfaces between two fluids were formed for which the semivertical angle was close to 49·3°.

An uncharged drop in a uniform electric field is pulled out into a nearly spheroidal form which becomes unstable when the field reaches a maximum value $1·62 (T/r_0)^{\frac{1}{2}}$ and the drop has become 1·85 times as long as its equatorial diameter.† Here the field was expressed in e.s. units and T the surface tension and r_0, the initial radius, are expressed in c.g.s. units. The same analysis predicts possible equilibrium shapes which are longer than 1·85 times the diameter though the corresponding value of the field is less than $1·62 (T/r_0)^{\frac{1}{2}}$. At higher fields the drop does not in fact pass through

---

* *de Magnete*, Book 2, chap. 2 (trans. by P. F. Mottelay).

† G. I. Taylor, *Proc. Roy. Soc.* A, CCLXXX (1964), 383 (paper 39 above).

approximately spheroidal shapes, but develops pointed ends at which narrow jets appear or small drops are torn off. Several authors* have described experiments with such jets, but have not given theoretical discussions of their mechanics, perhaps because they could not relate the electric field to the geometry of their apparatus, even before the jet appeared. The experiments to be described were undertaken in the hope of being able to supply a rational description of the mechanics of these jets but complete success in that direction has not been attained. Nevertheless, some points of interest have appeared.

Most of the experiments so far recorded have been peformed with apparatus sketched in fig. 1. Here A is an open-ended capillary tube of metal or glass containing

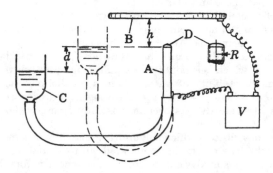

Fig. 1. Sketch showing the geometry of the single plate apparatus used by Zeleny.

a conducting fluid. Opposite it at a distance h is a metal plate B. A and B are maintained at potential difference V and the effects on the fluid surface at the open end of A observed as V increases. Under these circumstances no theoretical calculation of the force which the electric field exerts on the fluids at the end of tube A had been made. Zeleny, however, made measurements of the change in pressure which would keep the meniscus D (fig. 1) in approximately a constant shape as V changed. He measured the depth d through which a reservoir C connected with the fluid in A had to be moved downward in order to maintain the meniscus D at a constant height as V increased. He defined a mean electric field f, over the area of the top of the tube as being given by

$$f^2/8\pi = \rho g d, \tag{1}$$

and he thus measured f as a function of V, h and R the radius of the tube. This method depends on the assumption that the shape of the meniscus remains spherical and though the assumption could only be true if the normal electric stress over the whole meniscus were uniform, it seems to be a good approximation because, as will be seen later, the observed connection between f and V is very close to that calculated by Van Dyke, whose analysis is given in the appendix.

* J. Zeleny, *Phys. Rev.* III (1914), 69 and x (1917), 1; R. H. Margavey & L. E. Outhouse, *J. Fluid Mech.* XIII (1962), 151; B. Vonnegut & R. L. Neubauer, *J. Colloid Sci.* VII (1952), 616.

Van Dyke's calculation

Van Dyke has calculated the attractive force P between a long cylinder of length L and radius R and a perpendicular plane at distance h when a potential V is established between them. His calculations are given in the appendix* and his results are displayed in fig. 2 where P/V^2 is shown as a function of h/R for constant values

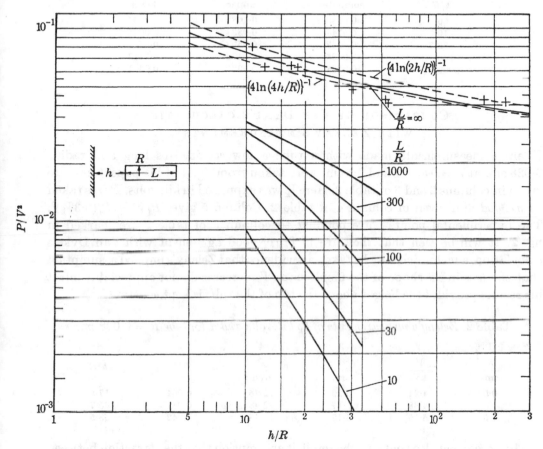

Fig. 2. Van Dyke's calculation of the attraction between a plate and a cylinder of radius R and length L when separated by distance h. $+$, experiments with single plate apparatus by Zeleny and by the present author. The upper broken line is $P/V^2 = \{4\ln(2h/R)\}^{-1}$ and the lower broken line $P/V^2 = \{4\ln(4h/R)\}^{-1}$.

of L/R. It will be noticed in fig. 2 that for a fixed value of h/R the attractive force increases as L increases and that it is still increasing when L/R is as great as 1000. When L/R becomes infinite the analysis was difficult but it was carried out by computer for $h/R = 5, 10, 33, 100, 1000$. For all values of h/R when L/R is infinite,

* *Editor's note.* The appendix by M. D. Van Dyke is not reproduced in this volume, since his results are described fully by fig. 2.

P/V^2 can be shown to lie between $1/\{4\ln(4h/R)\}$ and $1/\{4\ln(2h/R)\}$ which are fairly close limits when h/R is large. The values computed by Van Dyke and these upper and lower bounds are given in table 1.

Table 1. *Computed values of P/V^2 (column 2) for L/R infinite*

		bounds	
h/R	computer	upper	lower
5	0·0938	0·1086	0·0834
10	0·0744	0·0834	0·0678
33·33	0·0543	0·0593	0·0511
100	0·0435	0·0472	0·0417
1000	0·0308	0·0329	0·0301

COMPARISON OF VAN DYKE'S CALCULATION WITH ZELENY'S MEASUREMENTS

Zeleny's measurements made with a tube a few centimetres long and radius 0·028 cm. at $h = 0.5$, 1·0 and 1·5 cm. were taken from fig. 4 of his paper* and are given in columns 2 and 3 of table 2. Here V_k was expressed in kilovolts. P was taken as $\pi R^2 g \rho d$ and given in column 4 of table 2. Column 5 gives $P/V^2 = P/(0.3 V_k)^2$. The experimental points corresponding to column 5 of table 2 are marked in fig. 2. It will be seen that they are in fairly good agreement with Van Dyke's calculation, a fact which may be taken as evidence that Zeleny's implied assumption that changes in the shape of the meniscus as V increases do not appreciably effect his results provided the height above the top of the tube is kept constant.

Table 2. *Zeleny's measurements of V_k kilovolts and d for tube $R = 0.028$ cm.*

1	2	3	4	5	6
h	V_k	d	P	P/V^2	h/R
cm.	kV.	cm.	dyn		
0·5	4·24	5·29	12·75	0·064	17·8
1·0	4·88	5·17	12·48	0·047	35·7
1·5	5·15	5·17	12·48	0·042	53·5

The agreement also confirms the implicit assumption that the attraction between fluid and plate does not depend on the field far from the end of the cylinder. The fact that Van Dyke's calculation for a cylinder 1000 times as long as its radius differs very appreciably from that of a cylinder for which L/R is infinite is not surprising because Van Dyke's calculation was for the total force, that is the difference between the forces acting at the two ends. The force acting at the far end is still appreciable even when $L/R = 1000$. The force on the fluid is that which acts at the near end only and the implicit assumption that this depends only on the configuration in that neighbourhood can only be satisfied if the force is that on an infinite cylinder for the same value of h/R.

* *Phys. Rev.* III (1914), 69.

ZELENY'S MEASUREMENTS OF THE POTENTIAL NECESSARY
FOR INSTABILITY

In these experiments a pressure was applied to fluid in the tube A (fig. 1) till the meniscus rose to a definite height above the end of the tube and this height was maintained by reducing the pressure, p, as the potential of A was increased till breakdown occurred and it was no longer possible to maintain the meniscus steady. Assuming that the fluid is a conductor and wets the top of the tube, the equation for the equilibrium of the part of the drop above the top is

$$2\pi RT \cos \phi + W = P + \pi R^2 p, \tag{2}$$

where ϕ is the angle which the fluid surface makes with the tube axis at the point on the rim of the top where it leaves the tube and W is the weight of the fluid above the top. In Zeleny's experiments R was small enough to justify neglect of W. The results of experiments in which the height of the meniscus was equal to the radius, so that it was initially hemispherical, are given in table 3 which is taken from measurements displayed on p. 85 of Zeleny's paper* in his fig. 6. The values of the potential V_k at which breakdown occurs is given in column 3 of table 3 and the calculated value of P, taken from fig. 2, in column 4. To find p it is necessary to take the values given in Zeleny's curves from the initial value which, when the meniscus was initially hemispherical, was $2T/R$. Values of $\pi R^2 p$ found in this way are given in column 5. They were positive in all cases. According to (2)

$$\cos \phi = (P + \pi R^2 p)/2\pi RT$$

and the values of this at breakdown are given in column 6. Except in one case they are very close to 1·0 so that when the experiment was carried out in Zeleny's way with a positive pressure behind the meniscus, breakdowns occur when ϕ is close to 90° and in all cases where the meniscus was initially hemispherical the value of p at breakdown is small compared with its initial value.

Two-plate apparatus

Before Van Dyke's calculations had been made it was realised that a finite cylinder would have to be very long indeed before the total force on it was nearly the same as that on a semi-infinite cylinder. For this reason apparatus was constructed in which the electric field far from the tube containing the fluid would be known so that it would not be necessary to assume that the distant field does not affect appreciably the force on the fluid emerging from its end. This apparatus shown in fig. 3, consists of two parallel plates B and E separated by distance H and maintained at potential difference V_k kilovolts. Through the middle of the lower plate E passes the metal (hypodermic) tube A to a height L and the fluid was supplied either from a reservoir C through a flexible tube or, if it was necessary to measure the rate of delivery, from a graduated syringe S connected through a metal pipe with A.

* *Phys. Rev.* III (1914), 69.

Table 3. *Zeleny's measurements for critical voltage for instability when the meniscus was maintained at height R above the top of tube*

1	2	3	4	5	6
$\dfrac{R}{\text{cm.}}$	$\dfrac{h}{R}$	$\dfrac{V_k}{\text{kV.}}$	P(Van Dyke) dyn.	$\pi R^2 p$ dyn.	$\dfrac{P + \pi R^2 p}{2\pi R T}$
0·0543	27·6	5·9	21·3	3·3	$\dfrac{24\cdot6}{25} = 0\cdot99$
0·0420	35·6	5·45	17·4	2·8	$\dfrac{20\cdot3}{19\cdot5} = 1\cdot04$
0·0340	44·1	4·0	13·9	1·9	$\dfrac{15\cdot8}{15\cdot7} = 1\cdot00$
0·0281	53·4	4·7	11·8	0·6	$\dfrac{12\cdot4}{13\cdot0} = 0\cdot95$
0·0231	64·8	4·50	10·3	0·6	$\dfrac{10\cdot9}{10\cdot7} = 1\cdot02$
0·0200	75	4·05	8·2	0·8	$\dfrac{9\cdot0}{9\cdot3} = 0\cdot97$
0·0166	90	4·05	8·0	0·7	$\dfrac{8\cdot7}{7\cdot1} = 1\cdot23$
0·0146	103	3·55	5·9	0·6	$\dfrac{6\cdot5}{6\cdot7} = 0\cdot97$

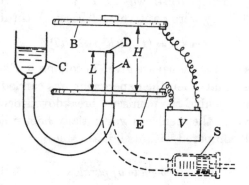

Fig. 3. Geometry of the two-plate apparatus.

In the two-plate apparatus the value of the force P exerted on a cylinder of length L is*

$$P = \frac{V^2 L^2}{4H^2} \frac{1}{\ln(2L/R) - \frac{3}{2}},$$ (3)

and (3) can be used in comparing experiments with theory.

* G. I. Taylor, *Proc. Roy. Soc.* A, ccxci (1966), 145 (paper 42 above).

EXPERIMENTS TO DETERMINE THE LOWEST VALUE OF V AT
WHICH FLUID IS DRAWN FROM THE TUBE

In these experiments the reservoir C (figs. 1 or 3) was first adjusted till the top of the fluid (D, fig. 3) could be seen to reflect as a plane mirror so that $p = 0$. No further changes in pressure were made. The voltage was gradually increased and the surface of the meniscus became convex till at a certain voltage, V_k kilovolts, static equilibrium could no longer be maintained. When water was used the break-down of equilibrium developed into violent oscillations just as it does when a soap bubble is subjected to an electric field.* When very bad conductors such as transformer oil or silicone fluids were used drops came from the tube at infrequent intervals, detached themselves and rose. Fig. 4, pl. 1, shows the outline of a water drop in the single plate apparatus when $R = 0.163$ cm., $h = 1.58$ cm. just before it became unstable at $V_k = 9.8$ kV. Using the curve of fig. 2 the calculated value of P was 85 dyn., and since $p = 0$ the equilibrium equation (2) is $P + W = 2\pi T \cos \phi$. The measured value of ϕ in fig. 5 is $29.5°$ so that taking $T = 74$ dyn. cm.$^{-1}$, $2\pi T \cos \phi = 66.4$ dyn. This is smaller than the calculated value of P, but part of the difference is due to the fact that W is not negligible when R is as large as 0.163 cm.; in fact an estimate of W is 8 dyn. so that the calculated value of P, namely 85 dyn. must be compared with $66.4 + 8 = 74.4$ dyn. It was thought that some of this discrepancy might be due to the fact that the plate used in the apparatus of fig. 1 was not large enough to justify the use of a theory envisaging an infinite plate but enlargement of the plate did not affect the result appreciably. The agreement with theory was better when the two-plate apparatus was used. Measurements were made with the two-plate apparatus (fig. 3) on a number of fluids and in each case the voltage at which drops first began to leave the top of the tube A when $p = 0$ was measured. It was found that when the fluid is conducting and sufficiently viscous the fluid does not vibrate but rises to a point. Figs. 5 and 6, pl. 1, show the appearance of the meniscus formed when a mixture of 95 % glycerine mixed with 5 % of a 10 % sodium chloride solution was used. In fig. 5 care was taken to ensure that $p = 0$ at the moment when the jet appeared at the top, but in fig. 6 p was a little higher. It will be seen that when $p = 0$ the meniscus is nearly conical with a semivertical angle close to the only possible equilibrium value $49.3°$† and it will also be noticed that the surface remains nearly conical at the outer edge of the top of the tube, though this cannot be an accurate result derivable from theory. This observation however suggests that an approximate expression for predicting the critical potential at which jets or drops can appear when $p = 0$ can be put forward by setting $\phi = 49.3°$ in (2) and substituting for P from (3). Since $2 \cos 49.3° = 1.30$ this gives

$$V_k^2 = \frac{4H^2}{L^2} \left(\ln \frac{2L}{R} - \frac{3}{2} \right) (1.30\pi RT) (0.09). \tag{4}$$

---

* C. T. R. Wilson & G. I. Taylor, *Proc. Camb. phil. Soc.* XXII (1926), 728 (paper 9 above).
† Paper 39 above.

The factor 0·09 is inserted to give the prediction in kilovolts. In table 4 details of a number of experiments conducted with $H = 7·66$ cm. are recorded. The fluid as well as R and L was varied but care was taken to ensure that $p = 0$. The results of the observed and computed values of V_k using (4) are given in the last two columns of the table. The agreement is good in cases where the conical form appears before the jet forms, but it is also good in cases like that of fig. 4 where the conical shape is not attained as a steady state.

Table 4. *Upper limit of statically stable meniscus in two-plate*
apparatus with $H = 7·66$ cm. and $p = 0$

	$\dfrac{R}{\text{cm.}}$	$\dfrac{L}{\text{cm.}}$	V_k(kV.)	
fluid			observed	calculated (eqn. (4))
1	0·05	2·1	13·0	13·4
1	0·05	3·0	9·5	10·0
2	0·05	1·5	17·5	17·6
2	0·05	2·5	11·5	11·6
2	0·162	3·0	13·0	14·2
3	0·12	2·1	17·5	17·8
3	0·12	3·0	14	13·2
4*	0·05	3·0	14	14·2
4	0·05	2·2	14	13·9
4	0·05	3·55	9·5	9·4
4	0·12	3·0	14	14·2
4	0·12	2·0	17·5	17·8
4	0·12	3·0	14	14·2
5	0·163	3·0	11·4	11·0

* Oscillating jet shown in fig. 7.

Fluids: 1. Glycerine, $T = 62·5$ dyn. cm.$^{-1}$.
2. 95 % glycerine, 5 % water, $T = 65$ dyn. cm.$^{-1}$.
3. 98 % glycerine, 2 % of 0·5 % NaCl solution.
4. Distilled water, $T = 73$ dyn. cm.$^{-1}$.
5. Transformed oil, $T = 37$ dyn. cm.$^{-1}$.

Fig. 7, pl. 1, for instance shows one of the cases listed in table 4. The oscillating drop is throwing off small drops. The agreement between the observed and calculated potentials is consistent with the experimental result* that there is little difference between the electrical field which makes an isolated drop become unstable and the probably slightly lower field at which it can be pointed at the end.

NON-CONDUCTING FLUIDS

When non-conducting fluids like transformer oil or silicone were used in either the single or double plate apparatus it was found that as the potential increased the meniscus was at first displaced in much the same way as for conducting fluids, but when the potential rose to the value at which a conducting fluid would become unstable and produce a fine jet, non-conducting fluids would throw off a drop and

* Paper 39 above.

then after a considerable time another. Raising the potential above the critical value increased the rate at which drops were thrown off and sometimes when the potential rose very considerably perhaps to twice the critical value the drops might coalesce into a steady stream through this condition could only be attained by increasing the flow of liquid.

Fig. 8(a), pl. 1, shows a meniscus of Tellus oil just before instability at the top of a steel tube 3·2 mm. diameter. Fig. 8(b) shows the same jet when the voltage has been raised sufficiently to produce a stream of drops. When fig. 8(a) was taken the pressure at the level of the top of the jet was atmospheric ($p = 0$). For 8(b), pl. 2 no doubt the pressure was slightly less owing to the hydraulic resistance of the fluid flowing through the tube. Calculation, however, showed that this pressure drop would be small. The two photographs are interesting because they show that the nearly conical equilibrium form with semivertical angle 49·3° is set up in non-conductors (no doubt owing to a very small conductivity) as well as in conductors even though the electric field cannot be that which would be necessary for the establishment of a complete conical meniscus covering the whole of the cylindrical tube.

FORMATION OF STEADY STREAMS

In the preceding pages the only force capable of moving the fluid which has been considered is the normal force which an electric field exerts on a conducting surface. A steady motion could not be produced in this way. In one of Zeleny's experiments* performed with apparatus shown in fig. 1, a jet of glycerine was produced which broke up into drops after traversing steadily a length of about 1·0 cm. It is possible, though perhaps unlikely, that such a jet might be drawn up by a normal force acting on the end of the continuous part of the fluid, but in experiments using glycerine in my two-plate apparatus (fig. 3) jets were formed which extended in steady motion the full distance $H - L$ between the end of the tube A (fig. 3) and the plate B, a distance up to 6 cm. These jets were so steady that an exposure of 1 sec. or more would reveal sharply a straight jet sometimes only 0·002 cm. in diameter. Fig. 9, pl. 2, shows the first 1·4 cm. of a jet of glycerine 4·66 cm. long issuing from a tube projecting 3 cm. into the space between two plates 7·66 cm. apart. This jet was quite steady up to a point less than 0·1 cm. from where it struck the upper plate. No doubt within a length comparable with its diameter it must have become unstable and piled up with an oscillating motion into a mound just as a thread of honey does when it falls on a plate, but this was too small to be visible.

Instability of viscous jets

Under some circumstances a jet was observed which appeared to rise steadily for a short distance and then to disappear suddenly. Fig. 10, pl. 2, shows a case of this kind. Here an upward pointing jet of glycerine disappears at a height of 0·8 cm. above the tube from which it originates. This disappearance can only be

* *Phys. Rev.* x (1917), 1.

due to an instability setting in at about 0·8 cm. at a frequency greater than the exposure time of 1 ms. Fig. 11, pl. 2, is a photograph of the same jet at the same magnification but with an exposure time of about 2 μsec. This photograph covers the blank space above the jet of fig. 10 and an arrow is drawn in both figs. 10 and 11 to mark the position 0·8 cm. above the orifice where the instability seems to begin. This instability does not increase violently like that observed by Zeleny and others. It increases fairly rapidly at first and then stops increasing. No doubt the difference is due to the difference in the electric fields set up in the single and the two-plate apparatus. The explanation of this type of bending of the jet seems to be that owing to viscous stress it is straight like a stretched string while it is extending but that at a certain stage the electrostatic forces cease to be able to support it against gravity or tangential air friction so that it begins to slow down and therefore to contract longitudinally. The viscous stress therefore reverses. It then becomes unstable with an instability analogous to that of Euler when an elastic column is under end compression. A small amount of crumpling, however, relieves the compression so that it does not become violently unstable. The kind of instability just described is well known in the case of syrup falling on a plate but in that case the part of the falling jet which is under longitudinal compression is limited to a short distance from the point in which its velocity is reduced to zero. Fig. 12, pl. 3, for instance, shows a jet of a viscous fluid (a silicone) falling under gravity. The diameter of the jet decreases downward till it reaches a short distance above the plate on which it spreads. It then thickens and begins to oscillate before it spreads out on the plate. When the jet is also pulled in the same direction as gravity by electric forces it may acquire a speed greater than the limiting velocity at which tangential aerodynamic stress exactly balances gravity. Any decrease in electric stress will then give rise to a compressive longitudinal stress in the jet. Fig. 13, pl. 3, shows a downward jet of glycerine in a vertical field, photographed with exposure time 1 msec. Fig. 14, pl. 3, shows the same jet when the exposure is 2 μsec. Measurements of the diameter of the steady part of the jets in these two photographs are shown in fig. 15. It will be seen that though they are not quite identical the diameter stops decreasing just before the jet begins to oscillate at about 0·8 cm. from the orifice. The fact that the diameter stops decreasing is evidence that the aerodynamic tangential drag has become equal to the sum of the forces due to electric attraction and gravity and one of the principal difficulties in attempting to give a quantitative theory of the electrically driven viscous jets is that of estimating the tangential aerodynamic drag even when the jet is quite steady as it is in fig. 9.

Experiment illustrating instability due to reversal of stress

Fig. 16, pl. 3, shows a stream of glycerine of density 1·26, made visible by a mixture of indian ink, falling into a stratified salt solution whose density in the upper half is approximately 1·24 and 1·25 in the lower half. The driving force 1·26 g. − 1·24 g. = 0·02 g. keeps the upper half in tension and therefore straight. In the lower half the driving force is 1·26 g. − 1·25 g. = 0·01 g. and this is not sufficient

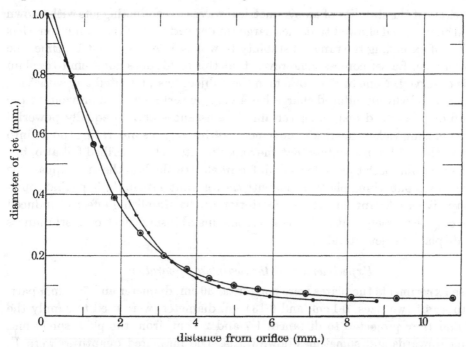

Fig. 15. Diameter of jets of figs. 13 (•) and 14 (⊙) (pl. 3).

to maintain the speed acquired in the upper half so the jet contracts longitudinally and becomes unstable owing to reversal of viscous stress. Fig. 16 can be compared with figs. 10 and 11. Fig. 16 shows the tank with the 1 mm. tube just projecting through the surface of the upper fluid. Above this is the top of the tank, held down by the screws into the Perspex sides which can be seen. The level of the interface between the two densities of salt solution shows faintly through dispersion of indian ink about half way down the jet.

Stability of jets of water and dilute salt solution

The effect of induced charges on water jets has often been studied. Plateau* showed that in the absence of an electric field a uniform column of fluid is unstable under the influence of surface tension for all axisymmetric waves of length greater than the circumference of the cross section. Rayleigh† also discussed the axisymmetrical instability of flowing jets and showed experimentally that the columns broke into drops corresponding with the wavelength of maximum growth rate. He also commented on the great effect which induced charges are observed to have in causing colliding drops to coalesce. Basset‡ and Melcher§ extended

* *Statique experimentale et theoretique soumis aux seules forces moleculaire*, II (1873), 254.
† *Proc. Lond. math. Soc.* X (1879), 473; *Proc. Roy. Soc.* A, XXVIII (1879), 406; *Nature* XLIV (1891), 249.
‡ *Am. J. Math.* XVI (1894), 93.
§ *Fluid-coupled Surface Waves*, M.I.T. Press (1963), chap. 6.

calculations of the stability of axisymmetrical waves to conducting jets with known induced charge, and showed that the charge (under certain limiting conditions) has the effect of extending the range of stability to waves of length about 1·6 times the circumference. Basset confesses, however, that this result does not seem to explain what is observed. None of the workers in the subject has presented any satisfying explanation of why an induced charge has a very powerful effect in preventing the breakup of jets into drops under certain circumstances and an equally powerful effect in causing violent unsteady movements ultimately disintegrating the jet into drops in others.* I must confess that the experiments to be described fail also, but since they exhibit both these effects at different electric fields in the same apparatus it seems worth publishing photographs with the relevant data in the hope that some-one may give a relevant analytical description of the stabilizing effect of induced charge on axisymmetric disturbances and the unstabilising effect on disturbances which displace the jet laterally.

Experiments with the two-plate apparatus

In these experiments the plates were horizontal 30 cm. diameter and 5·5 cm. apart. Steel tubes of two sizes, 0·1 cm. and 0·134 cm. diameter, were used to supply the water and they projected to distances 1·7 and 2·7 cm. from the plate sometimes pointing upwards and sometimes downwards. The measured quantities were V, the potential difference between the plates and Q the volumetric rate of flow of the water. Most of the pictures were taken with single flashes of about 2μsec. duration. In all cases when Q was small enough the jets exhibited the Rayleigh–Plateau instability before the electric field was applied, and the jets broke into drops (figs. 17a, 18a, 19a, 21a, pls. 3, 4, 6). As V or Q increased or R decreased the continu-ous part of the jet increased and at a certain potential a steady régime was estab-lished over the full length of the jet (figs. 17b, 18b, 19b, 20a, 21d, pls. 3 to 6). An increase in V beyond a second critical value makes the jet become unsteady and whip about, particularly at its lower end (figs. 17c, 17d, 18d, 18e, 19c, 19d, 19e, 20b, 20c, pls. 3 to 5). This unsteadiness develops into very violent motion as V increases. Such motions have often been reported, by Zeleny and by Magarvey & Outhouse for instance, but in most cases the threads break into drops which repel one another. In the present experiments the threads of fluid remain continuous through the whole length from the tube to the opposite electrode. The fact that at a definite V a steady stream becomes unsteady near its far end appears at first sight to be the same phenomenon as that observed with viscous conducting fluid, but in viscous jets the onset of instability seems to be associated with a reversal of the force over cross-sections near points where the sectional area is least. In the present experi-ments with water the effect of the field is to make the section decrease in area through the whole length of the jet and the onset of unsteady motion seems to be associated with a rapid decrease in jet diameter near the end where it strikes the electrode. To find whether the unsteadiness would occur if the cross-section were prevented

* J. Zeleny, *Phys. Rev.* x (1917), 1.

from decreasing at its lower end a fine silk thread (fig. 21 g, h, i, pl. 6) was led through the tube in such a way that its lower end was just clear of the lower electrode but dipped into the water deposited on it by the jet. This device effectively prevented the fluid from accelerating and kept the diameter nearly constant without impeding possible lateral motion. When the water was turned on drops ran down the wet silk thread (fig. 21 h). An increase in potential of about the same magnitude as that which made the drops coalesce into a steady jet when there was no thread had the same effect when the thread was present (fig. 21 i). Raising the potential up to and beyond the point where the free jet became violently unsteady had no effect on the threaded stream which remained steady till sparks jumped from the tube to the lower plate.

The effect of increasing Q when there was no thread was to increase the potential necessary for unsteadiness to occur (compare figs. 18 c and 19 c), but had little effect on the threaded jet (fig. 20 i). The water used had conductivity of order $10^{-8}\,\Omega^{-1}$ cm.$^{-1}$ and it might be low enough to make the convected charge comparable with the conducted charge. Experiments were therefore made with dilute NaCl solution of conductivities 10^{-5} and $2 \times 10^{-5}\,\Omega^{-1}\,\mathrm{cm}^{-1}$. The currents increased very greatly but there was little change in the potential at which the jet began to be unsteady in the lower half of its path. Figs. 19 b and 19 c show that unsteadiness with water began between 13 and 14·5 kV. Fig. 20 b shows that with the same value of Q and the same geometry the jet is steady at 12 kV. but unsteady at 13·5 kV. when the salt concentration was 10^{-5} g. cm.$^{-3}$. It seems therefore that it would be justifiable to discuss these experiments assuming that the potential at any point in the jet depends only on the distribution of cross sections. For a given geometry and a given Q the tangential stress on the jet depends only on the potential and the distribution of cross section along the jet so that the conductivity does not affect the stress which in turn determines the distribution of cross section along the jet.

After the stream had been made coherent and steady by raising the potential to a critical value it could remain steady when the potential was lowered. In fig. 21 a, b, c the potentials were 0, 3, 6 kV. At 7·8 kV. the drops coalesced (fig. 21 d) and on lowering the potential to 4 kV. (fig. 21 e) the stream remained steady and coherent though it became thicker at its lower end. When lowered to 3 kV. it broke into drops and looked like fig. 21 b. At 17 kV. (fig. 21 f) the jet had developed a considerable unsteadiness but remained coherent.

Upward jet

When the jet was aimed upwards the water just rose a very short distance above the top of the tube from which it flowed and then fell down outside it (fig. 22 a, pl. 5). It was not possible to produce a steady stream with the flow rate $Q = 0.37$ cm.^{3} sec.$^{-1}$ used for fig. 21 but at $Q = 0.466$ cm.^{3} sec.$^{-1}$ and with the same geometry as for fig. 21 a steady upward stream was produced at 11 kV. (fig. 22 b). The experimental difficulty encountered with upward jets was that the water collected on the upper electrode formed hanging drops which fell and disturbed the electric field. This difficulty was overcome by cutting a hole 1·3 cm. diameter in the upper plate on the axis of the jet (seen at the top of fig. 22*a*). The jet disintegrated into drops immediately after passing the hole, and a shallow conical copper box was placed on the upper electrode to catch them. In spite of this precaution, however, some hanging drops were liable to occur. Some can be seen in fig. 22*b*, but unless they coalesced and dropped off (fig. 22*d*) they did not disturb the field enough to have a visible effect on the jet when it became steady. It was noticeable that the potential required to make a steady upward jet was not very much larger than for a downward pointing jet, but the least possible value of Q was greater.

A POSSIBLE LINE OF THEORETICAL DISCUSSION

The instability of a charged liquid sphere, isolated in space, was shown by Rayleigh[*] to occur when the charge C is greater than $(16\pi r^3 T)^{\frac{1}{2}}$. This has been verified experimentally by Doyle, Moffett & Vonnegut.[†] The instability for the least critical value of C corresponds with a spheroidal form of the displaced surface. For the isolated sphere the critical charge coincided with the surface charge which was just sufficient to reduce the pressure inside the sphere to that of the surroundings. This result is not of more general application for it did not turn out to be true for the instability of an uncharged drop in a uniform field.[‡] Basset's calculation for the stability of a conducting cylinder of fluid of radius a at rest assumes a varicose disturbance,

$$r = a + b \cos mx\, e^{\kappa t}. \tag{5}$$

Using a form of associated Bessel function $K_n(z)$ which differs from that used by Watson[§] in the sign of $K_0(z)$, he finds the equation for κ

$$\frac{\kappa^2 \rho a I_0(ma)}{ma I_1(ma)} = -\frac{T}{a^2}(m^2 a^2 - 1) - \frac{E^2}{4\pi a^2}\left(1 + \frac{ma K_1(ma)}{K_0(ma)}\right), \tag{6}$$

where $I_0(z), I_1(z), K_0(z), K_1(z)$ are associated Bessel functions. E is the charge per unit length.

Using the same symbols defined as in Watson's treatise where these functions are all positive, I find

$$\frac{\kappa^2 \rho a I_0(ma)}{ma I_1(ma)} = -\frac{T}{a^2}(m^2 a^2 - 1) - \frac{E^2}{\pi a^3}\left(1 - \frac{ma K_1(ma)}{K_0(ma)}\right). \tag{7}$$

This differs from (6) in the factor $E^2/4\pi a^2$ which in my calculations is $E^2/\pi a^3$. The difference may not be important in qualitative discussion, but is of importance when comparing calculation with experiments.

As Basset points out, large charges tend to produce instability (κ^2 positive) when $ma > 0.6$ and stability (κ^2 negative) when $ma < 0.6$. It is only when

$$0.6 < ma < 1.0$$

[*] *Proc. Roy. Soc.* A, xxviii (1879), 406.
[†] *J. Colloid Sci.* xix (1964), 136.   [‡] Paper 39 above.
[§] *Bessel Functions*, Cambridge University Press (1922), p. 79.

that no values of E can stabilise the cylindrical fluid. Basset remarks that his result does not explain why it is that a very *slight* charge produces the stability described by Rayleigh. The experiment to which this remark refers was concerned with the large stabilising effect of a small difference in *potential* between the jet and its distant surroundings. The induced charge was not measured.

EXPERIMENTS ON THE STABILITY OF A CYLINDRICAL FILM

In view of the difficulties Rayleigh and others have found in accounting for the stability characteristic of a cylindrical liquid conductor in a radial field it seemed worth while to construct apparatus for producing a cylindrical soap film. Such a film can be produced when the total length is less than its circumference, and its stability under a radial field can be examined because the wavelength of a varicose instability is simply the length between the fixed supports of the film. The difficulty experienced in analysing the experiments with jets in the two-plate apparatus (that no method for calculating the charge induced on the jet has been found) can be overcome by surrounding the cylindrical film and the cylindrical conductors between which the film is held by a cylindrical conducting screen F, of radius R which is many times the radius a of the film. In this case the charge E per unit length is connected with V the potential difference between the cylinder and the screen by the equation

$$E = V/2\ln(R/a), \tag{8}$$

so that measurements of R and a determine E. The apparatus is shown in fig. 23. Two copper tubes A, A, sharpened at the ends, are held coaxially in non-conducting supports so that the distance λ between them can be controlled. The lower one is closed at its lower end. The upper one is connected by non-conducting tube to an open ended vessel, C which is inserted in a vacuum flask D containing water which can be raised or lowered by a jack J and it forms a simple means of controlling the volume of air in the film which is produced in the space λ (fig. 23) between the tubes. Soap films were laid over the ends of the copper tubes and a bubble is blown through the stopcock B till it touches the film stretched over the lower tube. A rod E (fig. 23) is then lowered to pierce the lower film. The stopcock is opened slightly and the bubble allowed to collapse till it is nearly cylindrical. The stopcock is closed and the fine adjustment provided by lowering the jack makes it possible to obtain a truly cylindrical film. The potential is then raised without further adjustment of the volume and was found that the film remains cylindrical till, at a certain potential which was measured, it bursts. Experiments were carried out in the range $1 \cdot 0 < ma < 3 \cdot 2$ and it was found that in the upper part of this range $ma > 1 \cdot 5$ the film oscillated violently before bursting but the point at which the burst occurred could be measured within a small percentage.

The diameter of the copper tubes AA was $1 \cdot 35$ cm. and that of the surrounding wire screen was 13 cm. so that for the apparatus used

$$V = E(2\ln 9 \cdot 6) = 4 \cdot 523E,$$

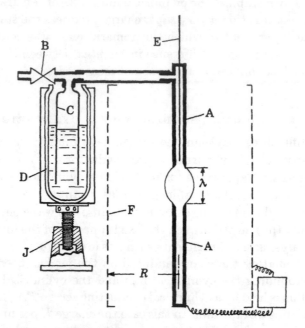

Fig. 23. Apparatus for producing a cylindrical soap film
and subjecting it to a radial field.

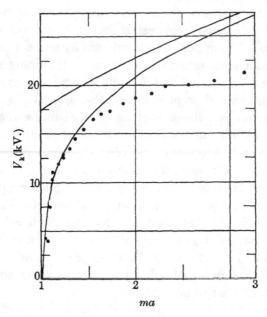

Fig. 24. Potential necessary to produce instability on a cylindrical soap film 1·35 cm. diameter
inside a wide screen 13 cm. diameter. Lower curve, axisymmetric disturbances. Upper curve,
disturbances with lateral displacement. •, experimental points.

expressed in electrostatic units. The values of V were measured in kilovolts, V_k, each of which is $1/0.3$ electrostatic units so that

$$V_k = 1.357E.$$

Using (7) the calculated value of V_k for neutral stability is therefore

$$V_k = 1.357(\pi a T)^{\frac{1}{2}} (m^2 a^2 - 1)^{\frac{1}{2}} \left(\frac{maK_1(ma)}{K_0(ma)} - 1 \right)^{-\frac{1}{2}}. \tag{9}$$

The surface tension of the soap solution used was $27\,\mathrm{dyn.\,cm.^{-1}}$ so that $T = 54$ and $(\pi a T)^{\frac{1}{2}} = 10.700$ so that

$$V_k = 14.50\,(m^2 a^2 - 1)^{\frac{1}{2}} \left(\frac{maK_1(ma)}{K_0(ma)} - 1 \right)^{-\frac{1}{2}}. \tag{10}$$

The values calculated from (10) and the observed potentials at which the cylindrical soap film burst are shown in fig. 24.

INSTABILITY IN MODE WHICH MOVES THE WHOLE JET LATERALLY

Referring to figs. 17 and 22 it seems that when the steady jet becomes unstable owing to large induced charge the displacement is one which moves the jet laterally rather than causing varicose swellings. Such a jet would be represented by

$$r = a + b\,o^{\kappa t} \cos m x \cos \theta, \tag{11}$$

where r, x, θ are cylindrical coordinates.

Applying the same methods that Basset used for discussing the varicose instability the equation for κ^2 analogous to (7) appears to be

$$\frac{\kappa^2 a \rho I_1(ma)}{ma I_1'(ma)} = -m^2 a^2 \left(\frac{T}{a^2} \right) - \frac{E^2}{\pi a^3} \left(1 + \frac{maK_1'(ma)}{K_1(ma)} \right). \tag{12}$$

Using Watson's definitions of $I_n(z)$, $K_n(z)$ and recurrence formulae

$$\frac{ma I_1'(ma)}{I_1(ma)} = \frac{ma I_0(ma)}{I_1(ma)} - 1, \quad 1 + \frac{maK_1'(ma)}{K_1(ma)} = \frac{-maK_0(ma)}{K_1(ma)}, \tag{13}$$

we can write (13) in a form convenient for use with Watson's tables, namely

$$\kappa^2 a \rho \bigg/ \left(\frac{ma I_0(ma)}{I_1(ma)} - 1 \right) = -m^2 a^2 \left(\frac{T}{a^2} \right) + \frac{E^2}{\pi a^3} \left(\frac{maK_0(ma)}{K_1(ma)} \right), \tag{14}$$

and the neutral stability condition is

$$E^2 = \pi a T \left\{ \frac{maK_1(ma)}{K_0(ma)} \right\}. \tag{15}$$

The interior is at atmospheric pressure when $E^2 = 2\pi a T$.

For small values of ma, the term $ma I_0(ma)/I_1(ma) - 1$ tends to the value 1.0, so that for long waves the left-hand side of (14) is $\kappa^2 a \rho$. When $E = 0$, $\kappa^2 = -m^2 T/a\rho$ thus the configuration is stable and waves travel at velocity $i\kappa/m = (T/a\rho)^{\frac{1}{2}}$.

Comparing this with the waves in stretched string whose velocity is (tensile force/mass per unit length)$^{\frac{1}{2}}$ the tensile force is $2\pi aT - \pi a^2$ (pressure) and the pressure is T/a so that the tensile force over each section is πaT and the mass per unit length is $\pi a^2 \rho$ so that the velocity of waves is $(\pi aT/\pi a^2 \rho)^{\frac{1}{2}} = (T/a\rho)^{\frac{1}{2}}$. Though this may serve as a confirmation of the correctness of the mathematics, it is of little practical value since the varicose instability relative to cylinders of fluid at rest prevents such waves from being observed when the length unsupported is greater than $2\pi a$.

For the neutral stability of disturbances of a cylindrical soap film of diameter 1·35 cm. in the mode represented by (11) the equations analogous to (10) is

$$V_{\mathrm{k}} = 14\cdot5 \left\{ \frac{maK_1(ma)}{K_0(ma)} \right\}^{\frac{1}{2}}. \tag{16}$$

The calculated values are also shown in fig. 24. It will be seen that the potential necessary for instability in this mode is always higher than that required for instability in the varicose mode when $ma > 1\cdot0$.

In the range $0 < ma < 0\cdot6$ the electric field could theoretically stabilise the Rayleigh varicose instability. Thus for the soap film cylinders of diameter 1·35 cm. the following potentials would be necessary to prevent the Rayleigh varicose instability from occurring:

ma	0	0·1	0·2	0·3	0·5
V_{k}	14·5	18·7	21·1	24·0	39

but this condition cannot be achieved physically with the apparatus shown in fig. 23, because axisymmetric waves whose lengths are submultiples of the distance between the copper tubes and are in the range $ma > 1$, require less potential to make them unstable than is necessary to stabilise them against the Rayleigh varicose instability in the range $0 < ma < 0\cdot6$.

My thanks are due to Professor Van Dyke for writing the appendix and to Mr Tillett for verifying my statement that Bassett's calculation is erroneous.

PLATE 1

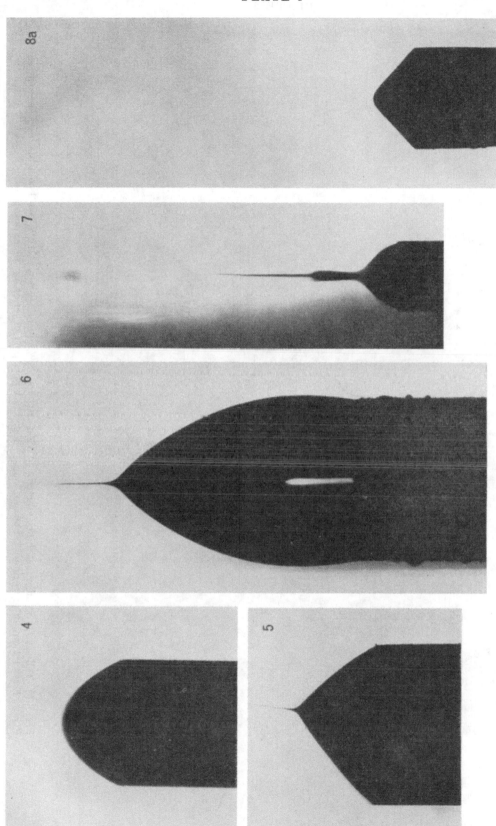

Fig. 4. Water surface just before instability set in when $p = 0$.
Fig. 5. Meniscus at the moment of jet formation. 95 % glycerine,
5 % dilute salt solution, $p = 0$.
Fig. 6. Same conditions as fig. 5 but p positive.

Fig. 7. Oscillating water jet in two-plate apparatus.
Fig. 8(a). Surface of Tellus oil ($\mu = 5$P) at the top of metal tube
($R = 1.62$ mm., $L = 3.0$ cm., $H = 7.66$ cm.) in two-plate apparatus just
before instability set in.

PLATE 2

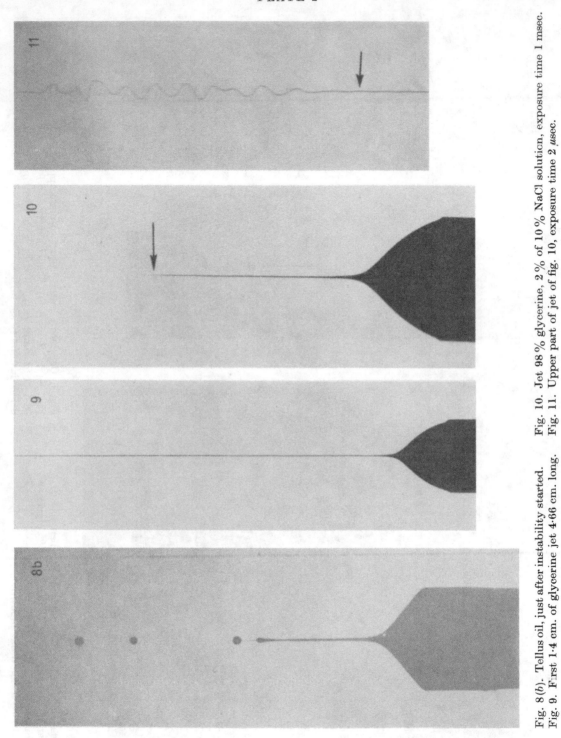

Fig. 8 (*b*). Tellus oil, just after instability started.   Fig. 10. Jet 98 % glycerine, 2 % of 10 % NaCl solution, exposure time 1 msec.
Fig. 9. First 1·4 cm. of glycerine jet 4·66 cm. long.   Fig. 11. Upper part of jet of fig. 10, exposure time 2 μsec.

PLATE 3

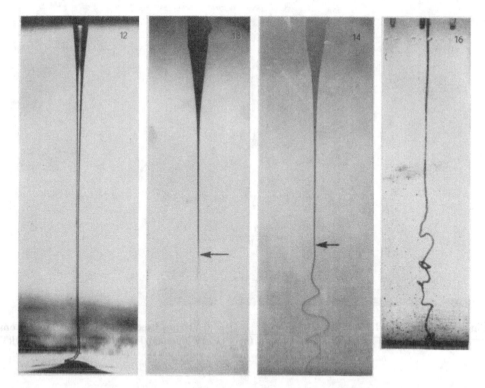

Fig. 12. Jet of silicone (100 P) falling on a plate under gravity.
Fig. 13. Downward jet of glycerine in a vertical field. Exposure 1 msec.
Fig. 14. Same jet as in fig. 13. Exposure 2 μsec.
Fig. 16. Glycerine jet made visible by indian ink falling in two layers of salt solution.

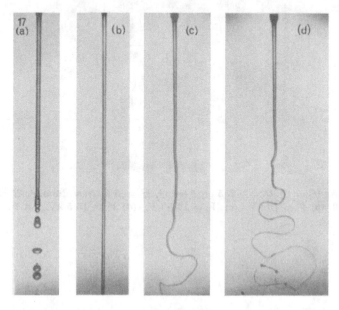

Fig. 17. Water jet pointing down in two-plate apparatus. $Q = 0.28$ cm.^{3} sec.$^{-1}$, $H = 5.5$ cm., $L = 1.7$ cm., $R = 0.053$ cm., $C = 0$, negative potential V on lower plate, upper earthed. (a) $V = 0$, (b), $V = 10.7$ kV., (c) $V = 15$ kV., (d) $V = 17.5$ kV.

PLATE 4

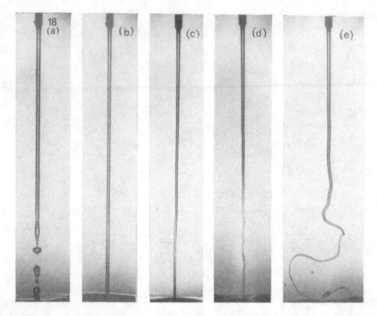

Fig. 18. Dilute NaCl ($C = 2 \times 10^{-5}$ g. cm.$^{-3}$), $Q = 0.316$ cm.^{3} sec.$^{-1}$, $H = 5.5$, $L = 1.7$ cm., $R = 0.053$ cm. (a) $V = 0$, (b) $V = 10$ kV., (c) $V = 18$ kV., (d) $V = 19$ kV., (e) $V = 20$ kV.

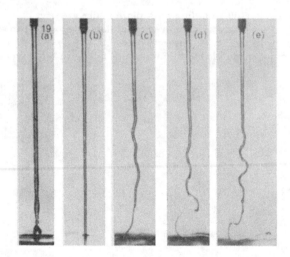

Fig. 19. Water ($C = 0$). $Q = 0.255$ cm.^{3} sec.$^{-1}$, $H = 5.5$, $L = 2.8$ cm., $R = 0.053$ cm. (a) $V = 0$, (b) $V = 13$ kV., (c) $V = 14.5$ kV., (d) $V = 15.5$ kV., (e) $V = 19$ kV.

PLATE 5

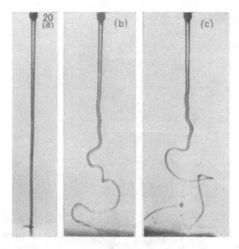

Fig. 20. Dilute NaCl ($C = 10^{-5}$ g. cm.$^{-3}$). $Q = 0 \cdot 255$ cm.^{3} sec.$^{-1}$, $H = 5 \cdot 5$ cm., $L = 2 \cdot 28$ cm. $R = 0 \cdot 053$ cm. (a) $V = 12$ kV., (b) $V = 13 \cdot 5$ kV., (c) $V = 15$ kV.

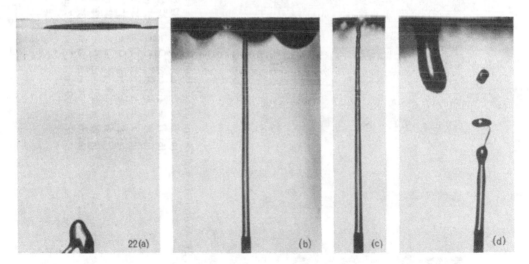

Fig. 22. Water jet upwards. $Q = 0 \cdot 466$ cm.^{3} sec.$^{-1}$, $H = 5 \cdot 5$ cm., $L = 2 \cdot 7$ cm., $R = 0 \cdot 067$ cm. (a) $V = 0$, (b) $V = 11$ kV., (c) $V = 12$ kV., (d) $V = 12$ kV.

PLATE 6

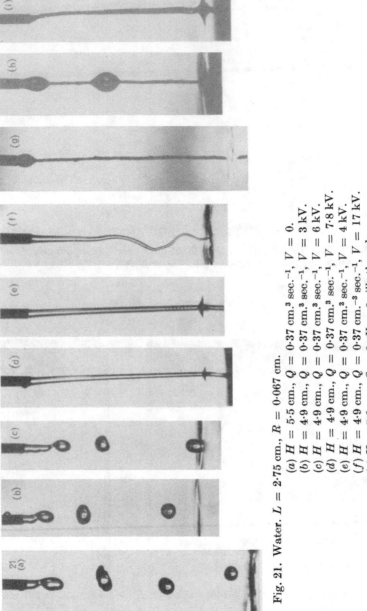

Fig. 21. Water. $L = 2.75$ cm., $R = 0.067$ cm.

(a) $H = 5.5$ cm., $Q = 0.37$ cm.^{3} sec.$^{-1}$, $V = 0$.
(b) $H = 4.9$ cm., $Q = 0.37$ cm.^{3} sec.$^{-1}$, $V = 3$ kV.
(c) $H = 4.9$ cm., $Q = 0.37$ cm.^{3} sec.$^{-1}$, $V = 6$ kV.
(d) $H = 4.9$ cm., $Q = 0.37$ cm.^{3} sec.$^{-1}$, $V = 7.8$ kV.
(e) $H = 4.9$ cm., $Q = 0.37$ cm.^{3} sec.$^{-1}$, $V = 4$ kV.
(f) $H = 4.9$ cm., $Q = 0.37$ cm.$^{-3}$ sec.$^{-1}$, $V = 17$ kV.
(g) $H = 5.2$ cm., $Q = 0$, $V = 0$ silk thread.
(h) $H = 4.9$ cm., $Q = 0.37$ cm.^{3} sec.$^{-1}$, $V = 0$ silk thread.
(i) $H = 4.9$ cm., $Q = 0.37$ cm.^{3} sec.$^{-1}$, $V = 17$ kV. silk thread.

CONSOLIDATED LIST OF PAPERS PUBLISHED
IN VOLUMES I–IV

The column on the right-hand side shows the volume in which a paper has been
reprinted and its number within that volume.

Statistical theory of turbulence. Part v. Effect of turbulence on boundary layer. Theoretical discussion of relationship between scale of turbulence and critical resistance of spheres. *Proc. Roy. Soc. Lond.* A, **156**, 307–17. ...... II, 33

The oscillations of the atmosphere. *Proc. Roy. Soc. Lond.* A, **156**, 318–26. ...... II, 34

Correlation measurements in a turbulent flow through a pipe. *Proc. Roy. Soc. Lond.* A, **157**, 537–46. ...... II, 35

Fluid friction between rotating cylinders. Part I. Torque measurements. *Proc. Roy. Soc. Lond.* A, **157**, 546–64. ...... II, 36

Fluid friction between rotating cylinders. Part II. Distribution of velocity between concentric cylinders when outer one is rotating and inner one is at rest. *Proc. Roy. Soc. Lond.* A, **157**, 565–78. ...... II, 37

1937 The emission of the latent energy due to previous cold working when a metal is heated (with H. Quinney). *Proc. Roy. Soc. Lond.* A, **163**, 157–81. ...... I, 26

Mechanism of the production of small eddies from large ones (with A. E. Green). *Proc. Roy. Soc. Lond.* A, **158**, 499–521. ...... II, 38

Flow in pipes and between parallel planes. *Proc. Roy. Soc. Lond.* A, **159**, 496–506. ...... II, 39

The statistical theory of isotropic turbulence. *J. Aeronaut. Sci.* **4**, 311–15. ...... II, 40

The determination of drag by the Pitot traverse method. *Rep. Memo. Aeronaut. Res. Comm.*, no. 1808. ...... III, 21

1938 Plastic strain in metals. *J. Inst. Metals*, **62**, 307–24. ...... I, 27

Analysis of plastic strain in a cubic crystal. *Timoshenko 60th Anniv. Vol.*, 218–24. ...... I, 28

Production and dissipation of vorticity in a turbulent fluid. *Proc. Roy. Soc. Lond.* A, **164**, 15–23. ...... II, 41

The spectrum of turbulence. *Proc. Roy. Soc. Lond.* A, **164**, 476–90. ...... II, 42

Measurements with a half-Pitot tube. *Proc. Roy. Soc. Lond.* A, **166**, 476–81. ...... IV, 13

1939 Stress systems in aeolotropic plates. Part I (with A. E. Green). *Proc. Roy. Soc. Lond.* A, **173**, 162–72. ...... I, 29

Determination of the pressure inside a hollow body in which there are a number of holes communicating with variable pressures outside. Paper for Aeronaut. Res. Comm. ...... III, 22

The propagation and decay of blast waves. Paper for Civil Defence Res. Comm. ...... III, 23

1940 Propagation of earth waves from an explosion. Paper for Civil Defence Res. Comm. ...... I, 30

Notes on possible equipment and technique for experiments on icing on aircraft. *Rep. Memo. Aeronaut. Res. Comm.*, no. 2024. ...... III, 24

Generation of ripples by wind blowing over a viscous fluid. Paper for Chem. Defence Res. Dept. ...... III, 25

Notes on the dynamics of shock-waves from bare explosive charges. Paper for Civil Defence Res. Comm. ...... III, 26

Pressures on solid bodies near an explosion. Paper for Civil Defence Res. Comm. ...... III, 27

The stagnation temperature in a wake. Paper for Aeronaut. Res. Comm. ...... III, 28

1941 Calculation of stress distribution in an autofrettaged tube from measurements of stress rings. Paper for Advis. Coun. Sci. Res. Tech. Devel. ...... I, 31

The propagation of blast waves over the ground. Paper for Civil Defence Res. Comm. ...... III, 29

Analysis of the explosion of a long cylindrical bomb detonated at one end. Paper for Civil Defence Res. Comm. ...... III, 30

The pressure and impulse of submarine explosion waves on plates. Paper for Civil Defence Res. Comm. ...... III, 31

1942 The plastic wave in a wire extended by an impact load. Paper for Civil Defence Res. Comm. ...... I, 32

PUBLICATIONS BY G. I. TAYLOR NOT REPRINTED IN THESE VOLUMES

1914   Report on the work carried out by the s.s. *Scotia*, 1913. H.M.S.O. Report, pp. 48–68.

1915   Report on the accuracy with which temperature errors in determining heights by barometer may be corrected. *Rep. Memo. Advis. Comm. Aeronaut.*, no. 239.

1916   Phenomena connected with turbulence in the lower atmosphere. *Rep. Memo. Advis. Comm. Aeronaut.*, no 304.

1917   Fog conditions. *Aeronaut. J.* **21**, 75–90.

1920   Navigation notes on a passage from Burnham-on-Crouch to Oban. *Yachting Monthly*.

1921   Experiments with rotating fluids. *Proc. Camb. Phil. Soc.* **20**, 326–9.

1924   Extracts from the log of *Frolic. Roy. Cruising Club. J.*, 85–105.

1925   Experiments with rotating fluids. *Proc. 1st Intern. Cong. Appl. Mech. (Delft, 1924)*, 89–96.

       Versuche mit Rotierenden Flüssigkeiten. *Z. Angew. Math Mech.* **5**, 250–3.

1927   Turbulence. *Quart. J. Roy. Met. Soc.* **53**, 201–11.

       Across the Arctic circle in *Frolic*, 1927. *Roy. Cruising Club J.*, 9–26.

1928   The force acting on a body placed in a curved and converging stream of fluid. *Rep. Memo. Aeronaut. Res. Comm.*, no. 1166.

       Report on progress during 1927–8 in calculation of flow of compressible fluids and suggestions for further work. *Rep. Memo. Aeronaut Res. Comm.*, no. 1196.

1929   The air wave from the great explosion at Krakatau. *4th Pacific Science Congress (Java, 1929)*, vol. II B, 645–55.

1930   Tour in the East Indies. *Proc. Roy. Inst.* **26**, 209.

       Strömung um einen Korper in einer kompressiblen Flüssigkeit. *Z. Angew. Math. Mech.* **10**, 334–45.

1931   The flow round a body moving in a compressible fluid. *Proc. 3rd Intern. Cong. Appl. Mech. (Stockholm, 1930)*, vol. I, 263–75.

       Round Ireland in *Frolic. Roy. Cruising Club J.*, 213–25.

1932   Note on review by Davies and Sutton of the present position of the theory of turbulence. *Quart. J. Roy. Met. Soc.* **58**, 61–5.

1933   L'onde ballistique d'un projectile à tête conique (with J. W. Maccoll). *Mem. de l'Art. Franc.* **12**, 651–83.

1935   The mechanics of compressible fluid (with J. W. Maccoll). Section H of *Aerodynamic Theory*, vol. III, pp. 209–50, edited by W. F. Durand. Springer.

1936   Well established problems in high speed flow. *R. Accad. Ital. Att.: 5, Conv. Sci. Fis. Mat. Nat.*, 198–214.

1937   The determination of stresses by means of soap films. Article in *The Mechanical Properties of Fluids*, pp. 237–54. Blackie.

1938   Turbulence. Chap. 5 of *Modern Developments in Fluid Dynamics*, vol. I, pp. 191–233, edited by S. Goldstein. Oxford.

       Method of deducing $F(n)$ from the measurements. Appendix to paper by L. F. G. Simmons & C. Salter, *Proc. Roy. Soc. Lond.* A, **165**, 87–9.

1939   Some recent developments in the study of turbulence. *Proc. 5th Intern. Cong. Appl. Mech. (Camb., Mass., 1938)*, pp. 294–310. Wiley.

1946   Some model experiments in connection with mine warfare. *Trans. Instn Nav. Archit.*, 165–6.

1948   Eiver 1948. *Roy. Cruising Club. J.*, 194–200.

       Explosives with lined cavities (with G. Birkhoff, D. P. MacDougall, and E. M. Pugh). *J. Appl. Phys.* **19**, 563–82.

1949   Aerodynamic properties of gauze screens. Appendix to *Rep. Memo. Aeronaut. Res. Coun.*, no. 2276.

1950 The 7th International Congress for Applied Mechanics. *Nature* **165**, 258–60.
Similarity solutions to problems involving gas flow and shock waves. *Proc. Roy. Soc. Lond.* A, **204**, 8–9.

1951 The mechanism of eddy diffusivity. *Proc. of General Discussion on Heat Transfer*, pp. 193–4. Inst. Mech. Eng.

1953 Dispersion of salts injected into large pipes or the blood vessels of animals. *Appl. Mech. Rev.* **6**, 265–7.
William Cecil Dampier 1867–1952. *Obit. Notices of Fellows of Roy. Soc.* **9**, 55–63.

1954 Diffusion and mass transport in tubes. *Proc. Phys. Soc. Lond.* B, **67**, 857–69.
George Boole 1815–64. *Proc. Roy. Ir. Acad.* 66–73.

1956 George Boole, F.R.S. 1815–64. *Notes Rec. Roy. Soc. Lond.* **12**, 44–52.

1960 Similarity solutions of hydrodynamic problems. Article in *Aeronautics and Astronautics (Durand Anniv. Vol.)*, pp. 21–8. Pergamon.

1961 Interfaces between viscous fluids in narrow passages. Article in *Problems of Continuum Mechanics (Muskhelishvili Anniv. Vol.)*, pp. 546–55. Society for Indus. & Appl. Math.

1962 Gilbert Thomas Walker 1868–1958. *Biog. Mem. of Fellows of Roy. Soc.* **8**, 167–74.

1963 Memories of Kármán. *J. Fluid Mech.* **16**, 478–80.
Scientific diversions. Article in *Man, Science, Learning and Education*, edited by S. W. Higginbotham, pp. 137–48. Rice Univ. Semicent. Publ.
Sir Charles Darwin (1887–1962). *Amer. Phil. Soc. Year Book*, pp. 135–40.

1964 Cavitation in hydrodynamic lubrication. Article in *Cavitation in Real Liquids*, edited by R. Davies, pp. 80–101. Elsevier.

1965 Note on the early stages of dislocation theory. Article in the *Sorby Centennial Symposium on the History of Metallurgy*, edited by C. S. Smith, pp. 355–8. Gordon & Breach.

1966 When aeronautical science was young. *J. Roy. Aeronaut. Soc.* **70**, 108–13.

1967 Low-Reynolds-number flows. 16 mm. colour sound film, produced by Educational Services Inc.

1969 Amateur scientists. *Michigan Quart. Rev.* **8**, 107–13.
Aeronautics fifty years ago. *Quest: J. City Univ.*, no. 8, 12–19.

1970 The interaction between experiment and theory in fluid mechanics. *Bull. Brit. Hydromech. Res. Ass.*
Some early ideas about turbulence. *J. Fluid Mech.* **41**, 3–11.

Printed in the United States
By Bookmasters